The Molecular Biology of Animal Viruses

Advisory Board

The Molecular Biology of Animal Viruses

VOLUME 1

edited by DEBI PROSAD NAYAK

Department of Microbiology and Immunology
School of Medicine
University of California
Los Angeles, California

MARCEL DEKKER, INC. New York and Basel

Library of Congress Cataloging in Publication Data

Main entry under title:

The Molecular biology of animal viruses.

Includes index.
CONTENTS: v. 1. RNA viruses.
1. Viruses. 2. Molecular biology. I. Nayak, Debi Prosad.
QR360.M55 576'.6484 76-29295
ISBN 0-8247-6533-8

MARCEL DEKKER, INC.
270 Madison Avenue, New York, New York 10016

Current printing (last digit):
10 9 8 7 6 5 4 3 2 1

PRINTED IN THE UNITED STATES OF AMERICA

Preface

In the last decade animal virology has become a major force in modern biology. During this period the major emphasis in animal virus research has been toward understanding the nature of animal viruses and the virus-host interaction at the molecular level. Furthermore, animal viruses have been used as a major tool in elucidating the basic regulatory mechanism involved in translation, transcription, and replication of infinitely more complex eukaryotic cells. It is not surprising to find that these basic regulatory processes of eukaryotic cells are essentially the same as those involved in the biology of animal viruses. After all, viruses must use the same host machinery, probably more efficiently, for their own survival.

Thus, the major assaults on both these fronts, i.e., viruses as pathogens and as probes in elucidating their host's function, have generated massive amounts of literature. Consequently, an ever increasing number of journals, specialized reviews, and monographs, as well as comprehensive multivolume treatises on virology are now being published.

This trend in modern biology and health care has created an increasing need for qualified virologists and has resulted in greater emphasis in graduate and postgraduate training in animal virology. Major universities now offer regular courses in the biology and chemistry of animal viruses. During the eight years I have been teaching courses in animal virology, I have felt the need of a suitable book covering the basic molecular biology of the major groups of animal viruses. The tremendous expansion of published literature in virology has made the need for such a book critical. The present treatise is an attempt to fulfill this need. This book (in two volumes) is therefore designed to serve as a text or a reference in advanced graduate courses in animal virology, with major emphasis on molecular biology.

Obviously, my experience in teaching animal virology has been a major influence in the organization of this book. I am fully aware that the design of each course and its emphasis will vary with individual instructors. This book is not an attempt to serve as a model for a course in animal virology, nor will it meet the specific needs of every instructor. Rather, the basic premise here is to present major groups of animal viruses. Each chapter is written by teacher(s), researcher(s) who are actively involved in their field of speciality. It is hoped that the basic information on the molecular biology of major groups of animal viruses presented here will be available and can be used by the students and the instructors regardless of their individual ways of synthesizing and formulating a course.

A few words are needed to explain the organization of the book's contents. As indicated above, the major emphasis is on elucidation of the nature of virus and host-virus interaction at the molecular level. Each chapter is not merely a chronicle of facts but represents an attempt to follow the logical progression of the life cycle of a major group of viruses. It includes the generality and uniqueness of their own group, as well as their relationships with other viruses outside the group. Chapters often present a hypothesis, raise critical questions about unsolved problems, and bring them to the attention of the readers, the future virologists. Although there is no attempt to explain viral pathogenesis in molecular terms, the relevance of each group of viruses to major animal and human diseases is indicated.

Some of the chapters are likely to become somewhat outdated soon. This is true for any fast-moving discipline, and even more so for animal virology. Indeed, this is a tribute to the dynamic nature of this field and to the unquenching curiosity of virologists. In spite of this inherently unsolvable problem, I hope this book will be of value in providing a source for basic information on the molecular biology of animal viruses to researcher and student alike. Newer information can always be found in recent journals to supplement the text as must be done in any advanced course in modern biology.

In addition, this book will serve as a reference for courses in biochemistry, molecular biology, genetics, or pathogenesis, where a basic understanding of the host-virus interaction at the molecular level is required. I hope it will be useful to present and future virologists as well as to other biologists.

I would like to thank the members of the advisory board for their help in reviewing one or more chapters, and to the contributors, who have done a remarkable job. Finally, I would like to thank my wife, Abantika, for her patience and understanding, and for help throughout this work.

Debi Prosad Nayak

Contents

Contributors to Volume 1

David H. L. Bishop, Department of Microbiology, The Medical Center, University of Alabama in Birmingham, Alabama

Charles E. Buckler, Laboratory of Viral Diseases, National Institute of Allergy and Infectious Diseases, National Institutes of Health, Bethesda, Maryland

Bernard N. Fields, Department of Microbiology and Molecular Genetics, Harvard Medical School, Boston, Massachusetts

Raymond V. Gilden, NCI Viral Oncology Program, Frederick Cancer Research Center, Frederick, Maryland.

David W. Kingsbury, Laboratories of Virology, St. Jude Children's Research Hospital, Memphis, Tennessee

Hilton B. Levy, Laboratory of Viral Diseases, National Institute of Allergy and Infectious Diseases, National Institutes of Health, Bethesda, Maryland

Carl F. T. Mattern, Laboratory of Viral Diseases, National Institute of Allergy and Infectious Diseases, National Institutes of Health, Bethesda, Maryland

Debi Prosad Nayak, Department of Microbiology and Immunology, School of Medicine, University of California at Los Angeles, California

Robert F. Ramig, Department of Microbiology and Molecular Genetics, Harvard Medical School, Boston, Massachusetts

*David M. K. Rekosh,** Department of Molecular Virology, Imperial Cancer Research Fund, London, England

Freddie L. Riley, Laboratory of Viral Diseases, National Institute of Allergy and Infectious Diseases, National Institutes of Health, Bethesda, Maryland

M. S. Smith, Department of Microbiology, The Medical Center, University of Alabama in Birmingham, Alabama

Ellen G. Strauss, Division of Biology, California Institute of Technology, Pasedena, California

James H. Strauss, Division of Biology, California Institute of Technology, Pasedena, California

*Present affiliation: National Institute for Medical Research, London, England.

Contents of Volume 2

Chapter 1

Symmetry in Virus Architecture

Carl F. T. Mattern

Laboratory of Viral Diseases
National Institute of Allergy and Infectious Diseases
National Institutes of Health
Bethesda, Maryland

I. Introduction

Since the several existing virus classification schemes are based in part on the symmetry elements exhibited by some viruses, this chapter is devoted largely to those symmetry operations which have counterparts in virus structure and to those virus groups the structure and symmetry of which have been firmly established. The historical development of this subject has been extensively covered in a number of published reviews, in which can be found accounts of the roles of x-ray diffraction and electron microscopy in uncovering the symmetry elements of viruses [19, 21, 35, 68, 83]. In fact, relatively little has been added to our understanding of the basic architectural plan of viruses in the last decade, as can be appreciated from the comprehensive review of this subject by Caspar in 1965 [19].

II. General Architecture of Viruses [20, 79]

A. Capsids

The capsid is the outer virus shell consisting of regularly arranged units of protein forming cylinders, spheres, icosahedra, and other forms. The capsid protects the core nucleic acid from degradation by nucleases and contributes to the determination of the host range of the virus through its specificity of interaction with host cell receptor sites.

The units of protein in the capsid are called structure units or biochemical subunits. Structure units may be regularly and uniformly distributed over the virus capsid or clustered into groups called capsomeres or morphological subunits as are seen in certain virus groups by negative-staining electron microscopy.

Some virus capsids consist of multiple structure units of a single type of protein; others contain 2 or more different protein types.

B. Nucleocapsids

The complete virus or virion consists of a capsid and a core containing a nucleic acid, either DNA or RNA, with or without additional core protein(s). The capsid with core is termed a nucleocapsid. Some viruses consist of a nucleocapsid with no additional coat and are referred to as naked nucleocapsids. Other viruses have their nucleocapsids enclosed in an additional envelope or peplos which is frequently derived in part from the host cell membrane, especially in those viruses which mature by the process of budding from the cell membrane. These are called enveloped nucleocapsids. Some envelopes are seen by electron microscopy to have surface projections or spikes, referred to as peplomeres.

C. Viroids

A unique system of free RNA, apparently devoid of protein, has been implicated as being the infectious "agent" causing potato spindle tuber disease [25]. In some respects this agent behaves as do a number of virus nucleic acids which, following their release from intact virions, maintain the capacity, albeit with low efficiency, of inducing susceptible cells to replicate complete nucleocapsids [98].

III. Viruses with Helical Symmetry

A. Symmetry Operations

Helical symmetry can be defined by 2 repetitive operations: rotation about an axis coupled with translation parallel to the same screw axis. In the case of a virus with helical symmetry one can consider the capsid constructed of protein units repeated by the rotation operation, r degrees, followed by the translation operation, t nanometers. These operations define a "cylindrical" surface with helical symmetry (Fig. 1.1). The number of subunits per helical turn (n) will be $n = 360/r$ and the pitch will be $n \cdot t$.

B. Tobacco Mosaic Virus

Tobacco mosaic virus (TMV) has been extensively studied and represents the classic naked helical nucleocapsid. Most of the architectural features were uncovered by x-ray diffraction studies [18, 19, 35, 68, 83] and the pitch was subsequently confirmed by electron microscopy [29].

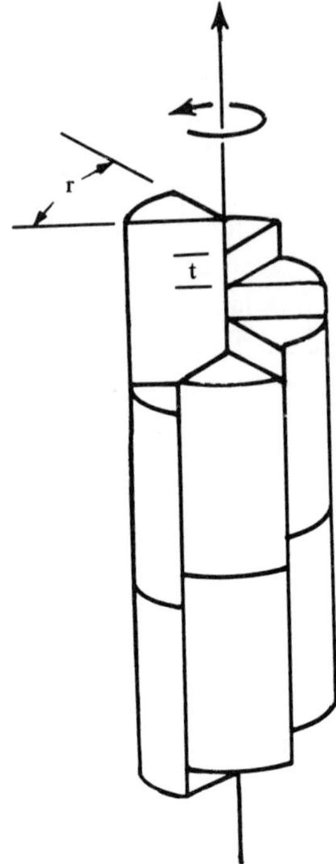

Fig. 1.1 A cylinder with helical symmetry is generated by repetitive operations rotation (r) and translation (t) along the screw axis.

The virus is about 300 nm long and 18 nm in maximum diameter. Figure 1.2 depicts a portion of the reconstructed structure consisting of helically arranged protein subunits of about 17,500 daltons. An integral number (49) of subunits is found in 3 turns of the helix, or 16 1/3 subunits in each turn of pitch 2.3 nm. The virus is hollow out to a radius of about 2nm and the RNA (about 2×10^6 daltons) courses a helical path between the protein subunits at a radius of about 4 nm, there being about 3 nucleotides associated with each protein subunit. Recently it has been determined that TMV is a right-handed helix [31].

x-Ray diffraction studies have probed the molecular structure of the TMV protein to a resolution of about 6Å. In these studies were employed either intact TMV [7, 54] or reaggregated disks [45] consisting of 2 rings each of 17 subunits, which in turn had been crystallized [34, 76].

Studies employing optical transforms of electron microscope images of negatively stained TMV have provided details of the protein packing [105] and the stacked disk polymer to an effective resolution of about 12 Å [110, 111].

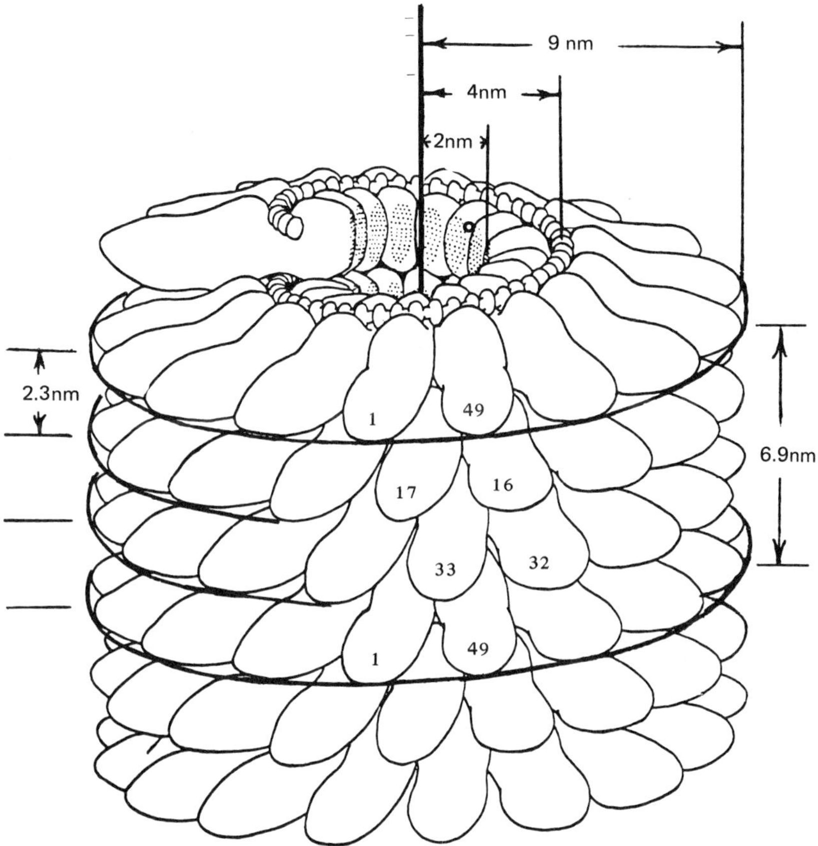

Fig. 1.2 Schematic representation of tobacco mosaic virus structure derived largely from x-ray diffraction studies (modified from Ref. [18]).

The significance of the double-disk polymer structure is discussed in relation to the single-disk-single-helix transformation [38, 69], which is thought to be initiated by TMV RNA and to play a role in the assembly of the virion [15, 16].

C. Other Helical Viruses

There is a substantial number of filamentous viruses among the plant and bacterial viruses. Some are known to be helical in structure; others probably will prove to be so.

In addition to TMV there are other similar, rigid helical viruses, such as dolichos enation mosaic virus [88], barley stripe mosaic virus [44, 51, 65], and lychnis ringspot virus [44] which resemble TMV by electron microscopy. An x-ray diffraction study of barley stripe mosaic virus [30] confirmed its helical architecture but revealed differences in diameter, pitch, and number of subunits per turn from those parameters of TMV. Tobacco rattle virus, studied by electron microscopy [50, 94, 97] and x-ray diffraction [30], is still another rigid helical virus with a number of interesting properties [53].

Among the flexible, filamentous plant viruses the sugar beet yellows virus has been seen to have a loose, hollow helical structure. It is smaller in diameter (about 10 nm) than TMV (18 nm) and has a pitch of about 3.0 nm [56, 103]. Several such flexible filamentous viruses whose diameters are 10-13 nm have been found to have helical features with pitches of 3.3-3.7 nm [112].

A substantial number of more or less flexible, filamentous bacterial viruses has been uncovered [80]. They seem to fall into 2 length groups, 600-900 nm and 1300-1400 nm, but their diameters are all about 6 nm. Their nucleic acid is a single-stranded circular DNA which courses a central channel in the protein coat. Recent x-ray diffraction studies [28, 81, 82, 114] reveal that class I filamentous viruses (fd, If1, IKe) are left-handed helices with a 1.5 nm pitch, 4.5 protein units per turn, an inner protein diameter of 2 nm, and an outer diameter of 6 nm. Protein subunits about 7 nm long are oriented with the virus axis and overlap as do shingles. Class II viruses (Pf1, xf) also have a 1.5 nm pitch, but there are 72 protein units in 5 turns of the helix in a more complex arrangement than those of class I viruses.

D. Enveloped Helical Nucleocapsids

A variety of viruses is now known to contain flexible helical nucleocapsids. The 2 main groups of such viruses are (a) the myxovirus (influenza and parainfluenza) group and (b) the rhabdovirus (bullet-shaped virus) group. Among the former [22] the diameters of influenza viruses (80-100 nm) are generally smaller than those of parainfluenza viruses (120-450 nm). Similarly, the diameters and lengths of the helical nucleocapsids of influenza viruses (9 and 50-130 nm, respectively) are smaller than those of parainfluenza viruses (18 and 1000 nm, respectively).

The nucleocapsids of parainfluenza SV5 [22], mumps, Sendai, and measles viruses [33, 116] have transverse periodicities of about 5.0 nm and these nucleocapsids appear quite similar to some of the flexible, filamentous plant viruses previously discussed.

Both influenza and parainfluenza viruses form elongated particles in which the helical nucleocapsid can be seen to be regularly coiled [22] into a higher order helix.

The envelopes of all of these viruses are seen to have surface projections (peplomeres) both on spherical and filamentous forms [3, 5, 22]. An interesting reticular pattern has been found in the membrane [5, 41]. In elongated forms it usually appears as a hexagonal network (7 nm along a hexagon edge) but occasionally as a square pattern [3] and in spherical forms both hexagons and pentagons have been observed.

The structure of the rhabdoviruses has been reviewed extensively [57, 59, 60]. These numerous viruses, of which vesicular stomatis virus is the prototype are characterized by having one rounded end and one blunt end, although bacilliform or double rounded ends are sometimes seen in these elongated viruses. Some viruses of this type are known to infect vertebrates: a large number of others produce disease in plants [42].

The envelope sometimes presents a surface consisting of a hexagonal network from which peplomeres project [52, 61, 99].

In the intact virion there is a relatively large coil, the nucleocapsid, about 50 nm in diameter, which runs almost the length of the particle (about 150 nm). The coiled nucleocapsid produces a transverse striation with a spacing of about 4.5 nm. This appearance, with minor variations, is seen uniformly in viruses of this group [60].

IV. Viruses with Icosahedral Symmetry

With the exception of some of the polyhedral bacteriophages which have octahedral (432) symmetry [12, 26], a possibility predicted by Crick and Watson [23], the only symmetry uncovered in the "spherical" viruses is icosahedral (532) symmetry.

A. Icosahedral Symmetry

The icosahedron is a platonic solid (Fig. 1.3) consisting of 20 equilateral triangular faces, 12 vertices, and 30 edges. The 6 lines through opposite vertices are axes of 5-fold rotational symmetry, 10 lines through the centers of opposite triangular faces are axes of 3-fold rotational symmetry, and 15 lines through centers of opposite edges are axes of 2-fold rotational symmetry. Because of the 5-, 3-, and 2-fold axes of rotational symmetry, the icosahedron is said to have 532 symmetry.

The icosahedron has planes of mirror symmetry perpendicular to each 2-fold axis and passing through the center of the polyhedron. Thus, as has been found to be of practical importance [83], the icosahedron seen on a 2-fold axis presents the only view in which faces (and capsomeres) on the upper and lower surfaces coincide in the "through" view as would be seen by negative staining electron microscopy.

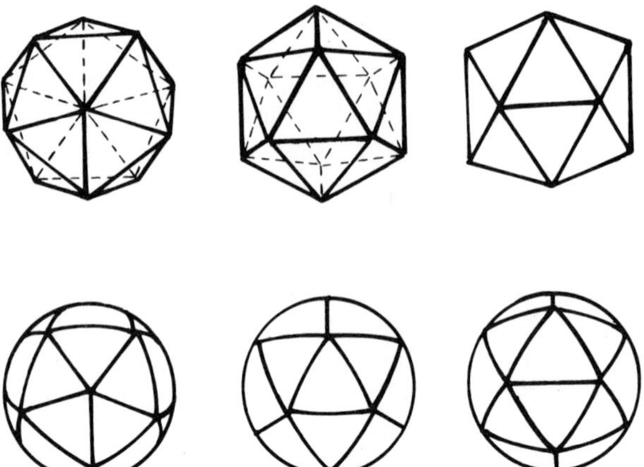

Fig. 1.3 Icosahedra (upper, left to right) seen on 5-, 3-, and 2-fold axes of rotational symmetry. The under surfaces are outlined in dashed lines. On a 2-fold axis the upper and lower edges coincide. Lower figures represent spherical models with icosahedral symmetry seen on the same axes as above. Lower surfaces not shown in any view.

An icosahedral model can be readily constructed by appropriate folding and cutting of an isometric net which consists of equilateral triangles [21]. Such a net (Fig. 1.4A) has axes, h and k, inclined at 60°. Selection of one vertex of a triangle at the origin (h = 0, k = 0) and a second vertex at point (1,0) defines a triangle which will form the characteristic face of the icosahedron folded from 20 such unit triangles (Fig. 1.4B, C). In this case the vertices of the icosahedral triangular face will coincide with the icosahedral vertices, hereafter called I-vertices to avoid confusion with triangle vertices. In fact the face vertices have determined the I vertices.

Actually the second I vertex can be selected at any point (h,k) on the net, generating an infinite number of different structures with icosahedral (532) symmetry and which are properly called deltahedra. The characteristic "icosahedral" face of each deltahedron will contain varying numbers of unit triangles, the number being called the triangulation number (T).

B. Icosahedral Symmetry and Virus Capsomeres

The correspondence of such deltahedra and their triangulation numbers to various icosahedral viruses and to the distribution of capsomeres on the virus surface was described by Caspar and Klug [19, 21].

The general triangulation number* is $T = h^2 + hk + k^2$, where h and k are the coordinates of the second I vertex of the characteristic icosahedral face of the deltahedron derived from the isometric net; the first I vertex is always at the origin (0,0).

If one considers each point of the isometric net which has been folded into an icosahedron as the locus of a virus capsomere, the total number of points of the deltahedron or capsomeres (C) of a virus are defined by $C = 10 (h^2 + hk + k^2) + 2$ [46]. C is thus a function of T; $C = 10T + 2$.

There are 3 classes of deltahedra which can be generated from the isometric net.

1. The first group (Fig. 1.5) results from the selection of the second I-vertex along the h axis where k = 0. The equation $T = h^2 + hk + k^2$

*For the interested reader the derivations of the equations for triangulation number and capsomere number are presented in the Appendix, Section I.VIII.A.

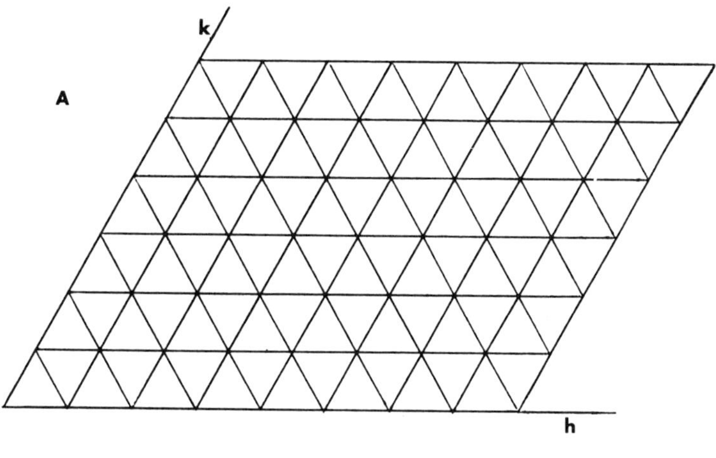

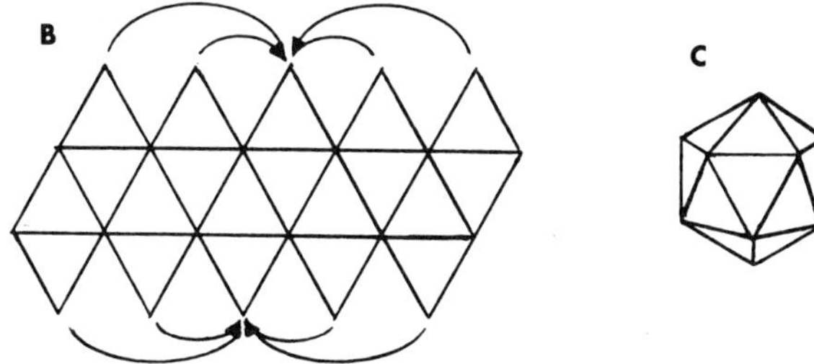

Fig. 1.4 The isometric net (A) with h and k axes inclined at 60°, and the folding of 20 triangles (B) into the icosahedron (C).

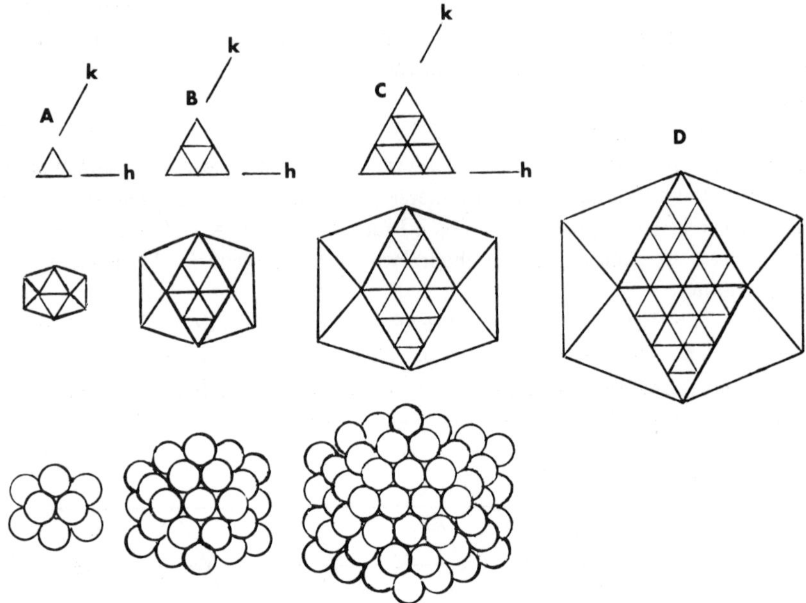

Fig. 1.5 Characteristic icosahedral faces determined by selection of the second icosahedral vertex at h = 1, k = 0 in A, (2,0) in B and (3,0) in C with corresponding icosahedral models (second row) having triangulation numbers T = 1, 4, and 9, respectively and T = 16 (D). Lower row, some models of corresponding icosahedral viruses.

simplifies to $T = h^2$ and $C = 10T + 2 = 10\,h^2 + 2$. The following series of deltahedra is thus generated (Fig. 1.5).

Coordinates, (h,k)	Triangulation no. ($T = h^2$)	Capsomere no. ($C = 10T + 2$)	Fig. 1.5
1,0	1	12	A
2,0	$4(1 \times 2^2)$	42	B
3,0	$9(1 \times 3^3)$	92	C
h,0	$h^2(1 \times h^2)$	$10h^2 + 2$	
	T = 1 series		

The significance of the triangulation number (T) is easily visualized in the icosadeltahedra of Fig. 1.5. The series is referred to as the T = 1 series.

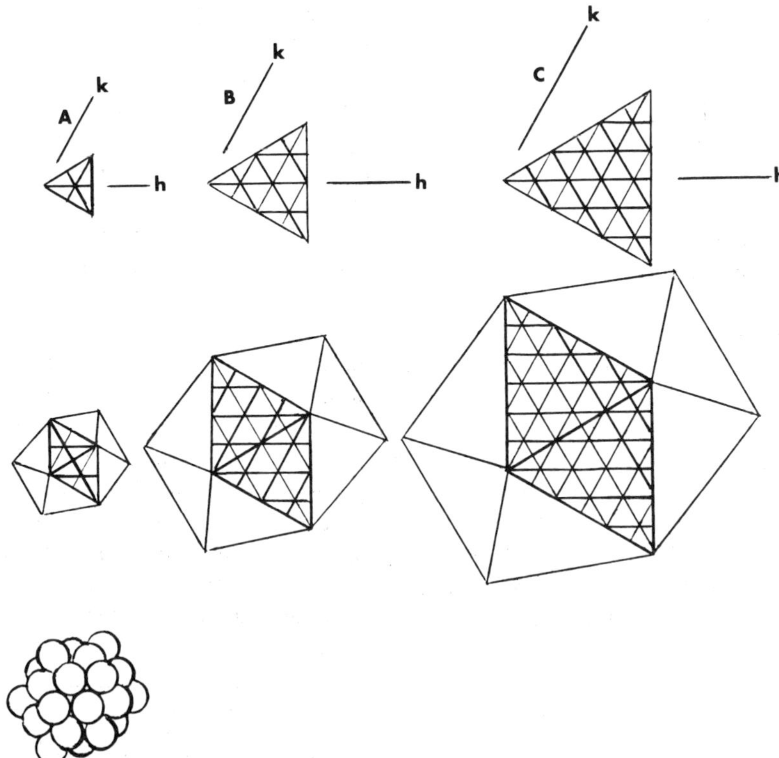

Fig. 1.6 Upper row, (h,k) indices of (1,1), (2,2), and (3,3) generate icosahedra (middle row) with T = 3, 12, and 27, respectively. Lower row, a T = 3, 32-capsomere virus model.

2. When the second vertex is selected along a line where h = k the equation for T becomes $T = 3h^2$ and $C = 30h^2 + 2$ and the following different series of deltahedra is generated (Fig. 1.6).

Coordinates (h,k)	Triangulation no. ($T = 3h^2$)	Capsomere no. (C = 10T + 2)	Fig. 1.6
1,1	$3(3 \times 1^2)$	32	A
2,2	$12(3 \times 2^2)$	122	B
3,3	$27(3 \times 3^2)$	272	C
h,h	$3h^2(3 \times h^2)$	$30h^2 + 2$	
	T = 3 series		

In this series a number of unit triangles is shared equally by adjacent icosahedral faces. In the first series T = 3 there are 6 half-unit triangles on each icosahedral face. The series has been called the T = 3 series.

In both cases described (a) where k = 0 and (b) where h = k the deltahedra generated have planes of mirror symmetry perpendicular to each 2-fold axis and passing through the center of the polyhedron. In viruses having a corresponding distribution of capsomeres there will be superimposition of capsomeres on the upper and lower surfaces of viruses seen on a 2-fold axis. Seen on 5- or 3-fold axes, viruses will not have superimposed capsomeres on the 2 surfaces, both of which are usually "stained" and contribute to the virus image to a greater or lesser extent with the conventional negative-staining procedure.

3. In the remaining deltahedra, the second vertex coordinates are such that $h,k \neq 0$ and $h \neq k$. Here the general equation for triangulation number applies, $T = h^2 + hk + k^2$ and $C = 10(h^2 + hk + k^2) + 2$. Such coordinates generate a series of "skewed" icosahedra (Fig. 1.7). In this group there are always 2 sets of coordinates (h = m, k = n and h = n, k = m) producing the same values of T and C. In the first of the series, for example, h = 2, k = 1, and h = 1, k = 2 both generate T = 7, C = 72. These 2 polyhedra are related by mirror symmetry. The convention for naming the hand in accordance with the isometric net is: for $h > k$, levo and for $h < k$, dextro. For convenience only one hand, levo, is considered further.

Coordinates (h,k)	Triangulation no. ($T = h^2 + kh + k^2$)	Capsomere no. ($C = 10T + 2$)	Fig. 1.7
2,1	$7(7 \times 1^2)$	72	A
4,2	$28(7 \times 2^2)$	282	
6,3	$63(7 \times 3^2)$	632	
2n,n	$(7n^2)$	$10(7n^2) + 2$	
	T = 7 series		
3,1	$13(13 \times 1^2)$	132	B
6,2	$52(13 \times 2^2)$	522	
9,3	$117(13 \times 3^2)$	1172	
3n,n	$(13n^2)$	$10(13n^2) + 2$	
	T = 13 series		
3,2	$19(19 \times 1^2)$	192	C
6,4	$76(19 \times 2^2)$	762	
3n,2n	$(19n^2)$	$10(19n^2) + 2$	
	T = 19 series		

$T = (h^2 + hk + k^2)$ series, where $h, k \neq 0$ (T = 1 series for h,k = 0), $h \neq k$ (T = 3 series for h = k), and h and k have no highest common factor (m) which would revert T to the lowest of the series with coordinates h/m, k/m

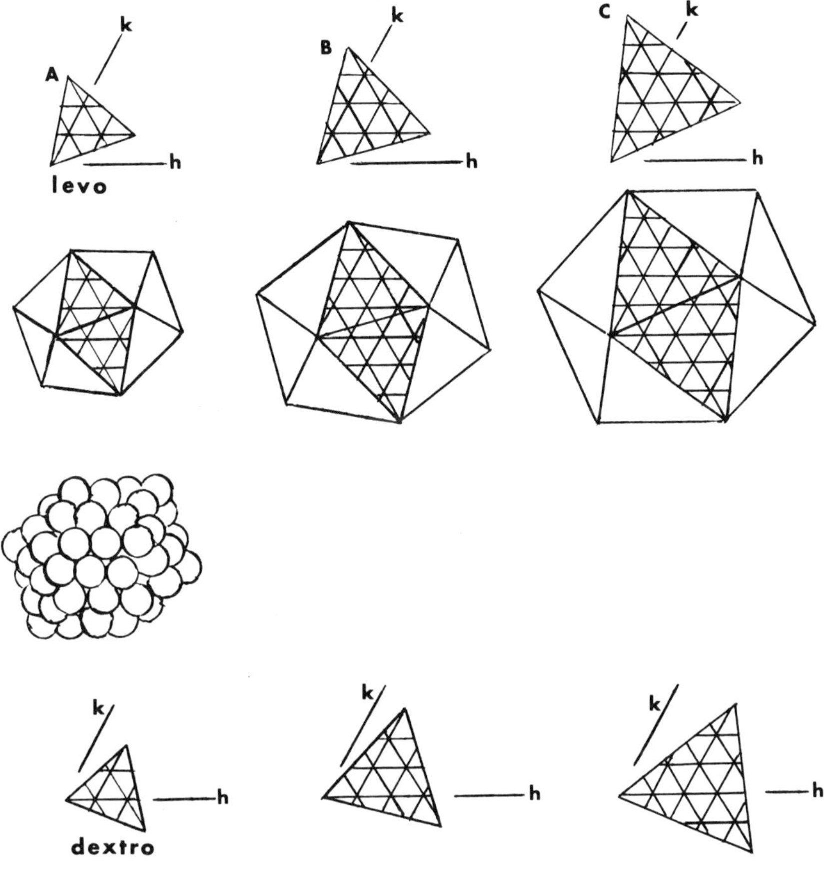

Fig. 1.7 Upper row, (h,k) indices (2,1), (3,1), (3,2) generate skewed icosahedra (second row) with T = 7, 13, and 19, respectively, all levo (h > k). A T = 7 levo, 72-capsomere model (third row). In row 4, the dextro counterparts of row 1 are shown.

In viruses corresponding to this group of deltahedra there is no view in which capsomeres on upper and lower surfaces coincide. Problems created by such nonsuperposition and the resolution of capsomeres distribution in the T = 7 group of papovaviruses have been reviewed elsewhere [83].

Figure 1.8A presents a schematic summary of characteristic faces of possible deltahedra with resulting capsomere numbers. Three subgroups are shown: (B) when h or k = 0, (C) when h = k, and (D,E) when h,k ≠ 0 and h ≠ k with the resulting dextro and levo forms. A list of viruses with established triangulation (T) and capsomere number is presented in Table 1.1.

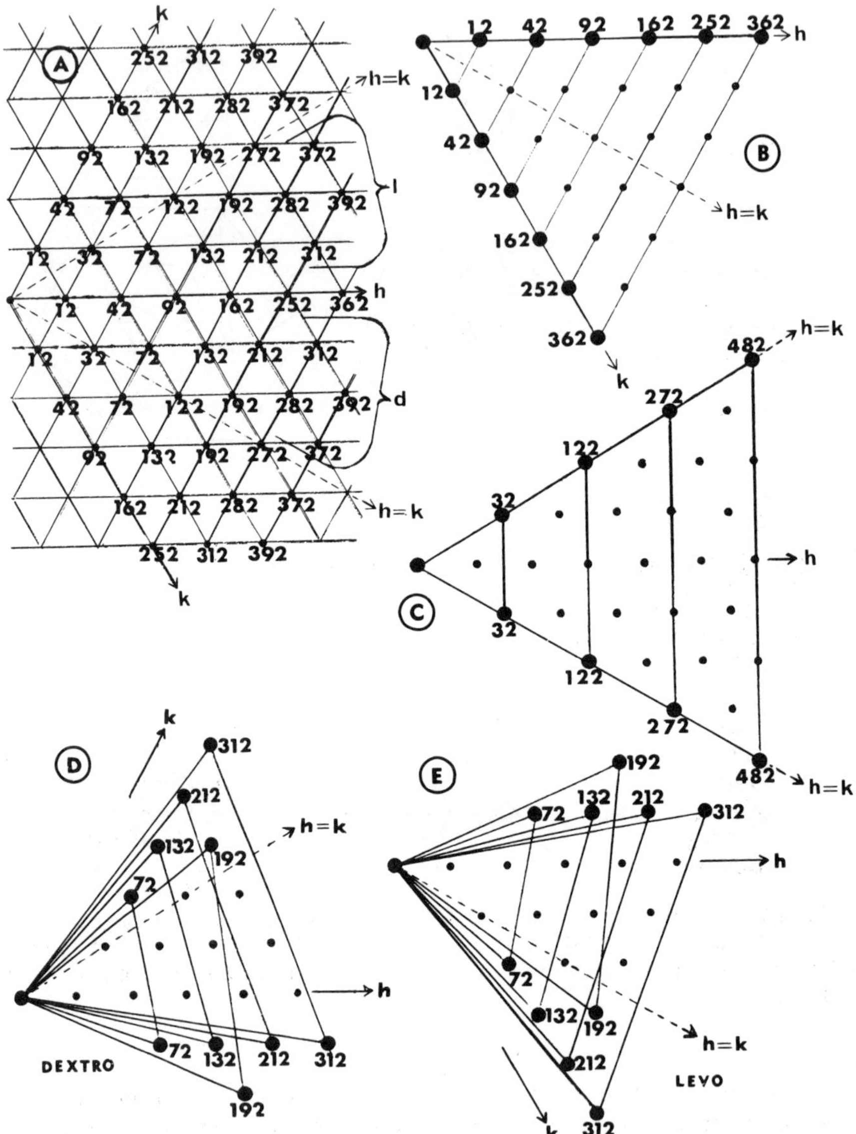

Fig. 1.8 The derivation of deltahedra from the isometric net (A) with capsomere numbers resulting from various (h,k) indices. Characteristic triangular faces are shown for T = 1 (B), T = 3 series (C), and the skewed icosahedra dextro (D) and levo (E).

Table 1.1 Icosahedral Viruses with Established Capsomere Number

T	Capsomere no.	Virus group	Virus	References
1	12	Bacterial parvo-viruses 1-DNA[a]	ϕX174	47, 109
			ϕR	64
3	32	Plant picorna-viruses 1-RNA	Turnip yellows mosiac,	36, 37, 62, 93
			Cowpea chlorotic mottle,	6
			Broad bean mottle,	37
			Cucumber mosaic	40
4	42	None known		
7	72	Papovaviruses		106
	dextro	2-DNA[a]	Human wart,	71, 72
			SV40,	4, 67
	levo		Polyoma	32
			Rabbit papilloma	72
			K	67, 106
9	92[b]	Phage 1-RNA	ZIK/1	11, 12
16	162	Herpesviruses enveloped 2-DNA	All	2, 101, 117
21	212 dextro	Phage head	Bacteriophage λ	8
25	252	Adenoviruses 2-DNA	All	58, 95, 96
81	812	Iridescent insect	*Tipula* IV	104
156	1562		*Sericesthis* IV	120

[a] 1-DNA single-stranded and 2-DNA double-stranded DNA; 1-RNA single-stranded and 2-RNA double-stranded RNA.
[b] For reovirus see Table 1.2.

It should be understood that viruses with 532 symmetry need not have the shape of an icosahedron with flat triangular faces. Their faces may be convex or even concave to varying degrees; the case where all capsomeres lie at equal radial distances from the particle center results in a spherical shell (Fig. 1.3) with 532 symmetry [19, 21].

The large viruses such as *Tipula* and *Sericesthis* iridescent viruses [104, 119, 120], adenovirus [55], and possibly herpes viruses [117] present rather flattened faces, whereas the somewhat smaller papovaviruses [32, 72] and still smaller turnip yellows mosaic virus [74] are probably best approximated by spheres with 532 symmetry. Recently ϕX174, previously described as having a 12 capsomere structure, may be more accurately represented as a 32 capsomere structure with 12 prominent vertex capsomeres and 20 deeply recessed "face" capsomeres [78].

C. Capsomere Substructure and Clustering of Structure Units

1. *Viruses with a Single Type of Protein Structure Unit (Hexamer-Pentamer Clusters)*

Along with the discovery of 532 symmetry in bushy stunt virus [17] it had been predicted that viruses with icosahedral symmetry would consist of 60 identical protein structure units (su), each having an identical environment by virtue of their being held by identical bonding sites [23, 68].

Turnip yellows mosaic virus (TYMV) is another small RNA plant virus with 532 symmetry confirmed by x-ray diffraction [70, 73, 74]. However, it was found to possess 32 morphological units (T = 3) or capsomeres [62, 93]. Biochemical data suggested that there were many more than 60 su [48] and more recently this number has been found to be nearly 180 [87].

The concept of quasi-equivalent positions of subunits [21, 68, 70] led to a solution of the probable structure of TYMV. If each of the 12 vertex capsomeres consisted of 5 su and each of the 20 face capsomeres consisted of 6 su, the entire particle would contain 32 capsomeres (T = 3) and 180 (T $\times$ 60) su and would satisfy 532 symmetry. The small difference in bonding angles between structure units in pentameric and hexameric capsomeres could be compensated for by flexion of identical bonds. Additional electron microscope studies have confirmed this structure [36]. Broad bean mottle virus [37] and cucumber mosaic virus [40] have fundamentally similar hexamer-pentamer clustering patterns. A scheme of the detailed structure fitting these 3 viruses is presented in Fig. 1.9 [36, 74].

It is also probable, although not altogether proved, that some of the other virus groups displaying large capsomere (6.0-8.0 nm in diameter) also are

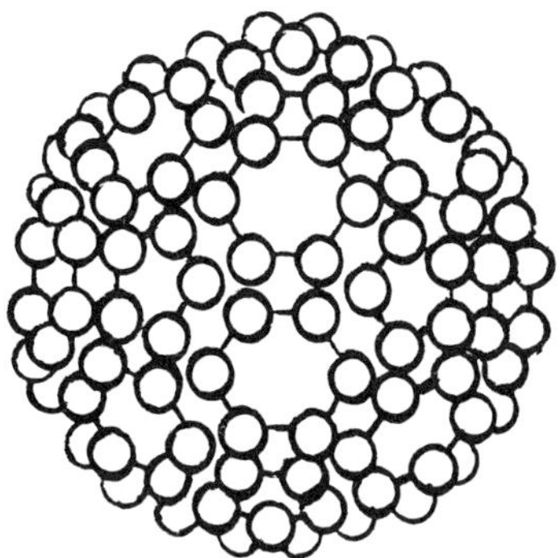

Fig. 1.9 A model of turnip yellows mosaic virus consisting of 32 capsomeres (T = 3) with pentamer-hexamer clustering, seen on a 2-fold axis.

constructed of pentamer-hexamer clusters. In such general deltahedral viruses the number of structure units in the virus will be 60T: 5 su in each of 12 vertex capsomeres and 6 su in each of 10(T − 1) nonvertex capsomeres.

2. *Viruses with Other Clustering Patterns*

A number of clustering patterns can be generated from the hexagonal net, in which net positions can be thought of as corresponding to positions of dimers and trimers (identical proteins) or nonidentical proteins. The net is thus employed to "resolve" structure unit organization.

An infinite number of structures can be systematically derived by creating holes which are symmetrically located in the net previously employed to locate capsomeres. For the derivation of such structures the reader may refer to the Appendix, Section I.VIII.C.

Not all icosahedral viruses as seen by negative staining possess a classic capsomere structure (pentamer-hexamer clusters). Three interesting examples of plant viruses are known to have other clustering patterns, and there are probably other examples among animal viruses (Table 1.2).

Such viruses may demonstrate systematically placed holes the triangulation number of which hereafter is referred to as t (hole pattern), as distinct from the more conventional triangulation number T (capsomere pattern).

Dimer clustering. Turnip crinkle virus and tomato bushy stunt virus [14, 35, 39] display morphological subunits in similar patterns, some located on axes of 2-fold rotational symmetry, which, in view of these viruses having verified 532 symmetry [17, 39], requires that all of their morphological subunits be dimers. A scheme of the construction of these viruses is presented in Fig. 1.10. They follow a t = 3 pattern but contain 90 morphological subunits (dimers) and 180 su (su = t × 60).

Table 1.2 Icosahedral Viruses with Morphological Units Consisting of Dimer or Trimer Clustering

t	No. morphological units	No. structural units	Virus	Reference
1	30 dimers	60	Small particle of turnip crinkle virus protein	39
3	90 dimers	180	Turnip crinkle Tomato bushy stunt	35, 39 39
9	180 trimers	540	Reovirus? Wound tumor?	63, 113 10, 89
	270 dimers	540	Reovirus?	83

An equally interesting small particle was formed by assembly of turnip crinkle protein [75] and shown to have icosahedral symmetry by x-ray diffraction [39]. This particle when studied by electron microscopy demonstrated dimer clustering with 30 morphological subunits (dimers) in a t = 1 pattern and 60 su [39] as shown in Fig. 1.11.

In all 3 of these dimer cluster particles a small bit of extra material was found in the center of the vertex hole [39]. Subsequently it was found that the 2 virus particles TCV and TBSV contained 2 types of protein [14]. One was of 28,000 daltons and present as 12 copies per virion; the second was 38,000

Figs. 1.10 and 1.11 The arrangement turnip crinkle virus (Fig. 1.10) showing a T = 3, dimer clustering pattern and a small turnip crinkle virus protein particle with similar dimer clustering and T = 1 pattern. Both seen on 2-fold axes.

daltons and there were 180 per virion. This distribution provided an interesting explanation for the additional small vertex material.

Trimer Clustering? Reovirus, an interesting double-stranded RNA animal virus, appears by negative staining to have 92 holes arranged in a t = 9 pattern [63, 113]. Wound tumor virus, a double-stranded RNA plant virus, may have a similar structure [10, 89]. Caspar [19] has suggested that reovirus may be an example of trimer clustering with 180 morphological units (clustered trimers) and 540 su in a t = 9 arrangement which would satisfy 532 symmetry and 92 holes. Similarly, dimer clustering of 540 su and 270 morphological units in a t = 9 pattern would satisfy the requirements (see Fig. 1.15, Appendix).

2. *Viruses with Multiple Polypeptide Chains (Nonidentical Structure Units)*

A number of viruses now appear to have capsids which consist of more than 1 protein. Problems involved in fitting multiple polypeptide chains with viruses with icosahedral symmetry have been addressed in polyoma virus [32] and some picornaviruses [100, 102]. In the former case it was suggested that viral protein No. 1 (VP1) could account for 360 su in 60 capsid hexamers and that VP2 and VP3 could provide 60 su to the 12 pentamers. The latter case [100] involves packing 4 different proteins into a structure assumed to have a T = 3 pattern (32 capsomeres) which has not yet been clearly resolved. Such numerical fitting, while of interest, must await more precise and reproducible data on the numbers of capsid proteins, their molecular weights, copies per virion, and morphological data, before the structure can be determined. The packing of multiple proteins will follow symmetry principles discussed elsewhere [27] and in the Appendix, Section I.VIII.C.

D. Core Structure

Relatively little is known about the structure of the nucleic acid-containing cores of viruses of icosahedral symmetry as compared to the organization of the nucleic acid in the helical viruses or helical nucleocapsids of other viruses discussed earlier [107].

There appears to be some relationship of the packaging of RNA of turnip yellows mosaic virus and the T = 3 arrangement of its 32 capsomeres [74].

Herpesvirus DNA has been described as being wound into a toroidal structure [43].

An unusual folding arrangement of the DNA of an amebal virus has been reported [84]. A model of the folded DNA, based upon these observations, has 222 symmetry which was previously predicted as the highest symmetry permissible for folded nucleic acid [70].

The presence of protein in the core of bushy stunt virus is well established [49], but it is not known whether it consists of capsid protein or a different protein, such as the basic core proteins found in papovaviruses [99a].

V. Viruses with Combined Symmetries

The T-even bacteriophages were hypothesized as having icosahedral caps and an elongated midportion so as to form "prolate icosahedra" [90]. Subsequently such structures were verified.

Bacteriophage T2 [13] consists of a slightly elongated structure with 2 polar caps (5 faces each) having icosahedral symmetry of T = 13 pattern (3,1). Inbetween the caps are 10 triangular faces with vertices (3,2) along the long triangle edge (Fig. 1.12). The model differs from T = 13 icosahedron by having an additional row of 20 capsomeres as indicated. The total number of capsomeres is 152 (132 + 20). Bacterophage T2 was found by freeze etching to have capsomeres consisting of subunit clusters compatible with hexamer-pentamer clustering [8].

Studies on bacteriophage T4 [92, 115] have revealed 3 classes of particle having shorter DNA molecules than the normal T4 phage. They represent 0.67, 0.77, and 0.90 the unit DNA length. Calculations of theoretical relative lengths of elongated phages with T = 13, 16, 19, 21, and 25 caps led to the suggestion that T4 phages might be constructed with T = 21 icosahedral caps, the smallest (nonelongated) particle being a T = 21 icosahedron. The others were postulated to contain, successively, 1 additional row of capsomeres producing elongated structures with normal T4 having 3 additional rows. More recently [1] it has been determined that T4 bacteriophages probably consist of a modified T = 13 surface lattice similar to that of bacteriophage T2 described above.

Bacteriophage λ was examined by freeze etching and found to have an icosahedral surface symmetry, apparently with a T = 21 dextro pattern [8]. More recently it has been proposed [118] that the capsid of phage λ consists of 2 types of protein in a T = 7 pattern. The more prominent capsomeres forming the T = 7 pattern consist of hexamer and pentamer (D protein) clusters. Centered between each 3 of these clusters there is a trimer cluster of the second and larger E protein. The low-resolution pattern is compatible with a T = 21 pattern but, in fact, the model is a unique and interesting composite T = 7 structure.

Combined elements of icosahedral and helical symmetry were found in the bacteriophage φCbK [77]. It is a rather long virus with 2 polar caps each

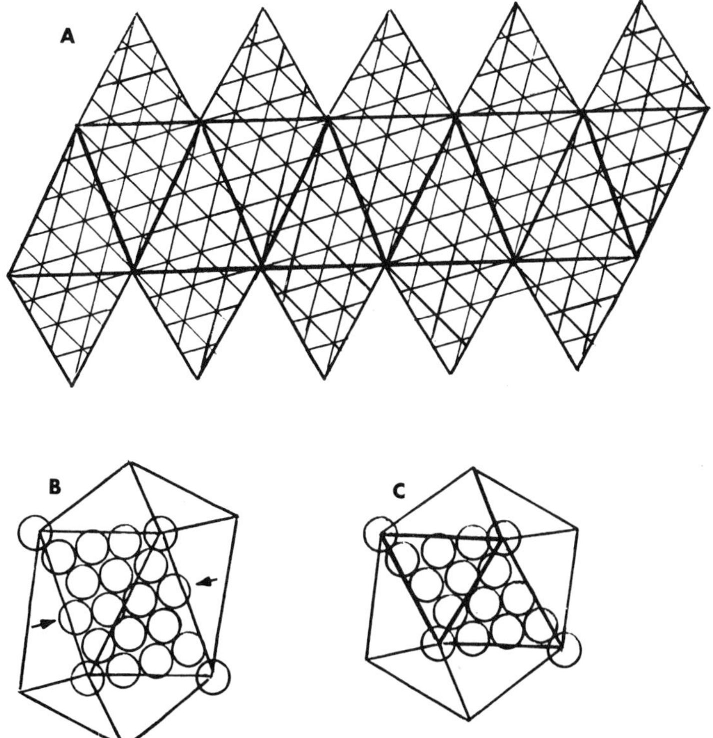

Fig. 1.12 The net pattern (A) shown by bacteriophage T2 folded into a model (B). It differs from a T = 13 icosahedron (C) by having an extra row of 20 capsomeres (arrows, B).

consisting of 5 faces with capsomeres in T = 7 icosahedral symmetry (Fig. 1.13). Between the 2 caps are 15 rows of capsomeres arranged with helical symmetry. The structure can be visualized as consisting of 10, T = 7 faces as would fold into a T = 7 (72 capsomeres) icosahedron plus 15 rows of additional capsomeres.

There are 3 other intriguing elements in the structure of this virus [77]. First, in addition to the capsomeres indicated there is a lattice of small projecting units seen 1 between each 2 capsomeres (Fig. 1.15).

Second, there are apparently 3 major proteins in the capsid: protein no. 6 with 36,000 MW and 1150 copies per virion; protein no. 7 with 35,500 MW, 130 copies/virion; and protein no. 8, 13,500 MW, 2452 copies/virion. It is speculated that protein no. 7 is associated with the vertices, 10 units per vertex, and nos. 6 and 8 are nonvertex capsid proteins.

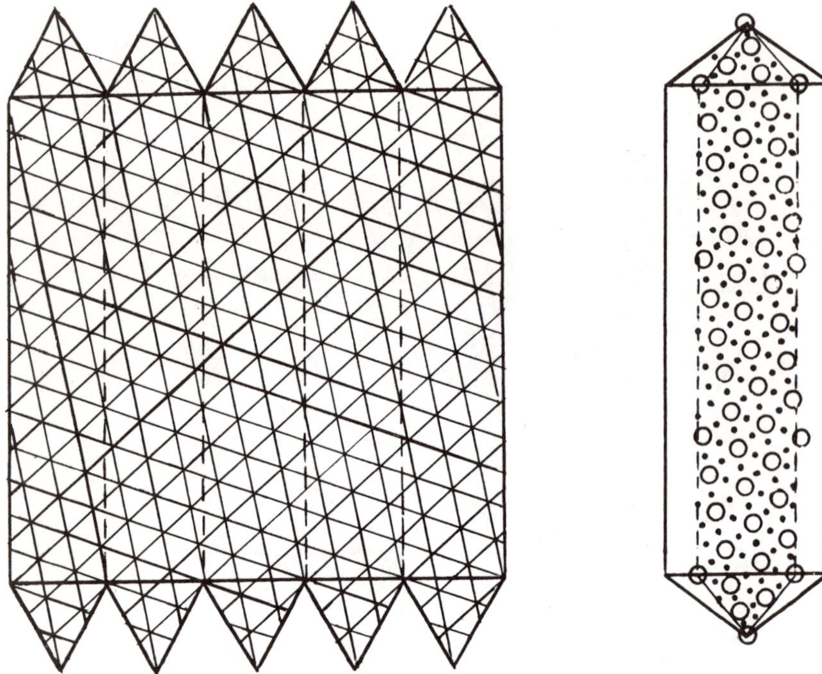

Fig. 1.13 The elongated phage ϕCbK, net pattern and model.

Third, the central vertex of the icosahedral cap seems to be the site of attachment of a helical tail (3 start helix) posing a problem in how a 3-fold symmetric tail is attached to a 5-fold symmetric vertex. It has been suggested that there may be a connector with 15-fold symmetry [90].

Another type of combined symmetry is that of bacteriophage G [26]. The head has the shape of an octahedron; it therefore has 432 symmetry. The tail consists of a helix with 6-fold symmetry attached to a vertex of 4-fold symmetry. Perhaps a 12-fold connector is involved.

VI. Tubular Variants of Viral Capsomeres

Tubular variants of bacteriophages are well known [24, 108]. In some instances the distribution of capsomeres has been studied in detail. Bacteriophage T4 polyheads [24, 121] have been found to consist either of hexamer clusters or to be in a pattern of 6 + 1 (hexagonal net). The helical structure of the sheath and polysheath of T4 tails has also been analyzed [91].

Many of the T = 7 papovaviruses produce tubular capsomere variants [66, 106]. A few of these, associated with human wart and rabbit papilloma viruses, have also been investigated in detail [66]. Wide tubes were found to consist of hexamer clusters arranged in a hexagonal lattice. Narrow tubes consisted of 2 unusual pentamer cluster arrangements with different helical parameters.

The significance of the papova filaments, some of which contain icosahedral caps or larger bulges at their ends, is not really known [66]. There is some evidence to suggest that these are not simply feasible arrangements of packaging capsomeres via the icosahedral-helical scheme, and that these structures may play a role in virus replication [85, 86].

VII. Discussion

As has been stressed elsewhere [19, 83] the purpose of the exercise with the isometric net is to show how a particular pair of coordinates of neighboring 5-fold axes (0,0) and any (h,k) define a unique structure (and generally a unique T). Therefore the structure of an icosahedral virus cannot be uniquely determined without the identification of (a) a pair of 5-coordinated capsomeres (on 5-fold axes) and (b) the pattern of 6 coordinated capsomeres inbetween to the extent that h,k coordinates, corresponding to those of the isometric net, can be assigned to the second 5-coordinated capsomere, the first taken as having coordinates (0,0). It is to be hoped that this minimum requirement for determining capsomere number and thus classifying icosahedral viruses can be more strictly met in the future.

Such viruses as reoviruses appear in electron micrographs to have holes filled with negative stain. The holes are surrounded by protein units of undetermined arrangement [19]. Some holes are 5 coordinated (surrounded by 5 holes) others are 6 coordinated. Since neighboring 5 coordinated holes have coordinates (0,0) and 3,0) this virus belongs to a T = 9 hole series. This arrangement is sufficiently different morphologically from viruses with typical capsomeres, that with some trepidation one may venture to suggest that the identifiable morphological aspect, namely, the holes with some surrounding protein of unresolved substructure, be termed coelomeres (with triangulation number t) rather than capsomeres (triangulation number T). Assuming that this pattern results from a specific and yet unresolved clustering of subunits, the term coelomere would describe the low-resolution morphological structure serve until the clustering pattern is resolved. A more sophisticated nomenclature indicating the type of clustering could be applied to define the high-resolution structure of coelomere just as hexamer-pentamer clustering defines the structure units of the capsomere.

In view of the obvious octahedral (432) symmetry of some phages [26] it seems worthwhile to point out that such symmetry can also be handled via the hexagonal net. The octahedron consists of 8 triangular faces, 6 vertices, and 12 edges and can be folded from 8 triangles. The faces can be subtriangulated and the triangulation number is $T = h^2 + hk + k^2$. The capsomere number, however, will be $C = 4T + 2$.

VIII. Appendix

A. Derivations of Equations for Triangulation and Capsomere Numbers (See Ref. 19)

1. *Derivation of* $T = h^2 + hk + k^2$

The base (b) of the general equilateral triangle (solid lines, Fig. 1.14), which represents the characteristic icosahedral face, coincides with the diagonal of the parallelogram (dashed lines). The length of the diagonal is given by

$$b = \sqrt{h^2 + k^2 - 2hk \cos B}$$

However, $B = 120°$ and $\cos B = -½$. Therefore

$$b = \sqrt{h^2 + hk + k^2}$$

The area of the general triangle is half the base times altitude $(b \cdot a/2)$.

In the equilateral triangle, $a = (\sqrt{3}/2)b$, therefore the area of the general equilateral triangle is $A = (\sqrt{3}/4)b^2$ and the area of the unit triangle (where $b = 1$) is $A\ \text{unit} = \sqrt{3}/4$.

The ratio of the areas $A/A\ \text{unit} = b^2 = h^2 + hk + k^2$.

The general triangle has an area $= b^2$ unit triangle areas. $b^2 = h^2 + hk + k^2$ is defined as the triangulation number (T) or the number of unit triangles within the general triangle.

2. *Derivation of* $C = 10T + 2$ $[C = 10(h^2 + hk + k^2) + 2]$

There are 20 faces on an icosahedron and therefore there are $20T = 20(h^2 + hk + k^2)$ unit triangles per icosahedron. Each unit triangle contributes to three capsomeres:

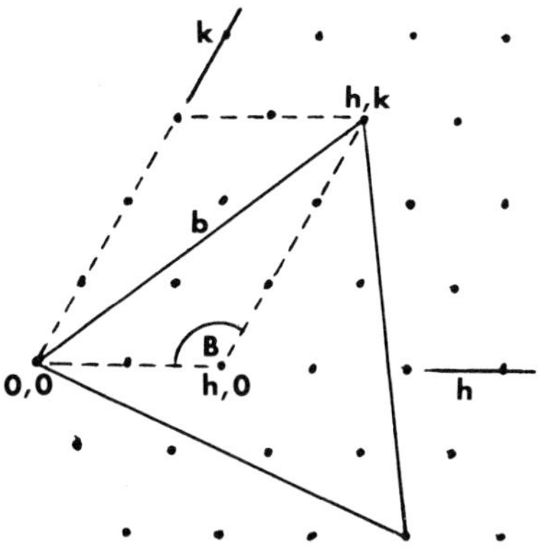

Fig. 1.14

1. Vertex unit triangles
   a. At the 12 vertices, each of 60 unit triangles contribute 1/5 to each 5-coordinated vertex capsomere or 60 × 1/5 = 12 vertex capsomeres.
   b. These same 60 unit triangles contribute 1/6 to their 2 nonvertex (6-coordinated) capsomeres; 60 × 1/6 × 2 = 20 nonvertex capsomeres. Therefore, vertex triangles contribute the equivalent of 12 + 20 = 32 capsomeres (except for T = 1, where there are no nonvertex capsomeres and C = 12 or 10T + 2).
2. Nonvertex unit triangles. Since there are a total of $20T = 20(h^2 + hk + k^2)$ unit triangles there are 20T − 60 unit triangles not associated with the 12 icosahedral vertexes. Each such triangle contributes 1/6 to the 3 (6-coordinated) capsomeres associated with each unit triangle vertex. Therefore there are (20T − 60) × 1/6 × 3 = 10T − 30 capsomeres contributed by nonvertex unit triangles.

The total of capsomeres from vertex and nonvertex unit triangles is therefore

$$C = 32 + 10T - 30$$

$$C = 10T + 2 = 10(h^2 + hk + k^2)$$

B. Topologically Unrelated Deltahedra with Same Capsomere Number

As pointed out by Goldberg [46] there are deltahedra which are topologically different but have the same capsomere number, the first being T = 49 which is generated by h = 7, k = 0 and h = 5, k = 3. Both deltahedra would contain 492 capsomeres.

There are an infinite number of such deltahedral pairs. Three infinite series which start from T = 49 are:

T series	(h,k)	or	(h,k)
49, 91, 147, 217 . . .	2n + 7,n		n + 5,2n + 3
49, 196, 441, 784 . . .	7(n + 1),0		5(n + 1), 3(n + 1)
49, 133, 259, 427 . . .	4n + 7,n		4n + 5,n + 3

Another set of infinite series is generated from T = 169:

T series	(h,k)	or	(h,k)
169, 273, 403 . . .	3n + 13,n		3n + 8, n + 7
169, 676, 1521 . . .	13(n + 1),0		8(n + 1), 7(n + 1)

C. Derivation of Dimer, Trimer and Multiple Protein Patterns

An infinite number of icosahedral structures can be systematically derived by creating holes with symmetrical patterns in the net previously employed to define capsomere patterns but now employed at "higher resolution" to define structure units. Consider any general triangle T; one can systematically delete points according to a second pattern the triangulation number of which is t. The vertices of the second icosahedral triangle (t) are selected to coincide with those of the first triangle (T). In order that the resulting holes be symmetrical T/t must be integral.

The pattern of holes so generated is defined by t, and the remaining points of the T pattern define a number of interesting structures (Figs. 1.15-1.17). Starting with a T series where h = k and subtracting general t values where T/t = n > 1 a series of structures results which can be useful models for dimer, trimer, and hexamer clusters as well as for determining the patterns of clusters of multiple proteins. The number of residual net points is 10(T + 2) – 10 (t + 2) = 10(T – t). If the number of structure units is not a multiple of 60 it must be multiplied by 2, 4, or 6, 1 of which will give n × 60 and define dimer, trimer, and hexamer clusters, respectively. This is required from the symmetry of the diagrams since in these situations net points are located on 2-fold, 3-fold, and both 2- and 3-fold axes, respectively. In the event that 10(T – t) is a multiple of 60 (60 × n) these patterns can be employed to define the distribution of n different proteins. In Figs. 1.15-1.17, 3 sets of

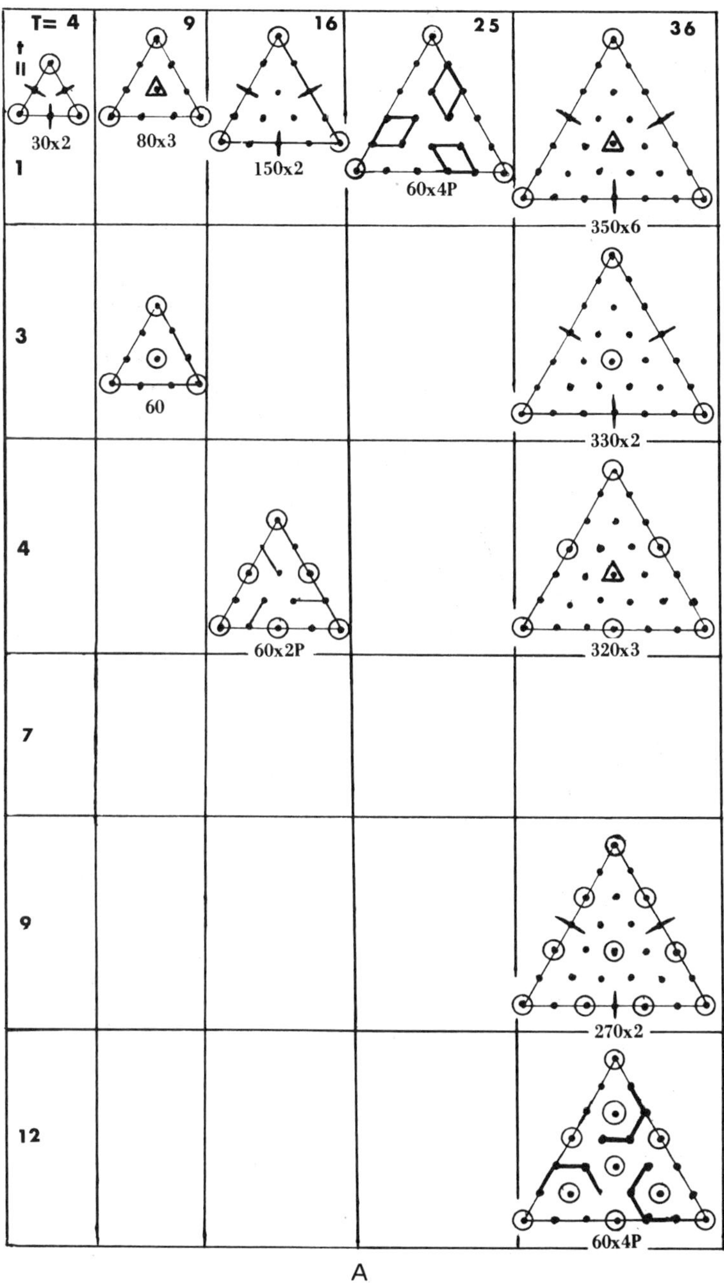

Fig. 1.15 (A,B) Systematic derivation of dimer, trimer and multiple protein clustering patterns by systematically deleting hole (circles) patterns ($t = h^2 + hk + k^2$) from $T = h^2$ net (dots) patterns (see Table 1.3).

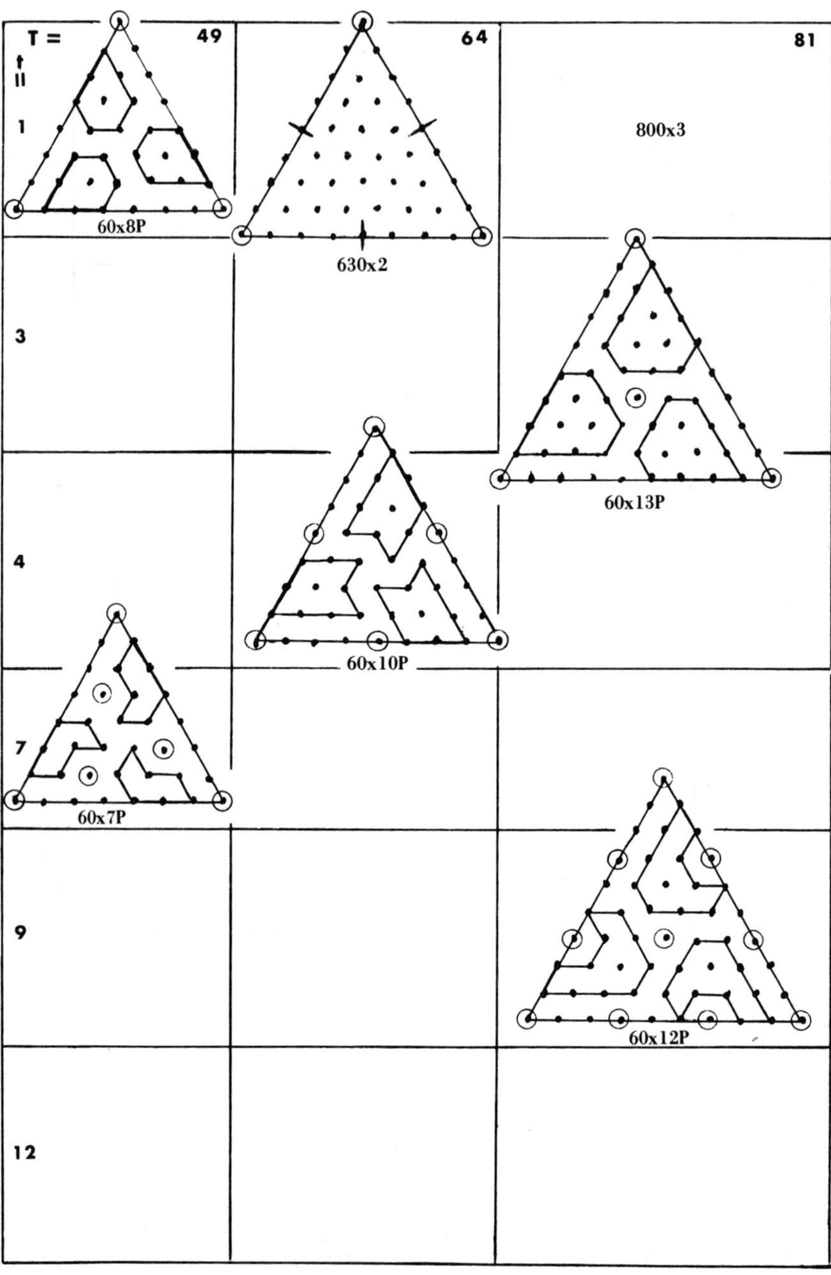

B

Fig. 1.15 (continued)

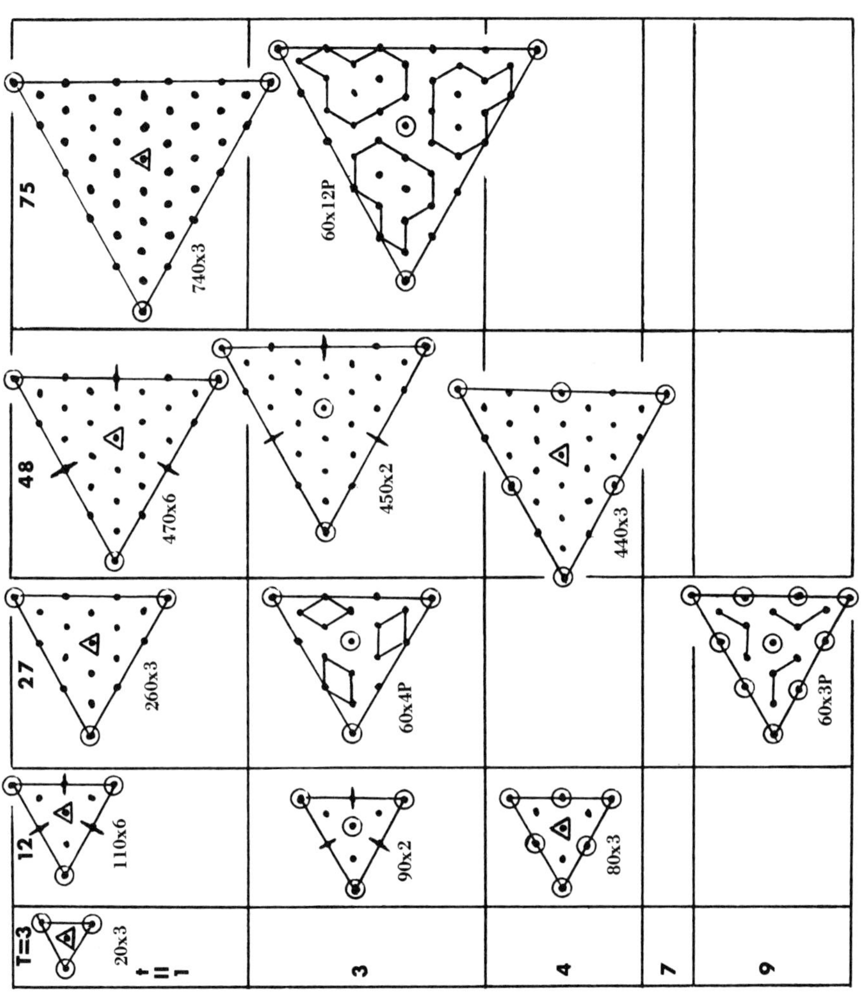

Fig. 1.16 As Fig. 1.15 but for T = $3h^2$ series (see Table 1.4).

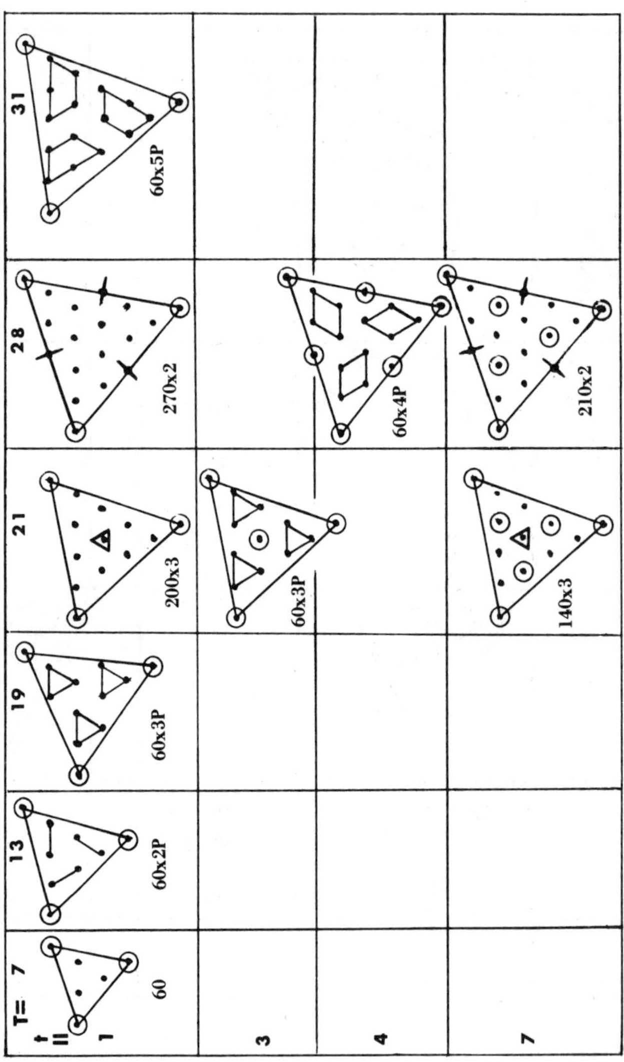

Fig. 1.17 As Fig. 1.15 but for the remaining $T = h^2 + hk + k^2$ patterns (see Table 1.5).

Table 1.3 Tabulation of Possible Patterns Containing Various Numbers of Morphological Units, Structure Units, and Different Proteins as Derived in Fig. 1.15

	$T = h^2$							
t	4	9	16	25	36	49	64	81
1	30 X 2 60 su	80 X 3 240 su	150 X 2 300 su	240 su (60 X 4P)	350 X 6 2100 su	480 (60 X 8P)	630 X 2 1260 su	800 X 3 2400 su
3		60			330 X 2 660 su			780 (60 X 13P)
4			120 (60 X 2P)		320 X 3 960 su		600 (60 X 10P)	
7						420 (60 X 7P)		
9					270 X 2 540 su			720 (60 X 12P)
12					240 (60 X 4P)			

Values of (T – t)10 where T/t is integral = number of morphological units. T = triangulation number of full triangle from which systematic deletions (holes) are made by superimposition of triangles of lower triangulation number = t.
(n × 2), morphological units (n) are dimers since there are units on the 2-fold axes and number structure units (su) = n × 2. (n × 3), trimer morphological units (units present on 3-fold axes) and su = n × 3. (n × 6), hexamer morphological units since there are units on both 2-fold and 3-fold axes; su = n × 6. (60 × nP), morphological units can be considered to consist of 60 clusters of n different proteins (P). Number of su = 60 × nP.

Table 1.4 Tabulation of Possible Patterns Containing Various Numbers of Morphological Units, Structure Units, and Different Proteins as Derived in Fig. 1.16

	$T = 3h^2$						
t	3	12	27	48	75	108	147
1	20 × 3 60 su	110 × 6 660 su	260 × 3 780 su	470 × 6 2820 su	740 × 3 2220 su	1070 × 6 6420 su	1460 × 3 4380 su
3		90 × 2 180 su	240 (60 × 4P)	450 × 2 900 su	720 (60 × 12P)	1050 × 2 2100 su	1440 (60 × 24P)
4		80 × 3 240 su		440 × 3 1320 su		1040 × 3 3120 su	
7							1400 × 3 4200 su
9			180 (60 × 3P)			990 × 2 1980 su	
12				360 (60 × 6P)		960 (60 × 16P)	

See Table 1.3.

Table 1.5 Tabulation of Possible Patterns Containing Various Numbers of Morphological Units, Structure Units, and Different Proteins as Derived in Fig. 1.17

	$T = h^2 + hk + k^2$ $(h; k \neq 0, h \neq k)$							
t	7	13	19	21	28	31	37	39
1	60	120 (60 X 2P)	180 (60 X 3P)	200 X 3 600 su	270 X 2 540 su	300 (60 X 5P)	360 (60 X 6P)	380 X 3 1140 su
3				180 (60 X 3P)				360 (60 X 6P)
4					240 (60 X 4P)			
7				140 X 3 420 su	210 X 2 420 su			
9, 12		No T/9 or T/12 integral						
13								260 X 3 780 su

See Table 1.3.

models are generated for T patterns where h or k = 0, where h = k, and where h, k ≠ 0; h ≠ k, respectively, and t is any allowable value ($h^2 + hk + k^2$). The results are presented in tabular form (Tables 1.3-1.5) where for a given t (hole pattern) one can find possible model structures.

For example, if reovirus consists of 92 holes (t = 9) the following patterns could define its structure.

1. (T36 – t9) with 270 dimers = 540 su
2. (T27 – t9) with 180 su and 3 different proteins
3. (T81 – t9) with 720 su and 12 different proteins
4. (T108 – t9) with 990 dimers = 1980 su

Thus, depending upon the size and number of protein units any of these symmetries would fit reovirus t = 9 pattern.

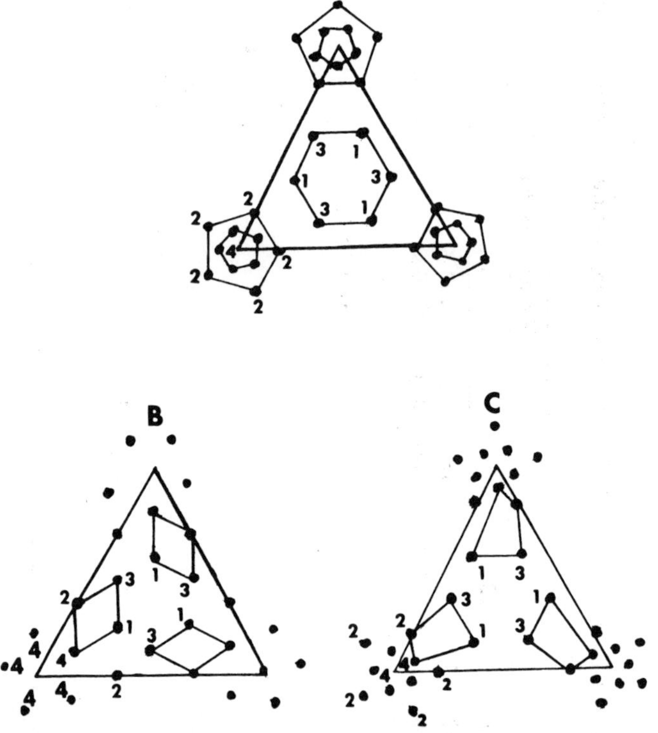

Fig. 1.18 A proposed model (A) for the clustering of 4 proteins in picornaviruses [100]. The (T27-t3) pattern (B) of Fig. 1.16 and the correspondence of the 2 (C).

A second example may be seen from (T27 – t3) with 240 su (60 × 4 different proteins). In Fig. 1.18 this may be seen to be the pattern of the 4-protein model of picornaviruses [100] under the assumption that they have a t = 3 pattern (32 holes).

It should be noted that these models define only symmetry patterns and that the relative sizes and shapes of different proteins and their absolute position within a given aggregate is not defined. If all 4 different proteins were approximately the same size and nearly spherical they would be arranged very nearly as in the model. Proteins of different size could, however, result either in more or less prominent holes or in small pentamer holes and large hexamer holes and vice versa. In this regard it is within the realm of x-ray diffraction and possibly high-resolution electron microscopy to work out the details, as it is the province of biochemistry to establish the relative sizes of proteins the number of different protein types and the number of copies per virion.

References

1. U. Aebi, R. Bijlenga, J. v. d. Broek, R. v. d. Broek, F. Eiserling, C. Kellenberger, E. Kellenberger, V. Mesyanzhinov, L. Müller, M. Showe, R. Smith, and A. Steven. J. Supramol. Struct. 2, 253-275 (1974).
2. J. Almeida, A. F. Howatson, and M. G. Williams, Virology, 16, 353-355 (1962).
3. J. D. Almeida and A. P. Waterson, in The Biology of Large RNA Viruses (R. D. Barry and B. W. J. Mahy, eds.), Academic, New York, 1970, pp. 27-51.
4. F. A. Anderer, H. D. Schlumberger, M. A. Koch, H. Frank, and H. J. Eggers, Virology, 32, 511-523 (1967).
5. K. Apostolov, T. H. Flewett, and A. P. Kendal, in The Biology of Large RNA Viruses (R. D. Barry and B. W. J. Mahy, eds.), Academic, New York, 1970, pp. 3-26.
6. J. B. Bancroft, G. J. Hills, and R. Markham, Virology, 31, 354-379 (1967).
7. A. N. Barrett, J. Barrington-Leigh, K. C. Holmes, R. Leberman, E. vonMandelkow, P. Sengbusch, and A. Klug, Cold Spring Harbor Symp. Quant. Biol., 36, 433-448 (1971).
8. M. E. Bayer and A. F. Bocharov, Virology, 54, 465-475 (1973).
9. A. J. D. Bellett, Adv. Virus Res. 13, 225-246 (1968).
10. R. F. Bils and C. E. Hall, Virology, 17, 123-130 (1962).
11. D. E. Bradley, J. Gen. Microbiol., 35, 471-482 (1964).
12. D. E. Bradley, in Comparative Virology (K. Maramorosch and E. Kurstak, eds.), Academic, New York, 1971, pp. 207-253.
13. D. Branton and A. Klug, J. Mol. Biol., 92, 559-565 (1975).
14. P. J. G. Butler, J. Mol. Biol., 52, 589-593 (1970).

15. P. J. G. Butler, Cold Spring Harbor Symp. Quant. Biol., 36, 461-468 (1971).
16. P. J. G. Butler and A. Klug, Nature, New Biol., 229, 47-50 (1971).
17. D. L. D. Caspar, Nature, 177, 475-476 (1956).
18. D. L. D. Caspar, Adv. Prot. Chem., 18, 37-121 (1963).
19. D. L. D. Caspar, in Viral and Rickettsial Infections of Man, 4th Ed. (F. L. Horsfall, Jr. and I. Tamm, eds.), Lippincott, Philadelphia, 1965, pp. 51-93.
20. D. L. D. Caspar, R. Dulbecco, A. Klug, A. Lwoff, M. G. P. Stoker, P. Tournier, and P. Wildy, Cold Spring Harbor Symp. Quant. Biol., 27, 49-50 (1962).
21. D. L. D. Caspar and A. Klug, Cold Spring Harbor Symp. Quant. Biol., 27, 1-24 (1962).
22. R. W. Compans and P. W. Choppin, in Comparative Virology (K. Maramorosch and E. Kurstak, eds.), Academic, New York, 1971, pp. 407-432.
23. F. H. C. Crick and J. D. Watson, Nature, 177, 473-475 (1956).
24. D. J. DeRosier and A. Klug, J. Mol. Biol., 65, 469-488 (1972).
25. T. O. Diener, in Comparative Virology (K. Maramorosch and E. Kurstak, eds.), Academic, New York, 1971, pp. 433-478.
26. G. Donelli, F. Guglielmi, and L. Paoletti, J. Mol. Biol., 71, 113-125 (1972).
27. A. K. Dunker, J. Virol., 14, 878-885 (1974).
28. A. K. Dunker, R. D. Klausner, D. A. Marvin, and R. L. Wiseman, J. Mol. Biol., 81, 115-117 (1974).
29. J. T. Finch, J. Mol. Biol., 8, 872-874 (1964).
30. J. T. Finch, J. Mol. Biol., 12, 612-619 (1965).
31. J. T. Finch, J. Mol. Biol., 66, 291-294 (1972).
32. J. T. Finch, J. Gen. Virol., 24, 359-364 (1974).
33. J. T. Finch and A. J. Gibbs, in The Biology of Large RNA Viruses (R. D. Barry and B. W. J. Mahy, eds.), Academic, New York, 1970, pp. 109-114.
34. J. T. Finch, P. F. C. Gilbert, A. Klug, and R. Leberman, J. Mol. Biol., 86, 183-192 (1974).
35. J. T. Finch and K. C. Holmes, in Methods in Virology, Vol. III (K. Maramorosch and H. Koprowski, eds), Academic, New York, 1967, pp. 351-474.
36. J. T. Finch and A. Klug, J. Mol. Biol., 15, 344-364 (1966).
37. J. T. Finch and A. Klug, J. Mol. Biol., 24, 289-302 (1967).
38. J. T. Finch and A. Klug, J. Mol. Biol., 87, 633-640 (1974).
39. J. T. Finch, A. Klug and R. Leberman, J. Mol. Biol., 50, 215-222 (1970).
40. J. T. Finch, A. Klug, and M. H. V. vanRegenmortel, J. Mol. Biol., 24, 303-305 (1967).
41. T. H. Flewett and K. Apostolov, J. Gen. Virol., 1, 297-304 (1967).
42. R. I. B. Franki, Adv. Virus Res., 18, 257-345 (1973).
43. D. Furlong, H. Swift, and B. Roizman, J. Virol., 10, 1071-1074 (1972).
44. A. J. Gibbs, B. Kassanis, H. L. Nixon, and R. D. Woods, Virology, 20, 194-198 (1963).

45. P. F. C. Gilbert and A. Klug, J. Mol. Biol., 86, 193-207 (1974).
46. M. Goldberg, Tohoku Math J., 43, 104-108 (1937).
47. C. E. Hall, E. C. MacLean, and I. Tessman, J. Mol. Biol., 1, 192-194 (1959).
48. J. I. Harris and J. Hindley, J. Mol. Biol., 3, 117-120 (1961).
49. S. C. Harrison, J. Mol. Biol., 42, 457-483 (1969).
50. B. D. Harrison and H. L. Nixon, J. Gen. Microbiol., 21, 569-581 (1959).
51. B. D. Harrison, H. L. Nixon, and R. D. Woods, Virology, 26, 284-289 (1965).
52. V. M. Hills and R. N. Campbell, J. Ultrastruct. Res., 24, 134-144 (1968).
53. L. Hirth, in Comparative Virology (K. Maramorosch and E. Kurstak, eds.), Academic, New York, 1971, pp. 335-360.
54. K. C. Holmes, G. J. Stubbs, E. Mandelkow, and U. Gallwitz, Nature, 254, 192-195 (1975).
55. R. W. Horne, S. Brenner, A. P. Waterson, and P. Wildy, J. Mol. Biol., 1, 84-86, 1959.
56. R. W. Horne, G. E. Russell, and A. R. Trim, J. Mol. Biol., 1, 234-236 (1959).
57. A. F. Howatson, Adv. Virus Res., 16, 195-256 (1970).
58. D. Howatson, I. A. Macpherson, M. C. Davies, Virology, 19, 418-419 (1963).
59. R. Hull, in The Biology of Large RNA Viruses (R. D. Barry and B. W. J. Mahy, eds.), Academic, New York, 1970, pp. 153-164.
60. K. Hummeler, in Comparative Virology, (K. Maramorosch and E. Kurstak, eds.), Academic, New York, 1971, pp. 361-386.
61. K. Hummeler, H. Koprowski, and T. J. Wiktor, J. Virol., 1, 152-170 (1967).
62. H. E. Huxley and G. Zubay, J. Mol. Biol., 2, 189-196 (1960).
63. L. E. Jordan and H. D. Mayor, Virology, 17, 597-599 (1962).
64. D. Kay and D. E. Bradley, J. Gen. Microbiol., 27, 195-200 (1962).
65. N. A. Kiselev, D. J. DeRosier, and J. G. Atabekov, J. Mol. Biol., 39, 673-674 (1969).
66. N. A. Kiselev and A. Klug, J. Mol. Biol., 40, 155-171 (1969).
67. A. Klug, J. Mol. Biol., 11, 424-443 (1965).
68. A. Klug and D. L. D. Caspar, Adv. Virus Res., 7, 225-325 (1960).
69. A. Klug and A. C. H. Durham, Cold Spring Harbor Symp. Quant. Biol., 36, 433-468 (1971).
70. A. Klug and T. J. Finch, J. Mol. Biol., 2, 201-215 (1960).
71. A. Klug and J. T. Finch, J. Mol. Biol., 11, 403-423 (1965).
72. A. Klug and J. T. Finch, J. Mol. Biol., 31, 1-12 (1968).
73. A. Klug, J. T. Finch, and R. E. Franklin, Biochim. Biophys. Acta, 25, 242-252 (1957).
74. A. Klug, W. Longley, and R. Leberman, J. Mol. Biol., 15, 315-343 (1966).
75. R. Leberman and J. T. Finch, J. Mol. Biol., 50, 209-213 (1970).
76. R. Leberman, J. T. Finch, P. F. C. Gilbert, J. Witz, and A. Klug, J. Mol. Biol., 86, 179-182 (1974).

77. K. R. Leonard, A. K. Kleinschmidt, N. Agabian-Keshishian, L. Shapiro, and J. V. Maizel, Jr., J. Mol. Biol., 71, 201-216 (1972).
78. S. E. Luria and J. E. Darnell, Jr., General Virology, 2nd Ed., Wiley, New York, 1967, pp. 51-89.
79. A. Lwoff, R. Horne, and P. Tournier, Cold Spring Harbor Symp. Quant. Biol., 27, 51-55 (1962).
80. D. A. Marvin and B. Hohn, Bacteriol. Rev., 33, 172-209 (1969).
81. D. A. Marvin, W. J. Pigram, R. L. Wiseman, E. J. Wachtel, and F. J. Marvin, J. Mol. Biol., 88, 581-600 (1974).
82. D. A. Marvin, R. L. Wiseman, and E. J. Wachtel, J. Mol. Biol., 82, 121-138 (1974).
83. C. F. T. Mattern, in The Biochemistry of Viruses (H. B. Levy, ed.), Dekker, New York, 1969, pp. 55-100.
84. C. F. T. Mattern, J. F. Hruska, and L. S. Diamond, J. Virol., 13, 247-249 (1974).
85. C. F. T. Mattern, K. K. Takemoto, and W. A. Daniel, Virology, 30, 242-256 (1966).
86. C. F. T. Mattern, K. K. Takemoto, and A. M. DeLeva, Virology, 32, 378-392 (1967).
87. R. E. F. Matthews and R. K. Ralph, Adv. Virus Res., 12, 273-328 (1966).
88. D. McCarthy and R. D. Woods, J. Gen. Virol., 2, 9-12 (1968).
89. S. Millward and A. F. Graham, in Comparative Virology (K. Maramorosch and E. Kurstak, eds.), Academic, New York, 1971, pp. 387-406.
90. M. F. Moody, Virology, 26, 567-576 (1965).
91. M. F. Moody, J. Mol. Biol., 25, 167-200 (1967).
92. G. Mosig, J. R. Carnighan, J. B. Bibring, R. Cole, H. O. Bock, and S. Bock, J. Virol., 9, 857-871 (1972).
93. H. L. Nixon and A. J. Gibbs, J. Mol. Biol., 2, 197-200 (1960).
94. H. L. Nixon and B. D. Harrison, J. Gen. Microbiol., 21, 582-590 (1959).
95. E. Norrby, J. Gen. Virol., 37, 565-576 (1969).
96. E. Norrby, in Comparative Virology (K. Maramorosch and E. Kurstak, eds.), Academic, New York, 1971, pp. 105-134.
97. R. E. Offord, J. Mol. Biol., 17, 370-375 (1966).
98. J. S. Pagano, Prog. Med. Virol., 12, 1-48 (1970).
99. D. Peters and E. W. Kitajima, Virology, 41, 135-150 (1970).
99a. D. M. Pett, M. K. Estes, and J. S. Pagano, J. Virol., 15, 379-385 (1975).
100. L. Philipson, S. T. Beatrice, and R. L. Crowell, Virology, 54, 69-79 (1973).
101. B. Roizman and P. G. Spear, in Comparative Virology (K. Maramorosch and E Kurstak, eds.), Academic, New York, 1971, pp. 135-168.
102. R. R. Rueckert, in Comparative Virology (K. Maramorosch and E. Kurstak, eds.), Academic, New York, 1971, pp. 255-306.
103. G. E. Russell and J. Bell, Virology, 21, 283-284 (1963).
104. K. M. Smith and G. J. Hills, Proc. 11th Intl. Congr. Entomol. (Vienna), p. 832 (1960).

105. R. Sperling, L. A. Amos, and A. Klug, J. Mol. Biol., 92, 541-578 (1975).
106. K. K. Takemoto, C. F. T. Mattern, and W. T. Murakami, in Comparative Virology (K. Maramorosch and E. Kurstak, eds.), Academic, New York, 1971, pp. 81-104.
107. T. I. Tikchonenko, Adv. Virus Res., 15, 201-290 (1969).
108. A. S. Tikhonenko, Ultrastructure of Bacterial Viruses, Academic, New York, 1970, pp. 1-294.
109. W. J. Tromans and R. W. Horne, Virology, 15, 1-7 (1961).
110. P. N. T. Unwin, J. Mol. Biol. 87, 657-670 (1974).
111. P. N. T. Unwin and A. Klug, J. Mol. Biol., 87, 641-656 (1974).
112. A. Varma, A. J. Gibbs, R. D. Woods, and J. T. Finch, J. Gen. Virol., 2, 107-114 (1968).
113. C. Vasquez and P. Tournier, Virology, 17, 503-510 (1962).
114. E. J. Wachtel, R. L. Wiseman, W. J. Pigram, and D. A. Marvin, J. Mol. Biol., 88, 601-618 (1974).
115. D. H. Walker, Jr., G. Mosig, and M. E. Bayer, J. Virol., 9, 872-875 (1972).
116. A. P. Waterson, J. C. Cruickshank, G. D. Laurence, and A. D. Kanarek, Virology, 15, 379-382 (1961).
117. P. Wildy, W. C. Russell, and R. W. Horne, Virology, 12, 204-222 (1960).
118. R. C. Williams and K. E. Richards, J. Mol. Biol., 88, 547-550 (1974).
119. R. C. Williams and K. M. Smith, Biochim. Biophys. Acta, 28, 464-469 (1958).
120. N. G. Wrigley, J. Gen. Virol., 5, 123-134 (1969).
121. M. Yanagida, D. J. DeRosier, and A. Klug, J. Mol. Biol., 65, 489-499 (1972).

Chapter 2

Interferon

Hilton B. Levy, Freddie L. Riley, and Charles E. Buckler

Laboratory of Viral Diseases
National Institute of Allergy and Infectious Diseases
National Institutes of Health
Bethesda, Maryland

I. Introduction

The phenomenon of viral interference, where the presence of 1 virus type in a cell population inhibits the growth of other virus types in the same cell population, has been known for a long time.* In 1957 Isaacs and Lindenmann [2] demonstrated that cells infected with influenza virus elaborated a soluble

*For a much more complete review of background material, see Ref. [1a]; for earlier review of the biochemical aspects of interferon, see Ref. [1b].

protein into the surrounding fluid that protected other cells from infection with the same or other viruses. They called this protein interferon. The production of interferon by cells in response to virus infection occurs not only in tissue culture, but also in lower animals and man. Interferon is a product of the cell's genome, not of the virus's genome. Thus, interferon is more or less animal species specific. Mouse interferon does not protect chicken cells, and vice versa. There is some cross-protection exerted between closely related species, such as mouse and rat, or monkey and man. However, interferon is not virus specific. That is, interferon induced in a given cell line by 1 virus is, as far as is known, the same as interferon induced in the same cells by another virus. The replication of different viruses shows a large range of sensitivity to the inhibiting action of interferon, with replication of some viruses being extremely strongly inhibited and that of others being relatively resistant. This range of sensitivity is not related to the virus that was originally used to induce the synthesis of interferon. No matter what the original inducer has been, the relative sensitivities of different viruses remain the same. In addition, different cell culture types of the same animal species show a rather wide difference in the amount of resistance to virus replication that can be induced in them by a given interferon preparation.

Interferon has never been obtained in a pure state. All the work to be discussed here has been done with preparations of a greater or lesser degree of impurity. Various biological and biophysical criteria [1] are used to establish that an effect is indeed attributable to interferon and not to impurities in the preparation.

Interferon is assayed by its ability to inhibit virus growth. One unit of interferon is defined as that amount of interferon which, when suitably incubated with proper test cells, causes a reduction of 0.5 $\log_{10}$ in the amount of a suitable sensitive challenge virus, as compared with the yield from cells not treated with interferon. Thus, a preparation containing 5000 units/ml can be diluted 5000-fold and still cause the required inhibition in the test cells. Since test systems are not uniform from laboratory to laboratory, international reference sources of interferon have been prepared and are used to facilitate standardization.

It is generally accepted that the interferon system plays a major role in natural recovery from virus infections. Soon after its discovery, it was realized that exogenously applied interferon might be a broad spectrum antiviral agent, useful in the treatment of disease in man. However, this potentiality has been very slow in realization. Some of the difficulties involved and attempts to solve these difficulties will be discussed in Section 2.II. A major problem has been that production of sufficient quantities of interferon has not been possible. Recent studies have shown that exogenous interferon is active in the control of virus disease in man [3], but the very large amounts needed precludes any immediate possibility of application. The chemical inducer of interferon, polyriboinosinic · polyribocytidylic acid [4] [poly(I) · poly(C)] is capable of acting as an antiviral agent [5] and an antitumor [6-8] agent in rodents, but is only marginally active in primates [9]. A nuclease-resistant derivative of poly(I) · poly(C) has been found to be effective in monkeys and chimpanzees [10] and is being tried in man.

Figure 2.1 represents, schematically, current thoughts on most of the cellular events that occur when a cell is caused to produce interferon and when that interferon acts on other cells. The rest of this chapter deals mostly with these events.

II. Induction and Production of Interferon

The production of interferon following exposure to an appropriate stimulus seems to be a general property of vertebrate cells. The ability of any non-vertebrate eukaryotic cells to produce an interferon is unknown, largely because of the lack of suitable systems for the detection of antiviral substances produced by such cells. In plants a substance with many of the properties of an interferon has been reported [11]. Interferons, however, have not been demonstrated among various prokaryotic systems studied [12]. While interferon induction has been studied both in vivo and in vitro, most of the information about the biochemistry of interferon production has come from studies utilizing tissue culture systems.

Interferons are products of cell genome rather than that of virus. Evidence supporting this is of 2 types. Interferons produced by a given cell seem to have the same biological properties, and usually many of the same physical properties regardless of the inducer used to stimulate interferon production [13]. Stronger support for interferon being a cell product comes from studies with metabolic inhibitors. Heller and Wagner estabished that production of interferons is inhibited by the treatment of cells with actinomycin D [14, 15]. Others have also shown inhibition by cycloheximide [16] or puromycin [17]. Therefore functional mRNA transcription and protein translation seems to be a prerequisite for interferon production. Two interpretations are possible for such studies. The more likely is that interferons are produced de novo and various inducers function by causing a derepression of a cell gene function. An alternate explanation for these observations with metabolic inhibitors is that cells produce, at a continuous rate, a prointerferon and that induction consists of stimulating the production of an activating system which converts this pro-interferon to an active form. In either case the appearance of interferon following induction does represent, either directly or indirectly, cell genetic control.

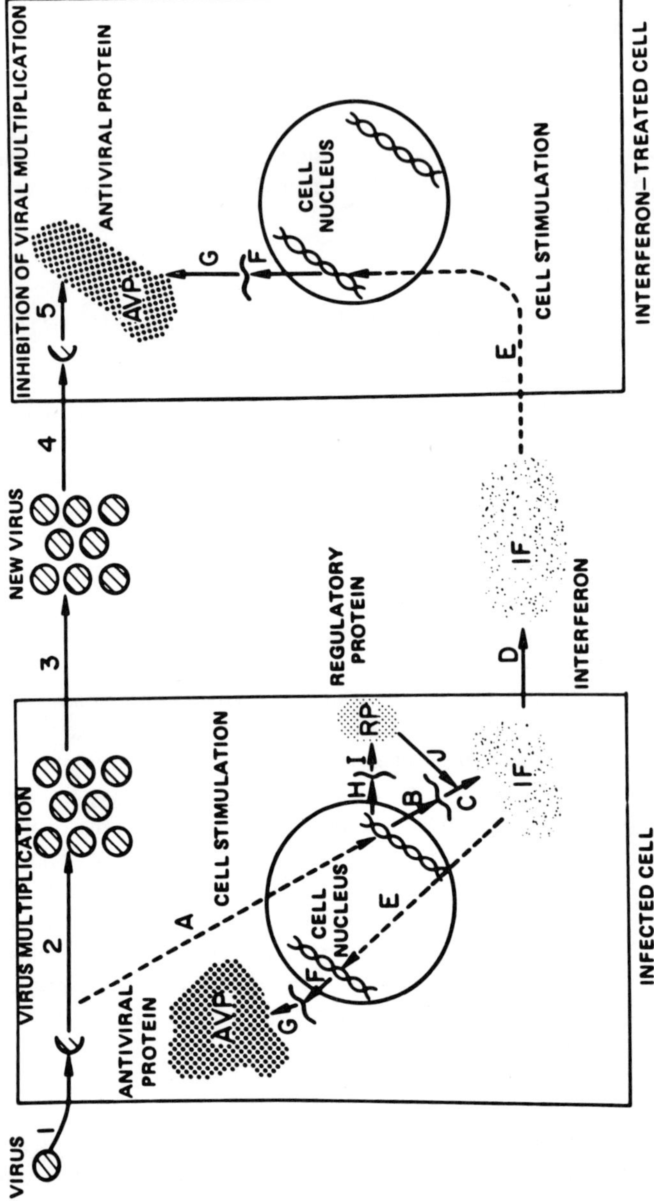

Fig. 2.1 Cellular events of the induction and action of interferon (IF). See text for explanation.

Interferon inducers can be divided into 2 classes: (a) viruses and some synthetic or naturally derived polyribonucleotides that are capable of stimulating both cells in culture and intact animals and (b) those inducers that so far have only been shown to stimulate interferon production in vivo. In the latter category bacteria and bacterial products [18, 19], tracoma-inclusion agents [20], rickettsia [21], protozoa [22], polysaccharides [23, 24], mitogens [25], and antibiotics [16] stimulate the appearance, in vivo, of interferon. Some organic chemical compounds also stimulate an in vivo interferon response: pyran copolymer [26], polyacrylic acid polymers [24], several vinyl copolymers [27], polyvinyl sulfate [28], and tilorone [29]. These substances may only stimulate certain cell types which are not present in tissue culture systems. There may be several mechanisms of interferon induction, one a specific response to virus infection and others a more general response to conditions that produce nonspecific alterations in cell metabolism and genetic control.

Viruses remain an important class of interferon inducers [29, 30] but the exact mechanism of induction is still unknown. Many studies have implicated viral RNA, both the input viral genome [31, 32] and the replicative form of the viral genome [33]. The ability of a given virus to stimulate interferon production depends on the nature of the virus-cell interaction. For example, Newcastle disease virus (NDV) is a potent stimulator of interferon production in many cells [34]. Usually little or no replication of NDV occurs in these cells. In chicken cells the virus grows well and stimulates very little interferon [35]. Even avirulent strians of NDV, which replicate in chicken cells without cytopathology are poor inducers in chicken cells, while they are good inducers in mouse cells. In general, virulent viruses rapidly induce an inhibition of host cell RNA and protein synthesis, thereby preventing the steps necessary for interferon production. Some evidence also supports the hypothesis that many of the virus infections which result in no detectable interferon synthesis induce specific inhibitors of interferon synthesis or induction [36]. Often, as with NDV, inactivation of infectivity by ultraviolet (UV) irradiation or heat will increase the interferon-inducing capacity of a given virus [37].

Some natural and synthetic double-stranded RNAs, such as polyinosinic · polycytidylic acid (In · Cn), have been found to be excellent inducers of interferon both in vivo and in cell culture [4, 38]. Neither DNA nor synthetic polydeoxyribonucleotides have been found to induce interferon [39], suggesting an important role of the 2′-hydroxyl group of the sugar phosphate backbone. A stable secondary structure seems to be related to the relative activity of various RNAs studied [40]. This may also reflect relative resistance to nuclease digestion, insuring a longer biological half-life for the most active double-stranded RNAs. Increasing the resistance of In · Cn to ribonuclease A digestion or to enzymes that degrade double-stranded RNAs, by complexing with poly cations, has been found to increase its activity in primates where high levels of RNAse III-like activity have been found [10, 41].

The site of the inductive action of the RNA inducers is unknown. They may function analogously to infectious viral inducers, presumably somewhere within the cell, or they may function by an interaction with the surface of the cell producing effects similar to hormones.

Various factors influence interferon production including age of the host animal or tissue culture cells, temperature, hormones, UV or x irradiation and carcinogens. The nature of the effect depends on the particular system and may be either stimulatory or inhibitory.

An interesting observation has come from studies of the superinduction of rabbit and human cells following exposure to In • Cn or UV-inactivated NDV [42, 43]. In these studies there is evidence for a complex mechanism of control of interferon synthesis. Thus, induction of cells with In • Cn in the presence of cycloheximide blocks translation and thus interferon production. If after 3-4 hr the translational block is reversed by the removal of cycloheximide there follows a period of rapid translation of large amounts of interferon. Furthermore, the addition of actinomycin D at this time results in an extended period of interferon production [44].

These observations have been explained as follows: During the initial period of induction in the presence of cycloheximide, the mRNA for interferon is transcribed but not translated. When protein synthesis is allowed to proceed, interferon is rapidly synthesized. Under ordinary conditions, in the absence of actinomycin, the production of interferon would lead to the synthesis of a new mRNA and its encoded protein which would act to block partially the further translation of mRNA for interferon. However, in the presence of actinomycin, mRNA for the regulator protein is not made, and interferon production goes on at a continued high rate.

III. Properties of Interferon

No interferon has as yet been isolated as a pure protein. For this reason it is difficult to make many definitive statements about the physical and chemical properties of any interferon. The contribution of contaminating substances to various measurements of interferon remains, to a large extent, unknown and a final statement of exact chemical and physical properties awaits successful purification. Therefore, the interferons are discussed as a class of proteins with emphasis on common properties.

The biological properties of interferons are shared by all substances in this class of active molecules and are the basis for the definition of an interferon. Interferons induce metabolic alterations in cells which result in the inhibition of viral replication [45]. There is no direct inactivation of the virus by interferon [30, 46]. There is a relative specificity of action of any interferon in that interferon produced in a given animal or cells derived from that animal tends to be most active only in cells from the same species [47-49]. Diminished activity can sometimes be demonstrated in cells of closely related (phylogenetic) species. There are a number of exceptions to this property, such as activity of human interferon on rabbit and mouse cells [50, 51], but in general this property has been very useful in the characterization of an antiviral activity as caused by interferon. Antiviral activity is demonstrated against a wide variety of viruses. Probably all viruses are sensitive to the antiviral action of interferon with some being much more so than others [52]. There is no relationship between interferons and their inducers other than 1 of an inducer and product, since immunological cross-reactivity does not exist between interferons and their inducers [45]. Various inducers, in a given animal or cell system, induce the production of biologically identical interferons. The differences in physical properties of such interferons, primarily molecular weights, may not be significant (see below).

A unit of activity, as yet, still has no universal meaning. It is usually defined by each investigator and is in general related to the minimum amount of material that will produce some reproducibly measurable reduction in the effect of virus replication. Reference interferon preparations derived from mouse, rabbit, and human cells are available to allow investigators to assess the relative sensitivity of their assay system [53]. The ultimate definition of a unit of activity will result from the availability of purified interferon.

Those physical measurements that have been made have been with interferon preparations with various amounts of contaminating substances. The reader should keep this in mind since the contribution of such contaminants is unknown. That interferons are proteins, or at least require proteins for their activity was demonstrated as one of the first observed properties [30]. Destruction of biological activity by proteolytic enzymes, such as trypsin, and not by other hydrolytic enzymes is well established. Thus, approaches employed to study other proteins have been utilized with the interferons.

Determinations of the isoelectric point of various interferons have been made [54] with values of from pH 3 to pH 10 reported. More recently, by utilizing the method of isoelectrofocusing, it has become apparent that most interferons have a considerable charge heterogenity. That this is in part caused by sialic acid residues on a glyprotein interferon has been demonstrated for rabbit and human interferon [55-58]. Removal of the terminal sialic acid residues results in the reduction of the number of bands formed to a relatively homogeneous preparation with most of the activity focusing at pH 6.7. The role of the carbohydrate portion is unknown and may reflect the mechanism of biosynthesis of interferon [59]. It seems to play no role in biological activity since careful removal of either the terminal sialic acid residues or more extensive removal of all of the carbohydrate residues have not caused loss in biological activity [56, 58].

In general, crude interferons tend to be more resistant to temperature and pH inactivation than many other biologically active proteins. Thus they tolerate incubation at pH 2 to pH 10 and subsequent neutralization without loss of activity [60]. Most crude interferons seem to be able to withstand heating to 37°C without loss of activity. Some interferons, such as those obtained from chicken cells, may be heated to 60°C without significant loss of activity [54]. The role of contaminating proteins in crude interferon preparations on these properties is unknown since a comprehensive study in parallel with attempts at purification has not as yet been made. As a result of studies with group-specific reagents on partially purified chicken interferon it was concluded that at least 1 sulfide group was essential for biological activity [61].

The effects of urea and guanidine HCl on interferons vary depending upon the type of interferon studied. Chicken interferon is stable in the presence of 8 M urea, while mouse interferon is not [62]. This may reflect the relative ease of reformation of hydrogen bonds during assay of chicken interferon rather than any lack of specific structure. Human interferon appears to be stable to 6 M guanidine HCl, treatment with guanidine seems to reduce the heterogeneity in molecular weight of crude preparation [63, 64].

Interferons exist as complexes with many proteins and may exist as oligomeric forms [65-67]. The biological significance of such behavior is unknown and may represent merely artifacts of preparation and isolation. In addition to binding to many proteins, interferon can form stable, biologically active complexes with carbohydrates, such as blue dextran [68]. Some of the loss in activity associated with purification attempts is thought to represent irreversible binding to glass or plastic, as protective, contaminating proteins are removed [62]. The implication is that the interferon molecule possesses areas rich in hydrophobic binding sites [69], and indeed this property has been used in some of the purification approaches used.

The reported molecular weights of the interferons are perhaps the most confusing property. Activity has been associated with proteins as small as 12,000 daltons for some forms of human interferon, to proteins greater than 100,000 daltons for most species, including human [54]. Such diversity in molecular weights could represent distinct species of interferon produced by various cells in response to different inducers. Alternately, various oligomeric forms of interferons could exist. Evidence to support this comes from the effect of urea on converting an 80,000 dalton rabbit interferon to what appears to be a 40,000 dalton monomer [70] and from observations of the effect of low ionic strength on converting what appears to be oligomers of human and mouse interferons into lower molecular weight monomers [71]. Another explanation for the molecular diversity observed with varous interferons may be specific or nonspecific binding to extraneous proteins [59]. The predominent active species of many interferons, when studied under conditions where binding and oligomer formation are controlled seems to be between 25,000 to 35,000 daltons with perhaps 5000 daltons being carbohydrate.

The purification of interferon is still an elusive goal. Most of the conventional approaches of protein isolation have been attempted with various success [54]. Precipitation, selective adsorption and elution, ion-exchange chromatography, and gel filtration have been tried. The most productive methods combine gel filtration, CM-Sephadex chromatography, and affinity chromatography [71-77]. Immunoadsorbants for affinity chromatography are prepared in rabbits or sheep by immunization with partially purified interferons. The isolated gammaglobulins are further purified by adsorption with cell and viral proteins in an attempt to remove noninterferon antibodies. By a combination of these procedures mouse, rabbit, and human interferons have been purified to specific activities between 1 and 10×10^8 units/mg proteins. In addition to immunoadsorbants, affinity chromatography under conditions that employ either hydrophobic interactions or lectin binding have been used [57, 69, 78].

The use of electrofocusing or gel electrophoresis as a final step usually achieves a considerable increase in purification [72, 75, 79]. Losses in activity during various purification steps depend, to a large extent, upon the particular procedure and the degree of purification achieved. Total recoveries range from as low as 5% to close to quantitative recovery [54, 80]. Thus, with crude interferon close to quantitative recoveries can be achieved by gel filtration or immunoaffinity adsorptoin with purification between 35- and 100-fold depending upon the nature of the crude interferon. However as specific activities increase above 1×10^6 units/mg of protein it is not unusual to obtain recoveries on the order of 5-10% in subsequent steps of purification [75].

Estimates of the specific activity of pure interferons are currently around 1×10^9 units/mg of protein [81]. This estimate is based on the assumption that interferons are homogenous proteins with molecular weights between 20,000 and 40,000 daltons. A further consideration in the estimate is the average number of cells that must respond in an assay system. Crude interferon preparations can be obtained with from 10^3 to 10^5 units of activity per milliliter as a starting titer. Therefore between 10 and 100 liters of crude interferon per milligram may be required for successful purification. With overall recoveries as low as 10% as much as 10,000 liters of crude interferon may be required in order to obtain a homogeneous isolation of sufficient material for further study. Sequencing and structural studies may therefore be a very optimistic goal.

The production of radiolabeled interferon is 1 possible way to provide an increased sensitivity for chemical studies. Such attempts have been made for mouse, rabbit, and human interferons [80, 82, 83]. However, in the case of mouse interferon control cells seem to be producing proteins with many similarities to interferon [84]. Removal of the last traces of these contaminating proteins has yet to be obtained, so that the usefulness of radiolabeled preparations is not known.

IV. Antiviral Action of Interferon

Inhibitors of both protein synthesis and mRNA synthesis prevent the development of the antiviral state in interferon treated cells [85-88]. It is therefore hypothesized that 1 or more new proteins are thought to be, at least in part, responsible for the inhibition of virus growth. This protein, has never been isolated or identified.

Interferon inhibits the multiplication of both RNA and DNA viruses. This has been shown in several virus-cell systems [89-93]. Interferon does not prevent the adsorption of virus to cells or its penetration into cells [94, 95]. Since interferon is effective even when infectious viral RNA is used as the challenge [96], it may be stated that its action occurs at some stage subsequent to the release of the viral genome. The synthesis of new viral RNA [97, 98] as well as replicative forms of viral RNA [99, 100] and the incorporation of input virus genome into the replicative forms [101] is inhibited in interferon-treated cells. Thus, interferon acts at a step between the time virus genome is released and new virus genome is synthesized.

While the final details of the mechanism by which interferon exerts its antiviral action have not been elucidated, 1 step of the virus replicative cycle that is inhibited seems to be clear.

A. Translation

Joklik [102] suggested that since interferon prevented the multiplication of both RNA and DNA viruses, the most likely locus of action would be at some common stage in their growth cycle, namely the translation into protein of information in the viral genome. In 1966, 3 laboratories presented data to show that indeed translation of viral RNA was impaired in interferon-treated cells.

Marcus and Salb [103] examined the interaction of Sindbis virus RNA with ribosomes from normal and interferon-treated chick embryo cells. With untreated cells, Sindbis viral RNA formed a viral polysome which broke down as it synthesized protein. Ribosomes isolated from interferon-treated chick cells, in contrast, bound Sindbis viral RNA less efficiently, did not disaggregate under conditions of protein synthesis, and did not incorporate amino acids into proteins.

Joklik and Merigan [104] studied vaccinia virus-infected mouse L-cells. This infection lead to the formation of viral polysomes in control cells. Formation of viral polysomes in this system was blocked by pretreatment of the cells with interferon. They found a higher rate of viral RNA synthesis in interferon-treated cells when compared to cells not treated with interferon. Protein synthesis, however, was found to be much lower in interferon-treated cells. They reported also that pretreatment of L-cells with interferon enhanced the rate of disaggregation of cell polysomes when subsequently infected with sarcoma virus, but not with Mengo virus. This work was confirmed by Metz et al. [105].

Levy and Carter [106-108] studied Mengo virus infection of L-cells in vivo. They were able to isolate on sucrose gradients, a 50 S particle from control cells which contained input Mengo viral RNA, and suggested that this particle might represent 40 S ribosome subunits in association with 37 S Mengo viral RNA. These particles were not found in cells pretreated with interferon. This step is considered to be a prerequisite for polyribosome formation in general. Metz et al. [109] have made analogous observations for vaccinia virus in L-cells. Carter and Levy [110] found that when ribosomes were prepared from L-cells exposed to homologous interferon, their capacity to associate with and translate [111] cell mRNA in vitro was preserved, while their ability to interact with Mengo viral RNA was markedly reduced. They concluded that the ribosomes from cells exposed to interferon were altered in a selective way which permitted only certain mRNAs to be bound and translated.

Kerr [112] confirmed these observations, working with cell-free fractions from chick embryo cells and encephalomyocarditis virus. Later, Friedman et al. [113] found that in addition to interferon treatment, mouse L-cells had to be infected with vaccinia virus to demonstrate the inhibiting effect of interferon on encephalomyocarditis RNA translation. Why infection with vaccinia virus is sometimes needed and sometimes not is not clear. These authors also showed that the products of the translation were indeed virus-specific polypeptides.

Falcoff et al. [114] also showed that L-cells exposed to interferon yield cell-free extracts with reduced activity for the translation of exogenous mRNAs whether viral (Mengo RNA) or cellular (rabbit reticulocyte hemoglobin RNA). Under the same conditons, however, endogenous mRNA translation was not impaired. Infection with vaccinia virus was not needed in these experiments to demonstrate inhibition.

Gupta et al. [115] studied extracts of interferon-treated Ehrlich ascites tumor cells and concluded that protein synthesis, directed by exogenous viral or cellular mRNAs was impaired but endogenous protein synthesis was not affected in these extracts. They concluded that the extracts from interferon-treated cells contained inhibitors which presumably were bound to ribosomes. They found that when they added a protein fraction from "interferon type" ribosomes to control ribosomes, they were able to convert the control ribosomes to "interferon type."

Samuel and Joklik [116] examined a cell-free system from Krebs II Ascites tumor cells. Extracts of untreated cells translated endogenous mRNAs, as well as exogenously added synthetic [poly(U)], cellular mRNA, and viral mRNA (reovirus and vaccinia). Extracts from interferon-treated cells, in contrast, translated endogenous and exogenous mRNAs but translated viral mRNA poorly. This finding is somewhat at variance with findings by both Gupta [115] and Falcoff [114] but is in accord with those of Levy and Carter.

There is general agreement that translation of viral RNA is selectively impaired in intact cells treated with interferon. Some differences among different laboratories have been found with regard to how well this difference is mirrored in cell-free systems. How the cell's translation machinery distinguishes between viral and cell mRNA is not clear. Levy and Riley [117] presented data that may bear on this last point. They reported finding modifications in certain restricted cellular RNAs after interferon pretreatment of the cells. It was shown that mRNAs and tRNAs from interferon-treated cells migrated slightly slower in polyacrylamide gels than did the corresponding components from control cells. These differences were found only in the mRNAs and tRNAs that were polysome bound (i.e., being translated). Ribosomal RNAs were not found to be altered, nor were nuclear RNAs. Additional data suggested that interferon-type mRNAs were larger than control mRNAs, while the tRNA differences appeared to be associated with secondary structure [117A]. They tentatively suggested that interferon treatment of cells leads to modification in cellular mRNAs such that the cell can distinguish between its own mRNAs and infecting viral RNA. No other reports of alterations in cellular molecules by interferon have at this time reached the literature.

Content and Revel et al. [118], working with uninfected mouse L-cell extracts, showed that the block in translation of mengo RNA or hemoglobin mRNA seen in cells pretreated with interferon could be eliminated by the addition of purified fractions of mammalian tRNA. These authors speculated that the discriminatory effect of interferon on cellular and viral mRNA translation could depend on the composition of the tRNA population of the cell. These observations are consistent with the observations made above by Levy and Riley. In addition, Gupta et al. [119], studying the extracts from control and interferon-treated Ehrlich Ascites tumor cells, reported nearly the same results. Mouse tRNA added to extracts from interferon-treated cells after the cessation of viral translation, restarts this translation at a rate close to normal. They suggested that the need for added tRNA may be the fact that in extracts from interferon-treated cells the amino acid acceptance by some of the endogenous tRNA species (but not added tRNA) is impaired.

B. Transcription

Several reports by different investigators support the idea that, in addition to its effect on translation, interferon may inhibit transcription of virion-contained nucleic acid.

Osman and Levin [120] studied lytic infection of monkey cells by a DNA virus, SV40. Treatment of these cells with interferon reduced the content of early virus-specific RNA, as well as the content of an early viral protein, the T antigen. They concluded that a possible explanation for these findings was that the mechanism of action of interferon, at least in the SV40-monkey cell system, might involve inhibition of transcription of virion DNA into early viral mRNA. However, the possibility existed that there was need for the synthesis of a very early viral protein for transcription to occur and that it was inhibition of synthesis by interferon of this early viral protein that was responsible for inhibition of viral RNA. This possibility appears to have been eliminated by experiments in which cycloheximide was used to show that transcription does not require prior protein synthesis (M. N. Oxman, 1976, personal communication).

Marcus et al. [121], studying vesicular stomatitis virus in chick embryo cells, also reported an inhibition by interferon of RNA synthesis mediated by virion-contained viral polymerase. They proposed that the action of interferon in this system might involve reactions that inhibit viral translation as well as viral RNA transcription or possibly some transcription-translation coupled system. Bialy and Colby [122] reported that pretreatment of chick cells with interferon inhibited the synthesis of early vaccinia mRNA. Their data suggested that the virion-bound transcriptase might be a target of interferon antiviral activity in this system. However, Metz and Esteban [105] showed that even in this system, translation is inhibited by interferon.

Clearly, there are 2 possible loci of virus replicating in which interferon may be active. It may be that both are operative in appropriate situations. A cell-free system in which interferon-induced inhibition of transcription can be demonstrated has not been found. Such a system would help clarify the reported effects on transcription. At the present time, the exact details of the mechanism by which interferon conveys its relatively specific virus-resistant state upon the cells still remains elusive.

V. Other Actions of Interferon

Interferon was originally thought of as specifically antiviral and without effect on normal animal or cell physiology. During the last several years it has become increasingly evident that interferon (as well as interferon inducers) can have several effects on cellular activities. These include: (a) anticellular action of the interferon system, including effects on tumors, and (b) effects on the immune system.

There have been several reports of an antitumor action of interferon [123] and of interferon inducers [6-8]. Tumors that were deliberately virus, or chemically induced and transplantable tumors without immediate virus association were studied. For example, Gresser has found that tumors in mice in ascites form (Ehrlich ascites, RC19, EL4, and L1210) were inhibited by exogenous interferon [127]. Friend leukemia in mice is also inhibited by an interferon inducer, Statolon [126] but it is not certain whether this inhibition is attributable to interferon or to enhancement of the animal's immune system. Treatment of L1210 leukemia in mice with the interferon inducer polyinosinic • polycytidylic acid had barely statistically significant effect on survival time in the mice [6], even though levels of serum interferon were obtained (3×10^4 units/ml) which were higher than those obtainable with exogenous interferon. However, poly(I) • poly(C) had strong inhibitory action on the growth of a number of other tumors [6-8]. The antitumor action of poly(I) • poly(C) may be partly ascribable to interferon, but also partly to immune enhancement [125] and partly to inhibition of tumor RNA and protein synthesis [124]. There is a recent report that administration of 3×10^6 units/week of human interferon to patients with osteosarcoma may lead to significantly enhanced metastases-free life [128]. This work requires confirmation and extension. Furthermore, it has been observed that interferon and interferon inducers, when given as one dose before a tumor transplant in mice, can enhance tumor growth [135].

The mechanism of growth inhibition by interferon appears to be complex. The growth of some tumors is inhibited in tissue culture as well as in animals. However, one tumor, the L1210R, is resistant in tissue culture but sensitive in the whole animal [136], indicating a host-mediated effect. There have also been reports that interferon inhibits the growth of nontumor cells in tissue culture [129]. Associated with findings of growth inhibition is a decrease in polysome formation [130]. There is not universal agreement that interferon inhibits cell growth. Levels of interferon that were strongly inhibitory to virus growth had no effect on L-cell growth [131] or macromolecule synthesis [132]. Conflicting observations on growth of L1210 leukemia have also been made [133].

Effects of interferon on an RNA and tRNA species of normal cells have been discussed in Section 2.IV. In summary, good data have been presented that interferon may inhibit the growth of tumors in animals and of cells in tissue culture, but there is not total agreement, at least on the magnitude of these effects. With the recently developed availability of relatively large amounts of human and mouse interferon of a reasonable degree of purity, these questions may soon be resolved.

There is more general agreement on a close link between interferon and the immune system. There is a 2-way relationship. Sensitized immunocytes, in tissue culture or in animals, when restimulated by the sensitizing antigen, produce interferon in addition to antibody [134, 137-139]. This interferon may be somewhat different from interferon induced by virus [140], at least in its pH sensitivity and immunological reactivity. In addition, cells in tissue culture or animals, when treated with interferon, show either increased [141, 142] or decreased antibody production as compared with controls not treated with interferon [141-144]. The exact response depends on the dose of interferon and the time relationship between interferon exposure and antigen stimulation, and possibly the test system used. The effects are very complex, and a detailed consideration is beyond the scope of this chapter. Both B- and T-cell functions are affected by interferon [146, 147]. Enhancement of macrophage activity by interferon has also been found [148].

The interferon system has also been reported to enhance [149] or to inhibit [150, 151] cell-mediated immunity. It is hard to understand the apparent contradiction of these observations. In our laboratories [151] in 6 independent, essentially identical experiments, 3 showed clearly that interferon exerts an inhibitory effect on graft versus host reaction in mice, while 3 showed a clear enhancing effect. It is hard to see how these diverse effects could be attributable to an effect on virus growth. It appears more reasonable to think of interferon more in the nature of a regulatory "hormone," possibly playing a role in selecting among different groups of mRNAs for translation. The alterations in mRNA and tRNA in uninfected cells treated with interferon, described in Section 2.IV., would be consistent with this idea (Fig. 2.1).

Virus comes in contact with the cell (1) and penetrates the cell membrane. The virus then releases its genetic material, and replication of the virus occurs (2). The new virus leaves the cell (3), enters the fluid around the first cell, and some of the replicated virus infects a second cell (4), where the release of the genetic material again takes place (5). During the early stages of infection of the first cell, some event stimulates a gene in the DNA which contains the stored genetic information for interferon (A). This leads to the production of an mRNA for interferon, which leaves (B) the nucleus, and is translated by the cells ribosomes (C), and the interferon protein is synthesized. Several events now occur more or less simultaneously. Some interferon leaves the first cell (D) and enters the surrounding fluid, where it comes into contact with and stimulates the second cell (E). The second cell is thereby induced to produce a new mRNA (F) which is translated to a new protein (G), the (hypothetical) antiviral protein (AVP). This in turn modifies the cell's protein-synthesizing machinery, such that cell mRNA is translated into protein, but viral RNA is poorly bound and/or translated. In the first cell processes E, F, and G may, in some instances, also operate to form AVP and thereby reduce the virus yield in the first cell. Shortly after interferon is synthesized in the first cell, another mRNA (H) is synthesized from the cell's DNA which is translated (I) into a regulatory protein (RP) (hypothesized). This regulatory protein combines with the mRNA for interferon, thereby preventing the further synthesis of more interferon (J).

It is difficult to anticipate the future of research on the interferon system, but 2 trends are clear. The next several years will in all likelihood see increased investigation of the clinical potential of the interferon system. Both interferon and interferon induces are being studied in man for possible antitumor and antiviral activity. It is reasonable to expect this work to expand. In addition, more and more laboratories are becoming involved in basic studies with the interferon system. Its antiviral action has been the subject of many studies, and significant understanding of this action has been developed. How interferon treatment modifies the uninfected cell so as to render it resistant to subsequent attack by virus is not clear and such studies are just beginning. The implications of the relationship between the interferon system and immunological defenses will, in all likelihood, furnish questions for several years. It is probable that interest in and support for interferon research will remain high.

References

1a. Interferon and Interferon Inducers (N. B. Finter, ed.), North Holland, Amsterdam, 1973.
1b. H. B. Levy, S. Baron, and C. E. Buckler, in Biochemistry of Viruses (H. B. Levy, ed.), Dekker, New York, 1969, pp. 579-612.
2. A. Isaacs and J. Lindenmann, Proc. Roy. Soc. Ser. B., 147, 258 (1957).
3. T. Merigan, S. Reed, T. Hall, and D. Tyrell, Lancet, 1, 563 (1973).
4. A. K. Field, A. A. Tytell, C. P. Lampson, and M. R. Hilleman, Proc. Natl. Acad. Sci. U.S., 58, 1004 (1967).
5. J. H. Parks and S. Baron, Science, 162, 811 (1968).
6. H. B. Levy, L. W. Law, and A. S. Rabson, Proc. Natl. Acad. Sci. U.S., 62, 357 (1969).
7. L. D. Zeleznick and B. K. Bhayan, Proc. Soc. Exp. Biol. Med., 130, 126 (1969).
8. H. Gelboin and H. B. Levy, Science, 107, 205 (1970).
9. H. B. Levy, S. Baron, and C. Gibbs, unpublished observations, 1969-1972.

10. H. B. Levy and G. Baer, S. Baron, C. Buckler, M. Iadarola, C. Gibbs, W. London, and J. Rice, J. Infect. Dis., 132, 434 (1975).
11. K. H. Fantes, in Interferon and Interferon Inducers (N. B. Finter, ed.), North Holland, Amsterdam, 1973, p. 263.
12. C. K. Mercer and R. F. N. Mills, J. Gen. Microbiol., 23, 253 (1960).
13. M. Boxaca and K. Paucker, J. Immunol., 98, 1130 (1967).
14. E. Heller, Virology, 21, 652 (1963).
15. R. R. Wagner, Nature, 204, 49 (1964).
16. J. S. Youngner, W. R. Stinebring, and S. E. Taube, Virology, 27, 541 (1965).
17. M. Ho and M. K. Breinig, Virology, 25, 331 (1965).
18. J. S. Youngner and W. R. Stinebring, Science, 144, 1022 (1964).
19. M. Ho, Science, 146, 1472 (1964).
20. T. C. Merigan and I. Hanna, Proc. Soc. Exp. Biol. Med., 122, 421 (1966).
21. H. E. Hopps, S. Kohno, M. Kohno, and J. E. Smadel, Bacteriol. Proc., p. 115 (1964).
22. M. W. Rytel and T. C. Jones, Proc. Soc. Exp. Biol. Med., 123, 859 (1966).
23. L. V. Borecky, V. Lackovic, D. Blaskovic, L. Masler, and D. Sikl, Acta Virol., 11, 264 (1967).
24. P. DeSomer, E. DeClercq, A. Billiau, E Schonne, and M. Claesen, J. Virol., 2, 886 (1968).
25. E. F. Wheelock, Science, 149, 310 (1965).
26. W. Regelson, Adv. Exp. Med. Biol., 1, 315 (1967).
27. N. A. Zeitlenok, L. M. Vilner, L. B. Trukhmanova, V. A. Kropachev, I. M. Rodin, T. A. Markelova, and M. M. Goldfarb, Vop. Virusol., 13, 401 (1968).
28. P. E. Came, M. Lieberman, A. Pascale, and G. Shimonski, Proc. Soc. Exp. Biol. Med., 131, 443 (1969).
29. G. D. Mayer and R. F. Krueger, Science, 169, 1214 (1970).
30. A. Isaacs and J. Lindenmann, Proc. Roy. Soc. Ser. B., 147, 258 (1957).
31. D. C. Burke, in Interferons and Interferon Inducers (N. B. Finter, ed.), North-Holland, Amsterdam, 1973, p. 107.
32. B. Lomniczi and D. C. Burke, J. Gen. Virol., 8, 55 (1970).
33. J. J. Skehel and D. C. Burke, J. Gen. Virol., 3, 191 (1968).
34. S. Baron and C. E. Buckler, Science, 141, 1061 (1963).
35. S. Baron, in Newcastle Disease Virus: An Evolving Pathogen (P. R. Hanson, ed.), Univ. Wisconsin Press, Madison, 1964, p. 205.
36. E. T. Sheaff, A. Meager, and D. C. Burke, J. Gen. Virol., 17, 163 (1972).
37. G. E. Gifford and E. Heller, Nature, 200, 50 (1963).
38. Z. Rotem, R. A. Cox and A. Isaacs, Nature, 187, 564 (1963).
39. G. P. Lampson, A. A. Tytell, A. K. Field, M. M. Nemes, and M. R. Hilleman, Proc. Natl. Acad. Sci. U.S., 58, 911 (1967).
40. C. Colby and M. J. Chamberlin, Proc. Natl. Acad. Sci. U.S., 63, 160 (1969).

41. H. B. Levy, G. Baer, S. Baron, C. F. Gibbs, M. Iadarola, W. London, and J. M. Rice, IRCS, 2, 1643 (1974).
42. J. Vilcek, Ann. N.Y. Acad. Sci., 173, 390 (1970).
43. Y. H. Tan, J. A. Armstrong, Y. H. Ke, and M. Ho, Proc. Natl. Acad. Sci. U.S., 67, 464 (1970).
44. J. Vilcek and M. H. Ng, J. Virol., 7, 588 (1971).
45. R. Z. Lockart, Jr., in Interferon and Interferon Inducers (N. B. Finter, ed.), North Holland, Amsterdam, 1973, p. 11.
46. A. Isaacs, Symp. Soc. Gen. Microbiol., 9, 102 (1959).
47. D. A. J. Tyrrell, Nature, 184, 452 (1959).
48. T. C. Merigan, Science, 145, 811 (1964).
49. S. Baron, S. Barban, and C. E Buckler, Science, 145, 814 (2964).
50. J. Desmyter, W. E. Rawls, and J. L. Melnick, Proc. Natl. Acad. Sci. U.S., 59, 69 (1967).
51. R. E. Levy-Koenig, R. R. Golgher, and K. Paucker, J. Immunol., 104, 791 (1970).
52. S. Baron, in Interferon and Interferon Inducers (N. B. Finter ed.), North Holland, Amsterdam, 1973, p. 12.
53. Reference interferons. For further information, see International Symposium on Standardization of Interferon and Interferon Inducers (F. T. Perkins and R. N. Rogamey, eds.), Karger, Basel, 1970.
54. K. H. Fantes, in Interferon and Interferon Inducers (N. B. Finter ed.), North Holland, Amsterdam. 1973, p. 171.
55. E. Schonne, A. Billiau, and P. DeSomer, in International Symposium on Standardization of Interferon and Interferon Inducers (F. T. Perkins and R. H. Regamey, eds.), Karger, Basel, 1970, p. 61.
56. F. Dorner, M. Scriba, and R. Weil, Proc. Natl. Acad. Sci. U.S., 70, 1981 (1973).
57. M. W. Davey, J. W. Huang, E. Sulkowski, and W. A. Carter, J. Biol. Chem., 249, 6354 (1974).
58. S. Bose, D. Gurari-Rotman, and C. Anfinsen, personal communication, 1974.
59. R. Weil and F. Dorner, in Selective Inhibiters of Viral Functions (W. A. Cater, ed.), CRC, Cleveland. 1973, p. 107.
60. J. Lindenmann, D. Burke, and A. Isaacs, Brit. J. Exp. Pathol., 38, 551 (1957).
61. K. H. Fantes and C. F. O'Neill, Nature, 203, 1048 (1964).
62. T. C. Merigan, C. A. Winget, and C. B. Dixon, J. Mol. Biol., 13, 679 (1965).
63. D. Gurari-Rotman, personal communication, 1974.
64. A. Davies, Biochem. J., 90, 290 (1964).
65. J. V. Hallum, J. S. Youngner, and W. R. Stinebring, Virology, 27, 429 (1965).
66. T. C. Merigan and W. J. Kleinschmidt, Nature, 208, 667 (1965).
67. W. A. Carter, Prep. Biochem., 1, 55 (1971).
68. S. Bose, personal communication, 1975.
69. J. W. Huang, M. W. Davey, C. J. Hejna, W. vonMuenchhausen, E. Sulkowski, and W. A. Carter, J. Biol. Chem., 249, 4665 (1974).

70. E. Schonne, A. Billiau, E. DeClercq, and P. DeSomer, in Colloque No. 6, Institut National de la Sante et de la Recherche Medicale, Paris, 1970, p. 195.
71. W. A. Carter, Proc. Natl. Acad. Sci. U.S., 67, 620 (1970).
72. S. Yamazaki and R. R. Wagner, J. Virol., 5, 270 (1970).
73. J. D. Sipe, J. DeMaeyer-Guinard, B. Fauconnier, and E. DeMaeyer, Proc. Natl. Acad. Sci. U.S., 70, 1037 (1973).
74. C. A. Ogburn, K. Berg, and K. Paucker, J. Immunol., 111, 1206 (1973).
75. Y. Kawade, Jap. J. Microbiol., 17, 129 (1973).
76. C. B. Anfinsen, S. Bose, L. Corley, and D. Gurari-Rotman, Proc. Natl. Acad. Sci. U.S., 71, 3139 (1974).
77. K. Berg, C. A. Ogburn, K. Paucker, K. E. Mogensen, and K. Cantell, J. Immunol., 114, 640 (1975).
78. M. A. Davey, J. W. Huang, E. Sulkowski, and W. A. Carter, J. Biol. Chem., 250, 348 (1975).
79. D. Stancek and K. Paucker, Acta Virol., 14, 125 (1970).
80. K. Paucker, B. J. Berman, R. R. Golgher, and D. Stancek, J. Virol., 5, 145 (1970).
81. M. H. Ng and J. Vilcek, Adv. Prot. Chem., 26, 173 (1972).
82. D. Stancek and K. Paucker, Ann. N.Y. Acad. Sci., 173, 427 (1970).
83. D. Stancek and K. Paucker, Appl. Microbiol., 21, 1067 (1971).
84. K. Paucker and D. Stancek, J. Gen. Virol., 15, 129 (1972).
85. J. Taylor, Biochem., Biophys. Res. Comm., 14, 447 (1964).
86. R. M. Friedman and J. A. Sonnabend, Nature, 203, 366 (1964).
87. S. Levine, Virology, 24, 586 (1964).
88. R. Z. Lockart, Biochem. Biophys. Res. Comm., 15, 513 (1964).
89. M. Ho, Proc. Soc. Exp. Biol. Med., 112, 511 (1963).
90. H. B. Levy, Virology, 22, 575 (1964).
91. J. Vilcek, in Virology Monographs, Vol. 6, Springer-Verlag, New York, 1969, pp. 1-115.
92. C. Colby and M. J. Morgan, Ann. Rev. Microbiol., 25, 333 (1971).
93. W. J. Kleinschmidt, Ann. Rev. Biochem., 41, 517 (1972).
94. N. B. Finter, Interferons, Saunders, Philadelphia, 1966.
95. E. DeClercq and T. C. Merigan, Ann. Rev. Med., 21, 17 (1970).
96. M. Ho, Proc. Soc. Exp. Biol. Med., 107, 639 (1961).
97. P. DeSomer, A. Prinzie, P. H. Denys, Jr., and E. Schonne, Virology, 16, 63 (1962).
98. S. E. Grossberg and J. J. Holland, J. Immunol., 88, 708 (1962).
99. R. M. Friedman and J. Sonnabend, Nature, 206, 532 (1965).
100. I. Gordon, S. Chenault, D. Stevenson, and J. Action, J. Bacteriol., 91, 1230 (1966).
101. R. M. Friedman, K. H. Fantes, H. B. Levy, and W. B. Carter, J. Virol., 1, 1168 (1967).
102. W. K. Joklik, Prog. Med. Virol., 7, 44 (1965).
103. P. I. Marcus and J. M. Salb, Virology, 30, 502 (1966).
104. W. K. Joklik and T. C. Merigan, Proc. Natl. Acad. Sci. U.S., 56, 558 (1966).
105. D. H. Metz and Esteban, Nature, 238, 385 (1972).
106. H. B. Levy and W. A. Carter, Fed. Proc., 25, 491; Bact. Proc., 119 (1966).

107. H. B. Levy and W. A. Carter, Ciba Foundation Symposium on Interferon (G. E. W. Wolstenholme and M. O'Connor, eds.), Churchill, London, 1967.
108. H. B. Levy and W. A. Carter, J. Mol. Biol. 31, 561 (1968).
109. D. H. Metz, M. Esteban, and G. Danielescu, J. Gen. Virol., 27, 197 (1975).
110. W. A. Carter and H. B. Levy, Science, 155, 1254 (1967).
111. W. A. Carter and H. B. Levy, Biochim. Biophys. Acta, 155, 437-443 (1968).
112. I. M. Kerr, J. Virol. 7, 448 (1971).
113. R. M. Friedman, D. H. Metz, R. M. Esteban, D. R. Rovell, L. A. Ball, and I. M. Kerr, J. Virol., 10, 1184 (1972).
114. E. Falcoff, R. Falcoff, B. Lebleu, and M. Revel, Nature, New Biol., 240, 145 (1972).
115. S. L. Gupta, M. L. Sopori, and P. Lengyel, Biochem. Biophys. Res. Comm., 54, 777 (1973).
116. C. E. Sameul and W. K. Joklik, Virology, 58, 476 (1974).
117. H. B. Levy and F. L. Riley, Proc. Natl. Acad. Sci. U.S., 70, 3815 (1973).
117a. F. L. Riley and H. B. Levy, Virology, 76, 49 (1977).
118. J. Content, B. Lebleu, A. Zilberstein, H. Berissi, and M. Revel, FEBS Lett., 41 (1974).
119. S. L. Gupta, M. L. Sopori, and P. Lengyel, Biochem. Biophys. Res. Comm., 57, 763 (1974).
120. M. N. Oxman and M. J. Levin, Proc. Natl. Acad. Sci. U.S., 68, 299 (1971).
121. P. I. Marcus, D. L. Engelhardt, J. M. Hunt, and M. J. Sekellick, Science, 174, 593 (1971).
122. H. S. Bialy and C. Colby, J. Virol., 9, 286 (1972).
123. Gresser, I., Adv. Cancer Res., 16, 97 (1972).
124. H. B. Levy and F. Riley, Proc. Soc. Exp. Biol. Med., 135, 141 (1970).
125. H. Cantor, R. Asofsky, and H. B. Levy, J. Immunol., 104, 1035 (1970).
126. E. F. Wheelock, L. N. Caroline, and R. D. Moore, J. Natl. Cancer Inst., 46, 797 (1971).
127. I. Gresser, D. Brouty-Boye, M. T. Thomas, and A. Macieira-Coelho, Proc. Natl. Acad. Sci. U.S., 66, 1052 (1970).
128. H. Strander, Proc. Workshop on Interferon in Malignancy, New York, 1975, in press.
129. K. Paucker and R. R. Golgher, Colloq. Inst. Natl. Sante Rech. Med. Interferon, 6, 119 (1971).
130. D. Brouty-Boye, A. Macieira-Coelho, M. Fiszman, and I. Gresser, Intl. J. Cancer, 12, 250 (1973).
131. S. Baron, T. C. Merigan, and M. L. McKerlei, Proc. Soc. Exp. Biol. Med., 421, 50 (1966).
132. H. B. Levy and T. C. Merigan, Proc. Soc. Exp. Biol. Med., 121, 53 (1966).
133. R. Adamson, S. Baron, and H. B. Levy, unpublished observations, 1975.

134. J. A. Green, S. R. Cooperband, and S. Kibrick, Science, 164, 1415 (1969).
135. A. F. Gazdar, A. D. Steinberg, G. F. Spahn, and S. Baron, Proc. Soc. Exp. Biol. Med., 139, 1132 (1972).
136. I. Gresser, M. T. Banden, and D. Brouty-Boye, J. Natl. Cancer Inst., 52, 553 (1974).
137. R. M. Friedman and H. L. Cooper, Proc. Soc. Exp. Biol. Med., 125, 901 (1967).
138. L. A. Glasgow, J. Bacteriol., 91, 2185 (1966).
139. L. B. Epstein, D. A. Stevens, and T. C. Merigan, Proc. Natl. Acad. Sci. U.S., 69, 2632 (1972).
140. J. Youngner and S. Salvin, J. Immunol., 111, 1914 (1973).
141. W. Braun and H. B. Levy, Proc. Soc. Exp. Biol. Med., 141, 769 (1972).
142. R. H. Gisler, P. Lindahl, and I. Gresser, J. Immunol., 113, 438 (1974).
143. S. Baron, H. M. Johnson, B. G. Smith, J. A. Bukovic, and A. F. Gazdar, J. Natl. Cancer Inst., in press.
144. M. P. Lindahl-Magnusson, P. Leary, and I. Gresser, Nature, New Biol., 273, 120 (1972).
145. E. DeMaeyer, L. Mobraaten, and J. DeMaeyer-Guignard, Compt. Rend. Acad. Sci. (Paris), 277, 2101 (1973).
146. B. R. Brodeur and T. C. Merigan, J. Immunol., 113, 1319 (1974).
147. K. Huang, R. M. Donohoe, F. B. Gordon, and H. R. Dressler, Infect. Immunol., 4, 481 (1971).
148. S. V. Skurkovich, E. G. Klinova, I. M. Aleksandrovskaya, N. V. Levina, N. A. Arkhipova, and T. I. Bulicheva, Immunology, 25, 317 (1973).
149. M. S. Hirsch, D. A. Ellis, P. H. Black, A. P. Monaco, and M. L. Wood, Transplantation, 17, 234 (1973).
150. L. E. Mobraaten, E. DeMaeyer, and J. DeMaeyer-Guignard, Transplantation, 10, 415 (1973).
151. H. B. Levy and R. Asofsky, unpublished observations, 1973.

Chapter 3

The Molecular Biology of Picornaviruses

David M. K. Rekosh*

Department of Molecular Virology
Imperial Cancer Research Fund
London, England

*Present affiliation: National Institute for Medical Research, London, England.

I. Introduction

Picornaviruses have been studied for over ¾ of a century. Originally, research was motivated by the fact that picornaviruses cause very serious diseases in both animals and man. In 1898, Loeffler and Frosch demonstrated that foot-and-mouth disease of cattle was caused by an agent [1], which we now know to be a picornavirus (foot-and-mouth disease virus, FMDV) and in 1909 Landsteiner and Popper showed that paralytic poliomyelitis of man was also caused by a picornavirus (poliovirus) [2].

Since poliomyelitis occurred in large epidemics and had such dread effects, an intensive research campaign was undertaken by medical scientists in an attempt to find a cure. This campaign, which has resulted in the virtual obliteration of the disease by the development of effective vaccines, has been called one of the major triumphs of modern medicine [3]. It involved intensive study of picornaviruses by hundreds of scientists and led to the isolation and characterization of over 200 distinct agents, which now comprise the picornavirus family. Over half of these agents are of human origin.

Research on picornaviruses did not stop with the development of a successful vaccine for poliomyelitis, however, for picornaviruses in more recent years have found their place in the laboratory of the molecular biologists. Picornaviruses have many technical virtues which make them ideal for detailed molecular investigations. They are among the smallest and simplest animal viruses known, they can be readily grown in cultured cells of many different origins, and they are extremely stable and easy to purify. During the past 15 years their simplicity has enabled them to be used as model systems in studies probing the complex molecular mechanisms of cellular and viral gene expression. Thus, many molecular biological tenets which today are taken for granted have first been suggested or more fully clarified in studies involving picornaviruses [4-6]. These include (a) the notion that polyribosomes are the site of protein synthesis and the concept of mRNA; (b) the concept of new enzymatic activities in virus-infected cells; (c) the ability of RNA to self-replicate and carry genetic information; (d) the precursor role of empty capsids in virus particle assembly; (e) the role of a special initiator tRNA in protein synthesis in animal cells, and (f) the role of proteolysis in protein synthesis and virus assembly.

It is the purpose of this review to provide an in-depth critical evaluation of recent work that has been carried out on the molecular biology of picornaviruses. As several recent reviews have provided detailed discussions of much of the earlier work, this review will not attempt to cover in detail material appearing before 1970. For the early work on picornavirus replication, the reader is referred to reviews by Baltimore [4] and Levintow [5], and to Rueckert [6] for the early work on picornavirus structure and assembly.

II. Classification

The classification of picornaviruses has been discussed at great length in Rueckert's review [6] and by Andrewes and Pereira [7]. The International Committee on Nomenclature of Viruses officially recognizes only 3 genera in the family Picornaviridae, namely the enteroviruses, the rhinoviruses, and the caliciviruses. However, a recent examination of the family by Newman et al. [8] makes a strong case for dividing the family into at least 5 subgroups, namely enteroviruses, cardioviruses, human rhinoviruses, equine rhinoviruses, and foot-and-mouth disease viruses, and reclassifying caliciviruses into a separate family altogether [8, 9]. The main distinction between the different subgroups is made on the basis of density in cesium chloride. It has been suggested that this difference is caused by differences in the ability of the various capsid proteins to bind cesium ions and exchange them with cations previously bound into the virion [10].

Enteroviruses include poliovirus types 1, 2, and 3; coxsackie viruses type A and B; bovine enterovirus; and the echoviruses. All of these viruses have a buoyant density of 1.33-1.35. They are all antigenically distinct. The cardioviruses (encephalomyocarditis, EMC; Maus-Elberfield, ME; and Mengo) also have a buoyant density of 1.33-1.34; however, unlike enteroviruses, they are unstable at pH 3 and at pH 6 in the presence of 0.1 M sodium chloride or bromide and thus they form a separate subgroup. The rhinoviruses, according to Newman et al. [8], can be further divided into 3 subgroups on the basis of density alone, consisting of human rhinovirus (density = 1.38 – 1.41), foot-and-mouth disease virus (density = 1.42 – 1.51), and equine rhinovirus (density = 1.45). The caliciviruses make up another group (density = 1.36 – 1.39).

All of the subgroups except caliciviruses contain 4 major types of polypeptide in their capsids. The caliciviruses contain only 1. Furthermore, caliciviruses appear different from the other picornaviruses in that they seem to have a fringelike appearance when examined in the electron microscope, are slightly larger, and contain less RNA by weight. Caliciviruses are not considered in this review.

Thus the differences between the various picornavirus subgroups are a function of specific interactions among the structural components which comprise the virions. Superficially all of the picornaviruses look the same and contain the same elements (4 different types of polypeptide chains and one molecule of single-stranded RNA). The only differences known to exist occur at the level of nucleotide sequences within the RNA molecule and, correspondingly, amino acid sequences within the proteins. As will be shown, the basic overall patterns of structure, genome organization, polypeptide species synthesized, RNA replication and assembly, seem to be conserved throughout the group, suggesting that at some stage a prototype picornavirus evolved and that the present picornaviruses all share a common ancestor. During the course of evolution, many mutations must have occurred in the structural components of the prototype virion particle, creating differences in the ability of the virus to grow on certain types of cells, and in the susceptibility of the virion particle to external influences (pH, ability to bind CsCl, etc.). Thus the present day multitude of picornaviruses and picornavirus-induced diseases.

For the purposes of this review, therefore, to stress the overall similarity of the molecular biology of all picornaviruses, the group will be reviewed as a whole. Specific differences between different virus-cell systems studied will only be referred to where I feel the differences are reflections of a basic mechanistic distinction. These times will be rare. For the most part analogies between different picornaviral systems are valid and will be made. The major virus-cell systems that will be reviewed here are poliovirus grown in HeLa cells and the cardioviruses (EMC, Mengo, ME), grown either in mouse ascites cells or HeLa cells. Some studies on bovine enterovirus, human rhinovirus, and foot-and-mouth disease virus will also be mentioned.

III. Disease Aspects

Picornaviruses are very widespread in nature. Many diseases are known to be caused by these viruses. For a detailed discussion of the overall disease aspects of picornaviruses Andrewes and Pereira [7] is very useful. Only the highlights are summarized here.

Enteroviruses are found primarily in the intestinal tract where they commonly cause little or slight illness. However, occasionally these viruses spread from the gut to other parts of the body, such as the heart or central nervous system, where destructive lesions occur. The devastating epidemic human disease, paralytic poliomyelitis, which is caused by poliovirus, has essentially been wiped out (only about 1 case per million now occurs in most parts of the world) by vaccines consisting of inactivated viruses developed by Salk [3] or live attenuated viruses developed by Sabin [11].

Human rhinoviruses are known to produce typical common colds and they are commonly isolated from the nose and throat. Infection occurs via the respiratory tract. Rhinoviruses are rarely found in the gut or feces, perhaps because of their inactivation at low pH making the intestinal tract a hostile environment.

Foot-and-mouth disease of cattle and swine vesicular disease are 2 diseases of livestock caused by picornaviruses. Swine vesicular disease virus seems to be a coxsackie-like virus, while foot-and-mouth disease virus is classified separately. Both diseases have caused great economic losses to ranches and farms and much research is currently under way to control their spread. Typically foot-and-mouth disease infection of cattle causes fever; vesicular eruptions on the mouth, tongue, and hooves; and myocardial damage. The diseases are extremely contagious. Several vaccines are being tested [7].

It is likely that several other diseases are caused by picornaviruses since the viruses are so widespread. Certainly, many inapparent infections of picornaviruses are known to exist and the viruses circulate readily in nature. Some recent lines of research implicate picornaviruses in diabetes. The literature on this subject has been recently reviewed [12]. It is clear that experimental diabetes can be induced in mice infected with EMC or Coxsackie virus. These viruses readily grow in the pancreas and cause diabetic symptoms. However it is not known to what extent model systems set up in the laboratory mimic the natural etiology of the disease in nature.

IV. Structural Components of the Virion Particle

A. Studies on RNA Structure

The picornavirus family consists of viruses which are small and icosahedral. The overall diameter of the particle has been measured to be between 20 and 30 nm, depending on the laboratory doing the measuring [4-6]. The particles contain 30% RNA by weight and the RNA exists as a unique single-stranded molecule having a molecular weight of between 2.5×10^6 and 2.8×10^6 daltons [4-6]. Many early studies have shown that RNA isolated from the virion is infectious [4-6].

Recently, the 5′ end of poliovirus RNA has been examined in detail by several groups [13-17]. As a result of these studies it is now clear that a distinction needs to be made between the RNA contained within the virus particle and the RNA associated with the polyribosomes of the infected cell. Although, previously, many workers have shown that virion-associated RNA can function as a messenger [18-24] and is virtually indistinguishable by base composition and size from polyribosomal RNA [25, 26], these two species now have been shown to differ at their 5′ terminus.

Available evidence suggests that poliovirus polyribosomal RNA contains pUp at its 5′ end [14-16]. This result is somewhat surprising, for recent data from many laboratories indicates that the 5′ termini of most mammalian mRNAs of cellular and viral origin consist of a 7-methyl guanosine residue, linked via a 5′5′-triphosphate linkage to an O-methylated nucleoside [27-30]. This "capped" structure appears to be essential for messenger translation in many cases [31, 33], yet it is clearly absent in poliovirus messenger RNA. Thus picornavirus RNA may prove to be exceptional in that it is translated without 7-methylguanosine at its 5′

end. Alternatively, 7-methylguanosine may be added to the RNA at some point during its translation and rapidly removed.

The 5′ end of the virion RNA has also been examined in detail. Early work by Wimmer suggested that the RNA 5′-terminated with pAp [13], but this conclusion was always considered unsatisfactory in that the amount of pAp detected was variable and in many instances a large amount of pUp was also found. More recent work, by the same group [17], has led to the discovery that the 5′ end of virion RNA is linked covalently to a small virus-coded polypeptide of molecular weight less than 7000. The nature of the linkage as well as the nucleoside to which the protein is attached is unknown. The protein may have a role in RNA replication, or morphogenisis.

The 3′ terminus of poliovirus virion RNA has also been examined in great detail. Using a variety of methods, Yogo and Wimmer have determined that this end of the RNA contains polyadenylic acid [34]. The poly(A) tract is heterogenous in size and has been estimated to consist of between 50 and 125 nucleotides [34-36]. Data obtained from RNase T_1 and RNase A digestion of [^{32}P]-labeled poliovirus RNA indicates that the sequence adjoining the poly(A) region is . . . YpGpGp(Ap)nA_{OH}, and periodate oxidation with subsequent borohydride reduction places the poly(A) at the 3′ terminus. Poly(A) has also been found in RNAs of other picornaviruses, although it has not been positioned and appears to be somewhat smaller [37, 38].

In a recent study Spector and Baltimore [39] removed most of the poly(A) from poliovirus RNA with ribonuclease H in the presence of poly(dT) and showed that infectivity of this poly(A) minus RNA was greatly reduced. Poly(A) must therefore play some important role in the infectious cycle of the virus, although this role is as yet undetermined. Poly(A) containing RNA has also been isolated from the polyribosomes and the replication complex of poliovirus-infected cells [40-43]. This subject is discussed more fully in Section 3.V.C.2.

Poly(C) tracts of 100-200 nucleotides in length have also been found in the genome RNA of some picornaviruses [44, 45]. Originally it was reported that poly(C) occurred only in the cardioviruses and foot-and-mouth disease viruses, but not in the enteroviruses or rhinoviruses [45]. Recently, however, poly(C) has been shown to be present in the genome of at least 1 enterovirus, bovine enterovirus, by the same group that originally reported it missing, so that the proposed correlation of subgroup and poly(C) in the genome is somewhat in question [46]. The function of poly(C) is at present a complete mystery.

RNA extracted from Mengovirus virions has been shown, by Salomon and Littauer, to be capable of being acylated by histidine when incubated in a cell-free preparation containing amino acids and crude mouse liver aminoacyl-tRNA synthetases [47]. *Escherichia coli* synthetases will not acylate the RNA. The Mengovirus RNA is cleaved to smaller fragments during the acylation reaction; however, this may be caused by contamination of the synthetase preparation with endonucleases and a specific cleavage has not been shown to be necessary for charging. It would be extremely interesting and important to determine the section of the Mengovirus RNA molecule in which charging occurs. Specific aminoacylation has also been detected for EMC RNA [48]. In this case serine seems to be the amino acid charged, although the charging detected appears to be only 2-3 times above background of the system. However, it has been reported by Öberg and Phillipson [49] that poliovirus RNA cannot be acylated. Whether this reflects a basic difference between poliovirus RNA (an enterovirus) and RNA from EMC or Mengovirus (cardioviruses) or differences in the charging systems used remains to be determined. The biological function of the amino acid acceptor activities of these RNAs is at present unknown, although a variety of plant virus RNAs also have this property [49, 50-55].

B. Studies on Protein Structure

As mentioned in Section II, the protein compositions of all of the picornaviruses that have been examined in detail are remarkably similar [6]. All of the virion particles contain 3 different species of polypeptide, ranging in molecular weight from 23,000 to 37,000 daltons (VP1 or α, VP2 or β, VP3 or γ) and a fourth polypeptide having a molecular weight in the range of 8000-9000 daltons (VP4 or δ) (see Table 3.1). In addition a small amount of a fifth polypeptide (VP0 or ϵ), having a molecular weight of 39,000-43,000 daltons, has sometimes been detected [6, 56-58]. As will be discussed more fully in Section V.C.1 this polypeptide is a precursor to VP2 and VP4 and is not genetically distinct. The virions are insensitive to ether and therefore do not contain any essential lipid [6]. Some reports indicate the presence of carbohydrate within the virion [6, 59]; however, other studies show very convincingly that in highly purified virus any carbohydrate which may be present certainly is not covalently bonded to protein [60, 61].

There is little doubt that polypeptides VP1, VP2, and VP3 are always present in equimolar amounts within the picornavirus particles. Polypeptide VP4, however, has been reported present in equimolar amounts in some picornaviruses [6] and in half-molar amounts in others [6, 62]. The significance of these data is difficult to interpret since such calculations are based on molecular weight determinations from SDS polyacrylamide gels. Such determinations are difficult to make accurately in the low molecular weight range. Thus the differences may be experimental rather than actual. If polypeptide VP4 is truly half-molar in amount, one would have to postulate some specific mechanism for dissociation of half of the VP4 polypeptide chains from

Table 3.1 The Molecular Weights of the Capsid Polypeptides of Various Picornaviruses[a]

Chain nomenclature						
1	2	EMCV[1]	Poliovirus[2]	Rhinovirus[1]	FMDV[2]	Bovine enterovirus[2]
ϵ	VP0	40,000	42,000	39,000	43,000	37,000
α	VP1	34,000	37,000	35,000	34,000	34,000
β	VP2	30,000	31,000	30,000	30,000	28,000
γ	VP3	23,000	26,000	25,000	26,000	26,000
δ	VP4	9000	8000	8000	13,500	9000

[a]Data from Refs. [64, 106, 153].
The superscript numbers refer to chain nomenclature.

the particle, since evidence which will be presented in Sections 3.V.C.1 and 3.V.D shows that the polypeptides of the virion are synthesized and assembled equimolarly.

The structural arrangement of the various polypeptide chains within the virion has been examined by electron micrographic methods, x-ray diffraction, and controlled chemical degradation [6]. A model for picornavirus structure consistent with all of the data, following established theories of symmetry, has been proposed by Rueckert and co-workers [6, 56, 57]. They suggest that the capsid of Maus-Elberfield virus consists of 58 mature structural units (mature protomers), 2 immature structural units (immature protomers), and 30-60 free chains of polypeptide δ. Each mature protomer is thought to consist of 1 chain each of polypeptides α, β, and γ, while each immature protomer consists of 1 chain each of polypeptides α, γ, and ϵ. Since polypeptides β and δ are derived by proteolysis of polypeptide ϵ during assembly of the capsid, the polypeptide chains present in each type of protomer have the same origin. This relationship will be elucidated more fully in the sections dealing with protein synthesis and virion assembly (Sections 3.V.C.1 and 3.V.D.).

It is further proposed that the protomers, both mature and immature, are held together by 2 sets of bonds. The first set binds 5 protomers together into a pentamer, while the second set holds 12 pentamers in place to form the icosahedral shell. The δ polypeptide chains are bond within this structure but the precise interactions involved are at present unknown. The proposed structure is illustrated in Fig. 3.1a.

The evidence for this proposal comes from data obtained during chemical disruption of the virion. Disruption with 2 M urea yields subviral particles which contain equal molar amounts of polypeptides α, β, and γ. These particles sediment at 5 S and have a molecular weight consistent with a protomer containing 1 molecule of each polypeptide. Thus urea seems to act by destroying the protomer-protomer bonds.

Evidence for protomer association into pentamers is obtained by mild acid dissociation of the virion. This generates subviral particles which also contain equal molar amounts of polypeptides α, β, and γ. However these particles sediment at 14 S and have a molecular weight consistent with that expected for a pentamer composed of 5 mature protomers. Polypeptide δ precipitates under these conditions and therefore does not appear to be associated with the pentamers. Also found in the precipitate in amounts consistent with the model presented are "immature pentamers" containing polypeptide ϵ as well as polypdptides α, β and γ. Thus, acid dissociation is thought to destroy pentamer-pentamer bonds but not protomer-protomer ones. A very similar model, based on similar types of studies as well as electron micrographic data, has recently been proposed for Mengovirus by Mak et al. [63].

Another degradation study using foot-and-mouth disease virus has been undertaken by Talbot and Brown [64]. They propose an identical arrangement

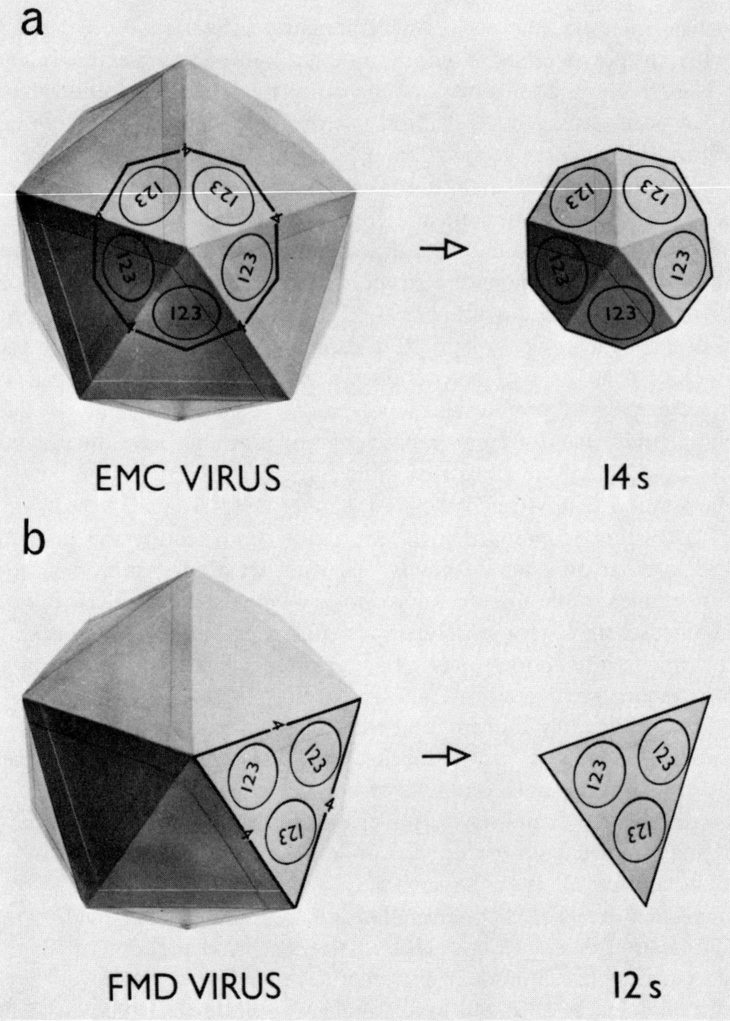

Fig. 3.1 Two models of picornavirus structure and particle dissociation. (a) The model proposed for EMC, ME, and Mengovirus showing dissociation of the capsid into 14 S pentamers. (b) The model proposed for FMDV showing dissociation of the capsid into 12 S trimers (By permission of Ref. [64].)

of the polypeptide chains within the virion particle to that presented above. However, they also propose that in foot-and-mouth disease virus the stabilizing bond holding the protomers together is trimeric rather than pentameric (see Fig. 3.1b). This is proposed because acid dissociation generates subviral particles having a molecular weight that is consistent with that of 3 protomers rather than 5. The particle sediments at 12 S. SDS gels have been known to yield anomalous molecular weight determinations for some proteins, so the significance of this difference is not clear until the molecular weight determinations are carefully checked by other methods. Another possible difference between FMDV and ME virus is that FMDV apparently does not contain any residual polypeptide ϵ and thus has 60 mature protomers rather than 58 mature ones and 2 immature ones.

All of the above studies present similar models for the structure of the picornavirus virion. In each case it is proposed that the icosahedral shell contains 60 more or less identical structural units arranged in a symmetrical manner. The role of the 2 immature protomers found in some virions is not known. Furthermore, the models are incomplete in that nothing is said about the interactions of the proteins with the RNA which is packed inside. This area awaits further investigation.

V. Intracellular Virus Replication

The virus multiplication cycle can more or less be subdivided into several events. Many of these events are interrelated and occur concomitantly. For purposes of analysis and rhetoric, however, they will be dealt with separately. The basic events which will be reviewed are:

A. Adsorption, uncoating, and penetration
B. Alteration of host cell functions
C. Synthesis of virus-specific protein and RNA
D. Virion assembly and release of virus

A. Adsorption, Uncoating, and Penetration

The early stages of picornavirus infection have not been well characterized. Different picornaviruses seem to have different host ranges, although non-susceptible mammalian cells can be infected by isolated RNA at a greatly reduced efficiency [4-6]. Polioviruses, for example, only infect cells of human or simian origin, while the cardioviruses are capable of infecting almost any mammalian cell type. Purified poliovirus RNA, however, has been shown to be infectious in many different cells types [4-6]. This difference presumably is caused by differential availability of specific receptors on the cell surface which are necessary for virus attachment. The necessity for a specific receptor can be bypassed by infectious RNA.

Recently, Green and co-workers [65, 66] have shown that the gene for the poliovirus receptor in human cell lines resides on chromosome 19. By hybridizing the susceptible human cells with nonsusceptible mouse cells these workers were able to follow the loss of chromosomes in several established lines and correlate this loss with susceptibility to virus infection. Lack of adsorption was distinguished from lack of replication by a radioactive virus adsorption assay. In addition to demonstrating a receptor-chromosome correlation they were also able to show that cells resistant to poliovirus still adsorbed several other enteroviruses, demonstrating directly that different receptors are involved for different viruses. This has also been shown by competition experiments carried out by other workers [4-6, 67].

Several workers have demonstrated that the small polypeptide, δ or VP4, present within the virion plays some role in the adsorption process. Crowell and Philipson [68] and Lonberg-Holm and Korant [69] have shown that both Coxsackie virus and rhinovirus eluted from cells, to which they had been briefly adsorbed, lack polypeptide VP4. The eluted viruses are not capable of reattachment. Furthermore, Cords et al. [70] and Briendl and Koch [71] have shown that Coxsackie viruses or polioviruses treated with low ionic strength buffers or diethylpyrocarbonate lose VP4 concomitantly with loss of infectivity.

Taken together the above results indicate that polypeptide VP4 is somehow involved in the adsorption of picornaviruses. The main problem that exists with these types of experiments is that they are only correlative and do not indicate reasons for the specific loss of polypeptide VP4. There are at least 2 models to explain this data which cannot be distinguished. The first model is that VP4 is directly involved in adsorption. It may bind tightly to a receptor and therefore would not be present in virus eluted from cells. The second model which has been proposed states that loss of VP4 from the particle results as a byproduct of adsorption. Some conformational change within the capsid may take place upon adsorption, causing VP4 to be lost. Studies fractionating rhinovirus into several populations, each with a presumed different conformational state, lend credence to this latter model [72]. It is also interesting to note that the structural studies mentioned above indicate that VP4 is loosely bound within the virion and therefore does not seem to form part of the major structural unit. Structurally, therefore, it is possible to lose VP4 and still retain the integrity of the particle.

B. Alteration of Host Cell Functions

After adsorption of the virus particle to the host cell, the next period of time in the infection is referred to as the eclipse period [1]. The length of this period, and of the entire infectious cycle for that matter, varies somewhat depending upon multiplicity of infection, virus strain used, and cell type. During this time no infectious virus can be detected and very few, if any, viral-specific products can be seen. In some little understood way, the attached virus penetrates the cell, becomes uncoated, begins to replicate its RNA, and synthesizes its proteins probably on membraneous structures which exist in the cytoplasm [73-76]. It is during the eclipse period, however, that the virus exerts its most significant effect on the metabolism of the host cell.

1. *Inhibition of Host Cell Protein Synthesis*

By 3-4 hr after infection in most systems studied, host cell protein synthesis is completely shut off and only virus-specific protein synthesis can be seen. This inhibition has been very well characterized in the earlier literature [1, 4, 5] and all of the data describing the phenomena will not be reviewed here. Briefly, however, inhibition and the switch from cellular protein synthesis to virus-specific synthesis is characterized by a gradual disaggregation of the heterogeneous host cell polyribosomes having an average sedimentation rate of 200 S, and a gradual increase of larger picornaviral polyribosomes which sediment more homogeneously at 350 S [4, 77]. During the course of the inhibition, the rate of cellular polypeptide elongation remains constant and cellular messenger RNA appears to be stable [4, 78].

Taken together, then, these data imply that inhibition takes place at the level of initiation. Any model which would hope to account for this shutoff must take this fact into account and also the fact that replication of the virus is not necessary for shutoff. At very high multiplicities of infection host cell protein synthesis shutoff efficiently occurs in the presence of drugs which inhibit RNA replication [4, 79]. In addition, it has been shown that some virus-specific protein factor is required [4, 80].

Mutants of poliovirus, defective in their ability to shut off host protein synthesis, have been isolated [81]. These mutants have been shown to map in the part of the genome which codes for the structural proteins. It has been proposed, therefore, that a structural protein is involved in the shutoff event perhaps by acting as a new initiation factor [82, 83]. There is conflicting evidence here, however, as work with a deletion mutant [84] has shown that although no stable coat protein is synthesized, protein synthesis can still be rapidly suppressed.

An alternative model of host cell shutoff has been proposed by Ehrenfield and Hunt [85]. They suggest that viral double-stranded RNA may be responsible for the inhibition of host cell protein synthesis. This suggestion is based on the fact that several different sorts of double-stranded RNA act as potent inhibitors of globin synthesis when added to a rabbit reticulocyte cell-free system. It has also been shown that double-stranded RNA can inhibit protein synthesis in vivo when added exogenously to cultured cells [86]. However, although the phenomena of shutoff in vitro is real, there is substantial evidence to indicate that this cannot be the mechanism responsible for protein synthesis shutoff in picornavirus infected cells. Some of this evidence follows.

1. Inhibition by double-stranded RNA in vitro is not specific. Viral as well as cellular protein synthesis is inhibited [87].
2. Host cell shutoff occurs early in infection and little, if any, double-stranded RNA is present at these times [4].
3. Host cell shutoff occurs during infection in the absence of RNA replication and presumably replication is needed for double-stranded RNA RNA formation [4, 77-79].

Recently, Lawrence and Thach, primarily on the data obtained in a cell-free protein-synthesizing system, have shown that the situation is much more complicated than either of the above models suggests [88]. They propose a mechanism based, at least in part, on the fact that viral mRNA seems to be capable of out competing cellular RNA for some factor necessary in the initiation of protein synthesis.

Cell-free protein-synthesizing systems were prepared from mouse plasmacytoma cells infected with EMC virus or from uninfected cells. If some factor exists in the infected cell which specifically inhibits translation of cellular mRNA, the infected cell-free system would be expected to be inactive when primed with cellular mRNA but active when viral RNA is added, while the uninfected cell-free system should be equally active for both. Such is not the case, however. Both systems are equally active with respect to their abilities to translate in vitro RNA of cellular or viral origin. These findings imply that there is not a stable factor made in the infected cell which specifically inhibits cellular message translation. Furthermore, simultaneous addition of viral RNA and cellular mRNA to the infected cell-free system results exclusively in translation of viral message. Cellular message translation is completely suppressed. Figure 3.2 taken from their paper, illustrates this point dramatically. Column b illustrates the product made in response to added cellular RNA. One of the major products made is a polypeptide with a molecular weight of 20,000 daltons. Column c illustrates that this polypeptide is not present when only EMC RNA is added and that the polypeptides synthesized are rather larger than in column b. Column d shows that simultaneous addition of the 2 RNAs results in the synthesis of only viral-specific products and thus appears identical to column c.

Additional evidence that translation of viral RNA is more efficiently initiated than cellular RNA translation has been put forth by Koch and coworkers [89, 90]. This is based on studies blocking initiation of protein synthesis with hypertonic medium. They have shown that poliovirus-infected HeLa cells exposed to high salt concentrations at early times in infection,

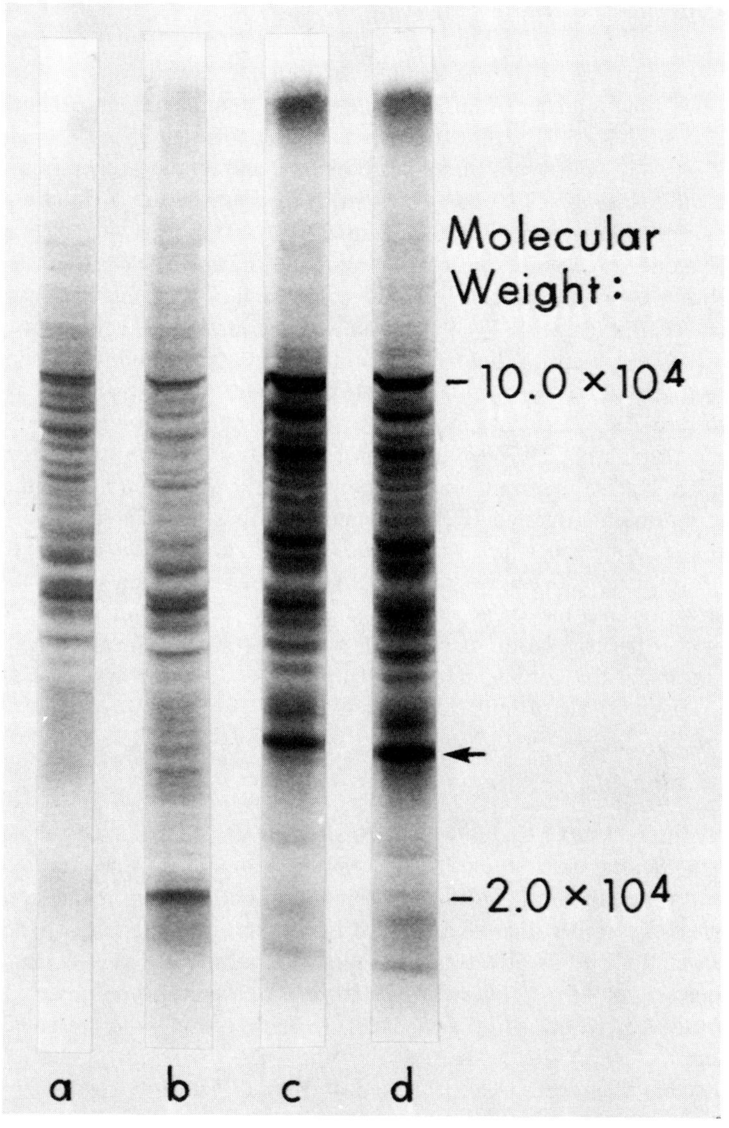

Fig. 3.2 SDS polyacrylamide gel of products synthesized in a cell-free system from mouse plasma cytoma cells in response to cellular and viral RNAs. (a) no RNA added; (b) 0.03 A_{260} 10 S cellular RNA; (c) 0.05 A_{260} EMC RNA; (d) 0.03 A_{260} 10 S cellular RNA plus 0.05 A_{260} EMC RNA. (By permission of Ref. [88].)

when both viral and cellular proteins would be synthesized, synthesize only virus-specific proteins. Given the right salt concentration cellular protein synthesis can be selectively and completely inhibited. Furthermore, they have shown that a higher osmolarity is required to completely inhibit polypeptide chain initiation in infected cells than in uninfected cells [90]. Thus these 2 very different types of studies both lead to the same conclusion: a picornaviral messenger RNA has a competitive advantage over a cellular one with respect to initiation of protein synthesis.

A model of host cell protein synthesis shutoff based exclusively on a competition between viral and cellular message cannot be completely correct, however. Earlier data [4, 77-79] as well as some presented in the above studies [88, 89] indicate that cellular protein synthesis is inhibited at times prior to the synthesis of significant levels of viral RNA. Furthermore, as mentioned previously, other data indicate that a virus-specific protein factor is involved in the shutoff, although no specific inhibitory factor could be found in the in vitro studies reviewed above. One possibility which puts all of the data together has been suggested [89]. The model proposes that a virus-specified protein affects a cellular factor and causes an alteration in the rate of initiation complex formation for all messenger RNA. Such an alteration would allow the translation of a messenger with a high affinity for initiation (viral RNA) but suppress the translation of one with a lower affinity (cellular RNA). This alteration might not be detectable in vitro, however, because in vitro systems have a very poor rate of initiation anyway.

2. *Inhibition of Host Cell RNA Synthesis*

Picornaviruses also exert inhibitory effects on the apparent rate of cellular RNA synthesis. Several early studies have suggested that a virus-specific protein is involved in the inhibition and that the inhibition occurs truly at the level of RNA synthesis, rather than being caused by an increased rate of RNA degradation. In addition to an effect on the actual synthesis of RNA, effects upon the processing of 45 S ribosomal precursor RNA have also been observed [91]. These studies are reviewed in great detail by Baltimore [4] and are not dealt with here.

Recently Schwartz et al. [92] and Miller and Penhoet [93] have attempted to analyze the molecular mechanism responsible for this shutoff. Much is now known about cellular RNA polymerases and these studies involve an analysis of the effects of virus infection on the varying species of RNA polymerases present in the cell nucleus.

Three classes of RNA polymerases exist in eukaryotic cells [94]. RNA polymerase I synthesizes ribosomal RNA (rRNA) and is associated with the nucleolus [95, 96]. RNA polymerase II synthesizes heterogeneous RNA

(HnRNA) and RNA polymerase III synthesizes 4 and 5 S RNA. The latter 2 polymerases are nucleoplasmic in origin [95-97]. All 3 enzymes exhibit differential responses to the inhibitor alpha-amanitin and thus they can be distinguished from each other in whole nuclei, using endogenous DNA as a template, or in soluble extracts with exogenously added DNA [97].

Both of the above studies demonstrate that nuclei isolated from Mengovirus-infected L-cells or EMC-infected ascites cells show inhibition of RNA polymerase II activity prior to inhibition of enzymes I and III. This differential effect implies that 2 separate mechanisms of inhibition may be involved. Attempts to further characterize these mechanisms were unsuccessful [92], however, and illustrate the problems that sometimes exist when in vitro systems are utilized to answer biological questions.

When soluble extracts of the individual RNA polymerases are made from infected or uninfected nuclei the difference in activities found in the whole nuclei vanishes. Enzymes from infected cells are equally active as those from uninfected cells at all times after infection. Furthermore, synthetic activity of the isolated enzymes from uninfected cells is unaffected by the addition of extracts of nuclei or cytoplasm from infected cells. Thus a specific inhibitory factor cannot be demonstrated in vitro.

A negative result of this sort is very difficult to interpret. The lack of observable inhibition in vitro makes any number of models for shutoff in vivo still possible. For example, inhibition may occur in whole cells by a virus-specified alteration of the DNA such that it no longer supports RNA transcription. Such an alteration would not be detected in vitro with solubilized enzymes and a DNA template of exogenous origin. Another possibility is that the virus-specified inhibitory factor functions by preventing the RNA polymerase from binding to the DNA. Such a factor also might not be active in vitro either because it is inactivated or lost in the solubilization procedure or because initiation in vitro occurs at nonnatural promoter sites which are refractile to inhibition by such a factor. Thus no conclusion is possible from these in vitro studies. The mechanism of host cell RNA synthesis shutoff remains obscure.

3. *Inhibition of Host DNA Synthesis*

DNA synthesis is also reduced in picornavirus-infected cells, but most of the available evidence is consistent with the view that this inhibition results as a secondary effect of protein synthesis inhibition [4, 5, 98, 99].

C. Synthesis of Virus-Specific Protein and RNA

Whatever mechanisms may be involved, it is nevertheless true that by the time virus-specific protein and RNA synthesis reach their maximum rate, host cell synthetic processes are completely suppressed in most picornavirus-infected cell systems studied. This usually happens between 2½ and 4½ hours after infection [1, 4, 5]. Any radioisotopically labeled precursor of protein (such as ^{3}H, ^{14}C, or ^{35}S amino acids) or RNA (such as ^{3}H uridine or ^{32}P phosphorus) added to the infected cell culture at this time will therefore only label virus-specified material, making observation of the various proteins and nucleic acid molecules which are synthesized during infection relatively straightforward. Thus protein synthesis and RNA replication can be easily studied. Virus growth occurs exclusively in the cytoplasm and it has been shown that the virus will grow quite well in enucleated cells [100, 101]. Much of the early work and methodology in this area has been discussed in detail in reviews by Baltimore [4, 102].

1. *Virus-Specified Proteins*

a. Mechanism of Synthesis. All of the virus-specific protein synthesis which goes on in picornavirus-infected cells occurs on large newly made polyribosomes [4, 5, 78-80]. There is much evidence which shows that these polyribosomes are tightly bound to cytoplasmic membranes [4, 5, 73-76]. Once dissociated from the membrane virus-specific polyribosomes have a much faster average sedimentation rate than cellular ones (350 S vs. 200 S) and furthermore appear to be more homogeneous [4, 5, 77]. This homogeneity probably arises from the fact that only 1 type of messenger RNA is present in the polyribosomes. The RNA, which sediments at 35-37 S is virtually indistinguishable from RNA contained with the virion, on the basis of size, base composition, and ability to hybridize to negative strands (see Section 3.V.C.2.) [4, 5, 25, 26].

The most direct evidence that virion RNA codes for viral specific protein comes from work with in vitro protein synthetic systems where various groups have shown that virion RNA isolated from several different picornaviruses can function directly as a messenger coding for the synthesis of many of the same polypeptides found in infected cells [18-24]. This is demonstrated in Fig. 3.3 where the tryptic peptides of the proteins synthesized in an E. coli cell-free system in response to poliovirus RNA are compared with tryptic peptides synthesized in the infected cell under conditions where synthesis is viral specific. From the position of elution of each peptide off of the Dowex column it is clear that all of the peptides synthesized in vitro have a counterpart in the infected cell. Similar studies carried out in cell-free systems of mammalian origin using EMC or poliovirus RNA have yielded essentially the same results.

There appears to be no temporal control over the synthesis of protein in picornavirus-infected cells. All of the virus-specific polypeptides present at early times of infection are present in approximately the same amounts at late

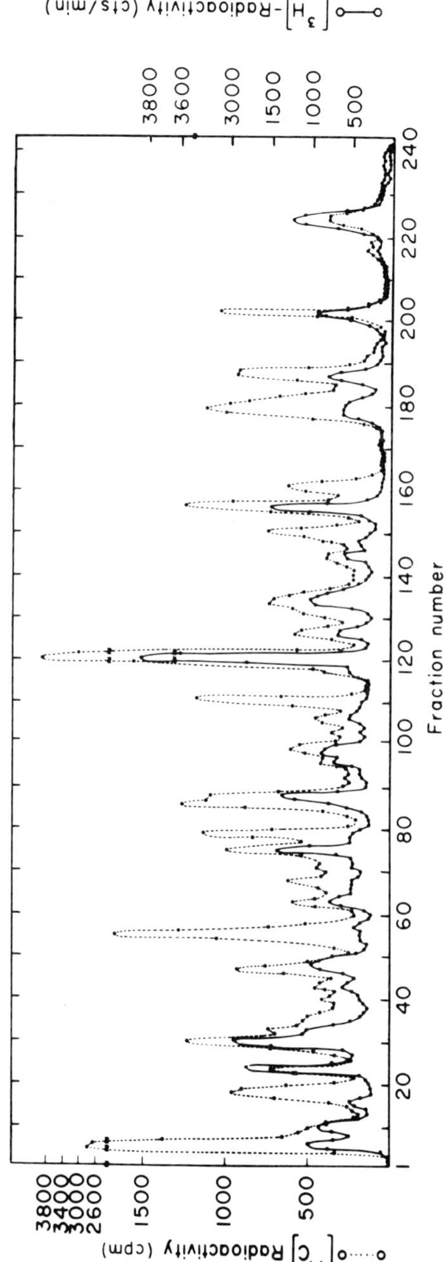

Fig. 3.3 Cation-exchange chromatography of tryptic peptides made in an *E. coli* cell-free system. Digest of the product made under the direction of poliovirus RNA (●—●) compared with a digest of total protein made in poliovirus-infected cells (●- - -●). (By permission of Ref. [18].)

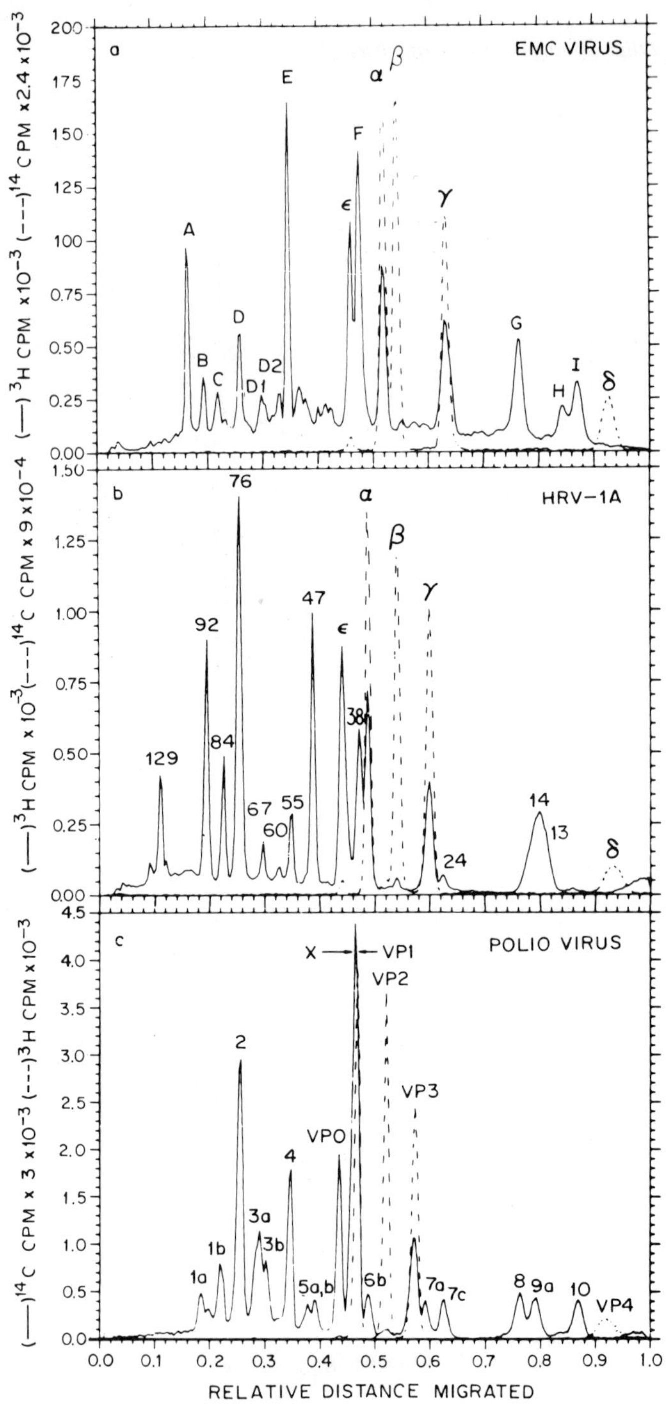
a
EMC VIRUS
b
HRV-1A
c
POLIO VIRUS
(—) ^{3}H CPM x10^{-3} (---) ^{14}C CPM x2.4 x10^{-3}
(—) ^{3}H CPM x 10^{-3} (---) ^{14}C CPM x 9 x10^{-4}
(—) ^{14}C CPM x 3 x10^{-3} (---) ^{3}H CPM x10^{-3}
RELATIVE DISTANCE MIGRATED
A B C D D1 D2 E ε F α β γ G H I δ
129 92 84 76 67 60 55 47 ε 38 α β γ 24 14 13 δ
1a 1b 2 3a 3b 4 5a,b VPO X VP1 VP2 VP3 6b 7a 7c 8 9a 10 VP4

times [4, 103, 104]. Furthermore, several lines of evidence indicate that each picornavirus messenger RNA molecule only contains a single site for the initiation of protein synthesis and that all of the virus-specific polypeptides found in the infected cell are generated by proteolysis, either at the level of the growing nascent polypeptide chain or subsequently in released precursor molecules. All of the initial species synthesized are made in equimolar amounts, in all instances studied except one [105, 106]. The significance of this apparently exceptional case is not known.

For exemplative purposes, only the model as it developed for poliovirus will be reviewed in detail, although references will be provided for the other viruses that have been examined. A slightly different nomenclature is used by each different laboratory, but for this review a nomenclature following that of Butterworth [106] will be adopted.

The first suggestion that some of the viral proteins are derived from others was made by Summers and Maizel [107] based on the observation that the combined molecular weight of the 14 virus-specific proteins which they could identify in poliovirus-infected HeLa cells exceeded 500,000 daltons, while the coding capacity of poliovirus RNA is only about 250,000 daltons. Figure 3.4c, taken from Butterworth [106], illustrates this point. In this case, because of increased resolution, over 20 different polypeptide species can be observed in extracts from poliovirus-infected cells, labeled for 45 min with [^{3}H]-amino acids and subjected to SDS polyacrylamide gel electrophoresis. For comparison, differentially labeled protein from purified poliovirus was also added. The molecular weight of each polypeptide, calculated from its position of migration within the gel, has also been determined and is listed in Table 3.2. In this case the total exceeds 700,000 daltons. For comparative purposes, although not to be discussed, Fig. 4, a and b, shows similar patterns for EMC virus and rhinovirus HRV-1A and the molecular weights of the virus-specific polypeptides for these viruses are also shown in Table 3.2.

Following Summers' and Maizel's observation, Jacobson and Baltimore [108] showed that during a pulse-chase experiment, radioactivity present in polypeptide 1a decreased as radioactivity in the capsid proteins VP0, VP1, and VP3 increased. Furthermore, the addition of parafluorophenylalanine to infected cells prevented both the disappearance of polypeptide 1a and the

Fig. 3.4 Structural and nonstructural virus-specific proteins synthesized in HeLa cells infected with various picornaviruses. (a) SDS polyacrylamide gel of proteins made in EMC-infected cells (—) vs. purified EMC virus capsid proteins (- - -). (b) SDS polyacrylamide gel of proteins made in human rhinovirus type 1a-infected cells (—) vs. purified rhinovirus capsid proteins (- - -). (c) SDS polyacrylamide gel of proteins made in poliovirus-infected cells (—) vs. purified poliovirus capsid proteins (- - -). (By permission of Ref. [106].)

Table 3.2 The Molecular Weights of Nonstructural Polypeptides of Various Picornaviruses[a]

EMC		Poliovirus		Rhinovirus	
Polypeptide	M.W.	Polypeptide	M.W.	Polypeptide	M.W.
A	100,000	1a	95,000	92	92,000
B	90,000	1b	85,000	84	84,000
C	84,000	2	77,000	76	76,000
D	75,000	3a	70,000	67	67,000
D1	65,000	3b	65,000	60	60,000
D2	59,000	4	57,000	55	55,000
E	56,000	5a	47,000	47	47,000
F	38,000	5b	37,000	38	38,000
G	16,000	X	37,000	24	24,000
H	12,000	7a	24,000	14	14,000
I	11,000	7c	22,000	13	13,000
A + C + F	222,000	8	14,000	92 + 84 + 47	223,000
		9a	13,000		
		10	10,000		
		1a + 1b + X	217,000		

[a]Data from Ref. [106]. The primary translation products are underlined.

appearance of the capsid proteins. Thus a strong kinetic argument can be made for polypeptide 1a being a precursor to the capsid proteins. Similar kinetic studies showed that VP0 is cleaved to VP2 and VP4 during virion maturation [109]. The ultimate proof that capsid proteins are derived by proteolytic cleavage from polypeptide 1a came with the demonstration that the tryptic peptides of 1a were almost identical to those found in the proteins of purified virus [110].

Experiments using very short periods of labeling with radioactive amino acids provide the evidence that 3 major polypeptides comprise the primary translation products of the genome [110]. These polypeptides, 1a, X, and 1b, are the major ones to appear after a very short pulse, although 1b is rapidly cleaved to form polypeptide 2. The relationship of 1b to 2 has been confirmed by tryptic peptide mapping. More recently similar kinetic studies have shown that polypeptide 2 is subsequently cleaved to polypeptide 4 [106].

The first evidence which suggested that even the 3 primary polypeptides are derived by proteolytic cleavage was obtained by exposing infected cells to several amino acid analogs [110]. In the presence of these analogs (para-fluorophenylalanine, ethionine, azetidine-2-carboxylic acid, and canavanine) it was observed that polypeptides larger than 1a accumulate. Several other groups reported shortly thereafter that these large polypeptides could be generated in several other ways: either by raising the temperature of infection from 37 to 43°C [111], by adding known protease inhibitors to the infected cell [110, 112, 113], or by adding Zn^{2+} [114]. The largest polypeptide observed has been called NCVP 00 [110]. This polypeptide has a molecular weight consistent with a polypeptide the length of which could be coded for by the entire genome, if translation initiated at one site near the 5′ end of poliovirus RNA and proceeded without interruption to the other end. Thus it is proposed that only 1 site for initiation exists on poliovirus RNA, and that there are 3 primary products of translation which are cleaved from nascent chains of the polyribosomes [110]. Each of these polypeptides is then subsequently cleaved to form 3 families of polypeptides which make up the 20-odd species seen in picornavirus-infected cells.

This suggestion has since been confirmed through the use of in vitro cell-free protein-synthesizing systems. Initiation of translation of either EMC or poliovirus RNA in an ascites cell-free system occurs at only 1 site [19-24]. The N-terminal sequence of the EMC polypeptide made in vitro has been examined in detail and shown to be Met-Ala-Thr . . . [24]. Furthermore, although premature termination of the growing nascent polypeptide chain occurs in all in vitro systems studied, probably because of ribonuclease digestion of the RNA, several groups have demonstrated the synthesis of a small but measurable amount of a polypeptide large enough to account for the entire viral genome. Thus all the available evidence is consistent with the model presented above.

b. The Biochemically Established Genetic Map. This novel mechanism of protein synthesis affords the opportunity for sequencing the picornavirus genome by specifically blocking initiation of protein synthesis in infected cells. Since protein synthesis initiates at one site near the 5′ end of the RNA, after initiation is inhibited synthesis of polypeptides encoded by the 5′ terminal end of the RNA molecule should be rapidly depressed followed in order by the polypeptides encoded by regions nearer to the 3′ end.

This approach, first suggested by Taber et al. [115] has since been followed by many groups, using either pactamycin [116-119] or hypertonic shock [90] to specifically block initiation. The results obtained using pactamycin are illustrated in Fig. 3.5. The top panel shows an SDS polyacrylamide gel comparing the polypeptides from poliovirus-infected cells radiochemically labeled in the presence of pactamycin (–●–) to those radiochemically labeled under normal conditions (- -○- -). To prevent the formation of capsid proteins, parafluorophenylalanine was added in each case. The data is normalized such that the amount of radioactivity in polypeptide 2 is equal in each case. In extracts treated with pactamycin synthesis of polypeptide 1a and 3a are the most inhibited, relative to the corresponding proteins in the untreated extract, followed in order by 3b, X, and 2 and 4.

Figure 3.5, bottom panel, shows the mapping of the proteins within the 1a region of the genome. In this case the relative amount of radioactivity within each polypeptide of purified virus synthesized in the presence of pactamycin is compared with polypeptides of purified virus synthesized under normal conditions. In the presence of pactamycin, synthesis of VP4 is the most suppressed, followed in order by VP2-VP3-VP1. Thus these experiments establish unambiguously the positions on the genome of many of the polypeptides found in poliovirus-infected cells. These data are summarized in Fig. 3.6 and are presented along with the similarly determined genetic maps of the other picornaviruses which have been examined [106, 118].

Two other slightly different biochemical approaches have also been employed to establish genetic maps for picornaviruses. One involves monitoring the appearance of each polypeptide as protein synthesis reinitiates after removal from hypertonic medium [90] while the other involves monitoring the kinetics of appearance of radioactivity in each capsid protein as pulses of radioactivity of increased length are administered to the infected cell [116, 119]. Both methods give results consistent with those presented above.

The striking similarity between the maps of each different picornavirus subgroup should be noted. In all 3 cases the capsid proteins map in the same relative positions at the 5′ end of the RNA molecule. Furthermore, the noncapsid half of the genome also appears to be processed in similar ways in all 3 systems.

In vitro translation of EMC RNA in an ascites cell-free system has provided some additional information which should be added to the above

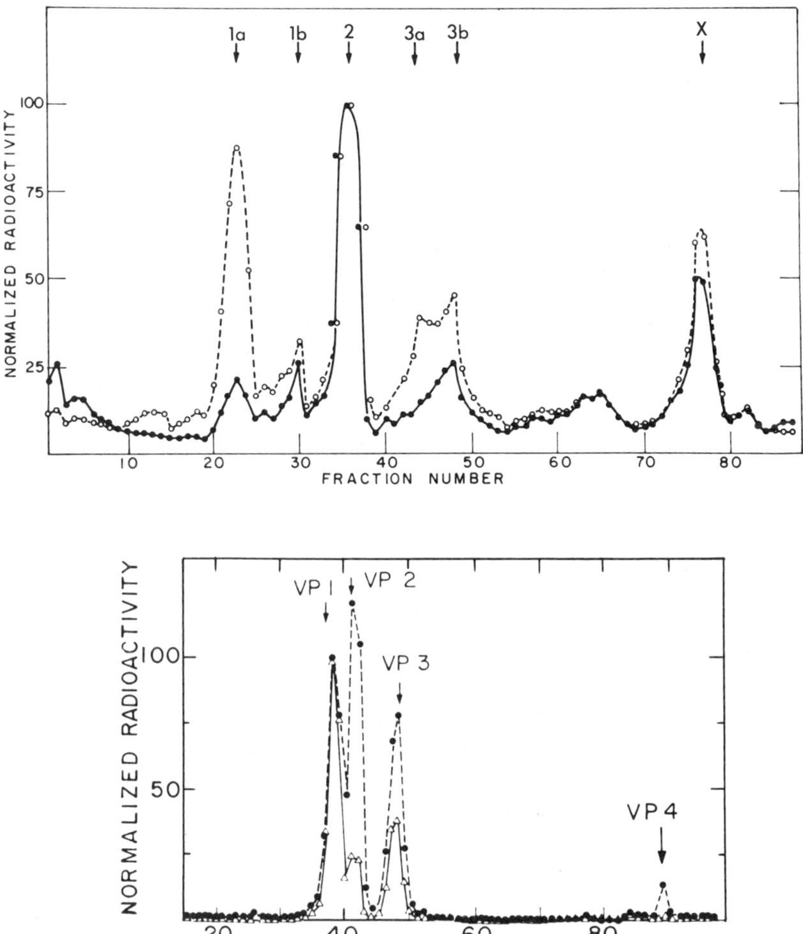

Fig. 3.5 Effect of pactamycin on poliovirus proteins. Top: Total proteins synthesized in the absence of pactamycin (○- - -○) compared with proteins synthesized in the presence of 10^{-7} M pactamycin (●—). Bottom: Proteins from purified virus synthesized in the absence of pactamycin (●- - -●) compared with proteins synthesized in the presence of 5×10^{-7} M pactamycin (△—△). (By permission of Refs. [115, 119].)

genetic maps. When tryptic peptides of the in vitro product were compared with the tryptic peptides of polypeptide δ or other EMC capsid proteins, the in vitro N-terminal peptide could not be found in the capsid proteins [120]. Furthermore this peptide is only found in infected cells during very short labeling times, indicating that it is rapidly turning over [121]. These results suggest that the structural protein gene starts some distance away from the ribosome

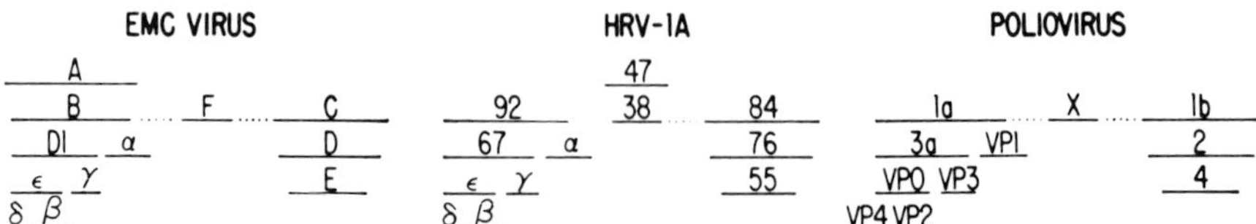

Fig. 3.6 Summary of the genetic maps of 3 picornaviruses. In each instance the lateral position of each protein corresponds to the relative location of each gene locus on the RNA (the 5′ end of the RNA is to the left). The vertical position represents precursor product relationships or alternative cleavage forms. Line lengths are proportional to molecular weights. (By permission of Ref. [106].)

binding site and that the polypeptide corresponding to the region between the two is synthesized and rapidly degraded in the infected cell. It has been suggested by Smith [120] that this region is important for maintaining the conformation of the ribosome binding site and not as a coding sequence. Hence the protein translated from it is degraded.

c. Correlation of Virus-Specific Functions with Specific Proteins. Although much is known about the physical map of the picornavirus genome and the processing of the virus-specific polypeptides, very little is known about the function of these proteins. Several virus-specific functions are known to exist in the infected cell and presumably a polypeptide is responsible for each. In addition to structural proteins and proteins which function as inhibitors of host cell synthetic processes (see Section 3.V.B.) a new RNA-dependent RNA polymerase activity exists in the cytoplasm of infected cells [4, 5] (see Section 3.V.C.2.) and it is also likely that the virus codes for its own protein-processing protease. Two approaches have been employed in attempts to assign functionality to specific polypeptides. The first approach is a genetic one. Cooper and co-workers, over the past decade, have isolated several temperature-sensitive mutants of poliovirus with which they have been able to demonstrate recombination and establish a genetic map [122, 123]. Correlation of this genetic map with the polypeptide map obtained biochemically is now possible and mutants known to possess a biochemical lesion when grown at restrictive temperatures can be assigned more or less to a specific polypeptide. The other approach involves the purification of a specific viral protein, whose activity can be readily assayed and the subsequent correlation of this activity with a specific poliovirus polypeptide, by comigration on SDS polyacrylamide gels.

The most obvious candidate to be examined in the above way is the viral RNA polymerase. Mutants defective in their ability to synthesize viral

RNA have been isolated by Cooper and co-workers [122-124] and these mutants have been shown to map at the opposite end of the genome from the structural protein region. Following the map shown in Fig. 3.6, such a map position suggests that at least some part of the viral RNA polymerase is contained in a protein derived from the 3′ end of the genome and belongs to the family of polypeptides derived from polypeptide lb. This suggestion seems recently to have been confirmed by a biochemical approach, in which viral RNA polymerase from infected cells was solubilized and isolated [125]. Polyacrylamide SDS gel electrophoresis revealed that the isolated polymerase activity contained only 1 virus-specific polypeptide, namely polypeptide 4 (NCVP 4). It is not known if the polymerase activity is complete, however, and capable of synthesizing both complementary as well as virion RNA. The involvement of additional virus-specific factors as well as host-coded ones is not ruled out. In support of this notion, a multicomponent poly(C)-dependent polymerase has been isolated from EMC infected BHK cells. It contains 5 polypeptide chains, 1 of which corresponds to the EMC equivalent of NCVP 4 (polypeptide E) [126].

It is not surprising that the purified polymerase activity seems to be associated with a stable polypeptide, such as polypeptide 4, rather than a larger precursor molecule, such as polypeptide lb or 2. Such precursors are unstable and relatively short lived. Any purification procedure of the sort carried out in the above study would result in their conversion to the stable polypeptide 4. Thus, the possibility cannot be ruled out that polypeptides lb or 2 possess some enzymatic activity themselves. They could represent forms of polymerase with the same or slightly altered activities. Alternatively, these precursor forms may represent inactive states of polymerase which become activated upon proteolytic conversion. The bits of polypeptide that are cleaved off could be necessary to maintain a certain structural conformation. Inactive forms of certain enzymes and hormones, which become activated by proteolysis, have been described [127, 128].

Other experiments have been carried out in an attempt to examine the specific proteolytic activity involved in the processing of virus-specified protein. Three types of proteolytic cleavage have been shown to exist in the infected cell. The first type generates the 3 primary products of translation and occurs within the nascent chain on the polyribosomes [110, 116]. The second type of cleavage occurs after release of these polypeptides from the polyribosome in the cytoplasm of the infected cell [110, 116], while the third type appears to be involved in the maturation of the viral particle [109]. It is not known how many enzymes are involved in these cleavages and only recently has some evidence developed which indicates that some of the proteolytic activities may be virus coded. The problem that existed until recently was that a decent assay system for studying virus-specific proteolytic cleavages did not exist. The precursor polypeptides which are necessarily needed as subsrates had only been identified as short-lived species, radiochemically, on SDS polyacrylamide gels. Any polypeptide isolated in this manner is not capable of acting as a substrate as it is totally denatured by the SDS.

Korant [129], however, has isolated a nondenatured form of polypeptide 1a from the cytoplasm of poliovirus-infected HeLa cells by taking advantage of the fact that the polypeptide is stable in the presence of iodoacetamide [112]. Cytoplasmic extracts of infected cells were subjected to zonal electrophoresis in a sucrose gradient and polypeptide 1a was isolated in an aggregate along with polypeptides 2 and X. These aggregates function very well as substrates for in vitro proteolytic cleavage assays. When cytoplasmic extracts from infected cells are added to these aggregates specific cleavage of polypeptide 1a to the capsid proteins occurs. Extracts from uninfected cells are inactive. Furthermore, consistent with the in vivo data, extracts from iodoacetamide-treated infected cells are also inactive. Thus, the protease which cleaves polypeptide 1a appears to be either virus-coded or virus-induced.

Another approach has been to use the polypeptides synthesized in an in vitro protein-synthetic system as substrate for specific cleavage activity [130]. When EMC RNA is added to an in vitro cell-free system from uninfected Krebs ascites cells, one of the products made is a polypeptide slightly longer than polypeptide A (polypeptide A in EMC is analogous to 1a in poliovirus). This polypeptide is identical to polypeptide A, by cyanogen bromide mapping, except that it appears to contain an extra 12,500 daltons of polypeptide at the amino terminal end. In an in vitro system from infected cells this longer version of polypeptide A is not found but an extra protein of 12,500 daltons is. Thus there seems to be a proteolytic activity, present in the infected cell-free system and not present in an extract from uninfected cells, that selectively cleaves off this extra bit of protein. Isolation and purification of this activity results in a polypeptide which comigrates with the viral capsid protein γ. Thus the suggestion is made that γ has proteolytic activity as well as a structural function, and it is tempting to propose a model of self-cleavage, similar to that known to exist for hormone and enzyme activation [127, 128], since polypeptide γ exists within the precursor polypeptide A.

Another type of cleavage activity may exist in the in vitro systems studied above since very few polypeptides longer than polypeptide A are synthesized. At present it is difficult to say whether this situation represents true cleavage or premature termination of synthesis caused by ribonuclease digestion which frequently occurs in in vitro protein-synthetic systems.

Thus at least 2 different types of protease activity have been assayed in vitro. The first activity is responsible for the cleavage of polypeptide A to the capsid proteins. It seems to be clearly virus coded or induced, implying that a specific viral gene product which has not been identified is involved. The second activity however, only trims a bit of polypeptide off a precursor of polypeptide A and polypeptide A is not degraded any further in its presence. This second activity has also been shown to be viral coded and seems to reside within a capsid protein. It is likely that the second activity functions on the polyribosomes, while the first is free in the cytoplasm. The 2 activities may account for all the posttranslational proteolytic events which occur in picornaviral systems. However, one cannot rule of the possibility that other enzymes are involved as well. Some of these additional enzymes may be host coded.

A functional map of the genome generalized to include the data obtained in all picornavirus systems is shown in Fig. 3.7. No doubt other functions will be shown to be viral specific and correlated with a specific protein as more molecular mechanisms are probed in depth. Very little genetic information exists in the virus to code for as many functions as there seem to be. A simple way to overcome this problem is to create a single polypeptide chain which has several different functions. This seems to be true in the case of polypeptide γ, where it appears to be both a structural unit and a protease.

More complex mechanisms of economy and control may exist as well. It is intriguing to postulate a mechanism whereby a precursor polypeptide has a different function than its cleaved products. Since cleavages seem to occur at different rates [119] they may be under strict control. Thus each 1 of the 20 or so polypeptides observed in picornavirus-infected cells may have a separate function, although genetically they are not distinct. If this were the case the appearance of certain activities would follow naturally from the destruction of others and very complex control mechanisms could exist.

2. *Replication of Viral RNA*

In addition to serving as a template for the translation of protein, an infecting RNA molecule functions as a template for RNA replication. Kinetic studies

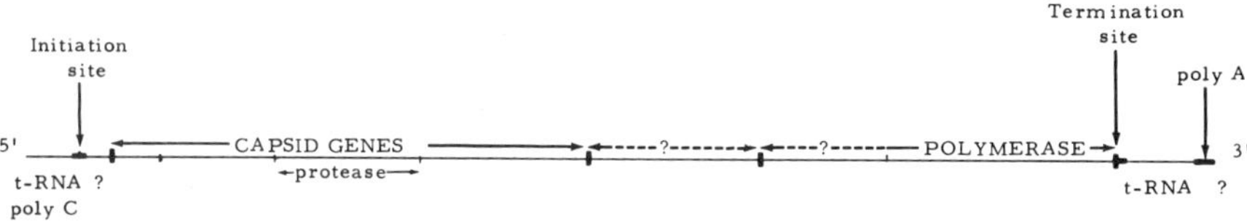

Fig. 3.7 Functional map of the genome of a picornavirus. Following Fig. 3.6 the position of each of the primary cleavage products is marked between the heavy bars (▪). The tRNA structure has not been positioned, but recent evidence places poly(C) near the 5′ end [165; 166]. Functions yet to be assigned to proteins include cell protein synthesis inhibition, cell RNA synthesis inhibition, another protease, and cell lysis factor.

have shown that replication begins within half an hour of infection and proceeds at an exponential rate for about 3-4 hr, after which synthesis becomes linear [4, 131] until it gradually stops at about 7-8 hr after infection. After a newly made viral RNA molecule is formed, some regulatory mechanism determines whether this new molecule is used for translation, transcription, or encapsidation. Control at this level would affect the rate of RNA synthesis. We know very little about this process.

Very little is also known about the initiation of RNA replication. It is known that initial transcription of the viral RNA into complementary RNA begins at the 3′ end of the viral RNA molecule; however, no mechanism exists to explain how the input RNA is cleared of ribosomes, which are moving 5′ to 3′, so that transcription can occur. We do know that in the *Qβ* RNA phage system the viral RNA polymerase acts as a repressor of protein synthesis by binding to the protein synthesis initation site [132, 133] and a similar mechanism may exist for picornaviruses. However, there is no evidence for this proposal, and there are some problems with it because we would then need to explain how the repressor activity of the polymerase is inactivated so that translation can proceed on the newly synthesized RNA strands.

Replication of RNA is known to occur exclusively in the cytoplasm on smooth membranes which make up a structure called the replication complex [74, 75]. This complex contains all of the viral RNA polymerase activity and a partially double-stranded and single-stranded RNA molecule called replicative intermediate RNA (RI). This molecule is heterogeneous in size, having a sedimentation rate of 20-70 S in sucrose gradients. Over the years, a wealth of evidence has accumulated indicating that RI is the site of RNA synthesis. The nature of this molecule and the overall mechanism by which replication occurs was nearly completely worked out by 1970. During the past 5 years there have only been a few significant advances in this area. Accordingly, the reader is referred to comprehensive reviews by Baltimore [4], Levintow [5], and Bishop and Levintow [134] for a detailed discussion of the earlier work. Only the essential facts of replication will be presented here, so that the more recent advances can be discussed.

A summary of replication is shown in Fig. 3.8. In synopsis, viral RNA functions as a template for complementary RNA synthesis. Several complementary RNA molecules are synthesized simultaneously, each by a separate polymerase molecule. These molecules then function as templates for the synthesis of viral RNA molecules. Although free single-stranded complementary RNA must be released from the replicative structure before a polymerase can begin to transcribe them (as shown in Fig. 3.8) such molecules are never found free from positive strands.

The model suggests that two different sorts of RIs exist–1 involved with the synthesis of minus strands and the other involved in plus strand synthesis.

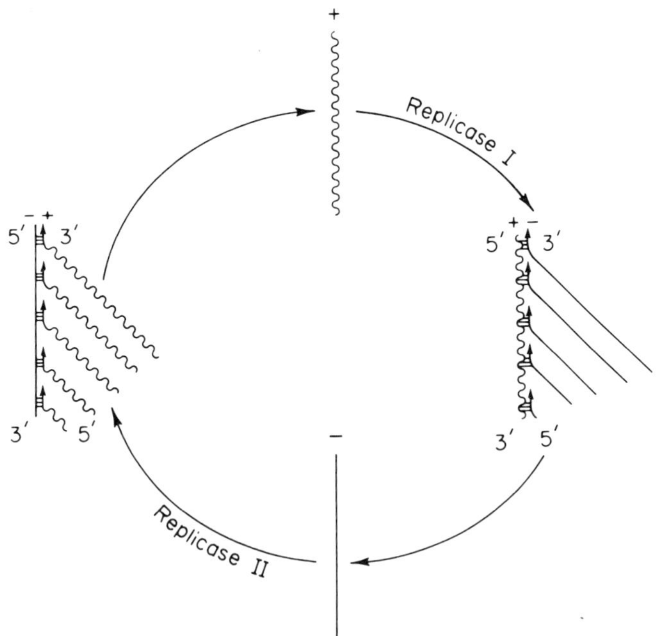

Fig. 3.8 Model of picornavirus RNA replication. (By permission of Ref. [1].)

In addition 2 different forms of polymerase may be involved. There is some genetic evidence to support this view as temperature-sensitive mutants can be isolated which synthesize either no RNA at all or only complementary RNA [122-124]. These different classes of mutants map in slightly different places within the genome.

Synthesis of RNA is asymmetric. Examination of total RI extracted from cells indicates that most of the complete single-stranded templates are complementary RNA. Thus more plus strands are made than minus strands (about 5-10 times as much). Each RI is thought to contain an average of 5 growing chains [4].

Recently, the exact nature of the in vivo RI has been reexamined. Originally, it was proposed that each of the growing RNA chains contained in the RI was extensively hydrogen bonded to the complementary strand. This was based on the fact that ribonuclease treatment of the molecule in vitro revealed extensive double-stranded regions. These double-stranded regions seem to be artifacts of isolation caused by deproteinization, however, as Öberg and Philipson [135] have found that treatment of RI with diethylpyrocarbonate prior to phenol extraction causes a great deal of the double-stranded structure to vanish. In this case diethylpyrocarbonate is thought to prevent single-stranded regions of the RI from reannealling during extraction. Furthermore, electron micrographic studies by Thach et al. [136] show that the in vivo structure of RI has very few double-stranded regions. Thus the in vivo RI present within the infected cell is a somewhat different molecule than its extracted deproteinized form. It is pictured as containing 4-6 nascent RNA molecules of varying length, each with its 5′ end free. Each growing chain is thought to be bonded to the template by a very small region of base pairing and a polymerase molecule.

Recently a membrane-bound replication complex other than the 250 S structure mentioned above has been found in poliovirus-infected cells [137]. This minor fraction sediments at 70 S and 66% of the newly synthesized RNA within it is complementary RNA, as opposed to the major complex which synthesizes only 8% complementary RNA. It has been suggested that the 70 S complex represents the primary site of complementary RNA synthesis. Thus a positional difference between the 2 replicative intermediate forms of RNA may exist. It would be interesting to know if the polypeptide containing polymerase activity present in each complex is the same and if the 70 S complex is enriched in the mutant defective in plus strand synthesis.

Also present in picornavirus-infected cells is a species of RNA that is entirely double stranded. This RNA has been shown to contain 1 viral RNA strand completely hydrogen bonded to its complementary RNA strand. Very little double-stranded RNA is present at the beginning of infection although it represents as much as 10% of viral-specific RNA by the end of infection. Its accumulation at late times suggests that it may be a byproduct of the RI [4, 5, 134].

The nature of the polyadenylate sequences on poliovirus RNA and the possible mechanism of their synthesis has been examined in detail [34-36, 40-43]. Since poly(A) is also found in Mengovirus, rhinovirus, and EMC RNA [37, 38, 46, 138, 139], the data for poliovirus can probably be generalized to other picornaviruses as well.

Polyadenylate sequences have been shown to exist on all species of poliovirus RNA present in the cell (virion RNA, RI, ds, polyribosomal, 35 S cytoplasmic). In addition polyuridylic sequences [poly(U)] have been identified on the 5′ terminus of the complementary RNA strand in both double-stranded molecules (DS) and replicative intermediate (RI) molecules. Thus the simplest explanation of poly(A) synthesis in poliovirus RNA would be that it essentially is a transcriptive process—that is, 3′ terminal poly(A) in the viral (plus strand) RNA is made by a polymerase copying of the poly(U) sequences present at the 5′ terminal of complementary (minus strand) RNA.

The entire picture is a bit more complicated than this simple model suggests, however, because of the following observations: (a) the poly(U) sequences identified in either double-stranded RNA or RI RNA are 150-200 nucleotides in length; (b) the poly(A) sequences found in double-stranded RNA are also very long, consisting of 150-200 nucleotides at all times in infection. However, (c) at 3 hr after infection (when RNA synthesis reaches its maximum rate) the poly(A) on virion RNA, replicative intermediate (RI) RNA, polyribosomal RNA, and total 35 S cytoplasmic RNA is heterogeneous in size with an average length of only 75 nucleotides. (d) At 6 hr after infection many of the intracellular RNAs have long poly(A)s over 150 nucleotides in length. At this time, however, the RNA found within the virion does not increase in size and remains between 50 and 125 nucleotides in length. (e) No nascent poly(A) can be detected in replicative intermediate (RI) RNA. The poly(A) found there is likely to be an artifact caused by binding of finished 35 S RNA during extraction. Thus a direct precursor product relationship between RI and 35 S RNA-containing poly(A), although not ruled out, cannot be demonstrated.

Little is known about the mechanism that controls poly(A) chain length. If poly(A) addition is primarily a transcriptive process, than at the midpoint of infection the newly formed molecules must rapidly leave the RI before the entire poly(U) region is copied. At latter times of infection the poly(A) made is longer. This longer poly(A) is likely to come from the copying of the entire poly(U) region of complementary RNA. Some AMP end addition after transcription cannot be completely ruled out and it has not yet been established whether long poly(A) sequences are the result of, the cause of, or unrelated to the decrease in rate of RNA synthesis.

It is interesting that, even at late times of infection when poly(A) of very long lengths is made, some mechanism exists which selectively encapsidates only molecules of short poly(A). Whether this mechanism involves a selection of molecules with the shorter poly(A) sequences or a cleavage of molecules with longer sequences is not known. Nevertheless, the fact that only short poly(A) stretches exist in virion RNA raises the question of how the long poly(U) stretches on complementary RNA are synthesized. It has been proposed [43] that a slippage mechanism in the viral replicase similar to the one proposed for *E. coli* DNA-dependent RNA polymerase is responsible [140]. Equally likely is the possibility that the poly(A) on the input strands of RNA is lengthened by an end addition mechanism before transcription occurs.

There have been several attempts to study RNA replication in in vitro systems. Crude preparations of virus-specific polymerase have been made from picornavirus-infected cells by isolating and solubilizing the smooth membrane-associated replication complex [141-143]. In the most carefully studied cases, intact viral RNA molecules are synthesized and released from the replication complex in vitro but initiation of new RNA chains does not seem to occur. To date only the completion of preexisting nascent RNA molecules which initiated in vivo has been demonstrated. There has been 1 isolation of an enzymatic activity which synthesizes poly(G) in response to exogenously added poly(C), after digestion of the native RNA template with micrococcal nuclease [126]. However, initiation on this unnatural template is not representative of the natural initiation process and the relationship of this enzyme to the native polymerase is not known.

Studies demonstrating the synthesis of poly(A) in vitro have also been carried out. Spector and Baltimore [144] have shown that poly(A) is added in vitro to all species of poliovirus-specific RNA known to exist in the infected cell. However, the crude polymerase preparation with which they were working also contained an enzyme capable of terminal poly(A) addition, as well as the viral polymerase. This activity exists in both uninfected and infected cells. Thus, from these studies it cannot be determined whether poly(A) is synthesized in vitro by an end addition mechanism or by a transcriptive process.

Poly(A) synthesis in vitro has been studied in more detail by Dorsch-Häsler et al. [145]. They have prepared a smooth membrane-associated polymerase complex from poliovirus-infected cells. This complex synthesizes both viral RNA and poly(A) in the presence of Mg^{2+}, like Spector's and Baltimore's preparation, but mostly poly(A) in the presence of Mn^{2+}. Treatment of the complex with 0.5% deoxycholate and subsequent rebanding of the enzymatic activity in a sucrose gradient, however, results in an activity which does not synthesize poly(A) by itself in the presence of Mn^{2+}. This "cleaned-up enzyme" is still capable of synthesizing viral RNA containing poly(A) in the presence of Mg^{2+}. One possible interpretation of this result is that the terminal addition enzyme (stimulated by Mn^{2+}) has been dissociated from the complex by treatment with deoxycholate. The viral polymerase (stimulated by Mg^{2+}), however, is still contained in the deoxycholate-treated complex and is capable of adding poly(A). This explanation implies that poly(A) is added by a transcriptive process carried out by the viral polymerase.

From the foregoing discussion it is clear that much is known about RNA replication in picornavirus-infected cells. Most of the picture which has evolved, however, has been based on data obtained in a steady-state situation. It is difficult to study the beginning and end of the process of replication but nevertheless it is here where the unanswered questions remain. We do not know (a) how or where the virus-specified replicase binds to the RNA, (b) what factor(s) control(s) the amount of complementary RNA synthesized, (c) what factor(s) may be involved in determining the fate of a newly made viral RNA strand, (d) what factors cause replication to shut down, (e) why poly(A) length increases late in infection, and last, (f) we do not know the role of double-stranded RNA, which appears to be the end product of replication. Does double-stranded RNA have a function, or is it a useless byproduct?

One can propose that the various molecules which associate with viral RNA probably have different binding efficiencies and thus many of the necessary controls are provided by competitive interactions between ribosomes, capsid precursors, and polymerase molecules. Such a statement is very vague and the specific interactions involved need to be carefully elucidated. Alternatively, specific proteins may function to actively control the rate of RNA synthesis and fate of the newly synthesized molecules. One approach to answer these questions would be through a detailed study of the molecular defects of some of the conditional lethal mutants which have already been isolated.

D. Virion Assembly and Release of Virus

Each picornavirus particle is a specific symmetrical aggregate of protein and nucleic acid containing 1 molecule of RNA and about 60 molecules of each of 4 polypeptides (Section 3.IV.B.). Assembly of the virion from its component parts involves a series of steps in which the individual polypeptides aggregate, in equimolar amounts, into structures of increasing size, which combine with RNA and ultimately assemble into a complete virus particle. There is evidence to suggest that assembly occurs on the smooth membranes which form part of the RNA replication complex [146]. Many of the steps in the assembly pathway have been elucidated for both EMC and poliovirus and are shown in Fig. 3.9. The nomenclature for EMC has been used in this figure (see Table 3.1).

All of the capsid proteins are translated from the genome into a single precursor polypeptide (1a in poliovirus, A in EMC). In the case of poliovirus it has been suggested that a single molecule of this precursor is cleaved by proteolysis into 3 polypeptides (VP0, VP1, VP3, or ϵ, γ, α) which aggregate together to form a structure that sediments at 5 S [147]. It has been proposed that this 5 S aggregate then combines with 4 other similar aggregates for form a larger 14 S structure [148, 149]. This 14 S structure is thought to contain 5 molecules of each polypeptide chain.

Although the above model has been taken as fact for the past 7 years, the evidence for this pathway is not very strong. 14 S aggregates containing these 3 polypeptides clearly do form after a short pulse of radioactive amino acid is given to the infected cell; however, the 14 S region of the gradient in the published reports always seems to contain some of the precursor polypeptide as well. Furthermore, a careful analysis of the data reveals that both the precursor polypeptide and its cleaved products are present in the 5 S region as well. A kinetic precursor-product relation of the 5 S aggregate to the 14 S aggregate has never been demonstrated.

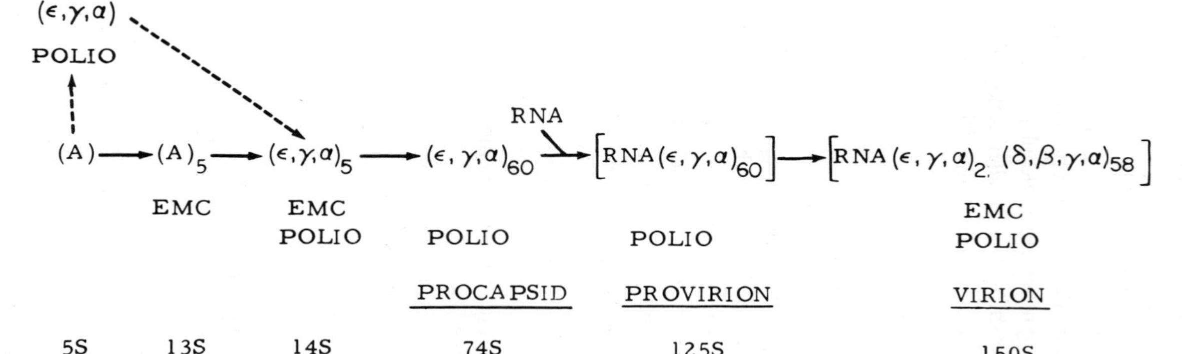

Fig. 3.9 Pathway of capsid assembly. The capsid precursor (A) is released from the polyribosome by a proteolytic cleavage at the start of this pathway. The path shown for polio (- - -), although reported in the literature, is likely to be incorrect. The nomenclature for EMC is used. (In poliovirus nomenclature A = 1a ϵ = VP0, α = VP1, β = VP2, γ = VP3, δ = VP4). The sedimentation rate for each intermediate is indicated. A procapsid or provirion has not been demonstrated in EMC.

Recently, McGregor et al. [150] have proposed a different mechanism for the generation of the 14 S aggregate in EMC. They propose that 5 molecules of precursor polypeptide A associate together into a 13 S structure *before* proteolytic cleavage occurs. Proteolysis then converts this 13 S structure into a 14 S one containing 5 molecules of each cleaved chain (ϵ, α, γ). The data for this model are a lot cleaner than those obtained in the older work with poliovirus. Furthermore, pulse-chase experiments indicate a true precursor-product relationship for the 13 S aggregate to the 14 S structure. It thus appears probable that the initial cleavages in assembly in poliovirus also occurs in a 13 S aggregate and that the 5 S aggregates of cleaved polypeptides found are breakdown products of the 14 S particle. The small amounts of the precursor polypeptide found in the 14 S peak, in the poliovirus case, probably arise from unresolved 13 S precursor particles which contaminate the 14 S preparations.

Once formed and stabilized (by the proteolytic cleavages?), the 14 S particle appears to self-assemble into a stable 74 S empty capsid, called the procapsid. As well as occurring in infected cells [109] this particle can be formed from 14 S particles in vitro [148, 149]. Next, RNA becomes associated with the procapsid and the faster sedimenting provirion is formed [151]. Although easily demonstrated for poliovirus, foot and mouth disease virus [152], and bovine enterovirus [153] a procapsid or provirion has never been found for EMC, presumably not because they do not exist but because they are unstable or shorter lived. The conversion of procapsid to provirion has been shown to take place in vitro as well [154]. The RNA contained in the provirion is insensitive to ribonuclease digestion, implying that it is on the inside of the particle; yet the particle is easily converted back to procapsid and free RNA by treatment with EDTA. Perhaps the RNA is held inside by some mechanism involving divalent cations. EDTA must remove these cations and allow the RNA to "leak out" of the polypeptide lattice which comprises the procapsid. The initial mechanism which "pulls" the RNA inside, however, is a complete mystery.

The last step in assembly involves the proteolytic conversion of most of polypeptide ϵ to polypeptides β and γ which results in the formation of a mature virus particle [109]. Some residual ϵ remains uncleaved. As the mature virion is now insensitive to a whole variety of agents, including SDS and EDTA, which denature the precursor structures, this latter cleavage must cause a vast conformational change in the shell of the particle. It is intriguing that polypeptide δ (or VP4) is one of the polypeptides formed by this cleavage. This polypeptide is lost on adsorption of the virion to the cell and another conformational change may happen there (see Section 3.V.A). Thus polypeptide δ must play an important stabilizing role, although structurally it does not form part of the basic repeating unit of the capsid.

There has not been very much recent work on the mechanism of release of the assembled virus particles. Older work, which has been reviewed by Baltimore [4], shows that, once assembly has occurred, the complete virions formed within the cytoplasm of the infected cell are released by actual rupture of the cell membrane. Release seems to be under some sort of viral control as different picornaviruses have markedly different cell lysis times and mutants of Mengovirus can be isolated which will induce lysis with different kinetics. Some virus-specified protein factor is likely to be involved.

VI. Miscellaneous Topics

A. Defective Interfering Particles

Defective interfering particles have been isolated and characterized from stocks of poliovirus passaged at high multiplicity [84, 155, 156]. Although easily obtainable for poliovirus there have been no reports of another picornavirus yielding such particles. It is difficult to say, however, if a systematic search with another virus has ever been carried out.

At least 3 different defective particles have been isolated from poliovirus. All seem to contain about 15% less RNA than wild type poliovirus, although each is a slightly different size [156]. A careful electron micrographic study of DI [1] RNA reveals that the RNA of this particle is 15.6% shorter than wild type [157]. The analysis is based on the measurement of length of defective double-stranded molecules compared with wild type double strands. These defective RNA molecules have been found to contain the same amount of poly(A) as wild type RNA [155].

Because of the deficit in RNA, defective particles are less dense than wild type ones and can be purified to homogeneity using cesium chloride density gradients. When purified DI infects a cell there is a normal amount of RNA synthesis and virus-specific protein synthesis [84]. However, no new particles are produced. Coinfection of cells with wild type virus is necessary for new DI particles to be produced.

The reason for the lack of virion production in cells infected with DI alone stems from the fact that DI RNA does not code for the polypeptide which is a precursor to the capsid proteins (polypeptide 1a). The RNA of this region appears to be deleted since there is no detectable coat protein found in the infected cell. A fragment of the capsid precursor does appear to be synthesized, but it is rapidly degraded after synthesis. Studies in in vitro protein-synthesizing systems primed with DI RNA or normal RNA indicate that the initiation site for protein synthesis on DI (1) RNA is the same as the one on normal RNA, as the initiation polypeptides formed in vitro in both systems are identical [151]. Thus the deletion present in DI RNA appears to be internal and within the capsid genes. It has not been mapped further.

Coinfection of cells with DI and wild type virus yields a mixed population of particles. The total amount of virus produced is lower in mixed infections than in infections with wild type virus alone and, in addition, the amount of DI present in the total virus population increases with each round of infection. These two processes are called interference and enrichment and are characteristic of all DI particles [155, 158, 159].

The mechanism of interference in this case seems to be a result of the inability of DI to code for the synthesis of capsid proteins. Because DI RNA is replicated equally well or better than wild type RNA and because it competes equally well for ribosomes, less capsid protein and consequently fewer procapsids are synthesized in coinfected cells than in cells infected with wild type virus alone. Thus fewer virus particles are able to assemble. The precise mechanism responsible for enrichment is not known in detail but seems to be the result of a preferential replication of DI RNA molecules [160].

The mechanism by which DI particles arise is not known. All the DI RNA molecules which have been examined in detail contain internal deletions. It can be argued that the initiation sites for protein synthesis, near the 5′ end of the molecule, and for RNA synthesis, near the 3′ end of the molecule, must be conserved if the RNA is going to be able to be translated and replicated. Thus any deletion which creates a DI particle with the properties of the above must be internal. There are at least 2 ways that an internal deletion can be generated. One mechanism would function after replication of the RNA has occurred by a chopping out and rejoining process. RNA rejoining enzymes (ligase) have been shown to exist in picornavirus-infected cells [161]. The other mechanism would generate the deletion during the replication process via a looping out of the negative-strand template and a skipping over a specific region by a polymerase.

Irrespective of the mechanism involved in the creation of a DI particle one must ask if DI particles are artifacts of in vitro culture of whether they play an important role in the natural life cycle of the virus. It has been argued that DI particles are important in the natural infectious cycle of a virus and are a way of regulating growth by inhibiting infection if the concentration of virus becomes too high [162]. The evidence for this assertion is not very compelling, however, and although the hypothesis is an attractive one, more work in this area is badly needed.

B. Selective Inhibition of Virus Replication by Guanidine and HBB

Specific inhibition of picornavirus replication by guanidine and 2-α(hydroxybenzyl)-benzimazole (HBB) has recently been reviewed at great length by Caliguiri and Tamm [163] and is also discussed in detail by Baltimore [4, 164]. Both compounds appear to be specific inhibitors of certain picornaviruses, leaving others unaffected. The spectrum of each is slightly different. Guanidine inhibits all enteroviruses, some rhinoviruses, and foot and mouth disease virus, while HBB only inhibits enteroviruses. Both drugs have no effect on cardiovirus replication. Mutants both resistant to and dependent on guanidine and HBB have been isolated.

There is general agreement that both drugs act by inhibiting viral RNA synthesis. More work has been carried out on the mechanism of action of guanidine than on HBB so that will be summarized here, as both mechanisms are likely to be similar.

Baltimore has proposed [4, 164] that the primary effect of guanidine is not to inhibit RNA synthesis per se but rather to inhibit the processing of RNA after it is synthesized. He proposes that newly made RNA is not released from the replication complex. This hypothesis is based on several observations: (a) guanidine does not inhibit the replicase in vitro but even the earliest detectable synthesis of viral RNA in the infected cell is prevented. (b) Protein synthesis is only secondarily inhibited, as a result of RNA synthesis. (c) Guanidine halts the encapsidation of newly made viral RNA into virion particles more rapidly than it inhibits RNA synthesis. (d) RNA made in the presence of guanidine is found in a structure which sediments more rapidly than the normal replication complex or viral polyribosomes. It has been called the guanidon.

Galiguiri and Tamm [163], however, feel that the guanidine acts primarily by inhibiting the initiation of synthesis of single-stranded viral RNA. Their results are at direct odds with Baltimore's. In their hands [unlike (c) and (d) above], some RNA appears to be released from the replication complex immediately after the addition of guanidine and a small amount is incorporated into virion particles and polyribosomes. They show that this process goes on for about half an hour before stopping. From these results they argue that in the presence of guanidine, preexisting RNA chains in the replication complex can be finished and processed but initiation of new ones cannot occur. Furthermore, since double-stranded RNA appears to be made at a normal rate in the presence of guanidine, they postulate that only the second step of replication is inhibited (see Fig. 3.8), i.e., the synthesis of positive-strand viral RNA. Also, since very little initiation of RNA synthesis occurs in the in vitro systems studied, they argue that it is not surprising that guanidine has no effect on the replicase in vitro.

Thus the exact mechanism of inhibition of picornavirus replication by guanidine (and HBB) is still controversial, although inhibition clearly occurs at the level of RNA synthesis. Until more is known about many of the controlling steps of RNA replication the question is likely to remain unsolved.

VII. Conclusion

Picornaviruses were originally studied by molecular biologists because of their apparent simplicity and the relative ease with which their growth cycle could be analyzed. They provided excellent model systems for the study of virus growth in a field that was just beginning to be established. During the late 1950s and early to middle 1960s studies on picornaviruses comprised a major part of the animal virus literature.

More recently, however, as animal virology has expanded with the development of new techniques and methods of analysis, the trend has been shifted away from picornaviruses toward the study of more complex viruses, whose genomes are many times larger and whose replication cycles are seemingly many times more complex.

While it certainly is true that largely because of their simplicity, picornaviruses cannot provide a model for every problem that exists in molecular virology today, it is equally ture that many aspects of their growth cycle are not well understood and need to be more fully elucidated. There are many major unsolved questions remaining in picornavirology which make their continued study worthwhile and the prospects for future research in this area exciting. Some of the remaining problems are reviewed below.

1. *Are there undiscovered viral functions?* We have just begun to correlate some of the known functions with certain virus-specified polypeptides. Some functions remain unassigned to a specific polypeptide and others need to be discovered. Temperature-sensitive mutants need to be analyzed more carefully. An analysis of the defect in each case should provide a more complete catalog of virus-specified functions.

2. *What is the detailed mechanism of proteolysis?* This is one of the most important areas that needs to be examined in detail. What enzymes are responsible for the cleavages? Are they virus or host coded? What are the recognition sites in the polypeptide which specify cleavage? Does proteolytic cleavage provide a means of control? (See Section 3.V.C.1.)

3. *How is the virion assembled?* Some steps in the assembly pathway have been described (see Section 3.V.D.) but little is known of the mechanisms causing aggregation to occur. What mechanisms allow the proteins to aggregate specifically? How is the RNA associated with the proteins of the particle and how does it get inside the procapsid?

4. *What are the early events of the virus cycle?* What is involved in adsorption, penetration, and uncoating? How does the input RNA strand initiation infection? (See Sections 3.V.A. and 3.V.C.2.)

5. *What are the specific mechanisms of the inhibition of host cell functions?* This has been discussed at length in Section 3.V.B. and although some partial models exist for protein synthesis shutoff they are clearly at present incomplete. The mechanism of RNA synthesis shutoff is even more obscure.

6. *What are the details of RNA replication?* Although the general pattern is established there are many questions remaining. How many replicases are there? One viral protein seems to be implicated in the replicase (Section 3.V.C.1.c.), are others involved? Do host-coded factors contribute to the replicase? How does guanidine work? What mechanism controls the fate of a newly synthesized RNA molecules? (Section 3.V.C.2.) How is poly(A) synthesized?

7. *What is the role of poly(A)?* It has been shown that poly(A) is necessary for infection but we do not know why.

8. *What is the role of membranes in infection?* It is clear that most virus-specific processes happen in association with cytoplasmic membranes, but we do not know why this association is necessary and the function that the membranes perform.

9. *What is the normal role of defective interfering particles in the virus infection cycle?* This has been discussed in Section 3.V.A.

10. *Where do picornaviruses come from and how did they evolve?*

At the end of his review in 1969 Baltimore [4] presented a similar list of remaining problems in picornavirus growth. Six years later, it is interesting to note that most of these questions remain unanswered and are included in the above list. In some areas the problems have been clarified a bit but there is still much work that remains to be done.

Acknowledgments

I thank Drs. D. Baltimore, K. Dorsch-Häsler, D. Frisby, M. Hewlett, D. Spector, N. Stebbing, R. Thach, and E. Wimmer for supplying unpublished information and preprints of manuscripts for this review; Drs. D. Frisby and E. Wimmer, for helpful discussions which clarified some of the more oblique areas of picornavirology; Drs. R. Thach and F. Brown for providing original photographs of published material; Academic Press, Cambridge University Press, and the American Society for Microbiology for permission to use published figures; I. Klavans-Rekosh for providing a broader perspective on replication; and C. Conway for deciphering and typing the manuscript.

This review was written in June, 1975, while the author was supported as a research fellow in the Department of Molecular Virology of the Imperial Cancer Research Fund, London, England.

References

1. F. Fenner, B. R. McAuslan, C. A. Mims, J. Sambrook, and D. O. White, The Biology of Animal Viruses, 2nd Ed., Academic, New York, 1974.
2. J. R. Paul, The History of Poliomyelitis, Yale University Press, New Haven, 1971.
3. J. Salk, in Viral and Rickettsial Infections of Man, 3rd Ed. (T. M. Rivers and F. L. Horsfall, eds.), Pitman Medical, London, 1959.
4. D. Baltimore, in Biochemistry of Viruses (H. B. Levy, ed.), Marcel Dekker, New York, 1969, p. 101.
5. L. Levintow, The Reproduction of Picornaviruses, in Comprehensive Virology Vol. 2 (H. Fraenkel-Conrat and R. R. Wagner, eds.), Plenum, New York, 1974, p. 109.
6. R. R. Rueckert, Picornaviral Architecture, in Comparative Virology, (K. Maramorosch and E. Kurstak, eds.), Academic, New York, 1971, p. 255.
7. C. Andrewes and H. G. Pereira, Viruses of Vertebrates, 3rd Ed., Bailliere Tindall, London, 1972, p. 3.
8. J. F. E. Newman, D. J. Rowlands, and F. Brown, J. Gen. Virol., 18, 171 (1973).
9. J. N. Burroughs and F. Brown, J. Gen. Virol., 22, 281 (1974).
10. K. C. Medappa and R. R. Rueckert, Binding of Cesium Ions to Human Rhinovirus-14, in Abstracts of the American Society of Microbiology Meeting, 1974, Chicago, Ill.
11. A. B. Sabin, Brit. Med. J., 1, 663 (1959).
12. T. H. Maugh II, Science, 188, 436 (1975).
13. E. Wimmer, J. Mol. Biol., 68, 537 (1972).
14. A. Nomoto, Y. F. Lee, and E. Wimmer, Proc. Natl. Acad. Sci. U.S., 73, 375 (1976).
15. M. J. Hewlett, J. K. Rose, and D. Baltimore, Proc. Natl. Acad. Sci. U.S., 73, 327 (1976).
16. R. Fernandez-Munoz and J. E. Darnell, J. Virol., 18, 719 (1976).
17. Y. F. Lee, A. Nomoto, B. M. Detjen, and E. Wimmer, Proc. Natl. Acad. Sci. U.S., in press.
18. D. M. Rekosh, H. F. Lodish, and D. Baltimore, J. Mol. Biol., 54, 327 (1970).
19. L. Villa-Komaroff, N. Guttman, D. Baltimore, and H. F. Lodish, Proc. Natl. Acad. Sci. U.S., 72, 4157 (1975).
20. I. Boime and P. Leder, Arch. Biochem. Biophys., 153, 706 (1972).
21. K. L. Eggen and A. J. Shatkin, J. Virol., 9, 636 (1972).
22. M. Esteban and I. Kerr, Eur. J. Biochem., 45, 567 (1974).

23. B. F. Öberg and A. J. Shatkin, Proc. Natl. Acad. Sci. U.S., 69, 3589 (1972).
24. A. E. Smith, Eur. J. Biochem., 33, 301 (1973).
25. J. M. Bishop, D. F. Summers, and L. Levintow, Proc. Natl. Acad. Sci. U.S., 54, 1273 (1965).
26. S. Penman, Y. Becker, and J. E. Darnell, J. Mol. Biol., 8, 541 (1964).
27. G. Abraham, D. P. Rhodes, and A. K. Banerjee, Cell, 5, 51 (1975).
28. J. M. Adams and S. Cory, Nature (London), 255, 28 (1975).
29. Y. Furuichi, M. Morgan, S. Muthukrishnan, and A. J. Shatkin, Proc. Natl. Acad. Sci. U.S., 72, 742 (1975).
30. C. W. Wei and B. Moss, Proc. Natl. Acad. Sci. U.S., 72, 318 (1975).
31. G. W. Both, A. K. Banerjee, and A. J. Shatkin, Proc. Natl. Acad. Sci. U.S., 72, 1189 (1975).
32. S. Muthukrishnan, G. W. Both, Y. Furuichi, and A. J. Shatkin, Nature (London), 255, 33 (1975).
33. J. K. Rose and H. F. Lodish, Nature, 262, 32 (1976).
34. Y. Yogo and E. Wimmer, Proc. Natl. Acad. Sci. U.S., 69, 1877 (1972).
35. J. A. Armstrong, M. Edmonds, H. Nakazato, B. A. Phillips, and M. H. Vaughan, Science, 176, 526 (1972).
36. D. Spector and D. Baltimore, J. Virol., 15, 1418 (1975).
37. R. L. Miller and P. G. W. Plagemann, J. Gen. Virol., 17, 349 (1972).
38. D. Spector and D. Baltimore, J. Virol., 16, 1081 (1975).
39. D. H. Spector and D. Baltimore, Proc. Natl. Acad. Sci. U.S., 71, 2983 (1974).
40. Y. Yogo and E Wimmer, Nature, New Biol., 242, 171 (1973).
41. Y. Yogo, M.-H. Teng, and E. Wimmer, Biochem. Biophys. Res. Comm., 61, 1101 (1974).
42. Y. Yogo and E Wimmer, J. Mol. Biol., 92, 467 (1975).
43. D. H. Spector and D. Baltimore, Virology, 67, 498 (1975).
44. A. Porter, N. Carey, and P. Fellner, Nature, 248, 657 (1974).
45. F. Brown, J. Newman, J. Stott, A. Porter, D. Frisby, C. Newton, N. Carey, and P. Fellner, Nature, 251, 342 (1974).
46. D. Frisby, C. Newton, N. H. Carey, P. Fellner, J. F. E. Newman, T. J. R. Harris, and F. Brown, Virology, 71, 379 (1976).
47. R. Salomon and U. Z. Littauer, Nature, 249, 32 (1974).
48. N. Stebbing, personal communication, 1975.
49. B. Öberg and L. Philipson, Biochem. Biophys. Res. Comm., 48, 927 (1972).
50. M. Pinck, P. Yot, F. Chapeville, and H. M. Duranton, Nature, 226, 954 (1970).
51. P. Yot, M. Pinck, A. L. Haenni, H. M. Duranton, and F. Chapeville, Proc. Natl. Acad. Sci. U.S., 67, 1345 (1970).
52. M. Pinck, S. K. Chan, M. Genevaux, L. Hirth, and H. M. Duranton, Biochimie, 54, 1093 (1972).
53. S. Litvak, A. Tarragó, L. Tarragó-Litvak, and J. E. Allende, Nature, New Biol., 241, 88 (1973).
54. T. C. Hall, D. S. Shin, and P. Kaesberg, Biochem. J., 129, 969 (1972).

55. I. Sela, Virology, 49, 90 (1972).
56. R. Rueckert, A. K. Dunker, and C. M. Stoltzfus, Proc. Natl. Acad. Sci. U.S., 62, 912 (1969).
57. A. K. Dunker and R. R. Rueckert, J. Mol. Biol., 58, 217 (1970).
58. T. R. Ziola and D. G. Scraba, Virology, 57, 531 (1974).
59. S. Halperen, H. O. Stone, and B. D. Korant, J. Gen. Virol., 20, 267 (1973).
60. A. T. H. Burness, I. U. Pardoe, and S. M. Fox, J. Gen. Virol., 18, 33 (1973).
61. R. Drzeniek and P. Bilello, J. Gen. Virol., 25, 125 (1974).
62. K. C. Medappa, C. McLean, and R. R. Rueckert, Virology, 44, 259 (1971).
63. T. W. Mak, J. S. Colter, and D. G. Scraba, Virology, 57, 543 (1974).
64. P. Talbot and F. Brown, J. Gen. Virol., 15, 163 (1972).
65. L. Medrano and H. Green, Virology, 54, 515 (1973).
66. D. A. Miller, O. J. Miller, V. G. Dev, S. Gashmi, R. Tantravahi, L. Medrano, and H. Green, Cell, 1, 167 (1974).
67. N. H. Levitt and R. L. Crowell, J. Virol., 1, 693 (1967).
68. R. L. Crowell and L. Philipson, J. Virol., 8, 509 (1971).
69. K. Lonberg-Holm and B. D. Korant, J. Virol., 9, 29 (1972).
70. C. E. Cords, C. G. James, and L. C. McLaren, J. Virol., 15, 244 (1975).
71. M. Breindl and G. Koch, Virology, 48, 136 (1972).
72. B. D. Korant, K. Lonberg-Holm, F. H. Yin, and J. Noble-Harvey, Virology, 63, 384 (1975).
73. K.-O. Habermehl, W. Diefenthal, and M. Buchholz, in Advances in the Biosciences, Vol. II, Workshop on Virus-Cell Interactions, Pergamon, 1973, p. 41.
74. L. A. Caliguiri, I. Tamm, and A. G. Mosser, in Advances in the Biosciences, Vol. II, Workshop in Virus-Cell Interactions, Pergamon, 1973, p. 65.
75. L. A. Caliguiri and I. Tamm, Science, 166, 885 (1969).
76. M. Roumiantzeff, J. V. Maizel, and D. F. Summers, Virology, 44, 239 (1971).
77. S. Penman, S. Scherrer, Y. Becker, and J. E. Darnell, Proc. Natl. Acad. Sci. U.S., 49, 654 (1963).
78. R. Leibowitz and S. Penman, J. Virol., 8, 661 (1971).
79. S. Penman and D. Summers, Virology, 27, 614 (1965).
80. M. Willems and S. Penman, Virology, 30, 355 (1966).
81. A. Steiner-Pryor and P. D. Cooper, J. Gen. Virol., 21, 215 (1973).
82. P. D. Cooper, A. Steiner-Pryor, and P. J. Wright, Intervirology, 1, 1 (1973).
83. P. J. Wright and P. D. Cooper, Virology, 59, 1 (1974).
84. C. N. Cole and D. Baltimore, J. Mol. Biol., 76, 325 (1973).
85. E. Ehrenfeld and T. Hunt, Proc. Natl. Acad. Sci. U.S., 68, 1075 (1971).
86. B. Cordell-Stewart and M. W. Taylor, J. Virol., 11, 232 (1973).

87. H. D. Robertson and M. B. Mathews, Proc. Natl. Acad. Sci. U.S., 70, 225 (1973).
88. C. Lawrence and R. E. Thach, J. Virol., 14, 598 (1974).
89. D. L. Nuss, H. Opperman, and G. Koch, Proc. Natl. Acad. Sci. U.S., 72, 1258 (1972).
90. J. L. Saborio, S. S. Pong, and G. Koch, J. Mol. Biol., 85, 195 (1974).
91. G. Contreras, D. F. Summers, J. V. Maizel, and E. Ehrenfeld, Virology, 53, 120 (1973).
92. L. B. Schwartz, C. Lawrence, R. E. Thach, and R. G. Rodder, J. Virol., 14, 611 (1974).
93. H. I. Miller and E. E. Penhoet, Proc. Soc. Exp. Biol. Med., 140, 435 (1972).
94. R. G. Roeder and W. J. Rutter, Nature, 224, 234 (1969).
95. R. G. Roeder and W. J. Rutter, Proc. Natl. Acad. Sci. U.S., 65, 675 (1970).
96. R. H. Reeder and R. G. Roeder, J. Mol. Biol., 67, 433 (1972).
97. R. Weinman and R. G. Roeder, Proc. Natl. Acad. Sci. U.S., 71, 1790 (1974).
98. R. Hand, W. D. Ensminger, and I. Tamm, Virology, 44, 527 (1971).
99. R. Hand and I. Tamm, Virology, 47, 331 (1972).
100. T. T. Crocker, E. Pfendt, and R. Spendlove, Science, 145, 401 (1964).
101. R. Pollack and R. Goldman, Science, 179, 915 (1973).
102. D. Baltimore, in Perspectives in Virology, Vol. VII (M. Pollard, ed.), Academic, New York, 1971, p. 1.
103. D. F. Summers, J. V. Maizel, Jr., and J. E. Darnell, Jr., Proc. Natl. Acad. Sci. U.S., 54, 505 (1965).
104. D. F. Summers, J. V. Maizel, Jr., and J. E. Darnell, Jr., Virology, 31, 427 (1967).
105. E. Paucha, J. Seehafer, and J. S. Colter, Virology, 6, 315 (1974).
106. B. E. Butterworth, Virology, 56, 439 (1973).
107. D. F. Summers and J. V. Maizel, Proc. Natl. Acad. Sci. U.S., 59, 966 (1968).
108. M. F. Jacobson and D. Baltimore, Proc. Natl. Acad. Sci. U.S., 61, 77 (1968).
109. M. F. Jacobson and D. Baltimore, J. Mol. Biol., 33, 369 (1968).
110. M. F. Jacobson, J. Asso, and D. Baltimore, J. Mol. Biol., 49, 657 (1970).
111. D. Baltimore, A. Huang, K. F. Manly, D. Rekosh, and M. Stampfer, in Strategy of the Viral Genome, Ciba Foundation Symposium, Churchill Livingstone, Edinburgh, 1971, p. 101.
112. B. D. Korant, J. Virol., 10, 751 (1972).
113. D. F. Summers, E N. Shaw, M. L. Stewart, and J. V. Maizel, J. Virol., 10, 880 (1972).
114. B. Butterworth and B. D. Korant, J. Virol., 14, 282 (1974).
115. R. L. Taber, D. Rekosh, and D. Baltimore, J. Virol., 8, 395 (1971).
116. D. Rekosh, J. Virol., 9, 479 (1972).
117. D. F. Summers and J. V. Maizel, Proc. Natl. Acad. Sci. U.S., 68, 2852 (1971).

118. B. E. Butterworth and R. R. Rueckert, J. Virol., 9, 823 (1972).
119. B. E. Butterworth and R. R. Rueckert, Virology, 50, 535 (1972).
120. A. E. Smith, in Control Processes in Virus Multiplication, Cambridge University Press, Cambridge, U.K., 1975, p. 183.
121. A. E. Smith, K. A. Marcker, and M. B. Mathews, Nature, 225, 184 (1970).
122. P. D. Cooper, E. Geissler, P. D. Scotti, and G. A. Tannock, in The Strategy of the Viral Genome, Ciba Foundation Symposium, Churchill Livingstone, Edinburgh, 1971, p. 75.
123. P. D. Cooper, in The Biochemistry of Viruses (H. B. Levy, ed.), Marcel Dekker, New York, 1969, p. 177.
124. P. D. Cooper, D. Stancek, and D. F. Summers, Virology, 40, 971 (1970).
125. R. E. Lundquist, E. Ehrenfeld, and J. V. Maizel, Jr., Proc. Natl. Acad. Sci. U.S., 71, 4773 (1974).
126. H. Rosenberg, B. Diskin, L. Oron, and A. Traub, Proc. Natl. Acad. Sci. U.S., 69, 3815 (1972).
127. J. Al-Janabi, J. A. Hartsuck, and J. Tang, J. Biol. Chem., 247, 4638 (1972).
128. R. A. Bradshaw, R. A. Hogue-Angeletti, and W. A. Frazier, Rec. Prog. Horm. Res., 30, 575 (1974).
129. B. D. Korant, J. Virol., 12, 556 (1973).
130. C. Lawrence and R. E. Thach, J. Virol., 15, 918 (1975).
131. J. E. Darnell, M. Girard, D. Baltimore, D. F. Summers, and J. V. Maizel, in The Molecular Biology of Viruses (J. S. Colter and W. Paranchych, eds.), Academic, New York, 1967, p. 375.
132. R. I. Kamen, in RNA phages (N. D. Zinder, ed.), Cold Spring Harbor Laboratory, 1975.
133. D. Kolakofsky and C. Weissman, Biochim. Biophys. Acta, 246, 596 (1971).
134. J. M. Bishop and L. Levintow, Prog. Med. Virol., 13, 1 (1971).
135. B. Öberg and L. Philipson, J. Mol. Biol., 58, 725 (1971).
136. S. Thach, D. Dobbertin, C. Lawrence, F. Golini, and R. E. Thach, Proc. Natl. Acad. Sci. U.S., 71, 2549 (1974).
137. L. A. Caliguiri, Virology, 58, 526 (1974).
138. D. S. Colby, V. Finnerty, and J. Lucas-Lenard, J. Virol., 13, 858 (1974).
139. C. N. Nair and M. J. Owens, J. Virol., 13, 535 (1974).
140. M. Chamberlin and P. Berg, J. Mol. Biol., 8, 708 (1964).
141. B. Dietz-Schold and R. Ahl, J. Gen. Virol., 8, 73 (1970).
142. E. Ehrenfeld, J. V. Maizel, and D. F. Summers, Virology, 40, 840 (1970).
143. M. Girard, J. Virol., 3, 376 (1969).
144. D. H. Spector and D. Baltimore, J. Virol., 15, 1432 (1975).
145. K. Dorsch-Häsler, Y. Yogo, and E. Wimmer, J. Virol., 16, 1512 (1975).
146. L. A. Caliguiri and R. W. Compans, J. Gen. Virol., 21, 99 (1973).
147. B. A. Phillips, D. F. Summers, and J. V. Maizel, Jr., Virology, 35, 216 (1968).
148. B. A. Phillips and R. Fennel, J. Virol., 12, 291 (1973).

149. B. A. Phillips, Virology, 44, 307 (1971).
150. S. McGregor, L. Hall, and R. R. Rueckert, J. Virol., 15, 1107 (1975).
151. C. B. Fernandez-Tomas and D. Baltimore, J. Virol., 12, 1122 (1973).
152. D. J. Rowlands, D. V. Sangar, and F. Brown, J. Gen. Virol., 26, 227 (1975).
153. M. D. Johnston and S. J. Martin, J. Gen. Virol., 11, 71 (1971).
154. C. B. Fernandez-Tomas, N. Guttman, and D. Baltimore, J. Virol., 12, 1181 (1973).
155. D. Baltimore, C. N. Cole, L. Villa-Komaroff, and D. Spector, in Mechanisms of Virus Disease (W. S. Robinson and C. F. Fox, eds.), Benjamin, Menlo Park, Calif., 1974, p. 117.
156. C. N. Cole, D. Smoler, E. Wimmer, and D. Baltimore, J. Virol., 7, 478 (1971).
157. M. Hewlett and D. Baltimore, personal communication, 1975.
158. C. N. Cole and D. Baltimore, J. Mol. Biol., 76, 345 (1973).
159. C. N. Cole and D. Baltimore, J. Virol., 12, 1414 (1973).
160. M. Hewlett and D. Baltimore, personal communication, 1975.
161. T. Linné, B. Öberg, and L. Philipson, Eu. J. Biochem., 42, 157 (1974).
162. A. S. Huang and D. Baltimore, Nature, 226, 325 (1970).
163. L. A. Caliguiri and I. Tamm, in Selective Inhibitors of Viral Functions (W. A. Carter, ed.), CRC, Cleveland, 1973, p. 257.
164. D. Baltimore, in Medical and Applied Virology, Proceedings of the Second International Symposium (M. Sanders and E. H. Lennette, eds.), Green, St. Louis, Mo., 1968, p. 340.
165. T. J. R. Harris and F. Brown, J. Gen. Virol., 33, 493 (1976).
166. K. M. Chumakov and V. I. Agol, Biochem. Biophys. Res. Comm., 71, 551 (1976).

Chapter 4

Togaviruses

James H. Strauss and Ellen G. Strauss

Division of Biology
California Institute of Technology
Pasadena, California

I. Introduction

The togaviruses are a family of animal viruses which consist of an icosahedral nucleocapsid (containing the single-stranded RNA of the virion) surrounded by a lipoprotein envelope. Most of the known members of this family are arboviruses, being transmitted in nature by blood sucking arthropods (primarily mosquitoes) and possessing the ability to grow in the arthropod vector as well as in the avian or mammalian host. The alphaviruses (formerly known as the group A arboviruses) contain about 17 known members [1], which cross-react serologically in complement-fixation assays but which can be distinguished by serological neutralization of infectivity. Of these, Sindbis virus and Semliki Forest virus have been the most extensively characterized, although a considerable body of literature has also been developed for western, eastern, and Venezuelan equine encephalitis viruses and for chikungunya virus.

The flaviviruses (or group B arboviruses) contain 33 known members [2], defined in a manner analogous to the group A arboviruses. These viruses have been less well characterized, although dengue virus has been studied in some detail and data are available for Japanese encephalitis virus, Kunjin virus, and St. Louis encephalitis virus.

The arboviruses, being insect mediated, are prevalent in tropical areas of the world, where insects abound and the problems associated with overwintering do not obtain. The alphaviruses cause a variety of sometimes painful but not generally fatal human diseases, which are usually characterized by a frank illness followed by complete recovery in 1-2 weeks. Several of the viruses, such as Sindbis virus and Semliki Forest virus, lead primarily to subclinical infections. Although many are of African origin, 3 equine encephalitis viruses, eastern, western, and Venezuelan, occur in the United States. Eastern equine encephalitis virus rarely infects humans but has been associated with several epidemics in the eastern United States which were characterized by a high percentage of fatalities [1]. Serological studies indicate that western equine encephalitis virus commonly infects humans, but the majority of these cases are subclinical. Periodic outbreaks of epidemic proportions occur in the western United States, however, with mortality rates of 2-15% of serologically confirmed cases [1]. Venezuelan equine encephalitis virus is reported to cause influenza-like symptoms in man and has been responsible for numerous laboratory acquired infections including 1 fatal case [3]. It caused a number of human cases in the recent epidemic of equine encephaltiis in Central America and the southwestern United States.

The flaviviruses, in general, cause more severe human diseases. The history of yellow fever virus and its effects on western civilization are well known. Dengue virus is epidemic in many regions of the world, in particular in Southeast Asia, and Japanese encephalitis virus causes periodic outbreaks of disease in Japan. St. Louis encephalitis virus is currently (July 1975) epidemic in the southern United States. The interested reader is referred to reviews by Casals and Clarke [1, 2] for further details of the occurrence of the arboviruses and their disease symptoms.

We should note that the togavirus family is defined by morphological criteria (the morphology of these viruses is discussed in detail in Section 4.II.), whereas arboviruses are defined by epidemiology and by serological cross-reaction. However, this chapter considers only those viruses formerly known as the group A and group B arboviruses, although rubella virus and lactic dehydrogenese virus are now thought to be togaviruses by virtue of their morphology. In fact most of the discussion centers on the alphaviruses, as a great deal more is known about their structure and mode of replication than for the flaviviruses. The togaviruses have been reviewed recently by Pfefferkorn and Shapiro [4].

II. Structure of Togaviruses

A. Composition

The overall composition of the togaviruses is given in Table 4.1. These viruses contain RNA, protein, phospholipid, cholesterol, and carbohydrate. Most of the carbohydrate is present in glycoproteins, although the viruses do contain a small amount of glycolipid. The alphaviruses contain 3 or 4 protein species, depending upon the virus; an acrylamide gel electropherogram showing the 3 proteins of Sindbis virus is shown in Fig. 4.1. Two of these proteins are glycoproteins and have been referred to either as the membrane proteins (or M) or as the envelope proteins (E1 and E2, with E2 migrating faster than E1 in the case of Sindbis virus). These can be labeled with glucosamine or other radioactive carbohydrates, as well as with amino acids. The third protein, the nucleocapsid protein, contains no carbohydrates. Semliki Forest virus contains, in addition to these 3 proteins, a small envelope glycoprotein referred to as E3.

Table 4.1 Composition of Togaviruses

	Alphaviruses		Flaviviruses	
	Value	References	Value	References
Protein				
% of total mass	57-61	[5, 6]		
Molecular weights × 10^{-3}				
Capsid protein	C ≃ 30-34	[7, 8, 9]	V-2 ≃ 13.5	[18, 19, 20, 21]
Envelope glycoproteins	E1 ≃ 50	[7, 10]	V-3 ≃ 53-63	[18, 19, 20, 21]
	E2 ≃ 50	[7, 10]		
	E3 ≃ 10	[10]		
Other proteins			V-1 = 7.7 (tick-borne viruses)	[18]
			V-1 = 8.5 (mosquito-borne viruses)	[18, 19, 20, 21]
RNA				
% of total mass	5.5-6.3	[5, 6]		
Molecular weight × 10^{-6}	4.0-4.6	[11, 12, 13, 14]	4.0-4.6	[13, 22]
GC content (%)	51	[5, 15]	48	[22, 23]
Carbohydrate				
% of total mass	6.4	[6, 16, 17]		
Lipid				
% of total mass	27-31	[5, 6]		

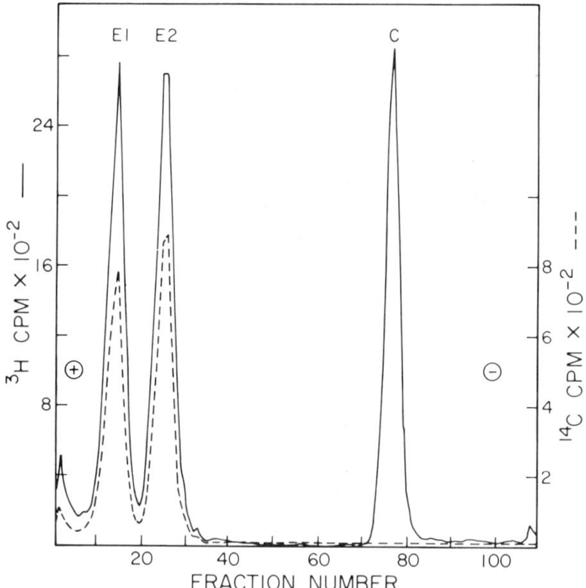

Fig. 4.1 Acrylamide gel electropherogram of Sindbis virion proteins. Purified Sindbis virions were disrupted with Triton X-100, reduced with mercaptoethanol, and alkylated with acrylonitrile. The gel system was a discontinuous system at pH 4.5 containing 2% Triton X-100 and 6 M urea. The solid line is [^{3}H]-amino acid label, the dashed line is [^{14}C]-glucosamine label. C is the nucleocapsid protein, E1 and E2 are the envelope glycoproteins (unpublished data of J. H. Strauss and E. G. Strauss, 1973).

The protein composition of the flaviviruses is somewhat different. There are 3 polypeptide species, of which only 1, the envelope protein V-3, is a glycoprotein. V-2 is the nucleocapsid protein. V-1 is not a component of the nucleocapsid, but its exact position in the virus is not clear.

The RNA and protein of the virus are synthesized de novo after infection [24, 25], whereas the phospholipids in the virion are derived from the preexisting host cell lipids [24].

A schematic representation of the structure of the alphaviruses and their mode of maturation is shown in Fig. 4.2. These viruses consist of a central nucleocapsid surrounded by a lipid bilayer, outside of which are the glycoproteins. The virion has an overall diameter of 70 nm [26, 27], a density of 1.21 g/ml [28], a sedimentation coefficient of 280 S [29, 30], and a particle molecular weight of approximately 70 × 10^6 (based either on the Svedberg equation using the sedimentation coefficient and density of the particle or upon RNA molecular weight and chemical composition).

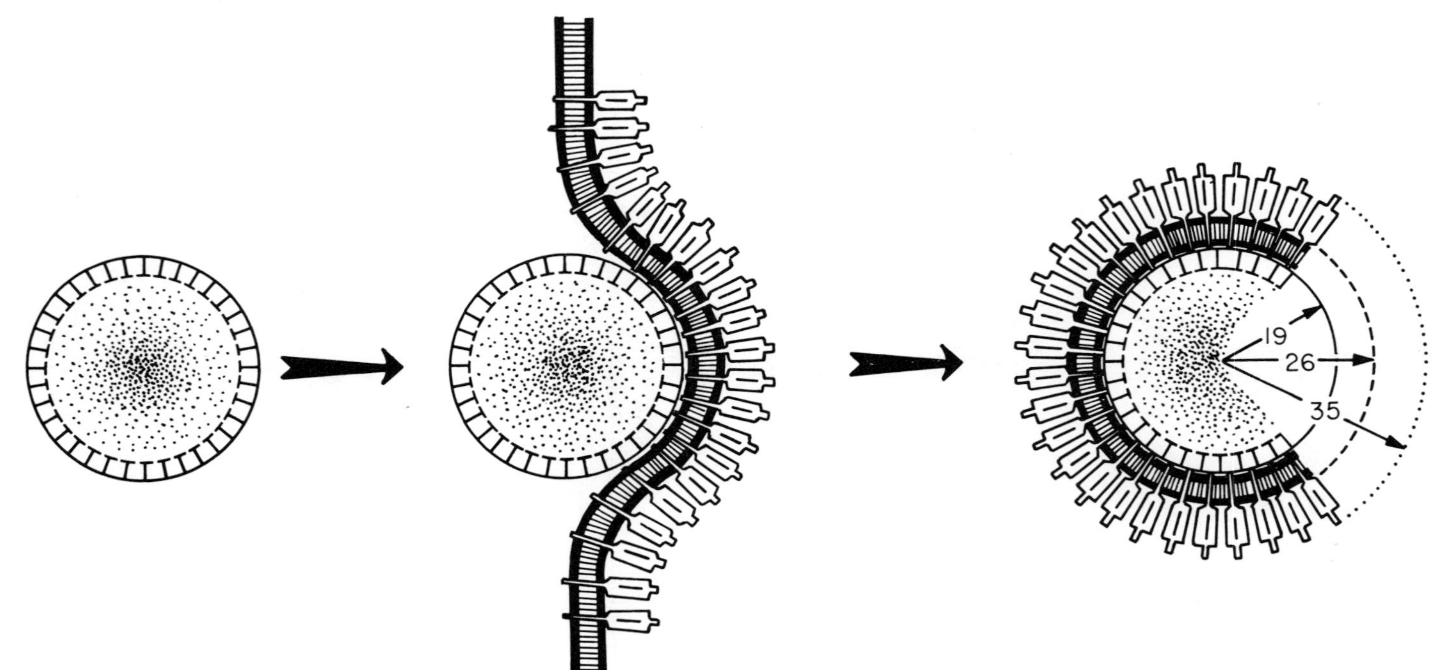

Fig. 4.2 Model of the structure and maturation of alphaviruses. Nucleocapsids are assembled in the cytoplasm and diffuse to the cell surface where they interact with the virus glycoproteins. The nucleocapsid buds through the virus-altered surface to produce the virion. Radii, in nm, determined from x-ray diffraction studies, are shown for the mature virion.

The flaviviruses appear to be somewhat smaller, with a sedimentation coefficient of 205 S [13] and a diameter, determined by negative staining in the electron microscope, of from 30 to 42 nm, depending upon the virus [32]. The density of the major infectious component of dengue virus is 1.24 g/cc [31].

The structure of the various components of the virus will now be considered in detail.

B. The Virion RNA

The alphaviruses contain single-stranded RNA of molecular weight approximately 4.5×10^6 as their genome. Estimates of $4.0\text{-}4.2 \times 10^6$ daltons have been obtained by acrylamide gel electrophoresis [11, 13, 33]. However, higher estimates have been obtained based upon the molecular weight of the double-stranded form of the RNA (8.8×10^6, giving a single-strand molecular weight of 4.4×10^6) [11] or upon electron microscope length measurements (4.6×10^6) [14]. This RNA has been reported to sediment between 42 S and 50 S. In our hands Sindbis RNA has an $S_{20,w}$ of 49 S in 0.2 M salt, as measured in the Model E Analytical Centrifuge. Although early experiments had shown that alphavirus RNA tended to break down under mild denaturing treatment, carefully controlled experiments with unnicked RNA molecules have now demonstrated that the alphavirus genome consists of a single polynucleotide chain of about 13,000 nucleotides [11, 12, 14]. It is interesting to note that electron microscopic studies have shown that the RNA isolated from virions exists primarily in the form of hydrogen-bonded circles [14], implying that there are regions of sequence complementarity near the ends of the molecule.

The viral RNA is plus stranded, serving as a messenger in vivo [34-37] and in vitro [38, 39]. As expected from its messenger functions the RNA contains 3′ terminal poly(A) tracts 60-80 nucleotides long [40-42, 91]. The naked RNA is infectious [43], and unlike many animal viruses no virion-carried enzymes are necessary to initiate the infection cycle.

The RNA of the flaviviruses has been less carefully characterized than that of the alphaviruses. It has a sedimentation coefficient of approximately 45 S [22, 44, 105, 106] and thus is similar in size to the RNA of the alphaviruses. In a comparative study, flavivirus RNA sedimented about 10% slower than alphavirus RNA in sucrose gradients, but both RNA species had the same mobility in acrylamide gel electrophresis (the latter method gave a molecular weight estimate of 4.2×10^6). The 45 S RNA of the flaviviruses is probably a single polynucleotide chain; it is infectious [44], and therefore "plus" stranded, and has been shown to contain poly(A) tracts [45, 46].

The base compositions of the various alphavirus RNAs are similar with about 29% adenine, 20-22% uracil, 25% guanine, and 25% cytosine [5, 15]. The flavivirus RNAs have a slightly different base composition, with about 30% adenine, 22% uracil, 27% guanine, and 21% cytosine [22, 23].

C. The Nucleocapsid

The RNA of the alphaviruses is complexed with about 300 molecules of a single protein species, the nucleocapsid protein or core protein, to form the virus nucleocapsid. This protein has a molecular weight of 30,000-34,000 [7-9] and is rich in basic amino acids [8].

The nucleocapsid is a stable structure, sedimenting at 140 S [29, 30], which can be isolated either from the cytoplasm of infected cells [47] or by treatment of the virion with mild detergents [29, 30]. The RNA within the isolated nucleocapsids is sensitive to RNase [48, 49]. Values of 35-40 nm have been reported for the diameter of the core based on electron microscopy; estimates have been made from thin sections of whole virus particles [50], from freeze-etched preparations of virions or of budding figures [52], and from isolated nucleocapsids visualized by negative staining [27, 51]. x-Ray diffraction studies of an alphavirus have been interpreted to indicate a diameter of 41 nm for the nucleocapsid [27].

Nucleocapsids of the alphaviruses are icosahedral in design [30, 51] and have been reported to have triangulation numbers of 3 [30], 4 (in this case inferred from the distribution of glycoproteins on the outer surface of the virion) [53], or 9 [54]. Further experiments will have to be done to clear up this point. At the current time we favor the T = 9 result of Brown and Gliedman [54]. These authors also found 2 smaller particles in addition to the virion after chronic infection of mosquito cells. These smaller particles had an envelope of normal thickness but smaller nucleocapsids. The authors assigned triangulation numbers of 9 to the normal virus nucleocapdis, 4 to the medium-sized nucleocapsid, and 1 to the small nucleocapsid. These 3 particles thus appear to form a series in the P = 1 class [55]; the construction of the smaller particles will be difficult to explain if the triangulation number of the nucleocapsid is 3 or 4 rather than 9. We might note that a triangulation number of 3 would require 180 subunits (this is probably too few). A triangulation number of 4 would require 240 subunits (and thus presumably 240 molecules of capsid protein in the nucleocapsid), while a triangulation number of 9 would require 540 subunits (and in this case presumably 270 molecules of capsid protein in the nucleocapsid, with each molecule filling the role of 2 subunits).

The nucleocapsid of the flaviviruses appears to be smaller, with a diameter of 26 nm [56], although the sedimentation coefficient of the nucleocapsid is the same as that of the alphavirus nucleocapsid, 140 S [13]. The nucleocapsid protein of these viruses has a molecular weight of 13,500 [18-21].

D. The Lipid Bilayer

The phospholipids and cholesterol of the alphaviruses are arranged in a bilayer centered at 23.2 nm and having a thickness of 4.8 nm; this was clearly shown by the x-ray diffraction studies of the virus by Harrison et al. [27]. The radial electron density of the virus derived from these studies is shown in Fig. 4.3. A minimum in the electron density profile occurs in the area of the lipid bilayer, since the electron density of hydrocarbon, such as the fatty acid portions of phospholipids, is less than that of protein or of RNA. The electron density at the center of the bilayer (r = 23.2 nm) is close to that of pure hydrocarbon, implying that this region contains very little protein (less than 10%). Other studies have shown, however, that glycoproteins do penetrate the lipid bilayer [57].

The small peaks centered at 21 nm and 26 nm are thought to represent the polar head groups of the phospholipids. Note also that the electron density rises toward the center of the virus, where the RNA and nucleocapsid protein are found. Outside the lipid bilayer are found most of the glycoprotein molecules; the electron density is high out to about 30 nm and then gradually falls to that of the solvent by 35 nm.

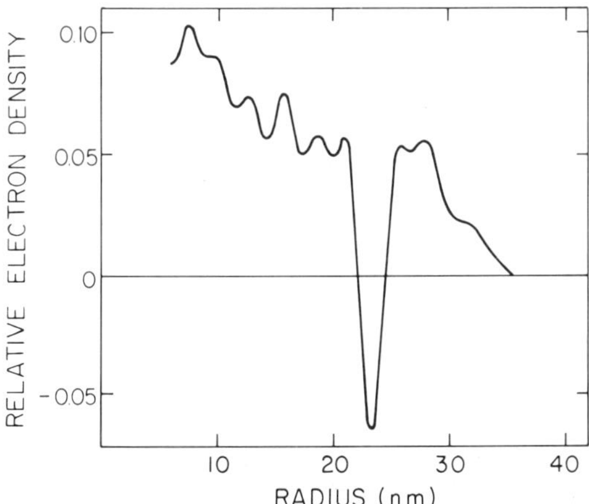

Fig. 4.3 Radial electron density distribution of Sindbis virus determined by x-ray diffraction. The units of the abscissa are approximately electrons/Å^3 relative to the electron density of the solvent. The minimum at 23.2 nm indicates the middle of the lipid bilayer. (Modified from Harrison et al. [27]. Reproduced by courtesy of Academic Press.)

Table 4.2 Phospholipid Composition of Alphavirus Virions (% of total)

Virus:	Semliki Forest virus	Sindbis virus		Venezuelan equine encephalitis virus	
Reference:	[58]	[59]		[61]	
Host cell:	BHK21-C13	BHK21	Chick fibroblast	L-Cell	Chick fibroblast
Phospholipid:					
Phosphatidylcholine	34	26	26	23	35
Phosphatidylethanolamine	26	35	32	29	28
Phosphatidylserine	12	20	21	15	17
Sphingomyelin	21	18	21	26	13
Phosphatidylinositol	1	Not detectable	Not detectable	0.5	2
Others	—	—	—	5	3

The composition of the phospholipids of the virus have been studied in some detail; representative data are given in Table 4.2. The major phospholipids are phosphatidylcholine, sphingomyelin, phosphatidylserine, and phosphatidylethanolamine. The phospholipid composition shows more resemblance to the composition of the host cell plasma membrane than to the membranes of other host organelles although the virus does appear to show some selectivity in phospholipid content [59, 61-63]. Semliki Forest virus grown in BHK cells is reported to contain 16,000 molecules of cholesterol and 17,000 molecules of phospho- and glycolipids per virion (molar ratio of 0.94) [6]; molar ratios of cholesterol to phospholipid of 0.68-0.80 have been found for Sindbis virus grown in chick embryo fibroblasts [5, 63]. The cholesterol to phospholipid molar ratio for the plasma membrane of chick cells was found to be 0.60 [63]. The distribution of the various phospholipids and cholesterol in the 2 halves of the bilayer is unknown, although it is interesting to note that because of the small size of the virus, the surface area of the outer leaflet is twice that of the inner leaflet [27].

In addition, the lipid bilayer of the virus appears to be less fluid than that of the cell [64]; this decrease in fluidity may result from the presence in the bilayer of those portions of the glycoproteins which span the bilayer and are anchored in position by the nucleocapsid (see Section 4.II.E. and 4.III.D).

The flaviviruses also possess a lipid bilayer, although the details of the bilayer composition and structure are not as well known as for the alphaviruses.

E. The Glycoproteins

The alphaviruses contain about 300 molecules each of 2 glycoprotein species of molecular weight approximately 50,000 [7, 10, 16, 65]. At least 1 virus, Semliki Forest virus, also contains equimolar amounts of a third glycoprotein of molecular weight 10,000 [10]. The amino acid analysis of each glycoprotein has been determined [10] as well as their respective carbohydrate compositions; the latter are shown in Table 4.3. The carbohydrates are found as polysaccharides of 2 size classes. The smaller chains are 6-8 residues long and contain only mannose and N-acetyl glucosamine. The longer chains are variable in size and composition, averaging 16-20 residues in length and containing mannose, galactose, N-acetyl glucosamine, fucose, and sialic acid [17, 66]; they can be resolved into at least 3 size classes [17]. The large glycoproteins, E1 and E2, consist of a polypeptide backbone with 2 covalently bound polysaccharide chains, 1 short chain which is relatively constant, and 1 longer residue. The average composition and distribution of the longer chains is dependent upon the host cell in which the virus is grown [16, 66]. The organization of the carbohydrate of the small glycoprotein E3, is not known.

Table 4.3 Carbohydrate Composition of Semliki Forest Virus Glycoproteins[a]

	Moles carbohydrate/mole protein		
	E1	E2	E3
N-Acetyl glucosamine	7	8	9
Mannose	5	12	4
Galactose	3	3	4
Fucose	1	1	2
Sialic acid	2	4	3
Total carbohydrate % by weight	7.5	11.5	45.1

[a] Data from Ref. [10].

Overall, polysaccharides account for 8-12% of the mass of E1 and E2 and 45% of the mass of E3.

These glycoproteins are, for the most part, external to the lipid bilayer, and can be removed from the virus particle by treatment with proteolytic enzymes, such as pronase [67], bromelain [68], or thermolysin [69]. Such treatment leaves a particle bounded by the lipid bilayer and having a density of 1.14 [68, 70] (the density of the intact virion is 1.21), which can be isolated and characterized. This naked particle is no longer infectious; contains all of the lipid, RNA, and nucleocapsid protein of the virion; and also contains small polypeptides with a molecular weight of about 5000 derived from the glycoproteins. These "roots" of the glycoprotein molecules are rich in aliphatic amino acids [70] and are thought to reside in the lipid bilayer, being soluble in the lipid phase, and thus to be protected from the action of the proteases. It is unclear at present which of the glycoproteins gives rise to these small polypeptide roots. Probably E1 and E2 both penetrate the lipid bilayer and E3 (if present) does not; it is possible, however, that only 1 glycoprotein is anchored in the bilayer or that all 3 are anchored.

At least 1 of the glycoproteins apparently spans the bilayer completely. Treatment of intact virions with dimethylsuberimidate, a cross-linking reagent, results in chemically coupling the glycoproteins to the nucleocapsid [57]. It is also possible to covalently link the glycoprotein to the nucleocapsid with formaldehyde [71]. Another experiment to show that at least 1 glycoprotein spans the bilayer completely is a labeling study with [^{35}S]-formyl methionyl sulfone methyl phosphate. When intact Semliki Forest virus is treated with this chemical, several tryptic peptides derived from the glycoproteins are labeled. When a membrane preparation from disrupted virions is used as a substrate, so that portions of the glycoproteins on both sides of the lipid layer are accessible to the reagent, 2 additional peptides, rich in basic amino acids, are found to contain ^{35}S [57]. Finally, studies with temperature-sensitive mutants of Sindbis virus imply that glycoprotein E1 can bind nucleocapsids to the inside of the plasma membrane of vertebrate cells [135].

Thus it appears that 1 terminus of at least 1 glycoprotein molecule, and probably of both E1 and E2, extends to the inner surface of the lipid bilayer and can interact with the nucleocapsid; there is a lipophilic region which spans the lipid bilayer and anchors the proteins; however, the major portion of the protein molecule and all the carbohydrate moiety are external to the bilayer and exposed to the aqueous environment.

A small glycoprotein corresponding to E3 of Semliki Forest virus has not been found in Sindbis virus, despite a careful search by several laboratories. E2 and E3 have been shown by tryptic peptide analysis to be the cleavage products of a precursor polypeptide, NVP62 [10, 72]. An analogous cleavage step occurs during the maturation of Sindbis virus; the precursor PE2 is cleaved to form E2 [73]. It thus appears that the small glycoprotein produced during this cleavage step is retained as an integral part of the Semliki Forest virion but is lost from Sindbis virus, either during virus maturation or during virus purification.

The glycoproteins occupy a region of about 10 nm outside the lipid bilayer [26, 27]. E1 and E2 also appear to interact in some way and perhaps a dimeric form (trimeric if E3 is present) is the principal structural component. Kinetic data on the appearance of the 2 glycoproteins in virus particles supports such an hypothesis [73], and treatment of intact virus with cross-linking reagents generates principally dimeric forms of the glycoproteins [74]. It is also known that 1 of the 2 glycoproteins, E2, is more accessible to iodination by lactoperoxidase than is the other [75].

The detailed structure of the glycoproteins on the surface of the virus particle is not clear. Recent electron microscopic evidence, however, suggests that the glycoproteins are arranged in an icosahedral lattice [53], perhaps because they are attached to the icosahedral nucleocapsid in a 1 to 1 binding ratio. The authors [53] concluded that the lattice of glycoproteins had a triangulation number of 4, and thus each E1-E2 dimer would fill the function of 1 subunit (240 subunits total). However, as discussed in Section 4.II.C., the triangulation number of the capsid may be 9, requiring 540 subunits, and in this case each E1-E2 dimer would fill the function of 2 subunits (270 dimers to form 540 subunits) or possibly, each glycoprotein of E1 and of E2 could be a separate subunit (540 total glycoprotein molecules excluding E3).

The flaviviruses appear to contain a single glycoprotein species, V-3, of molecular weight between 51,000 and 63,000 depending on the virus [18, 19, 21, 76]. In addition, a small, nonglycosylated protein species, V-1, of molecular weight 7700-8500, is present in the virion and may be associated with the viral envelope. The electrophoretic mobility of this protein for all the mosquito-borne flaviviruses is very similar, while the tick-borne viruses contain a V-1 species of reduced mobility [18].

III. Replication of Togaviruses in Vertebrate Cells

A. The Growth Cycle of the Virus

Togaviruses replicate in a wide variety of mammalian and avian cells. Cultures of chick embryo fibroblasts, BHK21 cells, and African green monkey kidney cells (VERO or BSC-1) are commonly used as laboratory hosts; other lines have also been used. The virus also grows well in mosquito cells; this subject is covered in Section 4.IV.

The replication cycle of an animal virus begins with the adsorption of the virus to specific receptors on the surface of a sensitive host cell. In the case of Sindbis virus the nature of the virus receptors is unknown, but chick or hamster cells will adsorb between 20 and 200 virus particles per square micron of surface area (or about 10^5 particles in all); the adsorbed virus particles are uniformly distributed over the surface of most cells, with different cells having different concentrations of virus receptors [77]. Following adsorption of the virus to a sensitive cell, the envelope of the virus probably fuses with the cell plasmalemma to initiate the infection cycle [78]. Another school of thought holds that pinocytosis is the initial event in virus penetration [79].

Sindbis virus has been shown to bind to protein-free lipid model membranes (liposomes) under conditions similar to those used for virus hemagglutination [80]. Liposomes made from a mixture of erythrocyte phospholipids would bind the virus; however, binding was much less efficient to liposomes lacking either phosphatidylethanolamine or cholesterol. Thus it is possible that the receptor for Sindbis virus consists primarily of phospholipid or glycolipid, and that host protein moieties are not involved in the initial binding step.

The alphaviruses grow well over a wide range of temperatures, probably because of their adaptations to growing in arthropods as well as in birds and mammals. The upper temperature for virus growth is 40-41°C. The lower temperature for growth is unknown, but a normal virus yield can be obtained at 25°C. A growth curve of the HR strain of Sindbis virus at 30°C in chick embryo fibroblasts is shown in Fig. 4.4. After an early exponential rise in virus yield, the cells yield virus at a constant rate of about 2000 plaque-forming units per cell per hour over a period of many hours. Infection with the alphaviruses is highly cytotoxic to the vertebrate cell, host RNA synthesis and protein synthesis being rapidly inhibited after virus infection [7, 81], and the

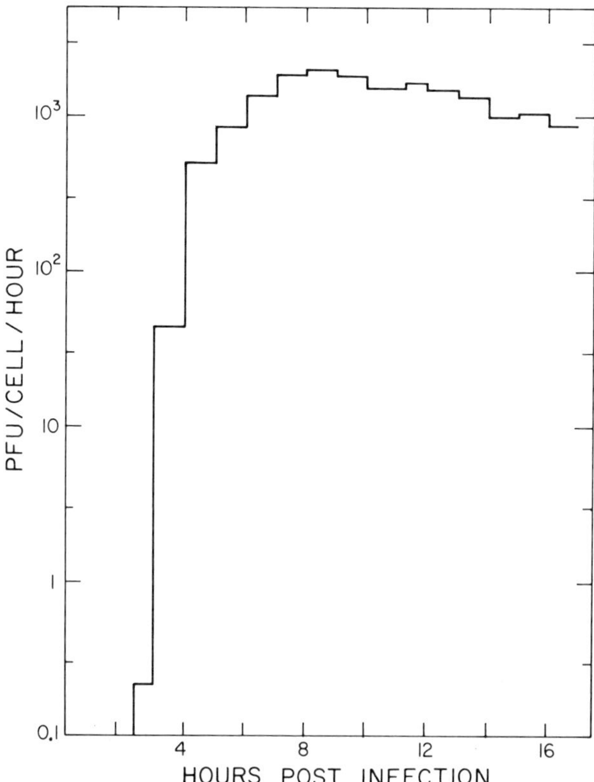

Fig. 4.4 Differential growth curve of the HR strain of Sindbis virus in chick embryo fibroblasts at 30°C. The medium on an infected roller bottle of cells was changed every hour and the virus yield titered. The results are expressed as plaque-forming units (PFU) produced per cell per hour. Note that the cells are still yielding large amounts of virus at 16 hr postinfection (unpublished data of J. H. Strauss and E. G. Strauss, 1973).

production of virus eventually ceases probably because of disintegration of the host cell.

The growth of the flaviviruses in most cultured mammalian or avian cells is slower in general than that of alphaviruses, and the infection is much less cytopathic with host cell macromolecular synthesis being inhibited only gradually and to a lesser extent than with alphaviruses. Several flaviviruses have been reported to establish persistent infections in some vertebrate cell lines. However, flaviviruses are strongly cytotoxic in certain cell lines. The cytopathology of flavivirus infection of vertebrate cells is discussed in more detail in the review of Pfefferkorn and Shapiro [4].

B. Transcription of Viral RNAs

1. *The Intracellular Viral Single-Stranded RNAs*

Two major species of single-stranded RNA are found in cells infected by the alphaviruses. One, the 49 S RNA, is identical to the RNA found in the virion. The second sediments at 26 S and has been referred to as "interjacent RNA" [82-84]. Early experiments suggested that 26 S RNA might be a different configurational form of 49 S RNA or perhaps a precursor to it [84, 85]. Recent hybridization-competition experiments have shown that 26 S RNA contains only 1/3 of the base sequences found in 49 S RNA; it has a molecular weight of 1.6×10^6 and, being of the same polarity as 49 S RNA, it represents a specific 1/3 piece of the alphavirus genome [11]. The exact location of the 26 S RNA sequences in the 49 S RNA sequences, whether 5′ terminal, 3′ terminal, or internal, is an important question; recent results indicate that 26 S RNA represents the 3′ end of 49 S RNA [88].

In addition to these major RNA species, minor RNA species sedimenting at 38 S and 33 S [86] and at 20-22 S [15] have been observed. The Sindbis 33 S RNA has been examined by Simmons and Strauss [34], using hybridization-competition experiments, and reported to share at least 90% of its base sequences with 26 S RNA and to be converted to 26 S RNA upon denaturation. These data indicate that 33 S RNA is a configurational variant of 26 S RNA. Cancedda et al. [87] have reported that the principal translation product of 33 S RNA in vitro is capsid protein as is the case for 26 S RNA; in their system, however, larger polypeptides not made in response to 26 S RNA were made in response to 33 S RNA and thus these authors believe that the 2 RNAs are not identical in their messenger function. The structure and function of these minor RNA species awaits further clarification.

Cells infected with flaviviruses appear to produce only 1 major species of single-stranded RNA, the 42-45 S RNA of the virion [22, 23, 103]. If a single-stranded molecule corresponding to 26 S RNA of alphaviruses is produced, it is made in only small quantities (see Section 4.III.B.3).

2. *Transcription of Alphavirus 49 S and 26 S RNA*

Transcription or replication of the RNA of single-stranded RNA-containing viruses occurs through the synthesis of the complementary RNA strand by a virus-specific RNA polymerase, using the infecting viral strand as a template. This complementary strand is then used as a template to produce more RNA of the virion polarity. The RNA strands which constitute the replication complex, both template and partial products, appear to exist as single strands inside the cell [89, 90], but upon extraction and deproteinization of the RNA, plus and minus strands collapse upon one another to form double-stranded structures

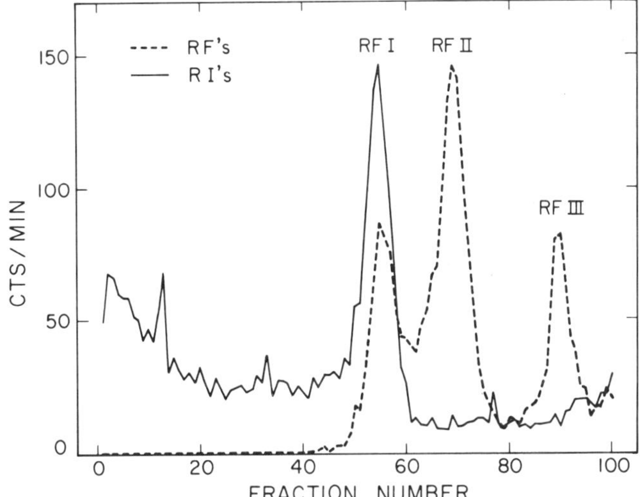

Fig. 4.5 Acrylamide-agarose gel electropherogram of the RFs and RIs of Sindbis virus. The RIs were labeled with [^{14}C]-uridine from 1 to 4 hr after infection. The RFs were labeled with [^{3}H]-uridine from 1 to 4.5 hr after infection and have been treated with pancreatic RNase. Migration is from left to right. Note that most or all of the RIs are as large as or larger than RF I. (Modified from Simmons [93]. Reproduced by courtesy of the author.)

which can be analyzed in order to determine the intermediates involved in RNA replication. Alphaviruses also replicate through production of complementary minus strands of RNA. Brief pulses of radioactive uridine label replicative intermediates (RIs) which are partially double-stranded and sediment as a broad peak at up to 30 S [92, 198]. Treatment with pancreatic RNase to remove single-stranded tails results in the formation of fully double-stranded molecules which sediment at about 20 S [83, 168, 172, 199] and will be referred to as replicative forms or RFs.

Cells infected with alphaviruses produce the 2 major species of single-stranded RNA, 49 S RNA and 26 S RNA. Simmons and Strauss [92] have presented evidence that both are transcribed from a full length minus strand, i.e., one having the same molecular weight as 49 S RNA. In these studies, pulse-labeled RIs sedimented between 23 and 30 S. Treatment of any fraction of these RIs with RNase resulted in the formation of 3 species of RFs. An acrylamide gel electropherogram of the RFs and RIs is shown in Fig. 4.5. The largest RF (termed RF I) has a sedimentation coefficient of 23 S, and a molecular weight of 9×10^6 and is a double-stranded form of Sindbis 49 S RNA. The smallest RF, RF III, sediments at 18 S, has a molecular weight of 3×10^6, and has been shown by hybridization-competition experiments to be a double-stranded form of 26 S RNA. The middle sized RF, RF II, sediments at 20 S, has a molecular weight of 5.6×10^6, and has been shown, again by hybridization-competition, to contain only base sequences not found in 26 S RNA; it is thus a double-stranded form of the remaining 2/3 of the genome [92]. The 2 smaller RFs are always found in equimolar ratios throughout the infection cycle. The fact that all RIs are as large or larger than the RF containing the full length double-stranded RNA and the 1 : 1 molar ratio of RF II and RF III implies that these 2 RFs are joined together prior to RNase digestion, and thus all RIs contain a full length minus strand; i.e., there are no small RIs which generate the smallest RF. Although they are not physically separable, we have designated the RI which generates RF II and RF III as RIb, the one generating RF I as RIa [92].

The RF II region of RIb is labeled very slowly (40 min to fully label this region compared to 1 min to label the RF III region of the same complex) [92], and we have been unable to find a free single-stranded RNA corresponding to the RF II region [34]. Thus this region does not seem to be involved in the production of single-stranded RNA.

These findings are summarized in the model for the transcription of Sindbis-specific plus stranded RNA (49 S and 26 S) from the minus strand shown in Fig. 4.6. The basic points of this model are as follows: Both 49 S and 26 S RNA are replicated from a full length minus strand. Synthesis of 49 S RNA is initiated at the 3′ end of the minus strand and the entire strand is transcribed. Twenty-six S RNA is produced independently by initiation at a second initiation site 2/3 of the way from the 3′ end of the minus strand, with transcription proceeding through to the 5′ end of the template. Although not absolutely essential to the model, we postulate a region of secondary structure in the RNA immediately preceeding the initiation site for 26 S RNA transcription. This RNA loop, perhaps upon being stabilized by a binding protein, reveals the initiation site for 26 S RNA transcription and prevents readthrough by a polymerase attempting to transcribe the entire molecule (i.e., producing 49 S RNA), thus converting the minus strand to exclusive production of 26 S RNA. Replicative intermediate RIa (which produces 49 S RNA) is convertible to replicative intermediate RIb (which produces 26 S RNA) by the formation (or stabilization) of such a loop at the second initiation site. Conversely, loss of the loop converts RIb to RIa and permits readthrough by the polymerase and synthesis of the entire 49 S RNA. The long labeling time of RF II implies that this interconversion of RIa and RIb is relatively slow.

Thus this model describes the production of only 3 species of single-stranded RNA during virus replication: the full length minus strand, the full length plus strand (i.e., 49 S RNA), and the 1/3 piece plus strand (i.e., 26 S RNA).

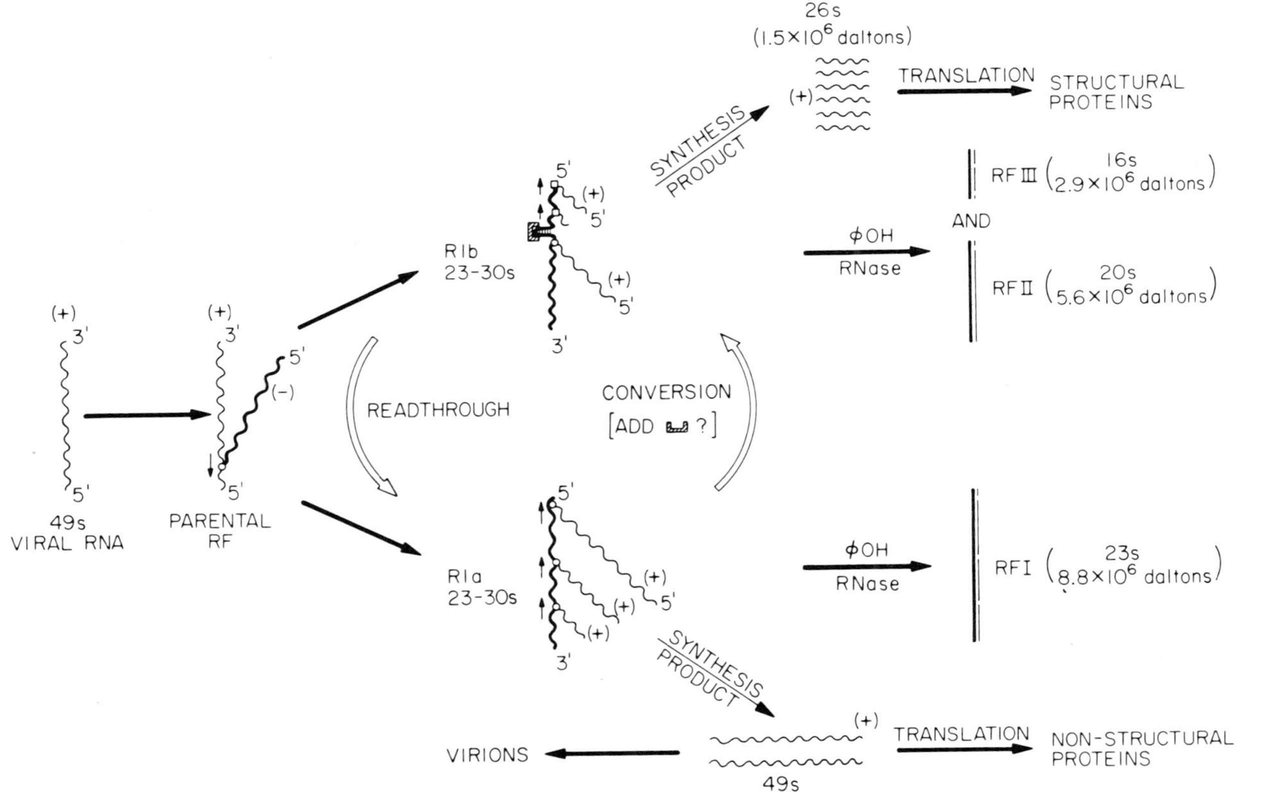

Fig. 4.6 Model for the replication of alphavirus RNA. Wavy lines indicate single-stranded molecules, straight lines represent double-stranded structures after RNase digestion. Minus strands are shown as heavy lines, plus strands as light lines. In this illustration RIb is depicted as synthesizing exclusively 26 S RNA; RIa contains the template for 49 S RNA synthesis. (Modified from Simmons and Strauss [92]. Reproduced by courtesy of Academic Press.)

There are several lines of indirect evidence which also lend support to this model. Waite [94] and Scheele and Pfefferkorn [95] have reported that 2 of their temperature-sensitive mutants (*ts*11 and *ts*24) cease to produce 26 S RNA but continue to produce 49 S RNA after a shift from the permissive to the nonpermissive temperature. The easiest interpretation of this result is that there are 2 viral RNA polymerases, 1 to produce 49 S RNA and 1 to produce 26 S RNA, although the hypothetical binding protein or other control function protein (such as a "sigma protein" to change the initiation specificity) could be involved.

Michel and Gomatos [96] have described an in vitro system which produces both 49 S and 26 S RNA. The RNA polymerase(s) are particulate and have not been separated from their templates. However, replication complexes synthesizing 26 S RNA could be partially resolved from those producing 49 S RNA. Thus 26 S RNA and 49 S RNA appear to be produced from different structures, perhaps by different polymerases. These authors reported the presence of only 1 RF in their in vitro system, however, having a molecular weight of approximately 5×10^6. The reason for the discrepancy between this result and that of Simmons and Strauss is not clear.

D. T. Brown [97] has examined Sindbis 49 S RNA by electron microscopy and found that a hydrogen-bonded loop exists in this RNA approximately 1/3 of the way from 1 end. This provides some evidence for the postulated control loop in the minus strand during RNA transcription.

Segal and Sreevalsan [98] also analyzed RNA from RIs and RFs and came to the conclusion that 26 S and 49 S RNA were made on different RI structures, although these were not separable by sedimentation. In addition, they found that both their RI preparations and RFs prepared from them were infectious, implying the presence of an intact plus strand.

It is also of interest to note that Hsu et al. [14] have found that most 49 S RNA molecules isolated from Sindbis virions exist as hydrogen-bonded circles. The negative strand, by definition, must also be able to circularize. The significance of this is not clear, but the hydrogen-bonded self-complementary region of such a circle could be the specific initiation site for the viral RNA polymerase responsible for transcription of the entire genome. However, formation of stable circles could conceivably block initiation of 49 S RNA production and result in an RI devoted to 26 S RNA production, although it appears difficult to reconcile this method of 26 S RNA production control with the data relating to the 3 species of double-stranded RNA found after RNase digestion. Further experimentation to locate 26 S RNA accurately on the 49 S RNA genome and further characterization of the in vitro RNA synthesis system are necessary to gain a greater understanding of the details of RNA transcription.

It should also be noted that RNA synthesis in vivo [99-101] or in vitro [96, 102] is associated with cytoplasmic membranes. Further experimentation is required to understand the significance of these structures.

3. *Transcription of Flavivirus RNA*

Double-stranded RNA, partially or completely RNase resistant, has also been found in flavivirus-infected cells [103-105]. This RNA sediments at 20 S and probably corresponds to RF I of the alphaviruses. Full length 43 S RNA has been isolated from this double-stranded molecule by denaturation with dimethylsulfoxide [104].

The major species of single-stranded RNA in flavivirus-infected cells is the 43 S virion RNA [22, 23, 103]. There does not appear to be a significant amount of another single-stranded mRNA corresponding to the 26 S RNA species of alphaviruses. Variable amounts of a molecule sedimenting at 26 S have been described [23, 103-105], but the nonreproducibility of this species and the variability of its ribonuclease resistance are problematic. This 26 S RNA may represent a heterogeneous population of RIs which have different degrees of RNase resistance depending on the infecting virus, the host cell, and the time after infection. It is also possible that a single-stranded 26 S mRNA is made in small quantities and its presence is masked by the RI peak.

Qureshi and Trent [105] have isolated particulate, membrane-bound "replication complexes" from cells infected with St. Louis encephalitis virus. These complexes contain RNA polymerase activity, 43 S single-stranded RNA, 20 S double-stranded RNA, and some partially RNase-resistant 26 S RNA. In addition these complexes contain viral antigens. These replication complexes in Japanese encephalitis virus-infected cells appear to be associated with the nuclear membrane rather than cytoplasmic membranes as in alphavirus-infected cells. A nuclear membrane fraction has been isolated which contains primarily 45 S RNA, 20 S double-stranded RNA, and 4 S RNA [106].

C. Translation of Viral Messenger RNAs

1. *Identification of the Viral Messenger RNAs*

Several laboratories have established that 26 S RNA is the major virus-specific messenger RNA in alphavirus-infected cells, comprising 60-90% of virus mRNA [34-37, 107, 108]. These experiments involve isolation of polysomes followed by release of mRNA using EDTA treatment or puromycin. In these experiments, the polyribosomes from alphavirus-infected cells appeared to be smaller than those of the uninfected cells, in contrast to the situation in poliovirus infection, where virus-specific polysomes are larger than the average cellular polyribosome [109]. The translation product of the 26 S RNA has been established as a precursor to the 3 structural proteins of the virion (see below).

The 49 S RNA of alphaviruses has also been reported to possess messenger activity [34-36] being present on functional polysomes. These experiments are complicated by the presence of viral nucleocapsids (which sediment at 140 S), by what appear to be nucleocapsid precursors in the polysome gradients, and by the fact that when 49 S RNA is released from polysomes with EDTA or puromycin it sediments as a ribonucleoprotein complex of 70-140 S, depending upon the conditions, and can thus be confused with nucleocapsids. Nevertheless, the combined data are quite convincing that 49 S RNA comprises 10-40% of virus messenger RNA during later stages of the infection cycle. In addition, it is known that 49 S RNA must serve as messenger early in infection since the naked RNA is infectious.

As is the case for most animal mRNA, poly(A) tracts have been found in both 26 S and 49 S RNA of alphaviruses [40, 42, 46, 91]. Eaton and Faulkner [110] found that the poly(A) tracts of virion 49 S RNA were of 2 size classes, and that 49 S RNA with large poly(A) tracts bound efficiently to Millipore filters, whereas 49 S RNA with shorter poly(A) tracts did not bind. Clegg and Kennedy [42] reported that the average size of the poly(A) tracts on 49 S RNA was about 100 nucleotides, while the poly(A) on 26 S RNA averaged 65 residues. Donaghue and Faulkner [111] reported that the poly(A) tracts were not 3′ terminal, in contrast to other mRNAs; this result is probably in error. Our results indicate that these tracts are 3′ terminal; we found the size distribution of poly(A) tracts in 26 S RNA and 49 S to be identical, averaging 60-80 nucleotides. We have also found that 10-20% of 49 S RNA molecules isolated from virions possess no detectable poly(A); these RNA molecules lacking poly(A) have a lowered specific infectivity for chick fibroblasts [91]. These results on specific infectivity are similar to those Spector and Baltimore [112] obtained with poliovirus RNA after partial removal of poly(A) with nucleases and may be a reflection of the short half-life reported for mRNA lacking poly(A) as compared to mRNA containing poly(A) [113]. Similarly, we have also found that 40-60% of 26 S RNA molecules lack detectable poly(A) tracts.

Several laboratories have also found a 33 S RNA associated with polysomes [34, 35, 107]. Since our results indicate that 33 S RNA is an altered form of 26 S RNA, however, it appears that 26 S RNA and 49 S RNA are the only 2 functionally distinct mRNAs in alphavirus-infected cells.

Nothing is known about the messenger RNAs of the flaviviruses. However, since the naked virion RNA is infectious [22] and contains poly(A) [45, 46] it may well serve as the primary message in the cell.

2. *Translation of Alphavirus 26 S mRNA in vivo*

During the replication of the alphaviruses, 26 S RNA is the major viral messenger RNA in the infected cells and is translated into the 3 (4) structural proteins of the virion. Thus most of the virus specific polypeptides seen in infected cells are structural proteins or their precursors.

Formation of the three structural proteins occurs through synthesis of large polypeptides which are processed by proteolytic cleavages. The scheme for protein processing is shown in Fig. 4.7. The 26 S mRNA is translated into a single, large precursor of molecular weight 130,000 (the *ts*2 protein or NVP130). This polypeptide is immediately cleaved (probably while nascent) into the capsid protein (molecular weight, 30,000) and a precursor to the envelope glycoproteins with molecular weight 100,000 (the B protein or NVP98). The B protein is further cleaved (again, probably while nascent) to envelope protein E1 (molecular weight of the polypeptide portion, 48,000) and to a precursor to the second envelope glycoprotein, PE2 or NVP62 (molecular weight of the polypeptide portion, 54,000). PE2 is cleaved, probably during virus maturation, to form E2 (MW 48,000) and E3. E3 remains associated with the virus in the case of Semliki Forest virus but is apparently lost from Sindbis virus. The evidence upon which this model is based is discussed in detail in the following paragraphs. An acrylamide gel pattern displaying these various polypeptides is shown in Fig. 4.8.

In one of the complementation groups of Sindbis virus (Group C, represented by *ts*2) the 130,000 dalton precursor is not further processed at

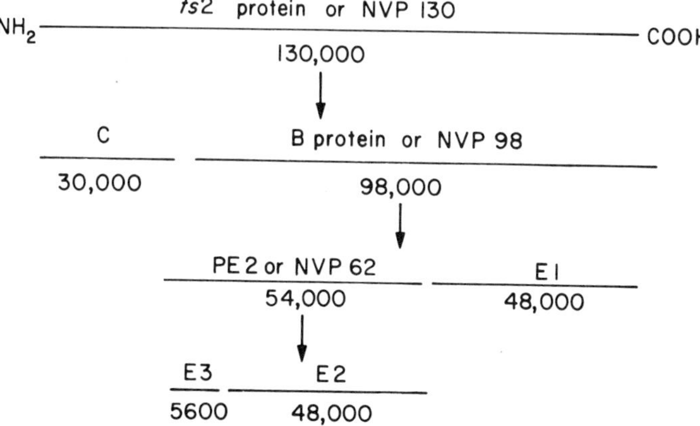

Fig. 4.7 Proteolytic cleavages in the formation of the structural proteins of the alphaviruses. Data from Schlesinger and Schlesinger [73, 115] and from Simons and collaborators [10, 72] have been used to construct the model. Molecular weights refer to the polypeptide moieties only of glycoproteins.

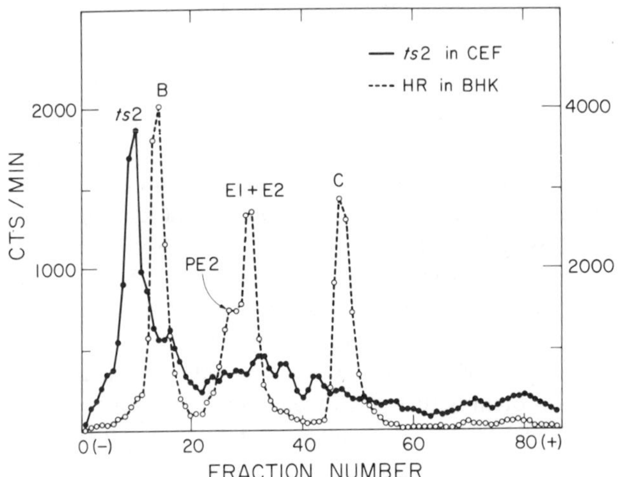

Fig. 4.8 Acrylamide gel electropherogram of Sindbis virus-specific polypeptides in infected cells. The solid line (left scale) is from chick embryo fibroblasts labeled with [^{14}C]-amino acids from 5 to 7 hr after infection with *ts*13 at 40°C (this mutant is in the same complementation group as *ts*2). The dashed line (right scale) is from BHK cells labeled with [^{3}H]-amino acids from 9 to 12 hr after infection with the HR strain of Sindbis virus at 37°C. The gel system is a continuous SDS-containing system which does not resolve the envelope proteins. (Modified from Strauss et al. [7]. Reproduced by courtesy of Academic Press.)

the nonpermissive temperature (Fig. 4.8) [7, 114]. Tryptic peptide analysis has revealed that this precursor polypeptide contains the amino acid sequences of all 3 structural proteins [115]. Similar mutants have been found for Semliki Forest virus [116].

The first cleavage step to produce the capsid protein and the B protein is very rapid and, in fact, occurs in vitro in lysates of rabbit reticulocytes (see Section 4.III.C.3). The capsid protein rapidly associates with 49 S RNA into nucleocapsids; radioactivity from amino acid label can be detected in nucleocapsids within 10-15 min [117]. In this manner the majority of the 49 S RNA in infected cells is found in nucleocapsids [34]. The B protein accumulates in BHK21 cells infected with Sindbis virus (Fig. 4.8) [7] and has been shown to contain the tryptic peptides of both E1 and E2 [115]. The B protein in these cells is not glycosylated [125]. An equivalent protein, NVP98, has been found in BHK21 cells infected with Semliki Forest virus [200] and in chick cells infected with certain temperature-sensitive mutants of this virus [118].

The precursor-product relationship between PE2 and E2 has been shown both by pulse-chase experiments and by tryptic peptide analysis [73]. The cleavage step requires about 30 min and is associated with virus maturation [140]. A similar process, cleavage of NVP62 to E2 and E3, occurs with Semliki Forest virus [10, 72].

The first 2 cleavage steps in this processing scheme appear to occur while the polypeptide is nascent; attempts to chase radioactive label from precursors into products have been largely unsuccessful. Thus the large precursor molecules (either the *ts*2 protein or the B protein), once released from their site of synthesis, can in general no longer be processed. However, Morser and Burke [124] have obtained data which support the cleavage scheme in Fig. 4.7 with a combination of pulse-chase experiments and experiments using amino acid analogs.

The order of the virion polypeptides shown in Fig. 4.7 has been established in 2 ways. Clegg [123] used very short pulses to establish the order of appearance of the polypeptides, and thus their sequence along the original precursor polypeptide. Lachmi et al. [118], working with temperature-sensitive mutants of Semliki Forest virus, found a polypeptide which contained the nucleocapsid protein-PE2 sequences, thus establishing the order (since the B protein contains the PE2-E1 sequences) as C-PE2-E1.

The current hypothesis for the site of synthesis of cellular plasma membrane glycoproteins suggests that they are synthesized on polyribosomes at the rough endoplasmic reticulum and migrate to the cell surface by way of the smooth endoplasmic reticulum and the Golgi apparatus, being glycosylated along the way, before finally appearing at the surface. By analogy, the B portion of the large precursor is probably synthesized at the rough endoplasmic reticulum, cleaved to E1 and PE2, and glycosylated during transit of the polypeptides along the smooth endoplasmic reticulum toward the cell surface. In BHK cells some of the B polypeptide escapes cleavage and is slowly degraded.

3. *Translation of Alphavirus 26 S mRNA in vitro*

The identification of 26 S RNA as the messenger RNA for the structural proteins has been established by translating the RNA in vitro. Several laboratories have established that capsid protein is produced in response to 26 S RNA, using lysates of rabbit reticulocytes, systems from Ehrlich ascites cells or L-cells, or a wheat germ system [38, 119-122]. Cleavage of the precursor to produce the capsid protein thus appears to be normal in vitro, at least in the reticulocyte, L-cell, and ascites cell systems [38, 119, 122]. Simmons and Strauss [38] also obtained production of the B protein in lysates of rabbit reticulocytes in response to 26 S RNA. These results are shown in Fig. 4.9. At high temperatures of translation (37°C), the B polypeptide is produced in equimolar amounts with capsid protein, and normal chain termination appears to be quite efficient; at lower temperatures (28°C) an excess of capsid protein

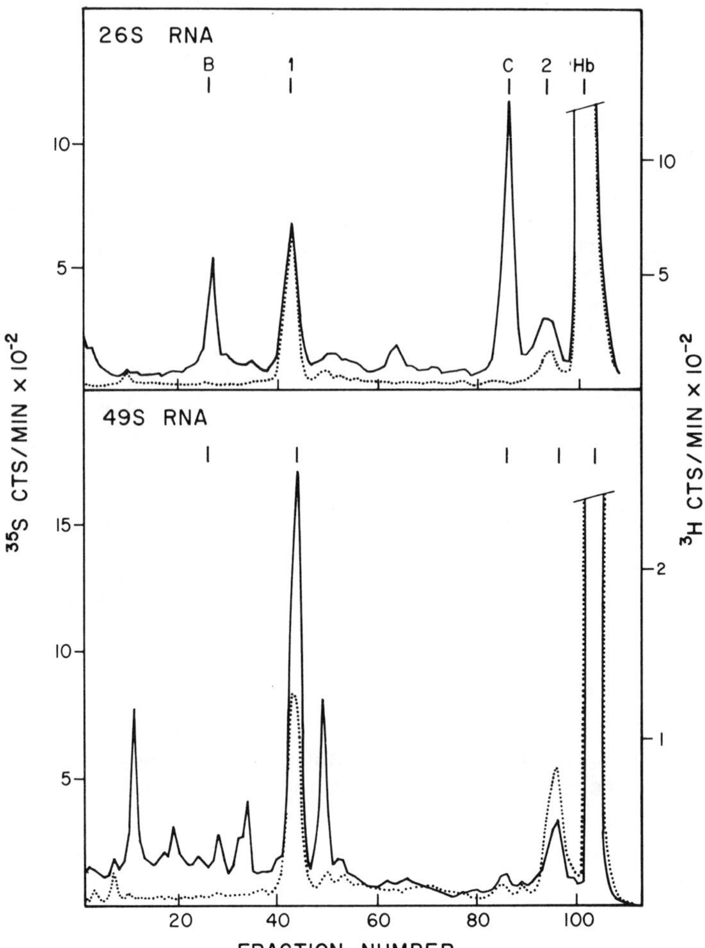

Fig. 4.9 Acrylamide gel electropherogram of polypeptides synthesized in vitro at 27°C in response to Sindbis virus 26 S and 49 S RNA. The upper panel shows the polypeptide products of wild type (HR strain) 26 S RNA translated in a rabbit reticulocyte cell-free system (solid line). B is the 98,000 molecular weight precursor to E1 and E2, C is the capsid protein. The dotted line is the endogeneous synthesis without added mRNA; the major polypeptide synthesized is hemoglobin (Hb), but significant amounts of 2 rabbit polypeptides, labeled 1 and 2, are also produced. The lower panel is a comparable experiment using 49 S RNA as messenger. Note that no detectable capsid protein is made in this case. (Modified from Simmons and Strauss [38]. Reproduced by courtesy of Academic Press.)

is produced, implying premature chain termination. Other workers have not obtained the B protein, but Glanville et al. [120] obtained smaller products which contained tryptic peptides of the envelope proteins as well as of the capsid protein.

Production of E1, E2, or E3 (either in glycosylated or nonglycosylated form) has not been seen in vitro. Processing of this portion of the polypeptide must require other cellular components. It is quite possible that these cleavage steps require prior glycosylation and/or membrane insertion.

We have also shown that translation of the 26 S RNA derived from cells infected with mutant *ts*2 gives rise in lysates of rabbit reticulocytes to the 130,000 precursor when translation occurs at 37°C (a nonpermissive temperature) but to the B polypeptide and capsid polypeptide when translation occurs at 28°C (a permissive temperature) [38]. These results are shown in Fig. 4.10 and support the cleavage scheme shown in Fig. 4.7. They are consistent with the hypothesis that the *ts* lesion in *ts*2 lies in the capsid region of the precursor polypeptide and prevents cleavage of the capsid protein from the precursor.

4. *Translation of Alphavirus 49 S mRNA in Vivo and in Vitro*

The second major viral messenger RNA of the alphaviruses is 49 S RNA. This RNA must be translated into all the nonstructural proteins seen in infected cells, including 1 or more RNA polymerases. These minor polypeptide products are difficult to see in the infected cell extracts because of the preponderance of virus structural proteins and their precursors [7, 126, 127]. Recently, however, Lachmi et al. [118] have observed 2 nonstructural, virus-specific polypeptides in cells infected by a *ts* mutant of Semliki Forest virus which have molecular weights of 86,000 and 78,000. This mutant makes only small amounts of 26 S RNA at a nonpermissive temperature [128]; the major messenger on polyribosomes is 49 S RNA [36].

We presume that translation of 49 S RNA begins at a single initiation site near the 5′ end and continues to a single termination site 2/3 of the way down the chain. The precursor polypeptide produced (molecular weight about 250,000) would be processed by cleavage to produce at least 4 and possibly as many as 6 or 8 functional polypeptides. We have shown that there are at least 4 complementation groups among RNA^- mutants (see Section 4.IV.B). We hypothesize that these 4 groups are located in the precursor polypeptide translated from 49 S RNA and that a missense mutation in any of them prevents proper cleavage of the precursor, leading to reduced amounts of functional RNA polymerase being produced.

Further evidence for this model comes from M. J. Schlesinger and collaborators. These workers have shown that there is 1 major initiation

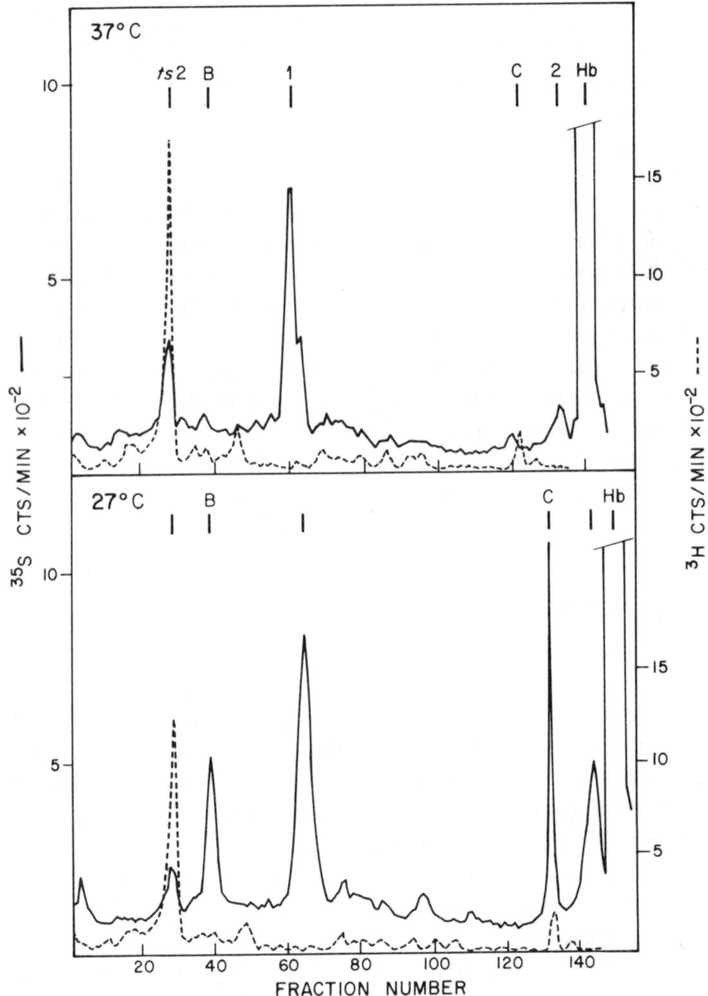

Fig. 4.10 Acrylamide gel electropherogram of polypeptides synthesized in vitro in response to 26 S RNA of *ts*2. The solid line in the upper panel shows the polypeptides synthesized at 37°C (a nonpermissive temperature) in a rabbit reticulocyte cell-free system primed with *ts*2 26 S RNA. The "*ts*2 protein" is the only prominent Sindbis-specific peak, since the peaks labeled 1, 2, and Hb are rabbit proteins (compare with Fig. 4.9). The dotted line is a marker from chick cells infected with *ts*2 at 40°C and labeled with [^{3}H]-amino acids. The lower panel shows the in vitro products when translation is at 27°C (a permissive temperature), using the same RNA preparation (solid line). A small amount of the "*ts*2 protein" is made but the major products are the capsid protein (C) and the B protein. The marker (dotted line) is the same as in the upper panel. (Modified from Simmons and Strauss [38]. Reproduced by courtesy of Academic Press.)

peptide when 49 S RNA is translated in vitro which is distinguishable from the initiation peptide when 26 S RNA is used [129], and that in temperature shift experiments with certain RNA$^-$ mutants a polypeptide with molecular weight about 200,000 is seen, which could be the primary translation product of 49 S RNA [142].

We have translated 49 S RNA in lysates of rabbit reticulocytes and observed the production of 6-8 polypeptides with molecular weights 60,000-200,000 [34]. The production of structural proteins was not observed. This result is shown in Fig. 4.9. The interrelationships among these polypeptides, whether they represent various cleavage products of a precursor, premature termination products, or products of internal initiation and termination, is not clear at present.

Smith et al. [39] have translated 49 S RNA from Semliki Forest virus in vitro in a cell-free system from mouse ascites cells. They obtained a large number of poorly resolved polypeptides, some larger than and some smaller than the structural proteins. By tryptic peptide analysis they showed the presence of peptides from both the capsid and the envelope proteins as well as tryptic peptides not derived from structural proteins. Wengler and Wengler [130] observed the production of small amounts of capsid protein translating 49 S RNA of Semliki Forest virus in vitro (5% of the amount of capsid protein they obtained with an equivalent amount of 26 S mRNA). The reason for production of virus structural proteins in these 2 cases is not clear. Although 49 S RNA contains the sequences of 26 S RNA at its 3′ end, one would not expect this internal initiation site to be used [201]. One possibility is that nuclease cleavage of the 49 S RNA might reveal the second initiation site which is normally read only from 26 S RNA.

5. *Translation of the Flavivirus mRNA*

Because flaviviruses are inefficient in shutting off host macromolecular synthesis after infection, it has been more difficult to identify the virus-specific polypeptides produced by these viruses. Trent and Qureshi [21] were able to visualize St. Louis encephalitis virus-specific proteins by treating infected BHK cells with actinomycin D, followed by a brief treatment with cycloheximide. Upon removing the cycloheximide and adding radioactive amino acids, these authors were able to resolve the 3 structural polypeptides of the virion and 5 additional nonstructural components. The same method was used by Shapiro et al. [20] to study protein synthesis in cells infected by Japanese encephalitis virus, with very similar results except that virion protein V-1 was not seen. The nonstructural proteins, termed NV-1 to NV-5, were assigned molecular weights of 10,500, 19,000, 45,000, 71,000, and 93,000, respectively. Westaway [131] conducted a comparative study of protein synthesis in cells infected by

Kunjin virus, dengue virus type 2, St. Louis encephalitis virus, and Japanese encephalitis virus, using actinomycin D alone and obtained comparable results.

In constrast to the situation in alphaviruses there is no evidence for large precursor polypeptides which are processed by posttranslational cleavage, with the exception of V-1, which will be discussed in Section 4.III.D.2 as a maturation event. The coding capacity of the 43 S viral RNA is about 400,000 daltons of protein and can easily accommodate the various polypeptide species observed. However, if the 43 S RNA is the primary message, then either posttranslational cleavage or independent initiation and termination must be involved. Eventually, perhaps, translation in vitro will answer some of these questions.

D. Maturation of the Virus

1. Maturation of the Alphaviruses

The maturation of the alphaviruses involves the appearance of virus glycoproteins in the cell plasma membrane, interaction of the nucleocapsid with the inner surface of the virus-modified membrane, followed by budding of the nucleocapsid through the altered surface.

Virus glycoproteins which have appeared at the cell surface can be visualized by treating the infected cells with antiviral antibody coupled to hemocyanin and examining them in the electron microscope using the surface replica technique [132]. By this method virus-specific polypeptides are first detected at the surface of a chick fibroblast at 2 hr after infection; the concentration of virion glycoproteins at the surface increases throughout the infection cycle, until by 6-8 hr after infection the concentration is quite high and viruses are budding at a maximal rate. By 3 hr after infection the concentration of glycoproteins is sufficient to show other characteristics of membrane modification, e.g., hemadsorption [133], agglutinability by Concanavalin A [134], and superinfection exclusion [145].

When the distribution of the viral glycoproteins was examined in the electron microscope, they were found to be randomly distributed over the cell surface, starting with the first detectable molecules. The only clustering of glycoproteins occurred on budding virus. Furthermore, the glycoproteins were free to diffuse laterally in the plasma membrane [132].

Virus nucleocapsids are assembled in the cell cytoplasm and probably migrate to the cell surface by simple diffusion. At the surface they interact with the internal terminus of a functional E1 glycoprotein in a noncovalent linkage and are then fixed in position. An illustration of this interaction is given in Fig. 4.11, which shows a cell infected with *ts*20 at the nonpermissive temperature with a striking array of nucleocapsids lined up beneath the surface of the cell, although no budding is seen. The *ts*20-infected cells synthesize a functional

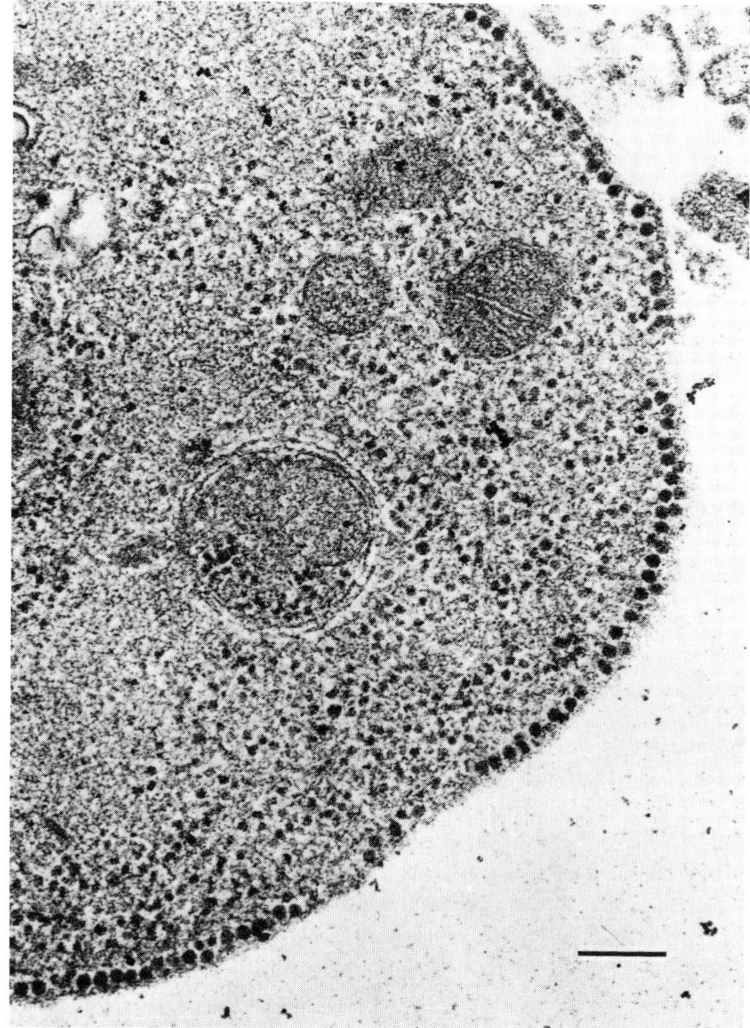

Fig. 4.11 Thin section of a BHK cell infected with *ts*20 for 8 hr at 39.5°C, the nonpermissive temperature. Note the array of nucleocapsids beneath the cell surface. × 54,800; scale bar is 200 nm. (Modified from Brown and Smith [135]. Reproduced by courtesy of the American Society for Microbiology.)

E1 since they can hemadsorb [47]; the failure to bud and produce mature virions at the nonpermissive temperature is probably caused by a defect in PE2 [135]. In cells infected with *ts*23 at the nonpermissive temperature, in contrast, no functional E1 is inserted into the cell surface and no hemadsorption occurs [47]; in this case the nucleocapsids are found dispersed throughout the cytoplasm [135].

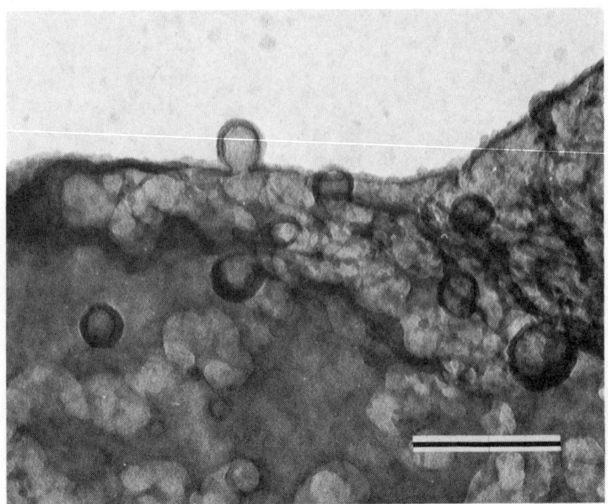

Fig. 4.12 Chick cell showing budding Sindbis virions. Note the virion budding from the edge of this embryonic chick heart fibroblast. The dark line surrounding the cell and the budding virus particle is the plasma membrane stained with osmium. The grey material outside the plasma membrane is vacuum evaporated carbon. The linear structures in the interior of the cell are microfilaments. X 91,000; scale bar is 200 nm. (Micrograph courtesy of Dr. P. Bell.)

Once at the cell surface the capsid can begin to bud through. During this process virus glycoproteins diffuse in and are anchored by attachment to the core [57]; host glycoproteins diffuse away to be replaced by more virus glycoprotein, which is bound in turn. This leads to envelopment of the nucleocapsid with the free energy for the budding event being supplied by the nucleocapsid-glycoprotein bonds formed during this process. The membrane surrounding the virus becomes less fluid [64] possibly because of the rigidity induced upon anchoring the glycoproteins to the core. The angles formed between the budding virus particle and the surrounding cell membrane can be quite acute [137], implying that the line of demarcation between "frozen" glycoprotein fixed to the core and membrane containing free virus-specific glycoproteins is quite sharp. An electron micrograph showing such a budding particle appears in Fig. 4.12. Eventually the virus particle pinches off and is released into the extracellular fluid. An upper limit for the total time required for budding is given by the fact that amino acid label can be detected in mature virions within 20 min of a pulse [117].

The interaction between the virus glycoproteins and the nucleocapsid is quite specific and complete; no host protein is detectable in released virions [24]. Because the ratio of proteins in the virion is 1 : 1 : 1 (: 1) it is probable that each capsid protein molecule in the nucleocapsid has a single site for binding a virus glycoprotein unit (consisting of 1 E1, 1 E2, and 1 E3 molecule, if present). Recent electron microscopic data have been interpreted as evidence for an icosahedral arrangement of the glycoprotein "spikes" of Sindbis virus [53], as expected for this model. It should be noted, however, that the specificity of the capsid for the virus glycoprotein is not absolute, as phenotypic mixing has been observed between alphaviruses [138, 139].

As discussed previously, E1 and E2 (and E3, if present) are thought to be bound together noncovalently so that the functional glycoprotein is a dimer (or trimer with E3). Probably an E1-PE2 dimer is integrated into cell surface as such and cleavage of PE2 to E2 occurs during the final maturation event. The cleavage of PE2 to E2 is known to require considerable time [73, 140] and may be the rate-limiting step. Furthermore, cleavage of PE2 to E2 fails to occur at the nonpermissive temperature with *ts* mutants in complementation groups D and E [97, 129], where no budding occurs. However, iodination experiments using lactoperoxidase [75] have shown that while E2 is readily iodinated in the virus or on the cell surface, PE2 is not iodinated in the reaction. Thus if PE2 is present at the surface its configuration must be such that it is not accessible to peroxidase until after cleavage to E2. In this regard we might note that Pedersen and Sagik [136], using ferritin-conjugated anti-Sindbis antibody, were able to detect Sindbis glycoproteins only in those regions of the plasma membrane where a nucleocapsid is present, whereas other experiments have found Sindbis glycoproteins over the entire surface of the cell [132]. It is possible that the conditions of Pedersen and Sajik detected only anchored glycoproteins and that the configuration of the glycoproteins after attachment to the core is significantly changed.

After cleavage of PE2 to E2, it appears that the smaller glycoprotein portion produced (E3) remains attached to the E1-E2 dimer in the case of Semliki Forest virus [10] but is only weakly bound in the case of Sindbis virus and is lost to the culture fluid.

Although virus glycoproteins diffuse freely and appear to cover the surface of the infected cell uniformly [132] the nucleocapsids often show preferred areas of budding [52, 136, 141]. In some cells Sindbis virus buds uniformly from the surface and no preferred areas are seen; but in other cells virus buds preferentially along the edges of the cell and in long processes (up to 7 μm in length) which originate from the edge of the cell [141]. These processes, which become more common as the infection proceeds, appear to be active sites of virus maturation with nucleocapsids budding through the sides of the processes as well as from the end. Some processes are only a single virus particle in diameter, others are several virion diameters in thickness. Two examples of processes are shown in Fig. 4.13.

We have isolated a mutant which is defective in maturation [28]. All of the particles produced by this mutant bud from processes, and the processes

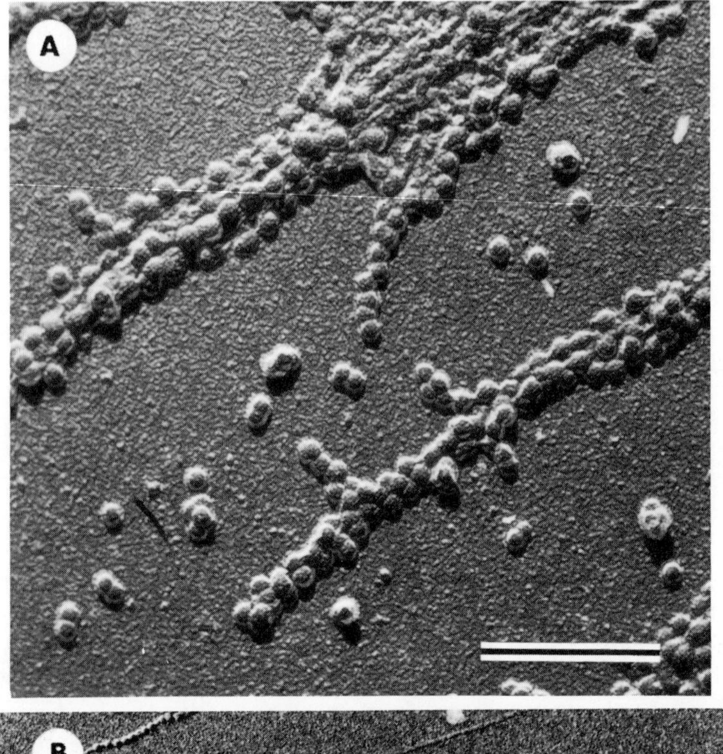

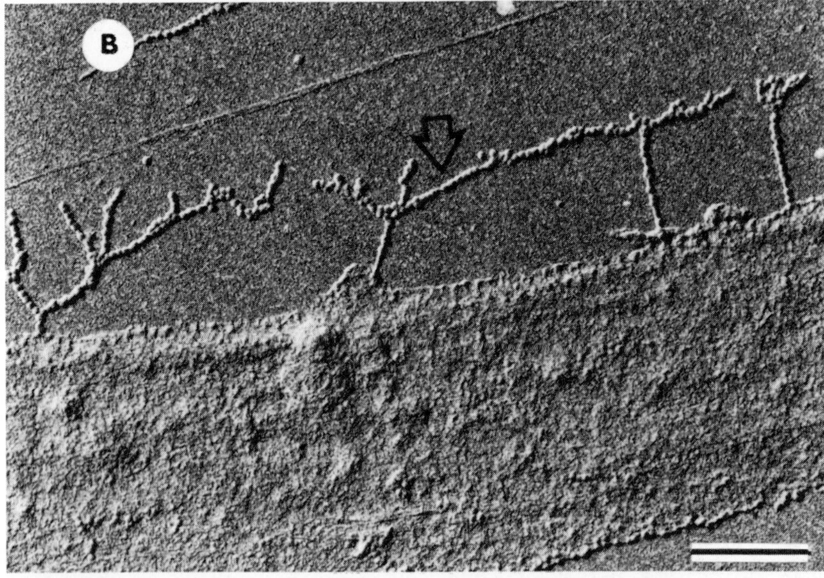

Fig. 4.13 Surface replicas of budding Sindbis virus. Panel A shows processes from a chick fibroblast at 7 hr after infection. X 57,000; scale bar is 500 nm. Modified from Birdwell et al. [141]. Reproduced by courtesy of Academic Press. Panel B shows a chick fibroblast 12 hr after infection with *ts*23 at 28°C (a permissive temperature). X 19,000; scale bar is 1 μm. (From Ref. [203], reproduced by courtesy of the author.)

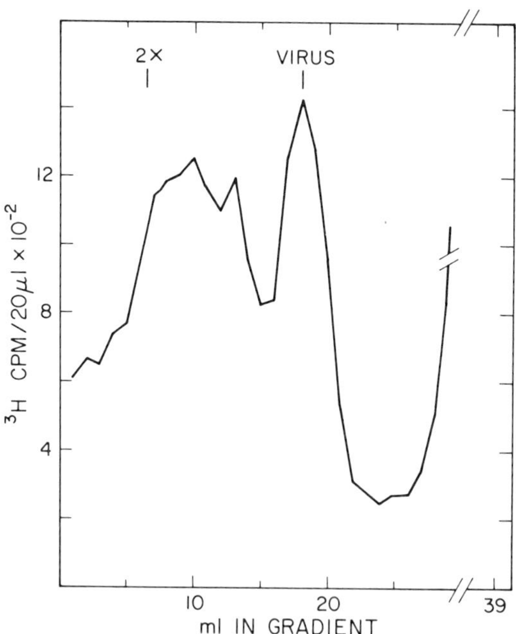

Fig. 4.14 Sucrose gradient of virions formed during replication of a maturation-defective mutant of Sindbis virus. Sedimentation from right to left. The position marked "virus" is the 280 S peak corresponding to wild type virions. "2x" indicates the position of particles sedimenting twice as rapidly (560 S) (unpublished data of J. H. Strauss and E. G. Strauss, 1973).

are unusual. They contain regions apparently devoid of nucleocapsids as well as balloon regions several nucleocapsids in width. The particles which are produced range in size from those of 280 S, which resemble wild type virions, to very large particles up to 3000 S. A sucrose gradient showing the distribution of particles sedimenting at less than 700 S in 1 preparation of this mutant is shown in Fig. 4.14. All of the particles larger than normal virions contain more than 1 nucleocapsid. These multinucleoid particles are very heterogeneous with respect to density; in some particles the nucleocapsids are packed tightly into the envelope, in others there is empty space inside the virion. These multiploid virions are fully infectious. Although no gross alterations have been observed in the structural proteins of this mutant, the defect appears to be in the organization of the nucleocapsid, which is therefore prevented from forming a proper association with the glycoproteins and completing the maturation step correctly [28]. A similar mutant which leads to multinucleoid particles has been reported by Hashimoto et al. [201], who has also found alterations in the nucleocapsids of the particle.

In the case of alphaviruses growing in chick cells, the terminal stage of virus budding is inhibited if the ionic strength of the medium is hypotonic [143-145]. The basis of this phenomenon is not understood. This effect does not occur with vesicular stomatitis virus or a number of other enveloped viruses; nor does it occur when Sindbis virus is grown in BHK cells [146].

2. *Maturation of the Flaviviruses*

Flaviviruses do not appear to mature at the cell surface in infected vertebrate cells. Maturation occurs by budding through the membranes surrounding cytoplasmic vacuoles [32, 147]. In the last stages of infection these vacuoles become full of mature virions packed in paracrystalline arrays. This method of maturation is at least superficially similar to the maturation of alphaviruses in arthropod cells (see Section 4.IV).

The final event in flavivirus maturation apparently involves cleavage of a membrane protein, similar to the PE2 to E2 proteolysis. Intracellular virions, or I-form, contain 2 polypeptides in their envelope, the large glycoprotein V-3 and a second glycoprotein called NV-2 [76, 148]. T-form particles, released from infected cells by treatment with tris, have the same polypeptide composition as I-forms. In mature Japanese encephalitis virions or dengue virions the envelope contains V-3 and a small component, V-1, which is not glycosylated. The final event in maturation, enabling the virions to be released from the host cell, is thought to be the cleavage of NV-2 to V-1 and the loss of the glycosylated moiety [76].

IV. Genetics of Togaviruses

A. Isolation of Mutants

Mutants of the alphaviruses have been extremely useful in studies of the replication of these viruses. Genetic studies of togaviruses began a number of years ago with the isolation and characterization of 24 temperature-sensitive mutants of Sindbis virus by Burge and Pfefferkorn [149]; prior to these studies an extensive catalog of animal virus mutants existed only for poliovirus [150]. The status of togavirus genetics has been reviewed recently by Pfefferkorn [151].

The most useful mutants of togaviruses have been temperature-sensitive (*ts*) mutants. Plaque morphology mutants exist, but no genetic tests can be carried out on these mutants at present and biochemical studies of the virus variants have not been conducted in most cases. In at least 1 case the difference in plaque morphology appeared to result from an altered surface charge of the virion, because separation of large and small plaque-forming variants was obtained on calcium phosphate columns [152]. Presumably a change in surface charge could lead to an altered diffusion rate because of the large negative charge of the amylopectin component of agar. In other cases, the difference in plaque morphology probably reflects differences in the kinetics of virus growth. Thus a naturally occurring small plaque variant of Sindbis virus (plaque assay on chick cells) has a selective advantage in BHK cells [146, 153], whereas the large plaque variant is selected in chick cells. Similarly, chronically infected *Aedes albopictus* cells were found to yield a small plaque variant of Semliki Forest virus [154] and a temperature-sensitive small plaque variant of dengue virus [155]; these small plaque formers probably have reduced virulence for the vertebrate cell.

Because of the great differences between the vertebrate and invertebrate hosts, host range mutants of the togaviruses should also exist and might prove useful in understanding virus replication. Such mutants have not been characterized as yet, however.

Temperature-sensitive mutants of the alphaviruses, in contrast, are relatively easy to isolate and work with. The virus replicates over a wide range of temperatures, 25-41°C, probably reflecting evolutionary adaptation to a poikilothermic vector as well as to homeothermic hosts. Viruses which are restricted to mammalian hosts, such poliovirus, generally can replicate only within a very narrow temperature range between 36 and 38°C, making *ts* mutants difficult to isolate as well as to work with.

The *ts* mutants of the alphaviruses have generally been isolated after treatment with mutagens. Mutagens used have included chemical agents which cause base modifications directly, such as nitrous acid, ethyl methane sulfonate, hydroxylamine, and N-methyl-N′-nitro-N-nitrosoguanidine, as well as base analogs, such as fluorouracil, 5-fluorouridine, and 5-azacytidine, which cause transcriptional misreadings during virus replication. In Table 4.4 are cataloged *ts* mutants of Sindbis virus and of Semliki Forest virus obtained by a number of laboratories, grouped according to the mutagen used to obtain the mutant and the RNA phenotype at the nonpermissive temperature. In addition to the mutants in this table, the isolations of 4 mutants of eastern equine encephalitis virus [159] and 2 of western equine encephalitis virus [160] have been described.

All of the mutants listed in Table 4.1 have been characterized as to phenotype with respect to RNA synthesis at the nonpermissive temperatures. The original 24 mutants isolated by Burge and Pfefferkorn [149] fell into two distinct classes: RNA^+ mutants, which made at least 50% as much RNA at the nonpermissive temperature as did the parental HR strain, and RNA^- mutants, which made less than 2% as much RNA at the nonpermissive temperature as did the parental strain. Later studies have found a continuum of

Table 4.4 Temperature-Sensitive Mutants of Group A Togaviruses

Virus	Mutagen[a]	Number of mutants	Phenotype[b]			References
			RNA^+	RNA^-	$RNA^{\pm}$	
Sindbis virus HR strain	HNO_2	5	2	2	–	
	NNG	17	4	13	–	[149]
	EMS	1	1	–	–	
Sindbis virus HR strain	NHO_2	16	7	4	5	
	NNG	32	17	5	10	[156]
	Aza-C	38	12	18	8	
	5-FU	2	–	2	–	
Sindbis virus wt, AR 339	HNO_2	20	–	15	5	
	NNG	19	–	17	2	[157]
	5-FU	15	4	9	2	
	EMS	23	3	20	–	
	HA	27	5	14	8	
Semliki Forest virus strain 25639	5-FU	11	1	8	2	
	HA	13	4	3	6	[158]
	NNG	14	11	2	1	
Semliki Forest virus	NNG	16	7	7	2	[128]

[a]Abbreviations used: HNO_2 = nitrous acid; NNG = N-methyl-N′-nitro-N-nitrosoguanidine; EMS = ethylmethane sulfonic acid; Aza-C = 5-azacytidine; 5-FU = 5-fluorouracil; HA = hydroxylamine.

[b]RNA^+ mutants make at least 60% as much RNA at the nonpermissive temperature as does the parental strain, $RNA^{\pm}$ make 10 to 60% as much RNA, and RNA^- mutants make less than 10% as much RNA as the parental strain.

RNA synthesis by various mutants. Tan et al. [158] used 3% as an upper limit for RNA synthesis at 40°C by RNA^- mutants and introduced an intermediate class, $RNA^\pm$, for one mutant. We have designated our new isolates of Sindbis virus as RNA^- if RNA synthesis is less than 10% of the parental HR strain and RNA^+ if synthesis is greater than 60% of HR; we have called the intermediate mutants $RNA^\pm$ [156]. For the purposes of comparison the mutants of Tan et al. and those of Atkins et al. [157] have been classified in Table 4.4 according to this convention as well.

More than half of the mutants isolated appear to be RNA^-. For reasons discussed in Section 4.IV.D., we believe this is a true reflection of gene frequencies. However, it is possible that this is an accident of isolation: RNA^- mutants tend to be more stable and fewer "leaky" mutants are seen, and thus a higher percentage of RNA^- mutants are probably retained during screening. Furthermore, the plaque enlargement technique employed for isolation by Atkins et al. [157] appears to enrich for RNA^- mutants. In this test, survivors of a mutagenized stock are tested by a plaque assay for 2 days at 30°C, followed by a shift to 40°C. Plaques which do not enlarge at 40°C are selected and retested. Because RNA^+ mutants kill the host cell at 40°C, a slight leakage would cause these plaques to enlarge at 40°C and many RNA^+ mutants might be missed.

B. Complementation

Recombination has not been demonstrated between mutants of alphaviruses. Among RNA viruses, true recombination has been found only with the picornaviruses [150], although viruses with fragmented genomes, such as the myxoviruses and reoviruses, undergo a reassortment process to produce recombinants. Failure to recombine means the sequence of genes will have to be determined biochemically rather than genetically.

Complementation analysis of alphavirus mutants has been extremely useful. Burge and Pfefferkorn [151, 161] were able to group their RNA^+ mutants into 3 complementing groups and established the presence of at least 2 groups among the RNA^- mutants. Keränen and Kääriäinen [128], however, were unable to obtain complementation with their 16 mutants of Semliki Forest virus, nor were Tan et al. [158] able to obtain complementation with their independently isolated series of mutants of Semliki Forest virus. Recently Atkins et al. [157] have reported they were unable to demonstrate complementation in 57 crosses involving their new catalog of Sindbis virus mutants isolated from the wild-type virus. In this last case, however, most of the mutants are RNA^- and it is our experience that complementation analysis of RNA^- mutants is much more difficult than RNA^+ mutants.

In our laboratory we have reproduced the results of Burge and Pfefferkorn and have been able to group a number of our new *ts* isolated into complementation

Table 4.5 The Complementation Groups of Sindbis Virus

Complementation group	Mutants of Burge and Pfefferkorn [161]	Mutants of Strauss et al. [156]
RNA⁻ Groups		
A	*ts*4, *ts*14, *ts*15 *ts*16, *ts*17, *ts*19 *ts*21, *ts*24	*ts*133, *ts*138
B	*ts*11	
F	*ts*6	*ts*110, *ts*118
G	*ts*7, *ts*18	*ts*134
RNA⁺ Groups		
C	*ts*2, *ts*5, *ts*13	*ts*106, *ts*112, *ts*113 *ts*114, *ts*127, *ts*128 *ts*131, *ts*152, *ts*153
D	*ts*10, *ts*23	*ts*104, *ts*127
E	*ts*20	

groups [156]. Our results, together with those of Burge and Pfefferkorn [151, 161], are summarized in Table 4.5. Among the RNA$^+$ isolates we have found new mutants which lie in groups C and D but we have not as yet found a new mutant for group E. We have divided the RNA$^-$ mutants of Burge and Pfefferkorn into 4 complementation groups and assigned new isolates to 3 of these groups. As yet, we have no new mutant corresponding to *ts*11 (group F), originally designated as group A′ by Burge and Pfefferkorn. Thus 6 of the groups are represented by multiple isolates, whereas 2 of the groups are represented by only a single mutant.

C. Functional Defects of RNA$^+$ Mutants

It is striking that, despite the number of new isolates examined, there are still only 3 complementation groups among the RNA$^+$ mutants. It is now clear that each group represents 1 of the 3 structural proteins of the virion, and that these functions are translated from the 26 S RNA message in the infected cell.

Group C mutants fail to cleave the large precursor polypeptide of the 3 structural proteins of the virion at the nonpermissive temperature as discussed in Section 4.III.C. Moreover, this polypeptide is inefficiently cleaved even at the permissive temperature. The mutants in this group fail to make nucleocapsids at the nonpermissive temperature, and this originally was taken as evidence that their defect lay in the capsid protein itself. Other evidence supports this conclusion. For example, particles made by these mutants at a permissive temperature are more heat labile at 56°C than is the parental strain [156, 162], implying that the defect does lie in a structural protein. In addition, we have shown that the virions produced by *ts*106 at the permissive temperature possess an electrophoretically altered capsid protein [205]. Thus the temperature-sensitive mutation in the capsid protein leads to failure of the precursor to be cleaved. Probably the missense mutation causes incorrect folding of the precursor so that it is not recognized as a substrate by the cleavage enzyme (whether cellular or viral in origin). It is possible, however, that the capsid protein itself possesses proteolytic activity and cleaves itself from the precursor; this enzymatic activity could be lost through the *ts* mutation.

Even though the defect of group C mutants is failure to cleave the precursor polypeptide at the nonpermissive temperature, some leakage does occur in most mutants. These mutants give rise to a small amount of hemadsorption by infected cells under nonpermissive conditions [47], indicating production and insertion of viral hemagglutinin, and, for the cases cited in Table 4.2, these mutants complement other groups of *ts* mutants.

Mutants having the same physiological defect as group C mutants of Sindbis virus have also been found for Semliki Forest virus. The *ts*3-infected cells accumulate the 130,000 molecular weight precursor at the nonpermissive temperature and fail to make capsids [116]. Several RNA^+ mutants isolated by Tan et al. [158] have the same phenotype, producing heat-labile particles and some membrane modification at 38.5°C, but failing to make nucleocapsids.

Mutants in group D have a defect in 1 of the envelope glycoproteins of the virion, the hemagglutinin, thought to be glycoprotein E1. Cells infected with group D mutants at the nonpermissive temperature do not hemadsorb [47]. In addition, 1 of the group D mutants, *ts*23, has been shown by Yin [163] to possess a temperature-sensitive hemagglutinating activity; virions (produced at 30°C) hemagglutinate at 30°C but not at 40°C. Finally, most group D mutants give rise to thermolabile virions [156, 162], implying a structural protein is involved in the mutation.

Mutant *ts*15 of Semliki Forest virus, described by Tan et al. [164], appears to correspond to the group D mutants of Sindbis virus. The *ts*15 mutant is RNA^+, makes heat-labile particles, and forms nucleocapsids at the nonpermissive temperature. At 30°C these nucleocapsids line the inner surface of the cell plasma membrane and intracellular vacuoles; however, at 40°C they are evenly distributed throughout the cytoplasm, as is the case for *ts*23 of Sindbis virus at the nonpermissive temperature (Section 4.III.D).

The single group E mutant, *ts*20, at the nonpermissive temperature produces a functional hemagglutinin (which leads to hemadsorption by infected cells [47]) and produces nucleocapsids, and these nucleocapsids line up under the cell plasma membrane (Fig. 4.11). No budding occurs at 40°C and the cleavage of PE2 to E2 [140] does not occur. By a process of elimination group E mutants are thought to be defective in the E2 or E3 region, and the failure of the PE2 to E2 cleavage may be the primary defect of this mutant. However, in group D mutants no cleavage of PE2 to E2 occurs either [97, 204] nor does the comparable event, cleavage of NVP62 to E2 and E3, take place with any of the RNA^+ mutants of Keränen and Kääriäinen [116]. As discussed in Section 4.III.D., this cleavage event is probably coupled to virus maturation.

D. Functional Defects of RNA^- Mutants

At least 4 complementation groups exist among the RNA^- mutants but the functions of these groups have not been identified. The RNA^- complementation groups probably represent those genes that are translated from 49 S mRNA, rather than 26 S mRNA. If the product of translation of 49 S mRNA is also a single large polypeptide, then any lesion which prevents proper cleavage could result in a nonfunctional RNA polymerase and therefore an RNA^- phenotype. Thus of the 4 RNA^- complementation groups only 1 or 2 might represent RNA polymerase(s) per se. Because RNA^- mutants are difficult to work with, it is also possible that not all RNA^- complementation groups have been identified. It is interesting to note that the majority of the RNA^- isolates so far belong to complementation group A; by analogy with the case for group C mutants, group A could represent the amino terminal portion of the precursor polypeptide.

Biochemical experiments to determine the nature of the RNA^- lesions have been difficult. Since none of the RNA^- mutants tested so far perform even the first step of the alphavirus infection cycle (formation of the minus strand of the RNA) at a nonpermissive temperature [165, 166], all experiments with these mutants must be conducted by a shift to the nonpermissive temperature after the infection is established under permissive conditions. This can lead to difficulties in interpretation because many *ts* mutations are only temperature-sensitive with respect to synthesis; once formed the proteins are stable and functional at either temperature. Furthermore, most of the proteins synthesized after togavirus infection are structural proteins or their precursors; all other gene products are only minor components and difficult to find in the infected cell. Nevertheless, several promising results have been obtained with RNA^- mutants. Two of the RNA^- Sindbis mutants make predominantly 49 S RNA after shiftup [94, 95] and these perhaps have an altered RNA polymerase. These mutants also lead to an altered pattern of protein synthesis after shiftup. Waite [94] found that *ts*11 gave rise to a large precursor polypeptide upon shiftup, which she believed to be identical to the polypeptide made after *ts*2 infection. M. J. Schlesinger [142] has observed a very large precursor polypeptide (molecular weight about 200,000) after shiftup with 1 of these mutants. Finally, Kääriäinen and collaborators [118, 128] have related findings with a Semliki Forest virus mutant, *ts*1. This mutant is classified RNA^+ because it makes RNA at a nonpermissive temperature, but at 40°C it makes predominantly 49 S RNA (10 times as much 49 S RNA as 26 S RNA); at 30°C the ratio of 49 S to 26 S RNA (2.5 : 1) is closer to the case with wild-type infection (0.4 : 1). Thus this mutant defect seems to be related to that of *ts*11 of Sindbis virus. At 40°C *ts*1 of Semliki Forest virus makes 2 large polypeptides which on the basis of tryptic peptides are nonstructural and nonoverlapping. These 2 polypeptides are thus probably translation products of the 49 S mRNA and together account for more than half of the coding capacity of this mRNA.

V. Defective Interfering Particles in Togavirus Infection

In most animal virus systems, continued passage of the virus at high multiplicity results in the accumulation in the stock of noninfectious particles called defective-interfering or DI particles. These contain incomplete genomes and are capable of interfering with standard virus replication during mixed infection. DI particles have been reviewed recently by Huang [167].

DI particles also arise during togavirus infection. In the case of Sindbis virus these particles have been shown to appear after 8-10 serial passages in either BHK21 cells or in chick cells [168-170]. In all cases these particles are noninfectious and reduce the yield of standard virus in a mixed infection by greater than 90%. However, there is some disagreement concerning the physical properties of the particles produced. The earliest description of DI particles of Sindbis virus by Schlesinger et al. [169] reported that they were inseparable from standard virus on the basis of buoyant density or sedimentation velocity. However, Shenk and Stollar [170] have described Sindbis DI particles of 2 classes, both separable from infectious virions, 1 class being lighter than standard Sindibs virus and the other denser. Thus the defective Sindbis particles isolated by these 2 groups seem to be different in a number of characteristics and may be the result of deletion of different segments of the genome. The RNA present in particles responsible for interference has not been thoroughly described, although the noninfectious light component of Shenk and Stollar [170] and the mixed virion-DI population both contain heterogeneous RNA which was smaller than the 49 S Sindbis genome [171, 173]. The protein complement of DI particles appears to be indistinguishable from that found in standard virus [171], although a new species of intracellular protein appears in the cytoplasm of cells infected with DI particles [172].

Most studies of DI particles have focused on the effect of these particles on RNA replication within the infected cell [168, 171, 173]. The synthesis of 49 S virion RNA and the number of nucleocapsids in the cytoplasm is much reduced during a mixed infection of DI particles and standard particles. In addition a new species of single-stranded RNA appears; this RNA sediments at 20 S and has a molecular weight of about 1×10^6. A new species of double-stranded replicative form with a sedimentation coefficient of 12 S also appears after mixed infection with DI and standard particles. This RF may contain the template for the 20 S single-stranded RNA.

The morphological variants found after repeated passages in arthropod cells [54] form another class of defective alphavirus particles. These particles contain a membrane of normal thickness but smaller nucleocapsids which have been assigned lesser triangulation numbers (discussed in Section 4.II.C.). It is not clear whether they should be designated DI particles. Although they are noninfectious for chick cells and therefore defective, it is not known if they interfere with the replication of the standard virus. However, the presence of a 12 S double-stranded RNA in chronically infected mosquito cells and the fact that these particles represent the majority of virus particles after the fifth day of persistent infection is suggestive.

VI. Growth of Togaviruses in Arthropod Cells

For a number of years there has been considerable interest in the replication cycle of togaviruses in their arthropod hosts. Early experiments proved that these viruses actively multiplied in the poikilothermic vector instead of being passively transferred to mammalian or avian hosts [175, 176]. Despite active viral replication, togaviral infection in insects appears to be symptomless and affects neither their life span nor their fecundity [176-179].

With the development by Singh in 1967 of tissue culture lines derived from mosquitos which can be readily maintained in the laboratory (reviewed by Yunker [180]), there has been renewed interest in pursuing the molecular biological aspects of togaviral infection in arthropods. The replication of a number of togaviruses, including such alphaviruses as Semliki Forest virus [181, 182], Middleburg virus [183], and Sindbis virus [174, 184-186]; several flaviviruses, including Japanese encephalitis virus [187], dengue virus type 2 [185, 188], and Kunjin virus [181, 184]; as well as the replication of a member of the California group viruses, California encephalitis virus [189], have been examined using cells derived from *Aedes albopictus* or *Aedes aegypti.*

In general, *A. albopictus* cells have given better results and higher virus yields, especially in the case of Sindbis virus [190].

There appear to be fundamental differences between the infection cycle in mammalian or avian cultured cells and the infection of arthropod cells. Although the replication of most togaviruses is rapidly cytocidal in chick or hamster cells, these viruses are capable of establishing chronic infections in arthropod cell lines [174, 181, 185, 187-189]. Chronically infected mosquito cells grow and divide normally for many generations while continually producing small amounts of virus.

The initial or primary infection of cultured insect cells by togaviruses, however, is established rapidly and the virus yield produced is comparable to the vertebrate cell culture system. This primary infection has been described in detail for dengue 2 virus [185, 188], Japanese encephalitis virus [187], Semliki Forest virus [181], and Sindbis virus [185]. During this initial phase some togaviruses exhibit sufficient cytopathic effect to form plaques [191], and it is possible to group the togaviruses by this response. All alphaviruses tested, except chikungunya, failed to make plaques in *A. albopictus* cells, while most flaviviruses which are known to be mosquito-borne in nature successfully formed plaques. Tick-borne flaviviruses or those whose natural vector is unknown failed to form plaques on mosquito cells. There are also reports of syncitial formation with flaviviruses as well as with vesicular stomatitis virus [190, 192]. It is interesting to note that most members of the Bunyamwera group of arboviruses are also cytopathic for insect cells, whereas most members of the Kemerovo group are symptomless [191].

Following this initial phase, a chronic infection develops; the resultant "carrier" mosquito cells, which shed 1-5 plaque-forming units (PFU) of virus per day (as opposed to up to 2000 PFU/hr in vertebrate cell infection) can be cultured and subcultured for many months [193]. It is unclear at present whether all cells in the culture yield low levels of virus or whether a few cells yield most of the virus produced. During this chronic infection there are gradual alterations in the virions produced: mutants temperature-sensitive in vertebrate cells arise [155, 194]; there are differences in plaque morphology which are only partially explained by reduced virulence for vertebrate cells [174, 185, 188]; changes occur in the antigenic properties of the particles [188]; and morphological variants arise (Section 4.II.C). In a study of mosquito cells persistently infected with Sindbis virus it was found that the majority of the intracellular double-stranded RNA produced sedimented at 12 S, as opposed to 20 S for replicative form RNA from either the initial infection of *A. albopictus* cells or Sindbis-infected chick or hamster cells [174]. In vertebrate cells the presence of 12 S double-stranded RNA is indicative of DI particles [168, 173] and DI particles could be responsible for the low yields in chronically infected cells.

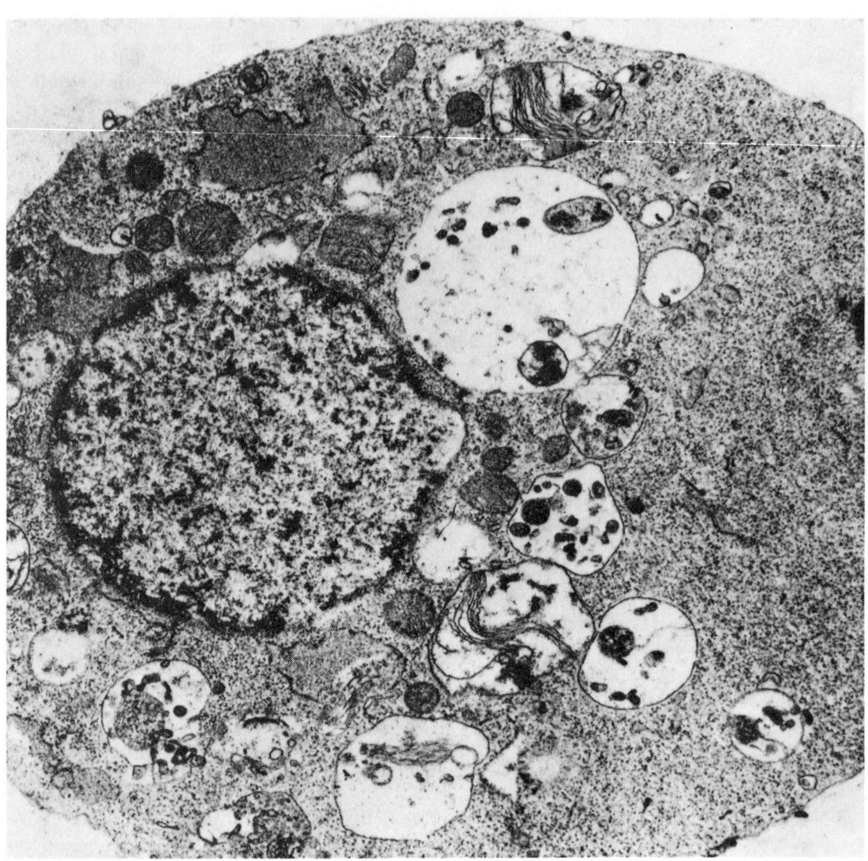

Fig. 4.15 Thin section of cultured *Aedes albopictus* cell 6 hr after infection with Sindbis virus. Note the large number of vesicles present, especially those showing a multilaminate membrane structure. X 18,500. (Micrograph courtesy of Dr. D. T. Brown.)

Why the togaviruses set up a persistent infection in arthropod cells and a cytolytic infection in vertebrate cells is not clear. It is possible that the defective particles containing smaller genomes become predominant and that these particles are responsible for the failure to shut off host macromolecular synthesis. It has also been suggested that interferon regulates this virus-host symbiosis [195]. A simpler, wholly mechanical explanation has been suggested by Gliedman et al. [196] from electron microscopic studies of Sindbis replication in *A. albopictus* cells. These authors found that in these cells Sindbis virus apparently replicates completely within membrane-bound "factories" in the

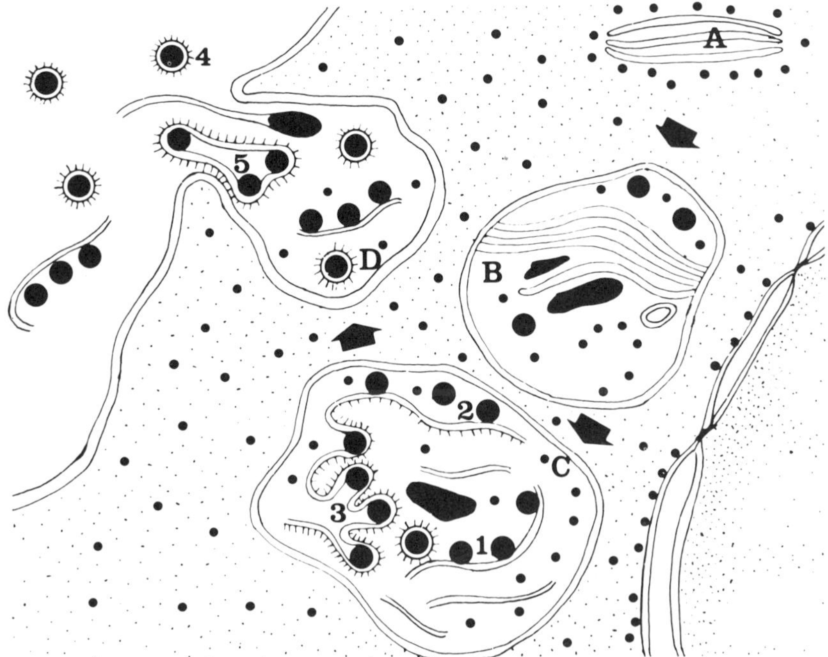

Fig. 4.16 Model for the replication of alphaviruses in arthropod cells. Membranes form within the cytoplasm (A) and become enclosed in a vesicle (B) in which are found ribosomes and nucleocapsids. The maturation of the virus is shown in C. The nucleocapsids attach to the membranes (1), which become modified by the appearance of the viral glycoproteins (2). The nucleocapsids become enveloped by the modified membranes (3) and mature to form virions. Finally, the vesicle fuses with the cell plasma membrane (D) and releases its contents. Immature forms of the virus are released in the process (5) as well as mature virions (4). (Model courtesy of Dr. D. T. Brown.)

cytoplasm. These factories contain ribosomes and laminate structures reminiscent of the Golgi apparatus. Virus matures within these vesicles, instead of budding through the cell surface, and when the viruses are completed the entire contents of the factory are released into the extracellular fluid.

An electron micrograph of Sindbis-infected mosquito cells is shown in Fig. 4.15, showing the numerous cytoplasmic "vacuoles" characteristic of these infected cells; note in particular the membranous structures in some of these vacuoles. A model for maturation of Sindbis virus in mosquito cells based on these studies is shown in Fig. 4.16. With this model the arthropod cell is effectively insulated from the virus after the initial infection and can be considered functionally uninfected. This may also imply that the cytocidal infection of mammalian cells is caused directly by viral encoded material and not merely by a competition between host and viral elements for the host macromolecular synthetic machinery.

Little is known at the present time about the differences at the molecular level between arboviruses produced in arthropod cells as compared to those from infection of mammalian or avian cells. Renkonen [197] has shown that the glycoproteins of Semliki Forest virus grown in *A. albopictus* lack sialic acid, as do the cellular glycoproteins of the mosquito cells. Brown and collaborators [97] have shown that the glycoproteins of mosquito-grown Sindbis virus are separable from those of BHK-grown virus on polyacrylamide gels whereas the capsid proteins migrate identically. Whether this is because of differences in the primary polypeptide sequence or reflects the altered carbohydrate content is unclear.

The lipid composition of mosquito-grown Semliki Forest virus resembles the composition of the plasma membrane lipids of *A. albopictus* cells. The virus contains 60% phosphatidylethanolamine, about 15% phosphatidylcholine, and very little cholesterol [62, 197], which is very different from the distribution of phospholipids in virus grown in vertebrate cells (Table 4.2). Mosquito-grown Semliki Forest virions appear to be less stable to physical manipulations and chemical treatments, which might be a result of this altered membrane.

VII. Concluding Remarks

The study of the togaviruses in general and of the alphaviruses in particular has been a very fruitful field in the past several years. A great deal has been learned about the molecular biology of virus replication and in the process much about cell surfaces and about macromolecular synthesis by animal cells. Obviously, however, much more needs to be learned from this system. Studies on the RNA polymerases of the virus are crucial to our understanding of virus replication. Sequencing of important areas of the 2 species of mRNA of the alphaviruses can tell us much about mRNA translation and function. Studies on the biosynthesis of the virus glycoproteins and their insertion into the cell surface will yield information on cell membrane biosynthesis. We might note that a paradox exists in this regard. Soluble proteins (such as the viral capsid protein) are normally synthesized on free polysomes, whereas membrane glycoproteins are synthesized by bound ribosomes attached to the rough endoplasmic reticulum. Do the ribosomes translating the 26 S mRNA exist free while translating the capsid protein and then attach to the endoplasmic reticulum when they reach the regions of the messenger corresponding to the glycoproteins? Studies to answer this question will prove very interesting.

Further studies of virus replication in insect cells are very important. Are there virus functions required in vertebrate cell infection but dispensable in arthropod cells or vice versa? Why is the virus cytolytic in vertebrate but not in insect cells? Studies along these lines might also tell us more about the physiological differences between vertebrate cells and invertebrate cells.

As has been repeatedly mentioned in this review, much less is known about the flaviviruses than about the alphaviruses. Although these 2 groups have been historically linked by both their mode of transmission and their morphology, further studies may show that their mode of replication is dissimilar. One question in particular is the identification of the messenger RNA(s) and their translation products in vivo and in vitro. If the genome RNA is the only message and the intracellular proteins are not generated by posttranslational cleavage of a large polypeptide precursor, then it is necessary to involve internal initiation and termination. Although common in prokaryotic systems, such internal signals have not been discovered as yet in any eukaryotic cell. Furthermore, since the flaviviruses appear to mature by budding into vacuoles and other organelles, they form a model system for intracellular membrane modification and biogenesis, just as the alphaviruses are a model for studying the plasma membrane.

Acknowledgments

We wish to acknowledge helpful discussions with Drs. D. T. Brown, M. J. Schlesinger, S. Schlesinger, S. I. T. Kennedy, D. C. Burke, L. Kääriäinen, K. Simons, and O. Renkonen. Many of these authors also supplied us with preprints describing their results prior to publication.

Supported by Grants AI 10793 and GM 06965 from the U.S. Public Health Service and by Grant BMS75-02469 from the National Science Foundation.

References

1. J. Casals and D. H. Clarke, in Viral and Rickettsial Infections of Man 4th Ed. (F. L. Horsfall and I. Tamm, eds.), Lippincott, Philadelphia, 1965, Chap. 26.
2. D. H. Clarke and J. Casals, in Viral and Rickettsial Infections of Man, 4th Ed. (F. L. Horsfall and I. Tamm, eds.), Lippincott, Philadelphia, 1965, Chap. 27.
3. R. P. Hanson, S. E. Sulkin, E. L. Buescher, W. M. Hammon, R. W. McKinney, and T. H. Work, Science, 158, 1283 (1967).

4. E. R. Pfefferkorn and D. Shapiro, in Comprehensive Virology, Vol. 2, Reproduction of Small and Intermediate RNA Viruses (H. Fraenkel-Conrat and R. R. Wagner, eds.), Plenum, New York, 1974, Chap. 4, pp. 171-230.
5. E. R. Pfefferkorn and H. S. Hunter, Virology, 20, 446 (1963).
6. R. Laine, H. Söderlund, and O. Renkonen, Intervirology, 1, 110 (1973).
7. J. H. Strauss, B. W. Burge, and J. E. Darnell, Jr., Virology, 37, 367 (1969).
8. K. Simons and L. Kääriäinen, Biochem. Biophys. Res. Comm., 38, 981 (1970).
9. N. H. Acheson and I. Tamm, Virology, 41, 321 (1970).
10. H. Garoff, K. Simons, and O. Renkonen, Virology, 61, 493 (1974).
11. D. T. Simmons and J. H. Strauss, J. Mol. Biol., 71, 599 (1972).
12. B. M. Arif and P. Faulkner, J. Virol., 9, 102 (1972).
13. R. W. Boulton and E. G. Westaway, Virology, 49, 283 (1972).
14. M.-T. Hsu, H.-J. Kung, and N. Davidson, Cold Spring Harbor Symp. Quant. Biol., 38, 943 (1973).
15. L. Kääriäinen and P. J. Gomatos, J. Gen. Virol., 5, 251 (1969).
16. J. H. Strauss, B. W. Burge, and J. E. Darnell, Jr., J. Mol. Biol., 47, 437 (1970).
17. B. M. Sefton and K. Keegstra, J. Virol., 14, 522 (1974).
18. D. Shapiro, D. Trent, W. E. Brandt, and P. K. Russell, Infec. Immun., 6, 206 (1972).
19. V. Stollar, Virology, 39, 426 (1969).
20. D. Shapiro, W. E. Brandt, R. D. Cardiff, and P. K. Russell, Virology, 44, 108 (1971).
21. D. W. Trent and A. A. Qureshi, J. Virol., 7, 379 (1971).
22. V. Stollar, T. M. Stevens, and R. W. Schlesinger, Virology, 30, 303 (1966).
23. D. W. Trent, C. C. Swensen, and A. A. Qureshi, J. Virol., 3, 385 (1969).
24. E. R. Pfefferkorn and H. S. Hunter, Virology, 20, 446 (1963).
25. E. R. Pfefferkorn and R. L. Clifford, Virology, 23, 217 (1964).
26. N. H. Acheson and I. Tamm, Virology, 32, 128 (1967).
27. S. C. Harrison, A. David, J. Jumblatt, and J. E. Darnell, Jr., J. Mol. Biol., 60, 523 (1971).
28. E. G. Strauss, C. R. Birdwell, S. E. Staples, and J. H. Strauss, Virology, in press (1977).
29. J. H. Strauss, B. W. Burge, E. R. Pfefferkorn, and J. E. Darnell, Jr., Proc. Natl. Acad. Sci. U.S., 59, 533 (1968).
30. M. Horzinek and M. Mussgay, J. Virol., 4, 514 (1969).
31. T. M. Stevens and R. W. Schlesinger, Virology, 27, 103 (1965).
32. F. A. Murphy, A. K. Harrison, G. W. Gary, Jr., S. G. Whitfield, and F. T. Forrester, Lab. Invest., 19, 652 (1968).
33. P. Dobos and P. Faulkner, J. Virol., 6, 145 (1970).
34. D. T. Simmons and J. H. Strauss, J. Virol., 14, 552 (1974).

35. D. Mowshowitz, J. Virol., 11, 535 (1973).
36. H. Söderlund, N. Glanville, and L. Kääriäinen, Intervirology, 2, 100 (1973).
37. G. Wengler and G. Wengler, Virology, 59, 21 (1974).
38. D. T. Simmons and J. H. Strauss, J. Mol. Biol., 86, 397 (1974).
39. A. E. Smith, T. Wheeler, N. Glanville, and L. Kääriäinen, Eu. J. Biochem., 49, 101 (1974).
40. B. T. Eaton, T. P. Donaghue, and P. Faulkner, Nature, New Biol., 238, 109 (1972).
41. R. E. Johnston and H. R. Bose, Biochem. Biophys. Res. Comm., 46, 712 (1972).
42. J. C. Clegg and S. I. T. Kennedy, J. Gen. Virol., 22, 331 (1974).
43. E. Wecker, Virology, 7, 241 (1959).
44. A. Igarashi, T. Fukunaga, and K. Fukai, Biken J., 7, 111 (1964).
45. T. A. Brawner, J. C. Lee and D. W. Trent, Ann. Meet. Am. Soc. Microbiol., Abstr., 202 (1973).
46. R. E. Johnston and H. R. Bose, Proc. Natl. Acad. Sci. U.S., 69, 1514 (1972).
47. B. W. Burge and E. R. Pfefferkorn, J. Mol. Biol., 35, 193 (1968).
48. N. H. Acheson and I. Tamm, J. Virol., 5, 714 (1970).
49. L. Kääriäinen and H. Söderlund, Virology, 43, 291 (1971).
50. R. W. Simpson and R. E. Hauser, Virology, 34, 568 (1968).
51. R. W. Simpson and R. E. Hauser, Virology, 34, 358 (1968).
52. D. T. Brown, M. R. Waite, and E. R. Pfefferkorn, J. Virol., 10, 524 (1972).
53. C.-H. von Bonsdorff and S. C. Harrison, J. Virol., 16, 141 (1975).
54. D. T. Brown and J. B. Gliedman, J. Virol., 12, 1534 (1973).
55. D. L. D. Caspar and A. Klug, Cold Spring Harbor Symp. Quant. Biol., 21, 1 (1962).
56. T. Matsumura, V. Stollar, and R. W. Schlesinger, Virology, 46, 344 (1971).
57. H. Garoff and K. Simons, Proc. Natl. Acad. Sci. U.S., 71, 3988 (1974).
58. O. Renkonen, L. Kääriäinen, K. Simons, and C. G. Ghamberg, Virology, 46, 318 (1971).
59. A. E David, Virology, 46, 711 (1971).
60. F. P. Heydrick, J. F. Comer, and R. F. Wachter, J. Virol., 7, 642 (1971).
61. J. P. Quigley, D. B. Rifkin, and E. Reich, Virology, 46, 106 (1971).
62. O. Renkonen, A. Luukkonen, J. Brotherus, and L. Kääriäinen, in Control of Proliferation in Animal Cells (B. Clarkson and R. Baserga, eds.), Cold Spring Harbor Laboratory, New York, 1974, p. 495-504.
63. C. B. Hirschberg and P. W. Robbins, Virology, 61, 602 (1974).
64. B. M. Sefton and B. J. Gaffney, J. Mol. Biol., 90, 343 (1974).
65. M. J. Schlesinger, S. Schlesinger, and B. W. Burge, Virology, 47, 539 (1972).
66. B. W. Burge and J. H. Strauss, J. Mol. Biol., 47, 449 (1970).

67. P. M. Osterrieth, Mem. Soc. Roy. Sci. Liege, Collect 8, 11 (1968).
68. R. W. Compans, Nature, New Biol., 229, 114 (1971).
69. C. G. Ghamberg, G. Utermann, and K. Simons, FEBS Lett., 28, 179 (1972).
70. G. Utermann and K. Simons, J. Mol. Biol., 85, 560 (1974).
71. F. Brown, C. J. Smale, and M. Horzinek, J. Gen. Virol., 22, 455 (1974).
72. K. Simons, S. Keränen, and L. Kääriäinen, FEBS Lett., 29, 87 (1973).
73. S. Schlesinger and M. J. Schlesinger, J. Virol., 10, 925 (1972).
74. H. Garoff, Virology, 62, 385 (1974).
75. B. M. Sefton, G. G. Wickus, and B. W. Burge, J. Virol., 11, 730 (1973).
76. D. Shapiro, K. A. Kos, and P. K. Russell, Virology, 56, 88 (1973).
77. C. R. Birdwell and J. H. Strauss, J. Virol., 14, 672 (1974).
78. C. Morgan and C. Howe, J. Virol., 2, 1122 (1968).
79. S. Dales, Bacteriol. Rev., 37, 103 (1973).
80. J. J. Mooney, J. M. Dalrymple, C. R. Alving, and P. K. Russell, J. Virol., 225 (1975).
81. M. Mussgay, P.-J. Enzmann, and J. Horst, Arch. Ges. Virusforschg., 31, 81 (1970).
82. J. A. Sonnabend, E. M. Martin, and E. Mécs, Nature, 213, 365 (1967).
83. R. M. Friedman, H. Levy, and W. B. Carter, Proc. Natl. Acad. Sci. U.S., 56, 440 (1966).
84. T. Sreevalsan and R. Z. Lockart, Proc. Natl. Acad. Sci. U.S., 55, 974 (1966).
85. P. Dobos and P. Faulkner, J. Virol., 5, 429 (1969).
86. J. G. Levin and R. M. Friedman, J. Virol., 7, 504 (1971).
87. R. Cancedda, R. Swanson, and M. J. Schlesinger, J. Virol., 14, 652 (1974).
88. S. I. T. Kennedy, J. Mol. Biol., 108, 491 (1976).
89. C. Weissmann, G. Feix, and H. Slor, Cold Spring Harbor Symp. Quant. Biol., 33, 83 (1968).
90. B. Oberg and L. Philipson, J. Mol. Biol., 58, 725 (1971).
91. T. K. Frey, unpublished data (1974).
92. D. T. Simmons and J. H. Strauss, J. Mol. Biol., 71, 615 (1972).
93. D. T. Simmons, Ph.D. Thesis, Division of Biology, California Institute of Technology, Pasadena, California, 1974.
94. M. R. F. Waite, J. Virol., 11, 198 (1973).
95. C. M. Scheele and E. R. Pfefferkorn, J. Virol., 4, 117 (1969).
96. M. R. Michel and P. J. Gomatos, J. Virol., 11, 900 (1973).
97. D. T. Brown, personal communication (1975).
98. S. Segal and T. Sreevalsan, Virology, 59, 428 (1974).
99. T. Sreevalsan, J. Virol., 6, 438 (1972).
100. P. M. Grimely, J. G. Levin, I. K. Berezesky, and R. M. Friedman, J. Virol., 10, 492 (1972).
101. R. M. Friedman, J. G. Levin, P. M. Grimley, and I. K. Berezesky, J. Virol., 10, 504 (1972).

102. T. Sreevalsan and F. H. Yin, J. Virol., 3, 599 (1969).
103. E. Zebovitz, J. K. L. Leong, and S. C. Doughty, Arch. Ges. Virusforschg., 38, 319 (1972).
104. V. Stollar, R. W. Schlesinger, and T. M. Stevens, Virology, 33, 650 (1967).
105. A. A. Qureshi and D. W. Trent, J. Virol., 9, 565 (1972).
106. E. Zebovitz, J. K. L. Leong, and S. C. Doughty, Infec. Immun., 10, 204 (1974).
107. S. I. T. Kennedy, Biochem. Biophys. Res. Comm., 48, 1254 (1972).
108. H. Rosemond and T. Sreevalsan, J. Virol., 11, 399 (1973).
109. S. Penman, K. Scherrer, Y. Becker, and J. E Darnell, Proc. Natl. Acad. Sci. U.S., 49, 654 (1963).
110. B. T. Eaton and P. Faulkner, Virology, 50, 865 (1972).
111. T. P. Donaghue and P. Faulkner, Nature, New Biol., 246, 168 (1973).
112. D. H. Spector and D. Baltimore, Proc. Natl. Acad. Sci. U.S., 71, 2983 (1974).
113. G. Huez, G. Marbaix, E. Hubert, M. Leclerq, U. Nudel, H. Soreq, R. Salomon, B. Lebleu, M. Revel, and U. Z. Littauer, Proc. Natl. Acad. Sci. U.S., 71, 3143 (1974).
114. C. M. Scheele and E. R. Pfefferkorn, J. Virol., 5, 329 (1970).
115. M. J. Schlesinger and S. Schlesinger, J. Virol., 11, 1013 (1973).
116. S. Keränen and L. Kääriäinen, J. Virol., 16, 388 (1975).
117. C. M. Scheele and E. R. Pfefferkorn, J. Virol., 3, 369 (1969).
118. B.-E. Lachmi, N. Glanville, S. Keränen, and L. Kääriäinen, J. Virol., 16, 1615 (1975).
119. R. Cancedda and M. J. Schlesinger, Proc. Natl. Acad. Sci. U.S., 71, 1843 (1974).
120. N. Glanville, J. Morser, P. Uomala, and L. Kääriäinen, Eur. J. Biochem., 64, 167 (1975).
121. J. C. S. Clegg and S. I. T. Kennedy, FEBS Lett., 42, 327 (1974).
122. G. Wengler, M. Beato, and B. A. Hackemack, Virology, 61, 120 (1974).
123. J. C. S. Clegg, Nature, 254, 454 (1975).
124. M. J. Morser and D. C. Burke, J. Gen. Virol., 22, 395 (1974).
125. B. M. Sefton and B. W. Burge, J. Virol., 12, 1366 (1973).
126. R. M. Friedman, J. Virol., 2, 1076 (1968).
127. H. W. Snyder and T. Sreevalsan, J. Virol., 13, 541 (1974).
128. S. Keränen and L. Kääriäinen, Acta Pathol. Microbiol., Scand., Sect. B, 82, 810 (1974).
129. R. Cancedda, L. Villa-Komaroff, H. Lodish, and M. Schlesinger, Cell, 6, 215 (1975).
130. G. Wengler and G. Wengler, Virology, 65, 601 (1975).
131. E. G. Westaway, Virology, 51, 454 (1973).
132. C. R. Birdwell and J. H. Strauss, J. Virol., 14, 366 (1974).
133. B. W. Burge and E. R. Pfefferkorn, J. Virol., 1, 956 (1967).
134. C. R. Birdwell and J. H. Strauss, J. Virol., 11, 502 (1973).
135. D. T. Brown and J. E. Smith, J. Virol., 15, 1262 (1975).
136. C. E. Pedersen, Jr., and B. P. Sagik, J. Gen. Virol., 18, 375 (1973).

137. P. Bell, personal communication (1975).
138. B. W. Burge and E. R. Pfefferkorn, Nature, 210, 1397 (1966).
139. E. Lagwinska, C. C. Stewart, C. Adles, and S. Schlesinger, Virology, 65, 204 (1975).
140. K. J. Jones, M. R. F. Waite, and H. R. Bose, J. Virol., 13, 809 (1974).
141. C. R. Birdwell, E. G. Strauss, and J. H. Strauss, Virology, 56, 429 (1973).
142. M. Bracha, A. Leone, and M. Schlesinger, J. Virol., 20, 612 (1976).
143. M. R. F. Waite and E. R. Pfefferkorn, J. Virol., 5, 60 (1970).
144. M. R. F. Waite, D. T. Brown, and E. R. Pfefferkorn, J. Virol., 10, 537 (1972).
145. J. S. Pierce, E. G. Strauss, and J. H. Strauss, J. Virol., 13, 1030 (1974).
146. E. G. Strauss and J. H. Strauss, unpublished observations, 1974.
147. T. Matsumura, V. Stollar, and R. W. Schlesinger, Virology, 46, 344 (1971).
148. D. Shapiro, W. E. Brandt, and P. K. Russell, Virology, 50, 906 (1972).
149. B. W. Burge and E. R. Pfefferkorn, Virology, 30, 204 (1966).
150. P. D. Cooper, in The Biochemistry of Viruses (H. B. Levy, ed.), Marcel Dekker, New York, 1969, Chap. 4, pp. 177-218.
151. E. R. Pfefferkorn, in Comprehensive Virology, Vol. 9, Regulation and Genetics (H. Fraenkel-Conrat and R. R. Wagner, eds.), Plenum, New York, in print, 1977.
152. H. R. Bose, G. Z. Carl, and B. P. Sagik, Arch. Ges. Virusforschg., 29, 83 (1970).
153. W. Schwöbel and R. Ahl, Arch. Ges. Virusforschg., 38, 1 (1972).
154. M. V. Davey and L. Dalgarno, J. Gen. Virol., 24, 453 (1974).
155. T. E. Shenk, K. A. Koshelnyk, and V. Stollar, J. Virol., 13, 439 (1974).
156. E. G. Strauss, E. M. Lenches, and J. H. Strauss, Virology, 74, 154 (1976).
157. G. J. Atkins, J. Samuels, and S. I. T. Kennedy, J. Gen. Virol., 25, 371 (1974).
158. K. B. Tan, J. F. Sambrook, and A. J. D. Bellett, Virology, 38, 427 (1969).
159. E. Zebovitz and A. Brown, J. Mol. Biol., 50, 185 (1970).
160. B. Simizu and N. Takayama, Arch. Ges. Virusforschg., 38, 328 (1972).
161. B. W. Burge and E. R. Pfefferkorn, Virology, 30, 214 (1966).
162. E. R. Pfefferkorn and B. W. Burge, in The Molecular Biology of Viruses, (J. S. Colter and W. Paranchych, eds.), Academic, New York, 1967, pp. 403-426.
163. F. H. Yin, J. Virol., 4, 547 (1969).
164. K. B. Tan, J. Virol., 5, 632 (1970).
165. E. R. Pfefferkorn, B. W. Burge, and H. M. Coady, Virology, 33, 329 (1967).
166. E. R. Pfefferkorn and B. W. Burge, Perspect. Virology, 6, 1 (1968).
167. A. S. Huang, Ann. Rev. Microbiol., 27, 101 (1973).

168. B. T. Eaton and P. Faulkner, Virology, 51, 85 (1973).
169. S. Schlesinger, M. J. Schlesinger, and B. W. Burge, Virology, 48, 615 (1972).
170. T. E. Shenk and V. Stollar, Virology, 53, 162 (1973).
171. B. Weiss and S. Schlesinger, J. Virol., 12, 862 (1973).
172. T. E. Shenk and V. Stollar, Virology, 55, 530 (1973).
173. T. E. Shenk and V. Stollar, Biochem. Biophys. Res. Comm., 49, 60 (1972).
174. V. Stollar and T. E. Shenk, J. Virol., 11, 592 (1973).
175. S. R. Pattyn and L. de Vleesschauwer, Acta Virol., 12, 347 (1968).
176. S. Ya. Gaidamovich, Ya. Ya. Tsilinsky, A. I. Lvova, and N. V. Khutoretskaya, Acta Virol., 15, 301 (1971).
177. F. Bras-Herreng, Ann. Microbiol. (Inst. Pasteur), 124A, 507 (1973).
178. H. G. Janzen, A. J. Rhodes, and F. W. Doane, Can. J. Microbiol., 16, 581 (1970).
179. S. G. Whitfield, F. A. Murphy, and W. Sudia, Virology, 43, 110 (1971).
180. C. W. Yunker, in Arthropod Cell Cultures and Their Application to the Study of Viruses (E. Weiss, ed.), Springer-Verlag, New York, 1971, pp. 113-126.
181. M. W. Davey, D. P. Dennett, and L. Dalgarno, J. Gen. Virol., 20, 225 (1973).
182. J. Peleg, in Arthropod Cell Cultures and Their Application to the Study of Viruses (E Weiss, ed.), Springer-Verlag, New York, 1971, pp. 155-161.
183. S. R. Pattyn and L. de Vleesschauwer, Arch. Ges. Virusforschg., 19, 176 (1966).
184. J. Řeháček, Acta Virol., 12, 241 (1968).
185. A. Stevens, Proc. Soc. Exp. Biol. Med., 134, 356 (1970).
186. V. Stollar, T. E. Shenk, and B. D. Stollar, Virology, 47, 122 (1972).
187. A. Igarashi, F. Sasao, S. Wungkobkiat, and K. Fukai, Biken J., 16, 17 (1973).
188. P. Sinarachatanant and L. C. Olson, J. Virol., 12, 275 (1973).
189. M. J. Lyons and J. Heyduk, Virology, 54, 37 (1973).
190. K. R. P. Singh, in Arthropod Cell Cultures and Their Application to the Study of Viruses (E. Weiss, ed.), Springer-Verlag, New York, 1971, pp. 127-132.
191. C. E. Yunker and J. Cory, Appl. Microbiol., 29, 81 (1975).
192. E. C. Suitor, Jr., and F. J. Paul, Virology, 38, 482 (1969).
193. J. Peleg, J. Gen. Virol., 5, 463 (1969).
194. J. A. Mudd, R. W. Leavitt, D. T. Kingsbury, and J. J. Holland, J. Gen. Virol., 20, 341 (1973).
195. P. J. Enzmann, Arch. Ges. Virusforschg., 40, 382 (1973).
196. J. B. Gliedman, J. F. Smith, and D. T. Brown, J. Virol., 16, 913 (1975).
197. O. Renkonen, personal communication (1975).
198. R. M. Friedman, J. Virol., 2, 547 (1968).
199. K. L. Cartwright and D. C. Burke, J. Gen. Virol., 6, 231 (1970).

200. M. Ranki, L. Kääriäinen, and O. Renkonen, Acta Pathol. Microbiol., Scand., Sect. B, 80, 760 (1972).
201. M. F. Jacobson and D. Baltimore, Proc. Natl. Acad. Sci. U.S., 61, 77 (1968).
202. K. Hashimoto, K. Suzuki, and B. Simizu, J. Virol., 15, 1454 (1975).
203. C. R. Birdwell, Ph.D. Thesis, Division of Biology, California Institute of Technology, Pasadena, 1974.
204. M. Bracha and M. Schlesinger, Virology, 74, 441 (1976).
205. E. G. Strauss, unpublished data, 1975.

Chapter 5

Rhabdoviruses

David H. L. Bishop and M. S. Smith

Department of Microbiology
The Medical Center
University of Alabama in Birmingham
Birmingham, Alabama

I. Introduction

The rhabdoviruses are a group of viruses obtained from vertebrate, invertebrate, and plant origins. Characteristically on phosphotungstic acid negative staining, the vertebrate and insect rhabdoviruses possess a bullet-shaped morphology. They are composed of an outer surface of glycoprotein spikes, a lipid envelope, and a central helical core consisting of a single RNA genome and various structural proteins. The plant rhabdoviruses appear to possess a similar composition although on thin section they are usually longer and bacilliform in shape even though they are often bullet shaped on negative staining [1]. Reviews on various aspects of the biology of different rhabdoviruses are available, including recent ones by Wagner [2], Murphy [3], Francki [1], and Knudson [4], as well as previous ones by Hummeler [5], Howatson [6], Matsumoto [7], and Hanson [8].

This review will provide a synopsis of the isolation and pathogenicity of certain vertebrate and invertebrate rhabdoviruses; then discuss the relatedness of the different isolates on the basis of genome, structural, and antigenic similarities; and finally delineate the known features of rhabdoviruses vis a vis their infection process and cell biology. The reader is referred to Francki [1] and Knudson [4] for a review of some of 16 plant rhabdoviruses and the relationships of those viruses to putative insect vectors, including evidence for the growth of plant rhabdoviruses within insects and in insect cell culture. This review will not include Marburg virus, since it differs structurally (by length, pleomorphism, and cross-section) as well as antigenically from all known rhabdoviruses [9].

II. Vertebrate and Invertebrate Rhabdoviruses: Origins and Isolations

A list of the major vertebrate and invertebrate rhabdoviruses is provided in Table 5.1 together with estimates of their length and whether cloned stocks of the virus are available. Most initial isolations of these viruses have been propagated by passage in newborn mouse brains or embryonated eggs before identification by electron microscopy, classification as rhabdoviruses, and cloning in chick embryo cells, VERO, or baby hamsters kidney cells. Several rhabdoviruses have yet to be cloned (Table 5.1).

Vesicular stomatitis disease was reported by Theiler in 1901 as a disease of horses and mules in the Transvaal, South Africa. He reported outbreaks occurring there in 1884 and 1897. Symptoms noted were vesicles on the gums, lips, tongue, and hooves. These vesicles were observed to frequently rupture or coalesce. Recovery began a week or so after onset of the disease and was both rapid and complete [10]. Similar disease signs have been reported in the United States among horses and cattle going back to the time of the Civil War [8]. The Indiana strain of vesicular stomatitis virus (VSV) was first isolated from tongue vesicles of young cattle shipped from Kansas City to Richmond, Indiana in 1925 [11]. VSV New Jersey has been isolated from many sources including the snout vesicles of an infected pig [30]. VSV Argentina was isolated by Garcia Pirazzi and collaborators [12], and VSV Brazil by Andrade and associates [13]. Both VSV Indiana and other VSV types have been repeatedly isolated from cattle diseases [8, 11, 13].

Cocal virus was isolated in 1961 from *Gigantotaelaps* mites removed from rice rats in Bush Bush Island near Trinidad and later isolates were obtained both from rodents and mosquitos in Trinidad and mites in Brazil [14].

Piry virus was obtained by R. E. Shope in 1960 from an opposum in Belém, Brazil [15].

Chandipura virus was first isolated from human sera obtained from patients with a febrile illness during an epidemic in India involving chikungunya and dengue viruses [16]. Serological surveys have shown evidence of Chandipura neutralizing antibodies on the Indian subcontinent involving camels, horses, donkeys, cows, buffalos, sheep, goats, and rhesus monkeys [16].

Klamath virus (M-1056) was isolated from a meadow vole (*Microtus montanus*) in Oregon [17].

Navarro virus (Cali 874) was isolated from the spleen of a vulture (*Cathartes aura*) in Navarro, Columbia [18].

Kwatta (Tr 58603) was obtained from *Culex* mosquitoes in Paramaribo, Surinam [19], while the antigenically related *BeAn 157575* was isolated by J. P. Woodall in Brazil [20].

Table 5.1 Vertebrate and Invertebrate Rhabdoviruses[a,b]

Group	Virus	Abbreviation or name used	Particle length (nm)	Cloned stock
1a	Vesicular stomatitis virus, Indiana	VSV Indiana	180	Yes
b	Cocal	Cocal	180	Yes
c	Vesicular stomatitis virus, Argentina	VSV Argentina	180	Yes
d	Vesicular stomatitis virus, Brazil, (Alagoas)	VSV Brazil		Yes
e	Vesicular stomatitis virus, New Jersey	VSV New Jersey	180	Yes
f	Chandipura (and IbAn 9978)	Chandipura	180	Yes
g	Piry	Piry	180	Yes
2a	Rabies	Rabies	180[c]	Yes
b	Lagos bat virus	LBV	180	Yes
c	Mokola	Mokola	180	Yes
d	Kotonkan (IbAr 23380)	Kotonkan	180	Yes
e	Obodhiang (SudAr 1154-64)	Obodhiang	170	Yes
f	Nigerian horse virus (H28)	NHV	–	No
g	Bolivar	Isolate no longer available	–	No

Table 5.1 (continued)

Group	Virus	Abbreviation or name used	Particle length (nm)	Cloned stock
h	Duvenhagé	Duvenhagé	180	Yes
3a	Bovine ephemeral fever virus	BEFV	145-185	Yes
4a	Kern Canyon virus	KCV	130	Yes
5a	Klamath (M1056)	Klamath	170	No
6a	Barur (I6235)	Barur	125-150	No
7a	Flanders	Flanders	220	No
b	Hart Park	HP	220	No
8a	Mt. Elgon bat (BP846)	MEBV	230	No
9a	Joinjakaka (MK7937)	Joinjakaka	200-250	No
10a	Navarro (Col. Cali 874)	Navarro	220-260	No
11a	Kwatta (Tr58603)	Kwatta	230-280	No
b	BeAn (157575)	BeAn 157575	ND	No
12a	Mossuril (Sa Ar1995)	Mossuril	225-305	No
b	Kamese (MP6186)	Kamese	ND	No
13a	Pike fry rhabdovirus	PFR	120-160	Yes
14a	Spring viremia of carp	SVCV	120-180	Yes
b	10/3 Swim bladder inflammation	SBI	120-180	Yes
15a	Egtved (viral hemorrhagic septicemia)	VHS or Egtved	120-180	Yes
b	VHS second serotype	VHS	ND	Yes

16a	Infectious hematopoietic necrosis virus	IHNV	120-180	Yes
b	Oregon sockeye disease virus	OSDV	180	Yes
c	Sacramento River chinook disease virus	SRCDV	160	Yes
17a	Sigma	Sigma	180	Yes
18a	European red mite virus	–	–	No
19a	Rhinoceros beetle virus	–	–	No
20a	Whirligig beetle virus	–	–	No
21a	*Dendrobium* virus	–	–	No

[a]No rhabdoviruses isolated from plants are included; of the insect viruses included in this table not all are believed to be involved in animal infections (groups 17-21) and several (groups 18-21) are not discussed further in this review (see Knudson [4] and Francki [1] for full discussions of these viruses).

[b]Lengths are rounded off to the nearest 5 nm and are based on the published values of several investigators (see text, Section 5.II). Where wide differences have been reported, ranges are given. For most viruses, the values are ±5%; however the influence of preparative procedures on the virus size may influence the observed lengths.

[c]Crick and Brown report rabies Flury (LEP) to have a particle length of 125 nm [212].

Mossuril (SA. Ar 1995) was obtained from *Culex* mosquitoes in Lumbo, Mozambique [21], and Kamese (MP 6186) which is antigenically related to it, came from *Culex* mosquitoes collected in Uganda [20].

Outbreaks of cattle diseases first described in East Africa in 1867, later in Rhodesia, Kenya, South Africa, Indonesia, India, Egypt, Australia, and Japan [22], have led to several isolations of bovine ephemeral fever viruses (BEFV) [23-28].

Kern Canyon virus was obtained from visceral tissues of *Myotis yumanensis* bats in Kern County, California in 1956 and 1961 [17].

Mount Elgon bat virus was obtained from the salivary gland of an insectivorous fruit bat (*Rhinolophus hildebrandtii eloquens*), collected from the Kiminimi caves on the slopes of Mt. Elgon in Kenya [29].

Joinjakaka virus (MK 7937) was obtained in 1966 from a pool of mixed *Culicines* insects in Joinjakaka, New Guinea [30], while Barur (I 6235) came from the spleen of a rat trapped near Barur village in India as well as from ticks off goats in the same district [30].

Flanders virus was isolated from mosquitos (*Culiseta melanura* and *Culex pipiens*) and an ovenbird (*Seiurus aurocapillus*) in New York State in 1961 and 1962 [31]. Hart Park virus has been isolated from mosquotos in California [31] as well as from a cardinal bird (*Richmondena cardinalis*) [32]. Many other isolates of both viruses have been obtained throughout the United States [32].

Sigma virus, discovered by L'Héritier, occurs in both wild fruit flies and laboratory strains of *Drosophila* [34, 98]. The virus is transmitted vertically and confers to the fruit flies a sensitivity to CO_2 exposure [33, 35, 36].

Among rhabdoviruses known to cause economically important disease of fish, the most recent isolate, pike fry rhabdovirus, causes an acute disease of young pike, "red disease of pike" [37].

Two carp fish diseases, swim bladder inflammation [38] and a carp dropsy-type disease [39] are caused by rhabdoviruses. The 2 diseases are quite different in their pathogenesis, seasonal variation, and geographical distribution which has led to speculation that the agents causing the disease are different; however, this does not appear to be the case, and it has been claimed that spring viremia of carp virus causes both diseases [38, 40].

A trout disease first observed in 1938 and later described by Schäperclaus [41] as "Infektiöse Nierenschwellung und Leberdegeneration" (INuL or viral hemorrhagic septicemia, VHS) was shown to be similar to a virus-caused disease of rainbow trout observed in Egtved, Denmark, around 1950 [42]. Many isolates of these viruses have been obtained and characterized [43].

Infectious hematopoietic necrosis virus and similar rhabdovirus isolates, such as Oregon sockeye disease virus and Sacramento River chinook salmon disease virus, have been obtained from diseased salmon in various locations on the west coast of the United States since 1950 [44, 45].

A large number of virulent rabies isolates have been obtained and a variety of attenuated, less virulent strains are available (see Table 5.2). Rabies as a disease has been known for more than 4000 years. In a reference in the pre-Mosaic Eshnunna Code there is a statement referring to mad dogs biting and causing death to man [46]. Rabies was isolated from a dog in 1882 and attenuated by Pasteur by passage in rabbit brain. In various parts of the world different animals transmit or carry the disease (wolves, jackals, mongooses, badgers, steppe foxes, dogs, cats, foxes, skunks, bats, raccoons, deer, etc.).

Nigerian horse virus was isolated by Porterfield and associates in 1956 in Ibadan, Nigeria, from the brain of an old horse exhibiting convulsions and paralysis termed "staggers" [47]. Bolivar virus was obtained from cattle and is probably another rabies-like virus [48]. A more recent rabies type virus is Duvenhagé isolated in South Africa [49].

Lagos bat virus was isolated in 1956 from a mixture of 6 brains of Nigerian fruit bats (*Eidolon helvum*), shot in Lagos, Nigeria [50].

Mokola virus (IbAn 27377) was obtained from pooled lungs, livers, spleens, kidneys, and hearts of shrews (*Crocidura* sp.), captured in and near Ibadan, Nigeria in 1968 [51]. Mokola isolates have also been made from children with central nervous system diseases [52, 53].

Kotonkan (IbAr 23380) was isolated from *Culicoides* gnats [54]. Serum samples of man, cattle, a cattle egret, giant rats, and African hedgehogs (all from northern Nigeria) have shown evidence of Kotonkan infections [54]. Obodhiang virus (Sud Ar 1154-64) was isolated from man-biting *Mansonia uniformis* mosquitos collected in the Sudan [55].

III. Rhabdovirus Transmission and Pathobiology

Clear-cut proof that insect or other vectors can transmit rhabdoviruses requires rigorous field and laboratory studies and, except for sigma and rabies virus, such evidence has either not been sought or is not absolutely conclusive. From this cautious standpoint, it is only fair to balance that statement by saying that for certain rhabdoviruses there is considerable *suggestive* evidence for transmission between vertebrates involving arthropods and some evidence that certain rhabdoviruses can multiply in arthropods.

Many rhabdovirus isolates listed in Table 5.1 have yet to be characterized by criteria other than serology and electron microscopy of infected material. Exceptions are the rabies isolates and putative rabies-related viruses (Mokola and Lagos bat viruses) as well as vesicular stomatitis virus (VSV Indiana) and some of its related viruses (VSV New Jersey, Cocal, Chandipura, and Piry). Also Kern Canyon virus, the fish rhabdoviruses, and sigma have been characterized to some extent (see below).

Some rhabdoviruses have been cloned only after successive mouse brain passages. Consequently, experimental pathobiological studies (or transmission studies) using uncloned or cloned virus should be considered in the light of possible selection of nonrepresentative strains. Use of uncloned stocks of low-passage isolates have been used in some pathobiology studies by Murphy and associates [3].

Evidence of virus transmission will be considered below as will various aspects of rhabdovirus pathobiology.

A. Vesicular Stomatitis Viruses, Piry, and Chandipura

In reports in 1862 by Major General McClellan to the Secretary of War of the United States, outbreaks of vesicular stomatitis in army horses were mentioned [8]. Mohler in 1904 [56] reported that the disease occurred in animals near water, being common in cattle pastured near ponds and streams rather than in cattle kept in the open plain. Also it was noted that outbreaks of the disease occurred after rain following a dry spell. Such conditions are suitable for the emergence of certain genera of diptera, such as stable flies, black flies, and horseflies, as well as mosquitos, the last 3 of which spend their immature stages in water. Disease incidence was found to decrease after a frost, which coincides with reduced biting by these insects (see Hanson [8] for a description of the natural history of vesicular stomatitis and of the various early reports of vesicular stomatitis disease which could have been caused by VSV). Although no virus was isolated by Mohler from those diseased cattle or the insects around them at the time, it has been shown that stable flies, mosquitos, deerflies, and horseflies can, in laboratory experiments, transmit VSV by bite from infected embryonated eggs to uninfected eggs [57].

In answer to a question of whether VSV can be transmitted from an insect to an animal and whether this transmission is passive or involves growth of virus in the insect vector, Mussgay and Suárez clearly showed that VSV fed to or injected into *Aedes aegypti* mosquitos could transmit VSV Indiana to newborn mice and that the virus grew in the mosquitos [58]. Virus persisted in the infected insects for a relatively long time (through 2 weeks). Tesh and associates [59] put phlebotomine sandflies (*Lutzomyia trapidoi*) on viremic (VSV) infant hamsters and showed an increase of the virus titer in the sandflies, as well as VSV transmission by bite as early as 3 days after the blood meal. Transovarian transmission of VSV has been demonstrated in such sandflies [213].

More recently, cell cultures of moths (*Antheraea eucalypti*) and mosquitos (*A. aegypti* and *A. albopictus*) have been productively infected with

VSV Indiana [60, 61]. Such infections produce low levels of released virus and can sometimes give rise to persistently infected cell cultures, although the exact nature of such persistent infections is unknown [62].

Despite this suggestive evidence, it has not been proved that insects are a major field vector in disease transmission to cattle, and either direct transmission and/or insects or other sources may be involved [63]. It has been postulated that disease transmission from South American enzootic regions may be responsible for the North American epizootics [8]. Whether disease carriers are imported cattle or other animals or windblown insect vectors is not known.

Signs of the disease caused by VSV in cattle involve a loss of appetite, loss of weight, cessation of milk production, abnormal salivation, and lesions on the mouth, tongue, teats, and hooves. Similar topical lesions occur in horses and infrequently the disease affects swine. It is doubtful if natural infections of sheep and goats occur [8], although a variety of animals are susceptible to experimental infection. It is quite possible that unrecognized mild infections occur in nature and that these act as a reservoir of the virus.

The cattle disease caused by VSV Indiana and other serotypes (VSV New Jersey, Alagoas, and VSV Argentina) is predominantly epitheliotropic [3], producing a cytopathic effect in the infected cells, spreading rapidly from the initial infection site to give characteristic vesicles, etc.

Replication of these viruses (as well as Piry and Chandipura) in young hamsters or mice is likewise primarily epitheliotropic but also occurs in visceral organs and tissues (liver, kidney, spleen, thymus, macrophages, and lung tissue). These infections give rise to a viremia and some subsequent brain and spinal cord involvement [3]. In older animals, there is less visceral organ involvement and more evident brain and spinal cord infections (see Murphy [3] for a discussion of the experimental infections of animals by VSV Indiana, VSV New Jersey, Piry, and Chandipura). When brain infections occur, extracellular virus is frequently observed.

In summary, VSV infection (both experimental and that observed in nature) is predominantly epitheliotropic. Disease transmission may be by direct contact and/or involve an insect or other vector. The reader is recommended to read the review by Hanson [8] for a discussion of the various documented epizootics in the United States.

B. Rabies

Rabies street strain virus infections are generally lethal in animals and man and are predominantly neurotropic. Transmission has been shown to occur via aerosols or by biting (see H. N. Johnson for a review [64] of this subject).

Transmission by biting is caused by the presence in saliva of rabies virus, derived by budding from the salivary gland secretory epithelium [65]. The aggressive nature of a rabid animal clearly aids in maintaining the transmission process. Many virulent isolates have been obtained; certain less virulent "attenuated" strains have been derived from street viruses (Table 5.2); these exhibit reduced pathogenicity for one or another host [66].

The experimental transmission of a rabies infection into and through

Table 5.2 Some Rabies Isolates[a]

Fixed strains	Example
Pasteur	Isolated from dog by Pasteur in 1882
CVS	From Pasteur strain, further passage in mouse brain
SAD	From rabid dog at CDC, Atlanta, Georgia in 1953
ERA	From SAD strain
LEP (Flury)	Low egg passage derived from brain of girl named Flury in 1939
HEP (Flury)	High egg passage of LEP
Fuenzalida S51	From brain of rabid dog in Chile, 1943
Fuenzalida S91	From human rabies case in Chile, 1943
Högyes	From rabid dog, obtained prior to 1888; generally similar to Pasteur strain
Kelev	From rabid dog obtained in Israel in 1950
Nishigahara	From Pasteur strain imported to Japan in 1915
Pasteur PV (PM)	Pitman-Moore official U.S. rabies strain for virus production in rabbit brains
DR19B	From vampire bat in Brazil
Mochalin	From human case, Moscow
NYC	From dog in New York City
R-205	From Czech badger in 1964
And many others	

[a]From Clark and Wiktor [66].

the nervous system and other organs has been followed by elegant electron microscopy and immunofluorescence experiments [3, 67-75]. In the experiments reported by Murphy and collaborators, it was shown that following an intramuscular injection of the hind leg of a newborn hamster, the first sign of virus replication was observed in nearby striated muscle cells 36-40 hr postinoculation [71]. Later, other muscle cells, the neuromuscular spindles within the inoculated limb muscle and their associated nerves, showed evidence of virus presence. By 60 hr, both peripheral nerves and ipsilateral spinal ganglia were positive in rabies immunofluorescence tests. The spinal ganglia and lumbar spinal cord neurons showed the presence of viral antigens 60-72 hr postinoculation, with subsequent rapid involvement of the brainstem. Internal budding of virus was observed within neurons and their processes; however, little extracellular virus was observed, nor was there much evidence of cytopathic effect in these infected cells. Normal brain structure was preserved despite the ascending infection of the brain and signs of inflammation were present only in the terminal phases of the disease [3]. These studies indicated that up through this phase of the infection process (the centripetal phase) apart from the initial inoculation site, viral products (antigens, inclusions, or virus budding) were essentially restricted to nerve cells and their processes and were not present in supportive tissue.

Other tissues shown to become involved during the terminal stages of the disease (the centrifugal stage [72, 75]) were the retina, autonomic nerves of the intestine, adrenal gland, brown fat, pancreas, heart, and many sensory nerve endings of the oronasal cavities. In these terminally infected tissues, viral products were not only associated with neural tissue. Specific organs were also shown to possess viral antigens [72].

Although in these hamster experiments little antigen was observed in salivary gland tissue, the neurotropism of rabies in hamster experiments clearly reflects the process as it occurs in nature [3]. Feral animal infections usually result in considerable amounts of virus and viral antigens present in salivary gland tissue and excretions [65].

The picture obtained in these studies is one in which the virus finds a nerve ending from its initial site of inoculation and then passes into the central nervous system of the animal. Why neuron supportive tissue does not become infected is an interesting aspect of this neurotropism. It is presumed that the quick transferral and temporal limitation of the virus to nerve cells results in a delay or suppression of the immune response by the animal [3].

Successful experimental transmission of rabies has been achieved by nasal infections using aerosols [75-77]. After a fatal rabies infection of 2 people who entered the bat-infested Frio cave in Texas, it was found that there was no evidence of bat bites on the subjects. Later it was shown that coyotes and foxes contracted rabies when kept in bat- and arthropod-proof cages in that cave [78]. Analyses of the cave, air, and infected bat nasal discharges also indicated that rabies was present [79, 80]. Transmissions from rabid bats or other animals to man following bites have been documented; see Sulkin and Allen [81] and Johnson [64] for reviews of the subject.

Orally administered rabies can induce disease [71, 82, 83], and oral vaccination of foxes has been reported ([84]; also see the recent review by Debbie [85]). Following oral infection by street strains, it has been shown that the virus is neurotropic, although the initially infected nerves varied from experiment to experiment (oral, intestinal, etc.).

There is no known arthropod vector playing a role in disseminating rabies disease. Although it is conceivable that arthropods can transmit the disease, all available evidence points to direct transmission from an infected animal to a susceptible subject either by aerosol inhalation, by bite, or possibly by oral ingestion of diseased tissue [64]. The slow progression of rabies infections through a population of wild animals also suggests that animal to animal is the principle form of virus transmission. It has been suggested that asymptomatic carriers of the disease may play a vector role (discussed in Ref. [81]), although whether or how frequently this occurs in nature is not known.

In summary, rabies infections are predominantly neurotropic in both experimental and field observations. Virus observed by Murphy and collaborators in infected brain tissue was mostly intracellular, although Iwasaki et al. [86] have reported virus budding out from neuronal cell membranes [87]. Disease transmission in nature is known to occur by direct animal to animal contact and association.

C. Lagos Bat and Mokola Viruses

The experimental diseases caused by these viruses clearly fit the rabies category. The infection is neurotropic and intracellular virus is observed in infected brain tissue [71]. There is, moreover, evidence of structural and antigenic similarities of these viruses to rabies itself (see Table 5.7).

Kemp and associates [88] showed that wild shrews subsequently inoculated with Mokola could transmit a disease to mice by bite. During the time course of those experiments not all the infected shrews apparently showed signs of illness, but it was shown that infected shrew brains possessed lesions with lymphocytic infiltration into the meninges, occasional perivascular cuffing, and necrosis of neurons close to blood vessels. Salivary gland cells, spleen, heart, lung, and kidneys also provided evidence of the presence of virus. Since salivary glands and mouth washings of these animals yielded virus and the animals violently attacked and bit mice, this was construed as evidence that a rabies-like mode of transmission could occur in Mokola infections. Kemp and associates [52] have also suggested that shrews may serve as asymptomatic carriers of the disease in nature, although definitive evidence for this suggestion is lacking.

In experiments using dogs and monkeys inoculated with LBV or Mokola, Tignor, Shope, and collaborators [89] showed that in the dog host (following intracerebral inoculation) there was intermittent severe depression, dyspnea, lack of coordination, periods of aimless wandering and barking, with death at 12 days in 1 case. In another case there was partial paralysis with progression to lateral recumbancy, champing fits, and intermittent convulsions. One monkey injected intracerebrally with LBV showed behavioral changes, including aggression, unresponsiveness, depression, and hydrophobia, as well as paralysis. Mokola and Lagos bat viruses have been obtained from experimentally infected monkey or dog brain tissues. In 2/3 of the monkeys infected with Mokola, virus was obtained from the saliva or submaxillary glands [89].

Murphy and associates, using immunofluorescent antibody techniques, obtained evidence that the infection process in hamsters injected with LBV or Mokola viruses was predominantly neurotropic, as in rabies infections, and that nearly all the virus observed was present within the cytoplasm of infected brain cells [71, 72].

D. Nigerian Horse Virus

The Nigerian horse "staggers" disease is generally a fatal one, fits the rabies neurotropic category, and is characterized by nervous derangement. Two forms of the disease have been observed, blind or frenzied staggers (in which there are severe convulsions and paralysis of the hind legs) and dumb or comatose staggers (in which the horse appears drowsy and paralysis starts in the hind legs and spreads over the body). No seasonal incidence has been noted. The disease causes death usually within 24-48 hr of the onset of symptoms. Virus isolated from such diseases has been inoculated into mice brains, causing paralysis and death in the mice. Virus has been recovered from the mouse brain, spinal cord, cardiac muscle, and in lesser amounts from the liver and serum [47]. In 1 experiment [47], a monkey was inoculated intracerebrally and after developing a fever became moribund. Various brain tissues were shown to contain virus; no electron micrographs have been published.

E. Duvenhagé and Bolivar Viruses

There is evidence that Duvenhage and Bolivar viruses are also neurotropic in a rabies-like fashion [48, 49].

F. Kotonkan and Obodhiang

Although there are apparently antigenic similarities between kotonkan and Mokola, the pathobiology of kotonkan (or Obodhiang) infections shows little similarity to rabies [54]. Mouse brain-adapted kotonkan injected intracerebrally into mice causes death by 14 days postinoculation. Electron micrographs of infected mouse brains [54] have shown virions budding from the plasma membranes of neurons and their processes and accumulating in extracellular spaces. Some infected cells contained inclusion bodies [54].

It has been suggested that kotonkan resembles bovine ephemeral fever virus, both physically (cone shape and overall dimensions) and possibly from a pathobiological viewpoint [54]. Serologically, however, they are distinct [90]. In a sampling of northern Nigerian cattle, 124 out of 129 trade cattle and 29 out of 29 dairy cattle, as well as other animals, were found to possess antibodies against kotonkan [54]. Southern Nigerian cattle lacking antibodies to kotonkan upon importation to the north have been observed to come down with a 3-day illness (as in BEFV infections) and to produce antibodies against kotonkan [54].

Since kotonkan was isolated from *Culicoides* gnats [54] and serum containing kotonkan antibodies has been obtained from various animals, there is a strong suspicion that an insect vector may be involved in virus transmission. Conclusive proof has not been obtained, however, although it has been shown that the virus grows well in *Aedes* mosquito cell lines [60]. The same may well apply for Obodhiang infections.

Injection of either kotonkan or Obodhiang by intracerebral or intramuscular routes into monkeys or dogs has been found to induce low antibody titers but no rabies-like disease (G. Tignor, personal communication, 1976).

G. Joinjakaka, Mossuril, Kamese, Kwatta, and BeAn 157575 Viruses

Very little is known about the transmission or pathobiology of these viruses isolated from insects. Mossuril has been found to be pathogenic only in newborn mice after intracerebral inoculations [21]. Infected brain cells showed typical rhabdoviruses budding from the cell membrane with virions present in interstitial spaces and occasionally within the cytoplasm of the cells [20]. Mouse brain adapted Kwatta inoculated intracerebrally into newborn mice [19] caused death in 3 days, although adult mice inoculated intracerebrally usually survived infection. Some adult mice so inoculated developed hind leg paralysis and died 10-13 days postinoculation. Newborn or adult mice inoculated intraperitoneally did not die. Karabatsos and associates found few cytoplasmic inclusions in Kwatta-infected mouse brain cells [20]. However, they did observe virus budding from cell membranes as well as virions enclosed in cytoplasmic vacuoles.

It has been shown that Joinjakala can multiply in *A. aegypti* mosquitos [30]. It also causes paralysis and death after intracerebral (but not intraperitoneal) inoculation in newborn mice (but not adult mice). No evidence of transmission from mosquitos to mice has yet been demonstrated [30]. It is not known if any of these viruses cause animal diseases in nature.

H. Klamath and Barur Viruses

Both these viruses have been isolated from rodents. Barur has also been obtained from ticks [30], which suggests that an arthropod may be involved in its transmission. Little is known about the pathobiology of either virus, although using mouse brain passaged virus it has been shown that Klamath gives rise to intracytoplasmic particles in infected mouse brain cells (as in rabies infections) and little extracellular virions [91]. Barur in newborn or adult mice has been observed to cause sickness and death [30]. No electron microscopy has yet been reported.

I. Mount Elgon and Kern Canyon Bat Viruses

The ability of these 2 viruses (isolated from Chiroptera) to infect mice has been studied. Again mouse brain-adapted virus was used in both cases. Mount Elgon bat virus (MEBV) injected intracerebrally into newborn mice or guinea pigs caused encephalitis and death, whereas similarly inoculated adult mice and newborn rats usually survived [29]. These authors claimed that MEBV inoculated into mosquitos can multiply in these insects although no evidence of transmission to mice was observed. In brain tissue of infected newborn mice, MEB virions have been observed only in extracellular spaces [92]. Essentially similar results have been observed with Kern Canyon virus and again in infected mouse brain tissue only extracellular virions were observed [93].

The mode of transmission to or from bats for either MEBV or Kern Canyon virus (KCV) is unknown; other than an insect vector, it is possible that aerosol infections may be involved in the transmission processes.

J. Flanders, Hart Park, and Navarro Viruses

Each of these viruses have been isolated from birds. Both Flanders and Hart Park have also been obtained from insects. Nothing is known about the transmission of any of these viruses or the pathobiology in the birds or the insects. In the initial isolation of Flanders virus, intracerebral inoculation into newborn mice gave rise to a disease characterized by lethargy, convulsions, spastic paralysis, and death [31]. As with vesicular stomatitis viruses, the presence of Flanders virus in brain tissue was found to be confined to extracellular spaces [32]. Jensen and collaborators have likewise shown that Hart Park gives rise to extracellular virus in infected mouse brain tissues [94].

Navarro-infected mouse brain cells have been shown [20] to exhibit mostly intracellular virus particles and prominent cytoplasmic matrices (as in rabies infections). Less frequently virus was observed protruding from cell membranes into interstitial spaces [20].

K. Bovine Ephemeral Fever Virus

Isolates of this virus have been obtained between 1967 and 1969 from cattle diseases in Australia, Japan, and South Africa. The virus causes a 3-4 day illness in cattle involving sudden fever with temperatures of up to 107°F, shivering, stiffness, loss of appetite, and lameness which sometimes becomes permanent but often leads into a rapid recovery. Serous nasal and ocular discharges as well as mouth drooling also occur, in addition to leukocytosis and neutrophila [95]. Reduction in milk production is commonly observed although it usually returns in about 2 weeks. Frequently the cattle appear utterly dejected [23].

The morphology of the virus is bullet to cone shaped, although how the cone shapes arise is not known [90].

Cattle have been experimentally infected by inoculation of blood from a diseased animal. Incubation periods following intravenous inoculations are usually of the order of 2-4 days depending on the amount of inoculum and the virulence or passage number of the stock [22]. The injection of crude white blood cell suspensions has been reported to be more effective than serum or red blood cells. Although injection of nasal secretions has been reported to induce disease, experiments attempting direct transmission by swabbing the gums or nasal membranes with saliva or nasal discharge from febrile animals have not been successful [22]. Except for Phillipine buffalos, other animals have not been experimentally infected (see Burgess [22] for a review).

There is some evidence for an insect vector involved in disease transmission. During epidemics of the disease it has been shown that disease incidence followed surface wind patterns and was influenced by rainfall and distribution of the rivers. However, direct transmission from insects or insect extracts has not been demonstrated [22, 96]. It is believed that the most likely site of multiplication of the virus is in the vascular epithelium. Virus in leukocytes is probably a result of their phagocytic activity. It has been suggested that the joint disorders (lameness, etc.) are caused by vascular damage, and that edema and emphysema may account for the respiratory distress of infected animals [22]. Whether nerve tissue damage is responsible for the occasional permanent disturbance of locomotion is not known.

Although newborn to 6-day-old mice succumb to bovine ephemeral fever virus (BEFV) infections (after either intracerebral or intraperitoneal inoculation), 21-day-old mice are not killed. Virus in brain tissue is seen in extracellular spaces but not in intracellular organelles [90].

L. Sigma

The only arthropod rhabdovirus which has been studied to any significant extent within the arthropod is sigma [99]. Sigma has been shown to be transmitted vertically in the germ lines of both male and female *Drosophila* [35, 36], and viral inclusion bodies and virions have been observed in association with germ cells but not in the somatic cells of male *Drosophila* [34, 97]. The sensitivity of *Drosophila* to carbon dioxide exposure has been studied by looking for a specific *Drosophila* genome site responsible for the phenomenon [98, 99]. The carbon dioxide sensitivity is not unique to sigma, since different strains of VSV adapted to *Drosophila* will also induce this effect [100, 101]. Recently a preliminary characterization of *in vitro* cultured sigma has been reported [310].

M. VHS and Egtved Viruses

Infections of rainbow trout by viral hemorrhagic septicemia (VHS) or Egtved viruses have been observed to peak in the spring and then slowly to dissipate in the summer, with only sporadic autumn or winter outbreaks [102]. Diseased fish exhibit a loss of appetite and become edematous with exophthalmia and subperitoneal hemorrhages, with the skin darkening to a purple or black, especially on the head and abdomen [102]. Other forms of the disease occur in which there are no hemorrhages, exophthalmia, or dropsy, but the fish become anemic and show motor disorders resulting in sudden violent twisting of the fish in the water [102]. Grayling (*Thymallus thymallus*), whitefish (*Coregonus* sp.), and brown trout (*Salmo trutta*) may also be naturally and/or experimentally infected with the virus [102, 103]. Outbreaks of the disease are lengthened by keeping fish in cold water (less than 8°C). At high temperatures the disease is either asymptomatic or disappears [102, 104, 105]. Virus has been isolated from the spleen and kidneys of infected young fry and fingerlings. At least 2 serologically distinct VHS or Egtved virus types have been reported [43].

N. Oregon Sockeye Disease, Sacramento River Chinook Salmon Disease, and Infectious Hematopoietic Necrosis Viruses

Various rhabdovirus isolates (from the sockeye salmon, *Oncorhynchus nerka;* chinook salmon, *Oncorhynchus tshawytscha;* and rainbow trout, *Salmo gairdneri*) have been obtained [103]. These isolates appear to be antigenically related to each other [106]. In the disease course for infectious hematopoietic necrosis virus (IHNV), initial lesions occur in the spleen and kidneys, and eventually necrosis is observed in liver, pancreas, and granular cells of the lamina propria of the gut [107, 108]. Typically fish become anemic. Kidney dysfunction may be the primary cause of death [108]. The potential for epizootics occurs maximally at 10°C and, although in cell cultures the virus replicates up to 18°C [109], the disease does not occur at water temperatures greater than 15°C.

In the case of Oregon sockeye disease virus (OSDV), the disease occurs in the anadramous and landlocked sockeye salmon and is characterized by normal fish behavior until near death, when about 90% of the fish become unresponsive and 10% become hyperactive, swimming on their sides in circles [103]. In the terminal stages there is abdominal swelling, pale gills, and hemorrhaging at the throat. The swim bladder and peritoneum become hyperemic, and the spleen pale. Kidneys, liver, spleen, and pancreas, as well as other tissues, are also affected [103].

The Sacramento River chinook disease virus causes an acute disease with high mortality in young chinook salmon. Rainbow and cutthroat trout (*Salmo clarki*) of up to 1 month old are also susceptible [103]. Epizootics of the disease occur between 8 and 13°C, but when the water temperature rises above 13°C the disease subsides. Diseased fish do not feed; they become darker from hemorrhaging and have massive vasular damage on the head, pale gills, distended abdomen, as well as necrosis of the spleen and kidneys. The pancreas becomes vacuolate and necrotic, while adrenal cortical tissue is always necrotic [103].

Darlington and collaborators [110] have shown by electron microscopy virions in interstitial spaces and occasionally in cytoplasmic vacuoles.

O. Spring Viremia of Carp Virus

Spring viremia of carp virus (SVCV) causes early spring epizootics of a dropsy disease. The pathological process starts in the carp central nervous system and peripheral nerves [111]. The fish develop distended abdomens, with protruding scales, pale gills, and dermal vesicles which rupture and ulcerate. The ventral areas are exanthematous, and some fish exhibit exophthalmia. The kidneys and spleen become enlarged, and there is hemorrhagic necrosis of the intestine [103].

Another disease is swim bladder inflammation of carp (aerocystitis). This disease occurs in early summer and its symptoms at 14-17°C resemble those of the dropsy disease, with strong swim bladder involvement. Along with edema, there is extensive petechiation in the swim bladder wall, brain, and pericard. Virus can be isolated in infected fish from the liver, swim bladder, intestine, kidneys, heart, muscle, and brain (in decreasing order of titer). It is possible that both diseases are caused by SVCV [40].

P. Pike Fry Rhabdovirus

Pike fry rhabdovirus causes an acute disease of the fry, involving hemorrhagic bilateral lesions on the trunk (hence the name "red disease"), hydrocephalus, necrobiosis of the kidney tubules, and hemorrhages of the connective tissue of the muscles [37]. The virus is probably present in the water in late winter and immediately infects the eggs when the pike spawn. Virus can be seen by electron microscopy in the hematopoietic tissue of the kidneys, but not in the skin, muscles, or nervous tissue. The mitochondria of the kidneys and blood cells can be seen to be swollen [112].

For some of the fish rhabdoviruses, the most probable transmission course is through water, whereby virus released from 1 animal infects another fish or its eggs. Other mens of transmission, such as feeding off diseased fish, may also be involved in disease transmission. Since most of the diseases are seasonal, the possibility exists that asymptomatic carriers of the disease are involved. The particular temperature optima for epizootics of the various virus types might reflect optimal temperatures for virulent growth of the viruses and, or an interrelationship with the efficiency of the host defense mechanisms (ability to produce interferon, antibody induction or cell mediated immunity, etc.).

Q. Summary of Rhabdovirus Transmission and Pathobiology

There is little definitive evidence on the transmission of rhabdovirus infections from subject to subject, except for rabies and sigma viruses. Unanswered questions remain concerning the possible existence of vectors (particularly arthropod ones) as well as asymptomatic carriers for almost every rhabdovirus type discussed above. What happens to the virus between outbreaks of vesicular stomatitis, bovine ephemeral fever, or the salmon, trout, pike, or carp diseases?

The observation that a disease in fingerling sockeye salmon in 1946 in fish hatcheries in Washington State reoccured 4 years later in 1950 suggests that fish from the initial outbreak may have carried the virus since this species of salmon migrates in 4-year cycles [113].

Concerning the pathobiology of the various rhabdoviruses, in experimental or field animals, it has been shown that the VSV group of viruses is primarily epitheliotropic and that, although visceral and neural tissue involvement occurs in experimental infections (depending on the route of inoculation, animal type and age, etc.), virions seen in those tissues are mostly extracellular. By contrast, rabies experimental infections can be characterized as primarily neurotropic with predominantly intracellular virions. Later stages of the infection result in virus production at plasma membranes of certain organs (notably salivary gland tissues).

The presence of eosinophilic inclusion bodies in rabies-infected cells (Negri bodies, especially in Ammon's horn, the cerebral cortex, medulla, and Purkinje cells of the cerebellum) often occurs in cells which exhibit little other cytological change. Most other rhabdoviruses also give rise to inclusion bodies at some stage of their growth cycle. It is believed that these inclusion bodies represent nucleocapsid material. Is there a defect in rabies virus maturation in differentiated nerve cells which leads to such nucleocapsid accumulation? The observation that rabies (but not VSV, Mokola, or Klamath) production is inhibited in enucleated cells [114, 312] suggests that there are particular cell functions required for rabies infections that are not required for VSV infections. If a particular cellular function is required, it is conceivable that differentiated cells may provide lesser or greater amounts of that function, depending on the cell type, etc.

Comparisons made from experimental pathobiology studies must be considered in the light of virus origin (e.g., adaptation by mouse brain passage), natural vs. experimental host (e.g., fox as opposed to mouse), route of inoculation, age of host, etc. Bearing these considerations in mind, Murphy [3] has suggested that the pathobiology of Mokola, Lagos bat, and Duvenhagé infections are essentially similar to rabies (Nigerian horse virus has not been studied in sufficient detail to determine whether it should be included). Klamath may also be similar to rabies [3]. However, until more is known about the molecular and cellular biology of each rhabdovirus, the meaning of these observed tropisms is obscure.

IV. Structural Components of Rhabdoviruses

A. Shape

As indicated in Table 5.1, rhabdoviruses vary considerably in length; all however appear to have the same basic cylindrical structure (Fig. 5.1). Electron microscopy has shown that the viruses possess outer spikes (6-10 nm in length) situated on a bilayer lipid membrane, and a spiral nucleocapsid core possessing about 35 turns. Similar basic structures have been obtained for both the infectious bullet-shaped particles (B particles) and defective truncated particles (T particles)—except in the case of the smallest T particles, which are almost round [115, 116].

Vesicular stomatitis viruses (Indiana and New Jersey serotypes) have been extensively studied by electron microscopy [117]; see the reviews by Howatson [6] and Knudson [4]. Both serotypes are 180 ± 10 nm long and 65 ± 10 nm wide with 1 rounded and 1 flattened end. An indentation is frequently observed at the flattened end; although how much this and other aspects of the structure are cuased by the preparative techniques involved in electron microscopy is a moot question (see Francki [1] for a discussion of this point).

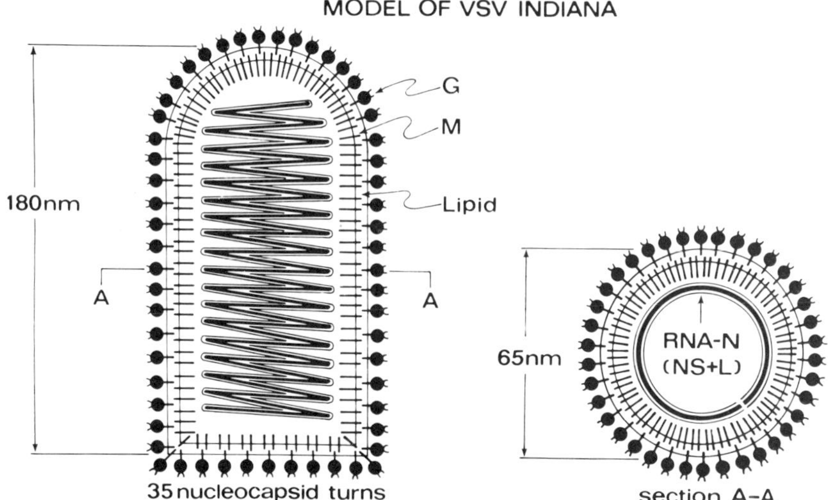

Fig. 5.1 A model of the structure of VSV Indiana.

B. The RNA Genome of Rhabdoviruses

The genome of vesicular stomatitis virus (Indiana serotype) has been shown by nuclease digestion to consist of a single-stranded RNA molecule of 3.8×10^6 daltons molecular weight [118]. The electrophoretic mobility or sedimentation coefficient of the viral RNA is not changed by boiling, formamide treatment, or self-annealing [119-121], indicating that it is a single-stranded, noncomplementary RNA. Of 9 other rhabdoviruses examined by polyacrylamide gel electrophoresis, VSV New Jersey, Cocal, Chandipura, Mount Elgon bat, Kern Canyon, rabies, Mokola, Egtved, and pike fry rhabdovirus, were all found to possess genomes indistinguishable in size to that of VSV Indiana [122, 123]. To that list can be added SVCV, Lagos bat, and Piry viruses (D. H. L. Bishop, unpublished observations, 1975). Schaffer and associates [124] have obtained a mutant of VSV New Jersey which apparently possesses a larger virion RNA as evidenced by gel electrophoresis. The significance of this apparently larger RNA is not known.

By the convention of describing messenger RNA as a plus strand, the viral RNA of VSV is thereby defined as a minus strand since it does not itself function as a messenger RNA; rhabdoviruses are thus described as negative-stranded viruses [125].

The 5′ terminal sequence of VSV Indiana has been shown to be pppApCpGp. . . consistent with the postulate that the RNA is linear and not circular [126]; the 5′ end is neither capped nor methylated. The base ratios of rabies, pike fry rhabdovirus, and VSV Indiana are given in Table 5.3 and, as can be seen, all contain a large amount of uridylic acid [118, 123, 127].

The RNA species of defective T particles of VSV and other rhabdoviruses (rabies, Mt. Elgon bat, Cocal, Chandipura, KCV, and VSV New Jersey) are

Table 5.3 Base Ratios of Some Rhabdoviruses

Virus	A	C	G	U
Rabies (HEP)[a]	26.4	22.5	21.6	29.5
Rabies (ERA)[a]	25.5	21.7	24.1	28.7
Rabies (PM)[a]	26.7	22.4	22.8	28.1
VSV Indiana[b]	27.4	21.5	21.0	30.3
Pike fry[c]	24.9	22.5	20.2	32.4

[a] Aaslestad and Urbana [127].
[b] From Repik and Bishop [118].
[c] From Roy and associates [123].

smaller than their corresponding B particle RNAs and possess various molecular weights [2, 122, 128].

No satisfactory nomenclature exists for categorizing the various T particles; in the literature there are classifications, such as long T (LT), T, short T (ST), heat-resistant virus T particles (HR T), VSV-11, and VSV-111. With one T-particle preparation (VSV-111), it has been shown that the RNA is unique and, by nuclease digestion, possesses a molecular weight of 1.1×10^6 [118]. Other T-particle preparations may be unique or mixtures of various sizes [119] or even contain plus and minus strand types [129]. Using elegant hybridization procedures, Leamnson and Reichmann [130] have mapped various T-particle RNA species to the various viral messenger RNA species.

C. Genome Homologies of Rhabdoviruses

The sequence homology between a variety of rhabdoviruses has been studied [131] by hybridizing labeled viral RNA of 1 virus to an excess of unlabeled viral complementary (vc) RNA of another virus and determining the ribonuclease resistance of the hybrids after digesting with ribonuclease A (nonstringent conditions) or ribonucleases A and T_1 (stringent conditions). The results obtained for a variety of rhabdoviruses are presented in Table 5.4 and are discussed next. For controls, labeled viral RNA was self-annealed or annealed to uninfected cell nucleic acids. For all the virus RNA preparations examined, self-annealing or annealing to uninfected cell nucleic acids gave essentially the same level of ribonuclease resistance (stringent or nonstringent conditions). For most viral RNA preparations this ribonuclease resistance did not significantly differ from the ribonuclease resistance of the unannealed viral RNA. However for some viruses (notably SVCV and rabies viruses), the ribonuclease resistance increased significantly upon self-annealing and this increase varied from preparation to preparation. Occasional preparations of rabies RNA gave only 4% ribonuclease resistance upon self-annealing, at other times as much as 40% resistance has been obtained (M. Smith and D. H. L. Bishop, unpublished results, 1975). Gel electrophoresis of unannealed SVCV RNA (or rabies virus preparations) have shown only 1 species of RNA (3.8×10^6 MW); electrophoresis of the self-annealed species has not been conclusive because of fragmentation which occurred during the long annealing at 60°C (D. H. L. Bishop, unpublished observations, 1975). These results suggest that under certain, as yet ill-defined, conditions virions containing plus and/or minus strands may be released from infected cells. Whether the plus strands represent replicative templates encapsulated into virions is not known but is an intriguing possibility. We would also like to know if the plus stranded

Table 5.4A Genome Homologies of Rhabdoviral RNAs: Ten Rhabdoviruses[a,b]

	Cell RNA species								
Viral RNA	Uninfected	Rabies ERA	SVCV	PFR	Chandipura	Cocal	VSV Indiana	VSV New Jersey	Piry
Rabies ERA	14	98(98)	15	19	18	19	19	ND	17
Lagos bat virus	11	27(4)	3(4)	5(3)	ND	5	13(10)	ND	5
Mokola	1	26	0	ND	ND	ND	ND	ND	ND
SVCV	26	29(22)	80	22	22	22	26	21	25
PRF	3(2)	6	5	100(96)	5	4	3	3	3
Chandipura	7(4)	8	5	5	100(96)	10(4)	15(6)	16(4)	ND
Cocal	11(4)	6	6	2	13(3)	99(97)	36(10)	24(5)	15(7)
VSV Indiana	5(3)	10	3	2	15(6)	22(10)	97(100)	17(5)	13(6)
VSV New Jersey	8(5)	6	12	12	14(4)	14(5)	20(5)	100(100)	16(7)
Piry	5(3)	2	17	3	14(5)		13(5)	10(5)	97(98)

[a] [^{3}H]-Viral RNA was annealed to uninfected or infected cell RNA and the ^{3}H resistance determined by pancreatic ribonuclease (nonstringent conditions). Values in parentheses are the resistancies after ribonuclease T_1 and pancreatic ribonuclease digestion (stringent conditions).
[b] Some of the data of Repik and associates [131] are included. ND not determined.

Table 5.4B Genome Homologies of Rhabdoviral RNAs: Various Fixed Rabies Strains[a]

	Viral RNA species				
Cell RNA	ERA	CVS	HEP	LEP	PM
Rabies ERA	98(89)	97(95)	97(100)	100(100)	94(100)

[a] Conditions as in part A.

virions or their RNA are infectious since VSV viral RNA (minus stranded) extracted by phenol and SDS and freed from protein is not infectious [132]. The presence of plus and minus strand RNA has been shown to occur with certain defective particle preparations of VSV Indiana [129].

The homologous annealing data (annealing labeled viral RNA to unlabeled cell RNA infected by the same virus type) is presented in Table 5.4 (underlined values). Other than SVCV, homologous annealing for all viruses gave essentially complete hybridization using either stringent or nonstringent digestion conditions (Mokola or Lagos bat cellular RNA have not yet been obtained). This result indicates that within the infected cells there is more viral complementary (vc) RNA than viral RNA present.

Among the heterologous annealing results, the following observations can be made. First, by comparison to the results with uninfected cell nucleic acids, annealing was obtained between Mokola or Lagos bat viral RNAs and rabies infected cell RNA. These results suggest some homology between these 3 viruses although until the reverse hybridization has been performed, the results can only be considered as tentative.

Second, all the attenuated rabies species examined were indistinguishable from rabies ERA (Table 5.4, part II).

Third, among the vesicular stomatitis, Cocal, Chandipura, and Piry viruses some homology was observed between Cocal and VSV Indiana (notably under nonstringent conditions of digestion) with lesser amounts between the other viruses of this group [131]. What do these results mean? Repik and collaborators have suggested that there is some partial sequence homology between these viruses. Such sequence homologies might be expected if there were a common ancestor for these viruses from which the current isolates have evolved. To support the idea that partial sequence homology does exist between these viruses, analyses of the heterologous hybrids by cellulose column chromatography [133], which separates single-stranded from double- or multistranded RNA complexes, have indicated that some form of complex is generated between VSV Indiana viral RNA and the vcRNA of VSV New Jersey, Cocal, or Chandipura viruses but not between VSV Indiana and uninfected cell RNA or between rabies viral and the vcRNA of VSV Indiana (P. Repik and D. H. L. Bishop, unpublished observations, 1975). Further studies (such as oligonucleotide sequence analyses) need to be performed to confirm those results.

D. The Structural Proteins of Rhabdoviruses

As indicated in Table 5.5, 3 major structural proteins are present in VSV Indiana, VSV New Jersey, Cocal, Chandipura, Piry, SVC, KCV, and PFR

Table 5.5 Molecular Weight ($\times 10^{-3}$) Estimation of Rhabdovirus Virion Proteins[a]

Virus	L[b]	G	N	NS[c]	M1	M2	Other[d]	Comments	References
VSV Indiana	149-160	65 ± 4 (	50 ± 4	38-54(P)[e]	–	27 ± 2	(50,48)		[134,135,136, 138]
Cocal	149-160	67 ± 4	50 ± 4	38-54(P)	–	25 ± 2	ND		[[a], 136,138]
VSV New Jersey	148-161	63 ± 4	50 ± 4	40 ± 4(P)	–	24 ± 2	ND		[135,136,138]
Chandipura	151-162	69 ± 2	50 ± 4	42 ± 4(P)	–	24	ND		[135,138]
Piry	150-160	74 ± 1	52 ± 2	38-54(P)	–	24 ± 1	ND		[135,138]
KCV	160 ± 10	70 ± 10(P)	50 ± 5(P)	–	–	25 ± 3	ND	N and G phosphoproteins are low specific activity (137)	[[a],137,138]
Rabies	160 ± 10	75 ± 10	62 ± 4(P)	–	40 ± 8	25 ± 4	50 ± 5	50 = Large fragment of N (137)	[[a],138,140,141, 164]
Mokola	160 ± 10	73 ± 5	64 ± 4(P)	–	40 ± 4(P)	25 ± 3	46(P)		[138,141]
LBV	160 ± 10	65 ± 5	61 ± 4(P)	–	40 ± 4(P)	25 ± 3	45(P)		[138,141]
Obodhiang	160 ± 10	87 ± 5	52 ± 4	–	32 ± 3	21 ± 2	38	No phosphoprotein analysis	([a])
Kotonkan	150 ± 10	87 ± 5	52 ± 4	–	32 ± 3	21 ± 2	ND	No phosphoprotein analysis	([a])

PFR	160 ± 10	85 ± 10	50 ± 5	40-50(P)	—	23 ± 5	ND		[123]
SVCV	160 ± 10	85 ± 10	50 ± 6(P)	40 ± 5(P)	—	23 ± 2	ND		[a,138,141]
IHNV	157 ± 10	72 ± 10	40 ± 4(P)	—	23 ± 3(P)	20 ± 1		f	[a,139]
Egtved (VHS)	157 ± 10	74 ± 10	41 ± 4(P)	—	22 ± 2	19 ± 1	ND	f	[a,139]

[a]Molecular weight estimates are the average of recent reported values where available, as well as unreported values obtained by J. F. Obijeski and D. H. L. Bishop. Since different gel systems give different results for the same proteins [135,136], the averages quoted also reflect these variations.

[b]Earlier reports of an L protein with a 190×10^3 MW for VSV Indiana are probably too high.

[c]In some systems the NS protein mobility varies considerably with the gel system used; hence a range is quoted.

[d]Only clearly defined minor or major protein bands are included in this column; minor bands that have not been clearly defined are referred to as ND.

[e]P stands for phosphoprotein.

[f]VHS and IHNV proteins recently reported by Hill and associates [210] differ from those found by McAllister and Wagner [139]; the latter's results are included in this table; we have observed similar results.

rhabdoviruses [123, 134-136]. These are a glycoprotein (G), a nucleoprotein (N), and a matrix or membrane protein (M). In addition, each possesses minor amounts of other proteins, a large protein (L) and (except possibly for KCV), the so-called nonstructural (NS) phosphoprotein [134, 135, 137, 138].

Rabies, Mokola, Lagos bat virus, and 2 fish rhabdoviruses (Egtved and IHNV) possess a glycoprotein (G), a nucleoprotein (N), and 2 putative matrix or membrane proteins (M1 and M2) [134, 135, 138, 139]. Two glycoproteins described for rabies preparations [140] have not been confirmed [141]. These 5 viruses contain minor amounts of other proteins, including an L protein [135, 138, 139, 141]. For rabies virus preparations a large fragment of the N protein (the nonphosphorylated N-LF) is frequently observed [137]. Other ill-defined minor proteins have been found in preparations of these viruses, although whether they are viral or host contaminants is not known. It appears that for all 5 viruses the nucleocapsid protein is phosphorylated, and for IHNV, Mokola, and Lagos bat viruses the M1 protein is also phosphorylated, whereas apparently for Egtved and rabies it is not [139, 141].

Obodhiang and kotonkan possess L, G, N, M1, and M2 proteins, although which protein is phosphorylated is not known (J. F. Obijeski and S. Bauer, personal communication, 1975). Other viruses have not been studied.

Although the structural polypeptides equate to almost all the genetic information of the virus (see Table 5.6), other proteins and various enzyme activities have been observed to be present in rhabdovirus preparations. In our laboratory we consistently find in VSV Indiana and other rhabdovirus preparations minor but distinct protein bands possessing an electrophoretic mobility slightly faster than that of the N protein. Such proteins, A and B, described by Bishop and Roy [142], are associated with the nucleocapsid when viral cores are made (i.e., lacking lipid and the G and M proteins). Other investigators have observed similar bands (J. F. Obijeski, personal communication, 1976). We do not know if these protein species are derived from other viral structural proteins or are host proteins packaged during maturation. Host proteins present in virus preparations could be considered as contaminants, but we believe that they should rather be considered as functional entities, the amounts and requirements for these entities may well be variable depending on the host cell type, its nutritional status, etc.

The presence and contribution of minor amounts of host or viral proteins should be considered in light of how many protein molecules are present per virion. Estimates of the molecular weight and numbers of virion protein molecules for VSV Indiana are presented in Table 5.6. For each protein, the estimate of molecular weight is based on its observed electrophoretic mobility in SDS polyacrylamide gels. For the phosphorylated NS protein, this mobility has been shown to vary with the electrophoretic system used [135] and it is possible that the true polypeptide size of the VSV Indiana NS protein is

Table 5.6 Proteins of Vesicular Stomatitis Virus[a,b] Indiana

Protein species	Percentage of total protein	Molecular weight	Daltons of protein per virion	Number of proteins per virion
L	2.5 ± 0.3	160,000	8.6×10^6	52
G	34.2 ± 1.0	67,000	120×10^6	1791
N	29.8 ± 1.0	52,000	105×10^6	2015
A	0.8 ± 0.2	50,000	2.9×10^6	58
B	1.6 ± 0.2	48,000	5.6×10^6	117
NS	2.3 ± 0.2	40,000[c]	8.0×10^6	200
M	28.8 ± 1.0	25,000	101×10^6	4052

[a]Average of 3 determinations.
[b]Revised for total virion RNA weight of 3.8×10^6 daltons [142], and average viral protein MW (Table 5.5).
[c]See text for a discussion of the MW value; the VSV Indiana NS protein is a phosphoprotein.

20-30 $\times$ 10^3 daltons [143]. That would bring it into the range of the Mokola, LBV, and IHNV phosphorylated M1 proteins. The mobility of the G protein probably also does not reflect its true polypeptide size because of the carbohydrate moiety which may account for as much as 11% of the G protein mass (see p. 198). The calculations shown in Table 5.6 of the number of protein molecules per virion is based on that described previously [142], a correction having been made for a revised molecular weight of the viral RNA [118]. It should be noted that if protein species are present in virions to the level of 1-20 for large protein species (e.g., greater than 50 $\times$ 10^3 MW) or 1-60 for smaller species, these would be undetected by the techniques of resolution currently employed. Also this point does not take into consideration the fact that not every virus particle is infectious [144].

It should be noted that the ratio of G to N to M is approximately 1:1:2. Although a plausible model of VSV virus structure and polypeptide ratios has been postulated based on other molar ratios of these 3 proteins [145], we have noted that the ratio of viral proteins varies from laboratory to laboratory, and in our laboratory variations have been observed using different types of media and different host cells for growing virus.

Defective particles of VSV and Chandipura viruses contain the same assortment of viral proteins as their respective infectious particles [135, 146].

E. Functional Considerations of the Structural Proteins of Rhabdoviruses

1. The G Protein

The viral G protein is located on the outer surface of the virus particle and constitutes about 1/3 of the viral protein mass. It can be specifically or partially removed by treating virus particles with proteases, leaving a non-glycosylated, hydrophobic tail embedded in the membrane [147]; bromelain, pronase, trypsin, and chymotrypsin (etc.) can be used to remove the G polypeptide [148, 149], although the G protein of some rhabdoviruses may not be removed completely by certain proteolytic enzymes (e.g., bromelain for Chandipura; P. Repik and D. H. L. Bishop, unpublished observations, 1975). Concomitant with the removal of the G protein there is a loss of infectivity (often to 0.0001% of the original titer, [150]).

Loss of 99% of VSV Indiana infectivity has been reported by neuraminidase removal of the terminal neuraminic acid from the G protein [2, 151]; resialylation using a sialyl transferase restores a considerable portion of the infectivity [151, 152]. Such results suggest that the carbohydrate is crucial for infectivity, perhaps by interacting with some positively charged receptor on the cell surface.

There are claims for 1 to 4, or more carbohydrate chains per G polypeptide ([153-155]; M. Kelley and S. U. Emerson, personal communication, 1976). The amount of carbohydrate per G polypeptide has been estimated to be about 11% of the G polypeptide mass [154]. McSharry and Wagner [156] determined for VSV Indiana grown in mouse L-cells that the following sugars were present on the G protein: neuraminic acid, 37%; glucosamine, 31%; glucose, 21%; galactose, 5%; mannose, 5%, with little detectable galactosamine or fucose. Etchinson and Holland [154] have found different amounts of various carbohydrates to those reported by McSharry and Wagner for VSV grown in BHK cells. Per polypeptide they report the presence of 1 fucose, 7-8 mannose, 6 galactose, 9-11 N-acetylglucosamine, 4-6 neuraminic acid, 1 N-acetylgalactosamine, and 3-5 glucose residues. Variation in carbohydrate composition might be caused by techniques employed in the analyses or could be interpreted to indicate that addition of carbohydrates to the G polypeptide is a variable, host-determined processing function. In support of this idea, Schloemer and Wagner [157] found that VSV grown in *A. albopictus* mosquito cells lacks sialic acid and possesses a low infectious to particle ratio as well as low hemagglutinin specific activity. They showed that in vitro addition of neuraminic acid increased 100-fold both infectivity and hemagglutination activity.

It is not known if every G polypeptide is equally glycosylated or what structural attribute of the G polypeptide determines glycosylation at one or another amino acid residue. The carbohydrate sequence on the various chains is also not known.

Low hemagglutinin activity, with goose erythrocytes (assayed optimally at 4°C and pH6), has been reported for several rhabdoviruses [158]. Hemagglutination is an attribute of the viral surface components. Specifically, the neuraminic acid part of the G protein carbohydrate appears to be involved [151, 157]. It has been claimed that isolated VSV glycoprotein also exhibits hemagglutination activity and that this can be blocked using VSV antiserum (McSharry and Choppin, quoted in Wagner [2]).

By preparing specific monocomponent antiserum against VSV G protein [159] or the various major structural polypeptides of VSV Indiana [160], it has been conclusively shown that the G protein is the only viral protein which induces neutralizing antibody. Antibody to glycoprotein was also shown to confer immunity to VSV infection in mice [160]. The same probably holds true for other rhabdoviruses, as indicated by the results of Wiktor and collaborators [161] using rabies virus.

Functional considerations therefore indicate that the viral glycoprotein is the outer component of the virus and that it is crucial for interacting with host cells, primarily through the initial involvement of the neuraminic acid residues of the G protein. Since there may be more than one way a virus may enter a cell, e.g., viropexis [162] or virus fusion [163], and differences could exist for entry into nerve cells as opposed to epithelial cells, more than 1 mechanism of virus interaction with a cell may be possible. We have observed that spikeless particles of VSV Indiana, Chandipura, VSV New Jersey, and Cocal viruses are infectious (about 0.0001% of the original virus infectivity) and that this infectivity is not inhibited by antiserum prepared against whole virus or purified glycoprotein [150]. Whether this infectivity is mediated by viropexis or by direct membrane fusion is not known.

2. *The M Proteins*

Very little is known about the function of the various M proteins of rhabdoviruses; although they constitute about 1/3 of the viral protein mass and are described as the matrix or membrane proteins, there is no certain information as to their site within the virus architecture. When spikeless particles of VSV Indiana [148], rabies virus [163], or the fish rhabdoviruses [123, 139] are prepared by protease treatment, only the G is removed.

The M1 protein of Mokola, Lagos bat, and IHNV are phosphorylated [138, 139, 141], although whether these proteins should be considered as NS-type proteins is a moot question. It has been claimed that labeling of intact VSV particles by radioactive iodine in the lactoperoxidase reaction [165] results in iodination of the G, L, NS, and M proteins. These results may be interpreted to indicate that some part of M, NS, or L are available to exterior iodination but since in these experiments the iodinated virus has not been repurified and it has been shown by several laboratories that the NS is tightly associated with the RNA-N protein complex [142] this interpretation may not be valid. Wagner and associates [166] also found that G and M are labeled in iodination experiments, but McSharry has obtained evidence which suggests that only G protein is on the outer surface of the particle (J. J. McSharry, personal communication, 1976).

Since protease treatment removes all but a tail of the G protein embedded in the lipid [147], one would like to know if the G protein penetrates to the interior of the virus and what molecular relationships exist between the G and M proteins. Also, one would like to know what relationship exists between the M protein and the lipid membrane and nucleocapsid of the virus [145].

3. *The N Protein*

The N protein, which accounts for about 1/3 of the viral protein mass, is closely associated with the viral RNA genome [167, 168]. The number of N protein molecules per RNA molecule (Table 5.6) is such that the RNA is probably completely covered with protein [142], which presumably accounts for the fact that the nucleocapsid is visible within the virus particle in a spiral configuration. The coverage of RNA by N is such that the 5′ terminus of the viral RNA is not accessible to nuclease removal (E. Hefti and D. H. L. Bishop, unpublished observations, 1975). What attributes of the N-RNA complex maintain the spiral configuration is not known.

The complex of RNA-N protein plus minor proteins NS and L is infectious, whereas the RNA-N complex minus NS and L is not infectious [167, 169, 170].

Removal of N protein from RNA can be obtained by SDS treatment, indicating that the N protein is hydrogen bonded and not covalently joined to the RNA. Although we would like to believe that the N protein is hydrogen bonded to the phosphate backbone of the RNA (as well as being hydrogen bonded to neighboring N molecules), there is no evidence to support this suggestion.

When antibody is raised to foot pad vesicle preparations of VSV (see p. 204), about 70% of the complement-fixing activity is directed against the viral G protein, 30% against the RNA-N (plus NS and L) complex, and less than 1.5% against the viral M protein [171]. No tests have been made to determine how much complement-fixing activity is directed against free NS or L proteins.

The N protein of rabies, Mokola, Lagos bat, Egtved, and IHNV appears to be phosphorylated [137-139, 141]. For rabies it has been suggested that the phosphorylation is confined to 1 end of the molecule, although it is not in fact known if rabies N protein molecules or other molecules of similar size are in fact phosphorylated.

Attempts to obtain soluble, undenatured N protein free from RNA have been unsuccessful, although low pH and guanidinium hydrochloride treatments can achieve removal (D. H. L. Bishop, unpublished data, 1975).

4. *The L Protein*

The L protein is not an aggregate of G, N, or M protein as shown by comparative analyses of tryptic or cyanogen bromide peptides [172, 173]. Together with the NS protein, in some as yet undetermined manner, it functions as the transcriptase [173, 174]. All rhabdoviruses so far examined possess an L protein (Table 5.5).

5. *The NS Protein of Certain Rhabdoviruses*

A virion NS phosphoprotein is present in some rhabdoviruses (VSV Indiana, VSV New Jersey, Cocal, Chandipura, Piry, PFR, and SVCV) but not others—unless one considers the phosphorylated M1 protein of Mokola, Lagos bat, and IHNV as an NS-type protein. Rabies, KCV, and Egtved do not appear to have a virion NS protein nor a phosphorylated M1 protein. Recent studies appear to indicate that the NS protein of VSV Indiana is an indispensable moiety of the transcriptase complex [174]. The fact that NS or a similar phosphoprotein is present in large amounts in infected cells is discussed on p. 258.

F. The Lipid Membrane and Possible Host Constituents of Rhabdoviruses

Lipids present in rhabdoviruses are derived from the host cell membrane from which the virus buds out, they constitute about 20% of the viral mass [175]. Apparently for rabies both the intracellular and extracellular virus possess similar lipid constituents [176]. Different rhabdoviruses grown in the same cell type have been shown [175] to possess essentially similar amounts of the various lipid components (phospholipids, neutral lipids, sterols, and glycolipids). The viral lipids were shown to differ somewhat from the average total lipid constitution of an uninfected cell and more closely resemble those of the host cell plasma membrane [175]. The same viruses grown in different host cells have been shown to contain different amounts of the various lipid constituents [175]. Since L cells synthesize dermosterol and not cholesterol, and since the synthesis of dermosterol can be repressed by cholesterol, a change in viral sterol composition was shown by growing VSV Indiana in mouse L-cells in the presence of cholesterol [177].

Suggestive evidence has been obtained by Dr. J. Lenard, Rutgers University (personal communication, 1975) that the distribution of phospholipids in VSV Indiana envelopes, like that of influenza virus [178], may be asymmetrical. If so, then, it is conceived that the viral G and M proteins possess different affinities for particular phospholipids or phospholipid configurations.

The presence of complement-fixing cellular antigens in preparations of VSV Indiana [179] has been demonstrated to result from a host glycolipid [180]. Cellular gangliosides have been found in virus preparations [181], and the mouse-specific histocompatability glycoprotein H2 antigen has also been observed in VSV grown in mouse cells [182].

Of various enzymes consistently observed in preparations of VSV and other rhabdoviruses (the transcriptase is discussed on p. 226), there is suggestive evidence that some are host derived. Enzyme activities which have been reported are a nucleoside triphosphatase [183], a nucleoside triphosphate phosphotransferase or nucleoside diphosphate kinase [183], a protein kinase [137, 184], a polyadenylation activity [185, 186], a 5′ nucleotide capping activity, and a nucleotide methylation activity ([187]; E. Hefti and D. H. L. Bishop, unpublished data, 1975). Of these enzyme activities, it has been clearly shown that the nucleoside triphosphatase and diphosphate kinase activities are released when a nonionic detergent is used to solubilize the viral membrane [188]. These results suggest that the enzyme may be part of the viral membrane. The possibility that cell membrane vesicles present in virus preparations contain these activities cannot as yet be excluded. Protein kinase activity has been observed in 11 out of 11 rhabdoviruses investigated (VSV Indiana, VSV New Jersey, Chandipura, Cocal, Piry, KCV, PFR, rabies, Mokola, LBV, and SVCV; unpublished data of F. Sokol, F. Clark, J. F. Obijeski, and D. H. L. Bishop, 1974). In each case the preferred substrate during an *in vitro* reaction was found to be the resident viral phosphoprotein(s) although the viral M proteins often also became phosphorylated (D. H. L. Bishop, unpublished data, 1974).

When VSV Indiana virion nucleocapsids are prepared from virus particles by nonionic detergent treatment, the protein kinase activity is not completely removed [189, 190]. In our hands, all of the endogenously templated activity remains with the nucleocapsids (D. H. L. Bishop, unpublished observations, 1974), even when 95% of the viral G, M, and lipids are removed by detergent and polyethylene glycol-dextran phase separation [142]. Since the viral protein kinase-specific activity varies depending on the host cell selected for virus growth and it is not associated with highly purified RNA-N or free L or NS protein preparations [189], this is suggestive evidence that the enzyme is probably a host protein component.

The minor proteins A and B of VSV Indiana [142] are good candidates for being the protein kinase, capping or methylation activities. Nothing is known about the proteins involved in the capping or methylation reactions; however, since they are found in a variety of viruses including cytoplasmic polyhedrosis virus [191], reovirus [192], and vaccinia [193], it seems likely that they are also host enzymes. The polyadenylation activity may be a viral transcriptase function (see p. 236).

G. The Viral Transcriptase Enzyme Activity of Rhabdoviruses

The known details of the synthesis of messenger RNA from the viral genome are discussed below. Suffice it to say that virion transcriptases have been demonstrated by Baltimore, Huang, and Stampfer [194] in preparations of VSV Indiana, this has been confirmed by others [119, 195] and extended to other rhabdoviruses including Kern Canyon virus (195); VSV New Jersey, Chandipura, Cocal, and Piry viruses [196]; pike fry rhabdovirus [123]; and SVC and Egtved viruses (P. Roy and D. H. L. Bishop, unpublished observations). We have obtained evidence of a low transcriptase activity in rabies virus preparations (5 pmole product/hr/mg virus protein, unpublished data) and conclusive evidence of transcription off the infecting viral genome in cycloheximide-treated rabies infected cells [197]. Other rhabdoviruses have not been examined. It is of interest to note that the optimal enzyme activity of the fish rhabdovirus preparations is around 18°C to 22°C ([123]; P. Roy and D. H. L. Bishop, unpublished observations, 1975). No transcriptase enzyme activity has been observed in standard defective T-particle preparations of VSV, e.g., VSV-111 [198, 199]; however, mixed T-particle preparations obtained from the HR strain of VSV Indiana [129] and VSV Indiana long T virions do appear to possess transcriptase activity [200].

V. The Relationships Between Rhabdoviruses as Shown by Serological Tests

A. Serological Tests, Reagents, and Antigenic Determinants

When antibody is raised against a virus or a viral component, the derived sera can be used for infectivity inhibition, complement fixation, and other tests to determine the relatedness of the initial virus antigens to antigens of other virus strains. As indicated in previous sections, for VSV Indiana, an infectivity determination by plaque assays probably usually involves an initial reaction between the neuraminic acids on the carbohydrate of the outer G protein and some host cell receptors. If antibody binds to VSV G protein and sterically inhibits this reaction, then it is believed that the ability of a virus to adsorb to a cell and initiate an infection will be blocked. Whether infection of all cell types by all rhabdoviruses involves the same molecular processes is not known, but, although without alternate evidence it seems probable, it should be borne in mind that the possession of neuraminic acid is a host-determined function and that VSV grown in *Aedes* cells lacks neuraminic acid [157]. Clearly for other virus types, other situations may also pertain.

It is generally assumed that the viral G polypeptide is a better antibody inducer than the G carbohydrate chains, since antibody produced against virus grown in 1 cell type is efficient in inhibiting the infectivity of the same virus grown in another cell type. However, exactly how different the carbohydrate chains might be is not known.

The intact virus particle in infectious and, to a lesser extent, so is a Tween-ether derived subviral skeleton of VSV Indiana described by Brown and collaborators [167, 201]. This skeleton is a particle from which the viral G, some lipid, and M have been removed (F. Brown, personal communication, 1976).

Deoxycholate, NP-40, or Triton N101 plus 0.4 M NaCl treatment of VSV results in the dissociation of the viral G, lipids, M, and some L protein from the RNA-N (NS + L) complex. This dissociated mixture can be resolved by gradient centrifugation into an infectious RNA-N (NS + L) entity and freed from the viral lipids, G, M, and free L proteins.

It has been mentioned above (p. 200) that for antibody raised against extracts containing live virus (e.g., VSV Indiana), most of the complement-fixing activity (70%) is directed against the G protein. Of the rest, some is directed against the M protein and the remainder toward the nucleocapsid complex of RNA-N (NS + L). The antibodies in such sera are probably also active with free L, NS, N, RNA, or RNA-N complexes. Although, for antibody raised against infected mouse brain extracts, the relative complement-fixing activities directed toward the individual viral components is not known, there are probably antibodies for most viral proteins.

Antisera have been used in neutralization tests with (a) infectious purified virus, (b) infectious structures derived by Tween-ether treatment of purified virus, and (c) infectious entities in mouse brain extracts (virus and possibly some infectious subviral structures).

With the availability of monocomponent antisera raised against individual viral proteins, clear-cut relationships between antigens of different viruses should be possible to establish. However, since many rhabdoviruses grow to low titer (making preparation of individual proteins a difficult task) and others have yet to be adapted to tissue culture, establishing such relationships with these viruses will perforce involve using less reliable reagents.

Other than using antisera for neutralizing infectivity (assayed by plaque reduction or newborn mouse killing tests), antisera can be used for hemagglutination inhibition and fluorescent antibody tests. Virus relatedness can also be determined by cross-vaccination challenge tests.

In studies to determine relatedness among the members of the putative VSV subgroup of rhabdoviruses, Dr. R. Shope and associates have used broad-spectrum antibody preparations (i.e., antisera prepared against infected mouse whole brain extracts) in experiments to determine the extent of inhibiting the killing of newborn mice by infected mouse brain extracts. Also such sera or ascitic fluids have been used in complement-fixation tests [202].

Cartwright and Brown [171] have reported similar studies using antisera developed against vesicle extracts (presumably containing intact virus but possibly also some subviral antigens as well). In their neutralization tests, they used virus from the supernatant fluids of infected BHK 21 cell cultures. These supernatant fluids probably contained only virus particles and not infectious subviral entities. The same antisera have been used to inhibit mouse killing by infectious subviral skeletons obtained by Tween-ether disruption of tissue culture virus. It is not clear if these Tween-ether skeletons contain much M protein or if, when antibody reacts with the M protein, the infectivity of those skeletons is affected. Probably most of the inhibition observed was directed against the RNA-N (NS + L) components of the skeleton since Brown and associates have found that deoxycholate-derived subviral preparations (which lack G and M) give essentially the same results (F. Brown, personal communication, 1976).

In order to discuss data on the relatedness of the VSV group of viruses, we should first delineate for each type of test the antigens which are being considered.

The major viral antigens reside on the G protein, the M protein, and the RNA-N (L + NS) complex. Cellular extracts used in some tests would be expected to contain different amounts of the same viral antigens and, in particular, considerable amounts of free NS as well as possibly some nonvirion antigens.

When investigating the cross-neutralization of infectivity of intact viruses (extracellular virus or free virus in cell extracts), the investigation is primarily being directed toward the relatedness of the G proteins of the different viruses. Similar investigations employing mouse brain cell extracts which may contain infectious subviral entities, as well as intact virus, would be expected to detect relatedness of the viral RNA-N (NS + L) complexes as well as that of the G proteins.

Potential problems to the utility of cross-neutralization data using intact virus would be the differing overall antigenicities of the various virus types or different particle to infectivity ratios of the various virus preparations. Even if 2 viruses possess similar G antigen masses per particle (which, as shown by

Sokol and collaborators [138] is doubtful) and these antigen masses are equally efficient at raising antibody, the effect of differences in particle to infectious unit ratios could seriously influence observed cross-neutralizations. Suppose 1 virus type with a particle to pfu ratio of 1:1 is compared with another in which the particle to pfu ratio is 100:1, then even if both viruses possess identical G proteins, the antibody required to neutralize the infectivity of the latter will be much greater. Even greater complications can be imagined when 2 viruses possess host antigens or only share one out of several antigenic determinants on a particular polypeptide. For reasons such as these one should probably interpret a quantitative difference in cross-neutralization tests as an indication of possible relatedness rather than its extent.

In investigations involving the antibody determined cross-neutralization of the infectivity of Tween-ether disrupted virus (i.e., extracellular virus from which the viral G, some of the lipids, and M protein have been removed), one is probably comparing antigenic similarities of the RNA-N (NS + L) complexes, although some M may contribute to the reaction. That these reactions give different results from those on intact virus was shown by Cartwright and Brown [171]. Here also the effect of different particle to infectious entity ratios and stability of these subviral entities might be expected to influence the results, although the results obtained by those investigators did show a reasonable degree of consistency.

By investigating the complement-fixing activity of deoxycholate-dissociated purified virus, the comparison is being made to all available antigens [i.e., RNA (NS + L) complexes and free G, M, and L proteins]. Complement-fixation tests with virus-infected mouse brain extracts are probably directed against viral G, M, free NS, L, and RNA-N (NS + L) complexes, as well as other cell-associated viral antigens.

It should also be borne in mind that the ability of 1 host to produce antibodies may not be the same as that of another and this may account for some of the variations observed by different investigators.

B. Serological Relationships Between VSV Indiana, VSV New Jersey, VSV Brazil, VSV Argentina, Cocal, Chandipura, and Piry Viruses

Early studies indicated that serological relatedness exists between VSV Indiana, VSV Brazil, VSV Argentina, and Cocal viruses, with a more tenuous relationship to VSV New Jersey [13]; see the reviews by Howatson [6] and Knudson [4]. The results to be discussed below are taken from the more detailed studies of Cartwright and Brown [171] and Shope [202].

More for convenience than as a judgment of the results, these seven viruses are commonly referred to as the VSV subgroup of rhabdoviruses, although as will be seen, VSV New Jersey, Chandipura, and Piry are on the outer borders of the group.

1. Neutralization Tests

Reciprocal neutralization tests have been done by Shope (Table 5.7), and Cartwright and Brown (Table 5.8). One initial point should be made concerning these tests. The neutralization tests of the latter start with a 1:100 dilution of antisera (therefore add 2.0 to their results to compare with Shope's results). The tests of Shope used undiluted sera and are of the sort used in initial studies to categorize a new virus isolate.

In the neutralization tests of Shope, some relationship was demonstrated between Cocal and VSV Indiana, and a significant but much less amount between VSV Indiana and VSV New Jersey, Chandipura, or Piry viruses. Cocal antiserum reacted with Piry and Chandipura but not with VSV New Jersey. Piry and Chandipura viruses and antisera, in their reciprocal crosses, showed that a relationship exists between these 2 viruses. Other reciprocal crosses (e.g., Cocal and VSV New Jersey; Piry and VSV Indiana; Piry and VSV New Jersey) gave indications of unidirectional relationships. Whether these results are significant is not known; however, the implications of such relationships should not be overlooked.

The reciprocal neutralization tests of Cartwright and Brown (Table 5.8A) employing intact virus, indicated that VSV Argentina and Cocal viruses are closely related. Like the results discussed above, some homology was found in the reciprocal tests using VSV Indiana, VSV Argentina, and Cocal antisera and viruses. A slight relationship was also observed between VSV Indiana and Chandipura, but almost none was seen between VSV Indiana and Piry viruses. Cartwright and Brown found essentially no relationship between Chandipura and Piry viruses. This latter result is not in agreement with the results of Shope. To resolve the conflict it may be necessary to obtain cloned virus stocks, make new reagents, and exchange stocks and reagents to repeat the analyses.

It is suggested from the results of Cartwright and Brown that VSV Brazil is closer to VSV Argentina, Cocal, and VSV Indiana than to Chandipura, VSV New Jersey, or Piry viruses. Whether one could establish nearest neighbor relationships in view of the probability that there are several antigenic determinants on the G proteins is doubtful.

In neutralization tests reported by Cartwright and Brown using Tween-ether disrupted viruses, the same basic pattern of homologies was evident except that the relationships appeared closer. This was particularly evident in

Table 5.7A Neutralization Tests[a,b] (Results of Dr. R. Shope [202])

Infected mouse brain extracts of[a]	Mouse hyperimmune sera										
	Titer-log LD_{50}	VSV Indiana	Cocal	VSV New Jersey	Piry	Chandi-pura	Hart Park	Flanders	Rabies	Kern Canyon	Mt. Elgon bat
VSV Indiana	7.1	6.2	4.2	1.9	0	2.6	0	0	0	0	0
Cocal	7.7	4.2	⩾7.0	2.7	1.7	1.8	0	0	0	0	0
VSV New Jersey	5.5	2.2	0	⩾5.0	0	0	0	0	0	0	0
Piry	7.9	2.3	2.6	2.5	5.7	3.4	0	0	0	0	0
Chandipura	7.3	2.5	2.7	0	4.6	⩾5.7	0	0	0	0	0
Hart Park	4.6	0	0	0	0	0	⩾4.1	⩾4.1	0	0	0
Flanders	4.4	0	0	0	0	0	⩾3.9	⩾3.9	0	0	0
Rabies	5.9	0	0	0	0	0	0	0	⩾5.4	0	0
Kern Canyon	5.4	0	0	0	0	2.2	0	0	0	⩾4.9	0
Mt. Elgon bat	5.8	0	0	0	0	0	0	0	0	0	⩾5.3

[a]Protocol: Mouse infected brain extracts injected into mice; boosted; serum or ascitic fluid recovered and used undiluted; 0 = 1.5 or less.

Table 5.7B Complement-Fixation Tests[c,d] (Results of Dr. R. Shope [202])

Infected mouse brain extracts of	Hyperimmune mouse sera or ascitic fluid												
	VSV New Jersey	VSV Indiana	Cocal	Piry	Chandi-pura	Flanders	Hart Park	Mokola	Rabies	Lagos bat	Mt. Elgon bat	Kern Canyon	Klamath
VSV New Jersey	256/512	0	0	0	0	0	0	0	0	0	0	0	0
VSV Indiana	0	256/512	32/128	0	0	0	0	0	0	0	0	0	0
Cocal	0	32/512	256/512	0	0	0	0	0	0	0	0	0	0
Piry	0	0	0	128/32	8/4	0	0	0	0	0	0	0	0
Chandipura	0	0	0	0	128/64	0	0	0	0	0	0	0	0
Flanders	0	0	0	0	0	64/16	32/4	0	0	0	0	0	0
Hart Park	0	0	0	0	0	32/16	256/64	0	0	0	0	0	0
Mokola	0	0	0	0	0	0	0	256/512	32/32	0	0	0	0
Rabies	0	0	0	0	0	0	0	32/128	1024/512	0	0	0	0
Lagos bat	0	0	0	0	0	0	0	64/128	64/128	8/128	0	0	0
Mt. Elgon bat	0	0	0	0	0	0	0	0	0	0	64/64	0	0
Kern Canyon	0	0	0	0	0	0	0	0	0	0	0	256/512	0
Klamath	0	0	0	0	0	0	0	0	0	0	0	0	256/64
Normal	0	0	0	0	0	0	0	0	0	0	0	0	0

[a]Extracts used to obtain antisera probably contained intact virus and cellular viral antigens; mouse brain extracts probably contain infectious virus and possibly also infectious subviral structures.
[b]To compare results with Table 5.8A, subtract 2.0 from all positive titers in this table. Reprinted from Ref. [202], by courtesy of Academic Press.
[c]See footnote a. Extracts probably contain intact virus and cellular virus-induced structures and antigens.
[d]Reciprocal of serum/antigen titers; 0 = no reaction at 1 : 4 serum dilution.

Table 5.8A Neutralization Tests (Results of Dr. B. Cartwright and F. Brown

Virus	Mouse hyperimmune sera						
	VSV Indiana	Cocal	VSV Argentina	VSV Brazil	VSV New Jersey	Piry	Chandipura
VSV Indiana	5.0	1.7	1.7	1.6	0.3	0.2	0.7
Cocal	1.4	4.9	3.2	0.7	0.9	−0.3	−0.1
VSV Argentina	1.9	5.1	4.4	1.9	0.7	0.3	0.1
VSV Brazil	0.5	1.8	1.7	5.3	0.4	−0.3	−0.1
VSV New Jersey	0.5	0.4	0.9	0.8	5.3	0.3	−0.3
Piry	0	−0.1	0.1	−0.3	−0.3	3.6	0.1
Chandipura	0.9	0.8	0.8	0.5	1.2	−0.5	5.3

Table 5.8B Neutralization Tests Against Infectivity of Tween-Ether-Disrupted Purified Virus (Results of Dr. B. Cartwright and F. Brown [171])

Virus skeleton	Mouse hyperimmune sera						
	VSV Indiana	Cocal	VSV Argentina	VSV Brazil	VSV New Jersey	Piry	Chandipura
VSV Indiana	3.6	3.0	3.4	3.1	2.4	0.1	0.5
Cocal	⩾1.5	⩾1.9	⩾1.9	⩾1.7	⩾1.7	−0.1	0.7
VSV Argentina	⩾2.7	⩾2.7	⩾2.7	⩾2.7	1.3	0.1	0.5
VSV Brazil	2.9	2.3	2.3	⩾3.7	2.9	0.5	0.7
VSV New Jersey	2.1	−0.2	1.7	1.9	4.2	−0.3	−0.1
Piry	−0.9	−1.1	−1.3	−0.9	−0.7	⩾3.5	0.4
Chandipura	−0.5	−0.7	−0.1	−0.3	0.3	0	⩾3.4

Table 5.8C Complement-Fixation Tests with Deoxycholate-Disrupted Purified Virus[c,d] (Results of Dr. B. Cartwright and F. Brown [171])

	Mouse hyperimmune sera						
Virus	VSV Indiana	Cocal	VSV Argentina	VSV Brazil	VSV New Jersey	Piry	Chandipura
VSV Indiana	100	26	26	13	10	<1	<1
Cocal	21	100	>70	20	2	<1	<1
VSV Argentina	24	>50	100	22	3	<1	<1
VSV Brazil	11	16	30	100	2	<1	<1
VSV New Jersey	3	<1	7	10	100	<1	<1
Piry	<2	<2	<2	<2	<2	100	<1
Chandipura	<1	<1	<1	<1	<1	<1	100

[a]Protocol: Guinea pig vesicle extracts injected into mice; boosted; hyperimmune sera used at 1/100 dilution against purified, intact (Table 5.8A), or Tween-ther-disrupted (Table 5.8B), or deoxycholate-disrupted (Table 5.8C) virus.
[b]To compare results with Table 5.7A, add 2.0 to all values.
[c]Expressed as percent of homologous activity.
[d]Antibody was used at a 1:50 dilution (i.e., antibody excess; F. Brown, personal communicatin). Reprinted from Ref. [171], p. 395, by courtesy of Cambridge University Press.

the various reciprocal crosses involving VSV Brazil or VSV New Jersey. Such results might be construed as indications that the *internal antigens of the viruses are much more related than the outer G antigens.* However, differences in nucleocapsid stabilities might bias the results substantially. In our limited experience, the specific infectivity of nucleocapsids of Cocal are always much less (1-2 logs) than those from similarly prepared VSV New Jersey, their infectivities being less (~ 1 log) than those of VSV Indiana (or Chandipura viruses; D. H. L. Bishop, unpublished observations, 1975). We have not studied VSV Argentina, Piry, or VSV Brazil nucleocapsid specific infectivities.

Whether such a stability problem may be relevant to the results of Cartwright and Brown is not known, although the fact that Cocal antiserum was less active on Cocal subviral structures than against those of VSV Indiana makes this possibility seem likely. As in their previous results, some slight relationship of Chandipura with the other viruses was detected, although this was most unidirectional.

2. *Complement-Fixation Tests*

The complement-fixation tests of Shope represent end point serum and antigen dilution results, involving mouse antisera and mouse brain extracts (Table 5.7). The results obtained also indicated that a relationship exists between VSV Indiana and Cocal viruses. There was no antigenic relationship detected between VSV New Jersey and VSV Indiana or Cocal viruses. However this may have been caused by low-titer antisera. Piry and Chandipura reciprocal tests gave evidence of only a slight unidirectional relationship. Since this test was a measure of different properties of infected mouse extracts than the neutralization tests described above, the results of Tables 5.7B and 5.7A are not necessarily conflicting; however, they could be interpreted to indicate that Chandipura and Piry G proteins are similar, whereas their internal antigens are dissimilar. Unfortunately the low antibody titers in the sera make such a conclusion premature.

The complement-fixation data of Cartwright and Brown (Table 5.8C) using a 1 : 50 dilution of antisera and varying antigen levels of deoxycholate disrupted virus indicated no homology between Chandipura and Piry antigens. Other ways of performing the test might have given alternate results; however, their results agree with the neutralization data they present (Table 5.8A and 5.8B). The other cross-complement-fixation results they obtained are also in fair agreement with the corresponding reciprocal neutralization results.

3. *Other Tests, Such As Defective Particle Interference*

Crick and Brown [203] have used defective-interfering particles to determine if homologies exist between the 7 putative members of the VSV subgroup of rhabdoviruses. This test is based on the observation that rhabdovirus-defective particles substantially interfere with the replication of their progenitor B virions [115, 204, 205]. It was previously shown by Huang and Wagner [115] that their VSV Indiana T particles interfere to only a small extent with those of VSV New Jersey. Greater interference with the replication of VSV New Jersey has been reported by Prevec and Kang [204] using the long T (LT) particles of the heat-resistant strain of VSV Indiana (this has been confirmed by C. R. Pringle, personal communication, 1976).

Crick and Brown [203], using T-particle preparations of VSV Indiana, showed that they interfered to equivalent extents with the replication of B particles of VSV Indiana, Cocal, VSV Argentina, and VSV Brazil, but only to a small extent with VSV New Jersey. Again, Pringle has obtained confirmatory results (personal communication, 1976). The replication of Piry and Chandipura viruses were not interfered with by the VSV Indiana-defective particles. Homologous interference and lack of heterologous interference was also obtained using Piry T particles tested against Piry, VSV Indiana, and Chandipura viruses. Similar homologous and lack of heterologous interference was obtained with Chandipura T-particle preparations. VSV New Jersey T particles interfered with VSV New Jersey B particles but not with VSV Indiana, VSV Argentina, Cocal, or VSV Brazil B particles.

The interference tests of Crick and Brown are probably measuring the ability of one virus T particle to use a B particle's replication machinery [206, 212]. In what way this is accomplished is not known.

In summary, it appears that there is some relatedness between VSV Indiana, VSV Argentina, VSV Brazil, and Cocal viruses and less between these viruses and VSV New Jersey, Piry, or Chandipura viruses. By Shope's tests Piry is related to Chandipura, whereas in Cartwright's and Brown's experiments, that relationship is tenuous or nonexistent.

C. Serological Relationships Between Flanders and Hart Park Viruses

These 2 viruses appear to be distinct from other rhabdovirus isolates as determined by reciprocal neutralization and complement-fixation tests ([93]; also see Shope's results, Table 5.7). By both tests, as well as by hemagglutination inhibition studies [158], Flanders and Hart Park appear to be very closely related to each other.

D. The Lack of Serological Relationships of Mount Elgon Bat, Klamath, Joinjakaka, Bovine Ephemeral Fever Virus, and Navarro Viruses to Other Rhabdoviruses

All these virus isolates appear to be serologically unrelated to each other and to other known rhabdoviruses (for MEBV see [92] and Shope's results, Table 5.7; for Navarro see [18] and [20]; for Joinjakaka see [30]; for Barur see [30]; for BEFV see [90]).

E. Serological Relationships Between Kwatta and BeAn 157575

These 2 viruses appear to be serologically related to each other but distinct from other known rhabdoviruses [20].

F. Serological Relationships Between Mossuril and Kamese Viruses

These 2 rhabdoviruses also appear to be serologically related to each other but unrelated to all other known rhabdoviruses [20].

G. Serological Relationship of Kern Canyon Virus to Other Rhabdoviruses

By complement-fixation tests, KCV appears to be unrelated to other rhabdoviruses; however slight cross-neutralization relationships to some members of the VSV subgroup have been reported, the meaning of which is unclear ([93] and Shope's results, Table 5.7).

H. Sigma Virus

It is not known if sigma is related to any other rhabdovirus [99].

I. Serological Relationships Between Rabies, Mokola, Lagos Bat, Nigerian Horse, Duvenhagé, Bolivar, Kotonkan, and Obodhiang Viruses

In all published reports, no relatedness has been detected between the rabies subgroup of rhabdoviruses and other rhabdovirus isolates (for example, see Tables 5.7A and 5.7B).

Although Nigerian horse virus is reported to be related to rabies [47], as was Bolivar virus [48], no extensive serological studies have been reported between these isolates and other members of the putative rabies subgroup of rhabdoviruses.

Using infected mouse brain extracts and sera raised against such extracts, Shope and collaborators have studied the relationships between rabies, Mokola, LBV, kotonkan, and Obodhiang viruses (see Table 5.7). Similar studies on rabies, Mokola, and LBV have been reported by Schneider and associates [207] using antisera raised against virus or purified ribonucleoprotein complexes (rabies RNA-N) and tested against purified virus or viral components (Table 5.9).

1. *Neutralization Tests*

In the cross-neutralization results obtained by Schneider and collaborators (Table 5.9A) some slight antigenic relationships were detected between rabies, Mokola, and LBV. The antibody titer in the LBV serum was not as high as those of the other 2 viruses. In similar unpublished results obtained by T. Wiktor, H. F. Clark, and F. Sokol (Table 5.10; personal communication, 1975) using either whole virus antisera or rabies glycoprotein monospecific antiserum of high titer, a definite antigenic relationship was detected between rabies and the other 2 viruses. Moreover, Mokola and LBV showed a somewhat closer relationship than in Schneider's tests.

The neutralization tests of Shope [51, 54] using infected mouse brain extracts also indicated that some relationship exists between LBV and Mokola but little relationship of either virus to rabies was detected.

In cross-neutralization experiments, kotonkan reacted only slightly with antisera raised against infected mouse brain extracts of Mokola, or LBV, although some slight neutralization by rabies antiserum was observed [54]. The sera used possessed relatively low antibody titers and no inhibition of rabies, Mokola, or LBV infectivity was observed with kotonkan antiserum.

2. *Complement Fixation Studies*

Schneider and collaborators [207] have used antisera raised against purified ribonucleoproteins of rabies, Mokola, and LBV in complement-fixations tests

Table 5.11 Complement-Fixation Studies on the Rabies Group of Viruses[a] (Results of Dr. R. Shope [202])

Infected mouse brain extracts	Mouse hyperimmune ascitic fluids				
	Rabies	Mokola	Lagos bat virus	Obodhiang	Kotonkan
Rabies	256	64	0	0	0
Mokola	32	256	4	16	0
Lagos bat virus	16	128	32	8	0
Obodhiang	0	4	0	128	0
Kotonkan	0	32	0	0	64

[a]Reciprocal of ascitic fluid titer; 0 = less than 4. Reprinted from Ref. [202] by courtesy of Academic Press.

Tignor and Shope [208] have vaccinated mice intracerebrally or intraperitoneally with mouse brain extracts of the attenuated rabies HEP strain or rabies Tr 5843 and challenged the mice subsequently with mouse brain extracts of Mokola or Lagos bat viruses. The mice vaccinated with rabies by the intraperitoneal route resisted challenge by LBV but not Mokola. Mice vaccinated intraperitoneally with LBV or Mokola resisted homologous but not heterologous challenge. These results indicate that some homology exists between LBV and rabies antigens although what antigens are involved is not known.

4. *Other Putative Rabies Viruses*

There is little published information concerning the relationship of the most recent rabies related isolate, Duvenhagé, to other members of the rabies subgroup [49]. In the laboratories of Dr. F. Murphy and Dr. R. Shope, using indirect immunofluorescence and cross-complement fixation, some relationship to rabies has been demonstrated. In tests on the neutralization of infectious material in mouse brain extracts, unidirectional results were obtained (Shope, Tignor, Murphy, and Bauer, personal communication, 1975).

5. *Summary of the Rabies Subgroup of Rhabdoviruses*

It appears that there is some slight antigenic relationship between rabies, LBV, Mokola, and kotonkan viruses as evidenced by neutralization tests (more between LBV and Mokola in the results of Wiktor and collaborators). These neutralization tests probably reflect slight antigenic similarities between the G proteins of the various viruses.

Considerable antigenic similarity has been reported for some of these viruses in complement-fixation tests. These results suggest that the internal antigens of rabies, Mokola, and LBV, as well as those of Mokola, Obodhiang, and kotonkan, are more closely related to each other than their corresponding G proteins.

Since the Bolivar isolate is not available, its relationship has not been studied extensively. However, the published reports of Bolivar, NHV, and Duvenhagé indicate that they are antigenically related to the rabies subgroup of viruses.

J. The Serological Relationships of the Fish Rhabdoviruses

Cross-neutralization studies performed by Bachmann and Ahne [40] or Hill and associates [209] with spring viremia of carp virus (SVCV) and isolates from carp swim bladder inflammation diseases indicate that these 2 viruses are indistinguishable. It has also been shown by cross-neutralization tests that SVCV is not related to PFR, IHNV, or Egtved viruses [209], although in more recent studies, Hill and collaborators [210] obtained some cross-neutralization between SVCV, PFR, and IHN viruses, but none with Egtved (VHS).

The 3 salmonid rhabdovirus isolates, OSDV, SRCSDV, and IHNV, have been examined by plaque reduction tests [106]. The results reported indicate that all 3 virus isolates are antigenically related and that OSDV and IHNV were indistinguishable by this test. No complement-fixation tests have been reported.

Jørgensen [43] showed that IHNV could be neutralized to some extent with an anti-Egtved serum, although fluorescent antibody staining of IHNV infected cells using anti-Egtved serum did not confirm this observation, and other investigators have not observed cross-neutralization between IHNV and Egtved [210, 211].

Jørgensen [43] has also investigated the antigenic relationships of 76 isolates of VHS (Egtved) from Danish, Norwegian, and Swedish rainbow trout and 1 isolate from Italian brown trout. Seventy-two of the isolates were found to differ only slightly in the neutralization tests. Four other isolates (3 Danish and 1 Norwegian) differed significantly and this result was interpreted to indicate that there is a second serotype of Egtved viruses.

VI. The Replication Process of Rhabdoviruses

Of the many rhabdoviruses isolated from different sources, only one (VSV Indiana) has been studied in sufficient detail to permit conclusions to be drawn concerning the molecular biology of its replication process. It should be understood in the following discussion that the replication process of VSV Indiana may be quite distinct by comparison to that of other rhabdovirus isolates, both in terms of the process and the timeliness of execution of viral-induced events, as well as the cellular interactions with these events.

It has already been pointed out that rhabdoviruses have been isolated from widely different sources including plants, insects, fish, birds, bats, and other mammals. It appears that many of these viruses can replicate in more than one host (e.g., both in insects and cattle for VSV Indiana). Consequently, it is possible that the details of the molecular biology of a virus in 1 host may be quite different from those occurring in another host. In terms of biological differences, it has been shown that whereas rabies is neurotropic, VSV Indiana is epitheliotropic [3]. Moreover, some rhabdoviruses produce large numbers of extracellular progeny virus while others produce very little. The list could be continued; however, it is sufficient to say that such differences are the result of different evolutionary developments of the various rhabdoviruses, so that it would be unwise to believe that all rhabdoviruses behave like VSV Indiana in terms of the process and proceeds of their growth cycles. This should be borne in mind in the following discussions.

In consideration of what we know about the replication process of a rhabdovirus, the progress of a productive infection of a cell by VSV Indiana can be delineated into several phases. These are depicted in diagrammatic form in Fig. 5.2.

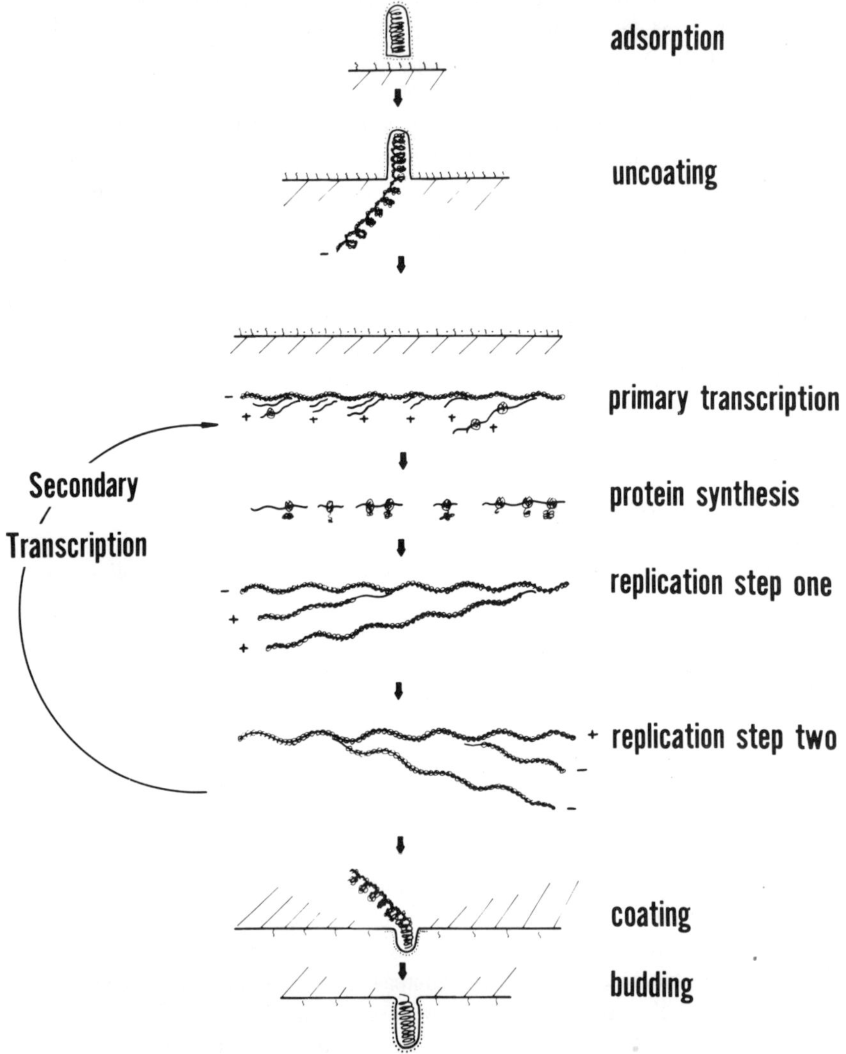

Fig. 5.2 A scheme of the essential steps in the infection process of VSV Indiana.

First there is attachment; this is followed by absorption, penetration, and uncoating. Second, there is the viral-directed synthesis of messenger RNA (mRNA) which, for rhabdoviruses, represents transcription of complementary (VC) RNA species from the viral genome by the transcriptase enzyme molecules which are present in the infecting virus. Third, there is protein synthesis. Fourth, there is replication of progeny genomes from the infecting parental genome. Fifth, there are further rounds of transcription, protein synthesis, and replication. Finally, there is progeny virus assembly and release.

In consideration of the various processes outlined above, we will delineate what is known about the attachment and penetration processes and then describe the viral-induced synthesis of RNA, drawing upon both in vitro and in vivo analyses. Finally, we will discuss viral protein synthesis and the location of the processes involved in progeny virus assembly and release.

A. Penetration

The attachment of VSV Indiana to cells can occur at 4°C as well as at higher temperatures. Using [^{3}H]-nucleoside-labeled virus and [^{32}P]-phosphate medium (to monitor dilution by extracellular liquid) and assaying the particle to plaque-forming unit (pfu) ratios in the inoculum and unadsorbed virus, Flamand and Bishop [144] showed that following 30 min of adsorption at 4°C: (a) very little of the virus in an inoculum adsorbed to a monolayer of cells and (b) the particle to PFU ratio of unadsorbed virus was the same as that of the initial inoculum even following 12 successive exposures of unadsorbed virus to cell monolayers. In their experiments, between 5 and 10% of the added virus adsorbed to a cell monolayer (up to 20% when DEAE-dextran was employed). Those results, as well as others they reported, indicated that adsorption is not just a monopoly of viable virus. It has been found that viruses adsorb to different cell types with different efficiencies. For VSV, BHK cell monolayers were shown to be better than mouse L-cells, which in turn were better than chick embryo fibroblast (CEF) cells [144, 214].

Not all the viruses which adsorb to cells establish an infection. Again Flamand and Bishop documented that, postadsorption, washed monolayers of cells subsequently desorbed about 20-40% of their virus within 20-30 min of incubation at 37°C [144]. The desorbed virus possessed particle to PFU ratios essentially similar to that of the original inoculum [144]. Desorption has been found to be influenced by the cell type, less occurring with BHK or L cells than with CEF cells [144, 214]. After 30 min of incubation at 37°C, it was shown that further release of desorbed virus was small in quantity and, in the absence of inhibitors of virus production, this release was masked by the production of progeny virus [144, 215].

It appears that multiplicities of up to 6000 adsorbed viruses per cell can be used to establish productive infections [144] and that in high-multiplicity infections there is no inhibition of the ability of any of the adsorbed virus to begin the initial phases of the infection process (see below).

The involvement of the viral glycoprotein in the adsorption process has been described on p. 198 of this review and is not discussed further.

Evidence has been presented that following attachment there is fusion of the viral and cellular membranes, leading to a release of the viral nucleocapsid (RNA-N, NS, and L) into the cell [163]. Alternate evidence has been presented that following attachment at 4°C, when cell monolayers are incubated at 37°C, virus can be taken up by viropexis [162]. Since both mechanisms appear to work in vitro, the conclusion to be drawn from these observations is that there is a possibility for more than 1 route of entry for a virus, although which mechanism occurs or predominates in the initial or subsequent infections of cells in an animal is not known.

It has been demonstrated that the infecting virion G and M proteins remain with the cellular plasma membrane when the nucleocapdis enter the cell cytoplasm [163]. We have confirmed these results and found that through the entire 6 hr of a productive infection by VSV Indiana in BHK 21 cell monolayers, the viral L, G, N, NS, and M proteins are conserved: they are not broken down nor, apparently, are they reutilized by progeny virions (A. Flamand and D. H. L. Bishop, unpublished observations, 1975). Likewise, it has been shown that the virion genome is also conserved intact [144].

Is virus uncoated at 4°C? In order to answer this question an experiment was performed in which [^{3}H]-nucleoside-labeled virus was allowed to adsorb to BHK cell monolayers at 4°C for 30 min, and the cells were mechanically sheared by passage through a syringe. When the homogenate was subsequently centrifuged on a gradient of sucrose, it was found that the majority of the [^{3}H]-labeled virus remained coated and sedimented in association with cell plasma membranes (Fig. 5.3). When these infected cell monolayers were incubated at 37°C in either the presence or the absence of cycloheximide (as an inhibitor of protein synthesis) and then mechanically sheared, most of the cell-associated, infecting parental virus was uncoated and recovered as 100-140 S structures (Fig. 5.3). While this argues for uncoating being a temperature-dependent process, it has been observed that for pike fry rhabdovirus, which possesses an RNA transcriptase capable of functioning at low temperatures, the transcriptase functions during adsorption at 4°C [123]. This observation suggests that for PFR, uncoating may occur at low temperatures.

B. RNA Synthesis

The productive infection of a cell by a virus involves the synthesis of viral-specified products in sufficient quantities, location, and timeliness to execute the necessary

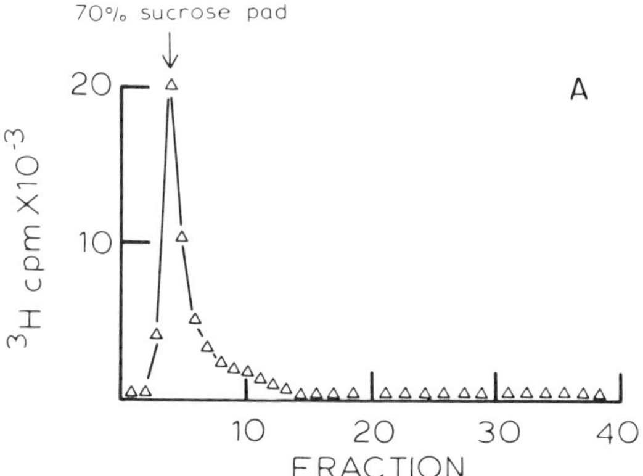

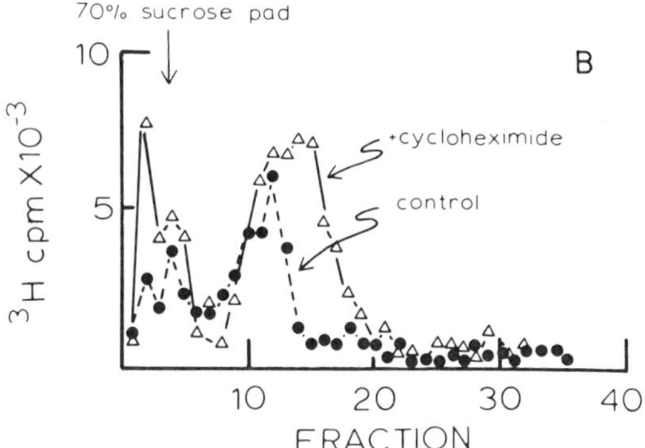

Fig. 5.3 The adsorption and uncoating of VSV Indiana. [^{3}H]-Uridine-labeled VSV Indiana (2×10^5 cpm) was allowed to adsorb 30 min at 4°C to monolayers of 3×10^6 BHK 21 cells. Following adsorption (A), the cells were washed, then mechanically sheared by passage through a 2 ml syringe (20 gage needle) and centrifuged at 35,000 rpm in a Spinco SW41 rotor for 2 hr in a 30-10% gradient of sucrose containing 0.01 M NaCl 0.2 M EDTA, 0.01 M Tris-HCl buffer, pH 7.4. A 0.5 ml pad of 70% sucrose was used to collect the residual complete virus and virus attached to cell membranes. For infected cell monolayers incubated for 1 hr at 37°C in Eagle's medium in the presence or absence of 100 μg cycloheximide (B) the procedure of analysis was identical (P. Repik and D. H. L. Bishop, unpublished results, 1975).

procedures to produce infectious progeny. For rhabdoviruses, which possess virion transcriptases, following uncoating, the critical second step to an infection process is the synthesis of viral messenger RNA species. This synthesis is described as transcription since the mRNA species (plus strands) are copied from the virion genome (minus strand). When, where, and in what form and quantities these mRNA species are synthesized is discussed in the next sections in relation to evidence gleaned from in vitro and in vivo studies.

Since some viral proteins are evidently required before others, a question which has to be asked is whether there is any regulation of viral messenger RNA synthesis to give unequal amounts of the various mRNA species so that through translation some protein species are available in greater amounts than others. Similarly, one can ask whether the regulation of transcription results in some mRNA species being produced at one time and others at another, thereby controlling the availability of certain proteins. Evidence pertaining to the answers of these questions is discussed next.

Transcription from the infecting virion genome is termed *primary transcription,* and transcription of mRNA from progeny replicas of the infecting genome is termed *secondary transcription.* In order to obtain progeny genome RNA species, replication must occur. Clearly, the process of mRNA transcription and replication must initially utilize the same template. However, these processes are not necessarily conflicting and this could be an important attribute of the infection process evolved by rhabdoviruses, as is proposed later in this review. The interrelationships of primary transcription, replication, and secondary transcription will be discussed using evidence gleaned from both in vivo and in vitro experimentation.

1. *In Vitro Studies on the Viral Transcriptases of Rhabdoviruses*

The properties of the transcription process of VSV have been studied extensively subsequent to the initial demonstration by Baltimore et al. [194] that VSV Indiana virions possessed an RNA-directed RNA polymerase. Since the virus genome is enclosed in a membrane, in order to assay the synthesis of RNA in vitro it is necessary to solubilize the structure to allow access of labeled nucleoside triphosphates to the enzyme-template complex [194]. Solubilization can be achieved by addition of a nonionic detergent, such as Triton N-101 or Nonidet P40 [119, 194]. Other requirements for RNA synthesis are all 4 nucleoside triphosphates, magnesium ions, and low concentrations of salt. The optimal concentrations of each have been determined [195, 216]. It is noteworthy that the optimal concentration of adenosine triphosphate is much greater than that of the other 3 ribonucleoside triphosphates [216]. Manganese ions in a reaction mixture inhibit the transcription process [216]. Although a reducing agent, such as dithiothreitol or 2-mercaptoethanol, is not required for the VSV enzyme, one is required for the in vitro assay of Kern Canyon viral transcriptase [195].

The optimal pH for an in vitro assay is around pH 8.0 ± 0.2 and, for all rhabdoviruses so far investigated except the fish rhabdoviruses, the optimal temperature for the in vitro transcription reaction is between 28 and 32°C [195, 217]. This is in clear opposition to the results obtained in vivo where the optimal temperature for primary transcription was shown to be 38°C [215]. Why is the in vitro temperature optimum so low? The reason is not definitely known. It has been shown that the in vitro transcription process is not sustained at 37°C, and that it is also selective and incomplete [200]. Within an infected cell, the primary transcription process is both sustained and complete at 38°C, or even 40°C [215]. Some evidence has been presented which suggests that the observed in vitro inhibition may result from the presence of a nonionic detergent [218], despite the fact that for other enzyme systems, such as that of an oncornavirus, Rous sarcoma virus, the RNA-directed DNA synthesis in the presence of Triton N101 is greater in vitro at 38°C than at 31°C [219].

The optimal temperature for the in vitro assay of the virion transcriptase of pike fry rhabdovirus is around 20°C [123], whereas that of SVCV or Egtved viruses is around 18°C (P. Roy and D. H. L. Bishop, unpublished observations, 1975).

Rhabdovirus preparations possess nucleoside diphosphate kinase enzyme activities (nucleoside triphosphate phosphotransferases). These activities are activated by a nonionic detergent and are capable of catalyzing such reactions as GTP + CDP $\rightleftharpoons$ GDP + CTP [183, 220]. It has been shown that provided an excess of at least 1 nucleoside triphosphate is supplied, the in vitro reaction can function in the presence of the 3 other nucleoside diphosphates following their conversion into triphosphates. Reaction mixtures containing ribonucleoside monophosphates in lieu of the corresponding diphosphates or triphosphates, or reaction mixtures only containing all 4 nucleoside diphosphates, do not support the in vitro transcription process [188], presumably since no triphosphates are synthesized.

Nucleoside triphosphatase enzyme activities (ATPase, etc.) are also present in all rhabdovirus preparations so far investigated [183]. These enzymes appear to be inhibited by Triton N101 and probably do not affect the in vitro transcription process unless limiting concentrations of triphosphates are present [183].

The transcriptase enzyme specific activities (assayed in terms of picomole nucleoside monophosphate incorporated into product RNA per hour per milligram viral protein, or per milligram viral RNA) vary considerably from 1 rhabdovirus to another. The highest rhabdovirus enzyme specific activity we have observed was obtained with preparations of VSV Indiana, giving values of about 30,000 pmole GMP/hr/mg protein [216]. Other rhabdoviruses exhibit lower in vitro specific activities (VSV New Jersey, Cocal, Chandipura, Piry [196], KCV [195], PFR [123], SVCV, Egtved, and rabies viruses; P. Roy and D. H. L. Bishop, unpublished observations, 1975), although such lower activities do not necessarily correlate with the observed in vivo primary transcription rates (P. Repik and D. H. L. Biship, unpublished observations, 1975). As an example, rabies virus preparations exhibit very little, if any, in vitro transcriptase activity (usually less than 5 pmole UMP incorporated/hr/mg protein), while the in vivo primary transcription rate is much greater ([197], and D. H. L. Bishop, unpublished observations, 1975). It is not known if the variation of enzyme specific activity is a function of the lability of the enzyme or a genetic attribute. Neither is it known if the greater intracellular activity reflects an interaction with some specific factors in the cell which facilitate transcription. Notwithstanding these considerations, it has been shown that the VSV Indiana transcription process functions sufficiently well in vitro to give product RNA which can serve as messenger RNA in in vitro translation systems (see p. 239).

2. *Product Analyses of the In Vitro Transcription Reaction*

It has been shown that the in vitro transcription process for VSV Indiana is complete at 31°C under optimal conditions [221]. It has also been shown that the process is repetitive, sequential, and disproportionate in that some sequences are transcribed more frequently than others [198]. This has been demonstrated by hybridizing the product RNA either to VSV viral or to VSV defective particle RNA (representing a unique 1/3 of the viral genome). By such studies it was shown that (a) much less than 1/3 of the product could hybridize to the defective particle RNA for any timepoint examined and (b) those product sequences which did hybridize could only be detected late in a reaction timecourse and not at all in reactions conducted at 37°C [200]. Product RNA species are, for the most part, smaller than the viral genome [200].

a. Transcription Initiation, Capping, and Methylation. The direction of product RNA synthesis has been shown to occur in a 5′ to 3′ mode [222]. A variety of 5′ initiation sequences have been identified in the in vitro transcription process involving either pppAp . . . or, to lesser extents, pppGp . . . sequences [197, 222]. No product initiations have been demonstrated to commence with pyrimidine nucleotides, and this has been documented for VSV Indiana, VSV New Jersey, Chandipura, Cocal, and Piry viruses [196]. Of the product RNA sequences initiated in vitro with purine nucleotides, 4 have been partially sequenced for VSV Indiana, these are pppApCpGp . . . , pppApApPypXPGp . . ., pppGpCp . . ., and pppGpGpPyp . . . [196, 222]. What role do these initiation sequences have in the process of in vitro mRNA transcription or viral RNA replication? Before we answer this question, we will first discuss evidence for specific posttranscriptional modification of VSV mRNA species.

It has been shown that virions of the double-stranded RNA viruses, cytoplasmic polyhedrosis virus (CPV), and human reovirus, in addition to their RNA polymerases, possess a methylation activity [191, 192]. Subsequently it was shown that on the product mRNA synthesized by either of these viruses, there is both nucleotide capping and methylation of nucleotides at the 5′ end. For reovirus mRNA transcripts, the modified 5′ sequence is $m^7G^{5'}ppp^{5'}GmpCp$ . . ., where m^7G is a methylated capping guanosine and Gmp is a 2′-O-methyl, 3′-phosphate derivative of guanosine [233]. These observations have been extended by Banerjee and associates and shown to apply to VSV Indiana transcription reactions [187, 224]. The effects of a methyl donor (S-adenosyl-L-methionine, SAM) upon the reaction rates of VSV Indiana, PFR, or SVCV virion transcriptases are shown in Fig. 5.4. Although the effects on VSV Indiana and PFR transcription rates were marginal, for SVCV (like CPV), the stimulation was significant. The results obtained by Banerjee and associates indicate that the VSV Indiana methylation involves a posttranscriptional modification of RNA synthesized by the

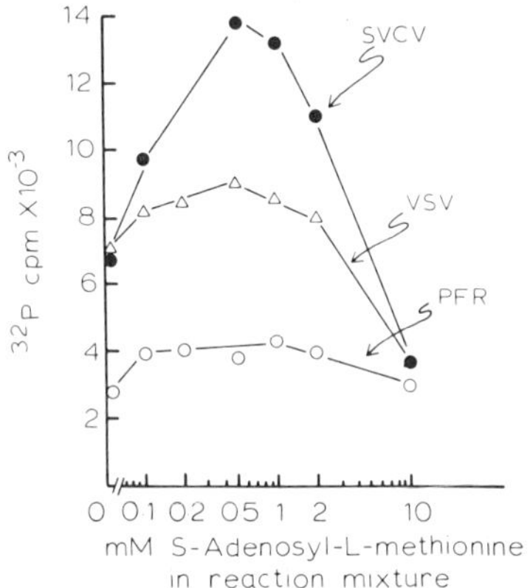

Fig. 5.4 Transcriptase assays in the presence of S-adenosyl-L-methionine. Standard reaction mixtures [195] containing [α-^{32}P]-UTP and different concentrations of S-adenosyl-L-methionine were templated by either VSV Indiana, spring viremia of carp virus (SVCV), or pike fry rhabdovirus (PFR) and incubated at 31°C. The hourly increase in incorporation of [^{32}P]-UMP was determined. The specific activity of the UTP in the VSV reaction was 1/10 that in the SVCV or PFR reaction mixtures (P. Roy and D. H. L. Bishop, unpublished results, 1975).

virion transcriptases [224]. Whether this modification occurs while product RNA is being completed or after it is released is not known, although the observations for SVCV (Fig. 5.4) that the reaction rate was stimulated by the presence of SAM argues, at least in some cases, for the former proposition.

The transcription process of RNA can now be delineated into several phases: (a) initiation, (b) chain elongation; (c) 5′ modification involving capping and methylation, and (d) release and polyadenylation of the 3′ terminus.

The results obtained on transcription initiation with cores of VSV Indiana utilizing [^{32}P-γ]-ribonucleoside triphosphates [222] indicate that initiation involves purine nucleotides (Fig. 5.5). The [^{32}P]-labeled terminal nucleotides can be recovered from the product RNA by alkali hydrolysis or, for [^{32}P-γ]-GTP-labeled product RNA, by ribonuclease T_1 digestion [222]. The label in these termini is sensitive to alkaline phosphatase (Fig. 5.5).

When the [γ-^{32}P]-ATP- or [γ-^{32}P]-GTP-labeled product was digested by pancreatic ribonuclease and resolved by DEAE cellulose column chromatography, 4 distinguishable termini were identified (Fig. 5.6). Nearest neighbor analyses using [α-^{32}P]-ribonucleoside triphosphates as well as analyses involving ribonuclease T_1 digestion have allowed us to characterize these termini as pppApCpGp . . ., pppApApCpXpGp . . ., pppGpCp . . ., and pppGpGpPyp . . . ([197, 222] and E. Hefti and D. H. L. Bishop, unpublished observations, 1975).

The termini analyzed in these experiments represent only those sequences which have retained their γ-phosphates. Are there other sequences which lose their γ-phosphates? Following the demonstration that reovirus, CPV, and VSV contain capping and methylation enzyme activities [187, 191, 192, 224], we have been able to answer this question and have found that in addition to the termini containing triphosphates there are also capped termini, and that in the presence of S-adenosyl-L-methionine these capped termini possess the sequence m^{7}GpppAmpApCpXpGp . . . and represent the capped and methylated derivatives of the pppApApCpXpGp sequence.

When [^{32}P-α]-ribonucleoside triphosphates were used to synthesize product RNA (±SAM) and the termini isolated from the total product by alkali digestion then subjected to phosphatase treatment, the following observations were made [314, 315].

First, as expected, both [^{32}P-α]-GTP and [^{32}P-α]-ATP labeled the initiating nucleotides (Fig. 5.7). Second, without SAM in the reaction mixture, alkaline phosphatase released phosphate and identified a phosphatase-resistant nucleotide fragment which eluted from DEAE cellulose columns with a -3 charge. From experiments involving nearest neighbor analyses, this nucleotide was found to correspond to a capped nucleotide of sequence GpppA. The distribution of label between inorganic phosphate and the capped nucleotide was 1 : 3 for the [^{32}P-α]-GTP product and 3 : 1 for the

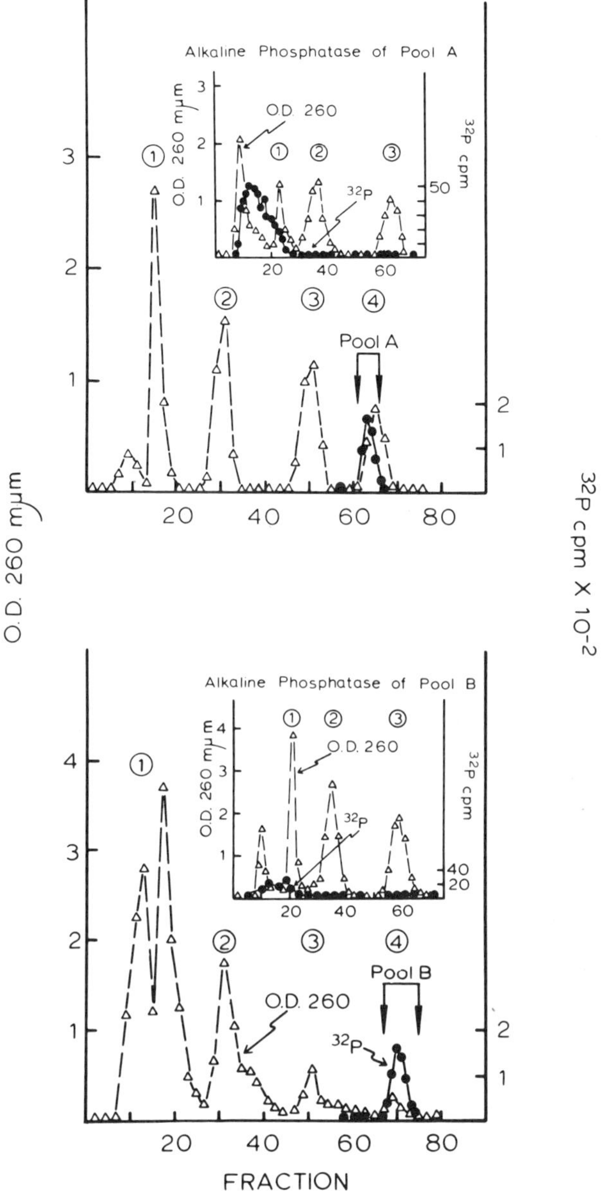

Fig. 5.5 Analysis of (A) [γ-^{32}P]-ATP- or (B) [γ-^{32}P]-GTP-labeled VSV Indiana transcription product RNA. Purified samples of 8 $\times$ 10^2 cpm of VSV Indiana transcription product RNA [222], were digested with (A) alkali or (B) ribonuclease T_1 and the nucleotides resolved by DEAE cellulose column chromatography [222] with suitable oligonucleotide markers obtained from a pancreatic ribonuclease digest of 1.5 mg RNA. In either case the tetranucleotides were recovered [126], treated with alkaline phosphatase and (inserts) rechromatographed (E. Hefti and D. H. L. Bishop, unpublished results, 1975).

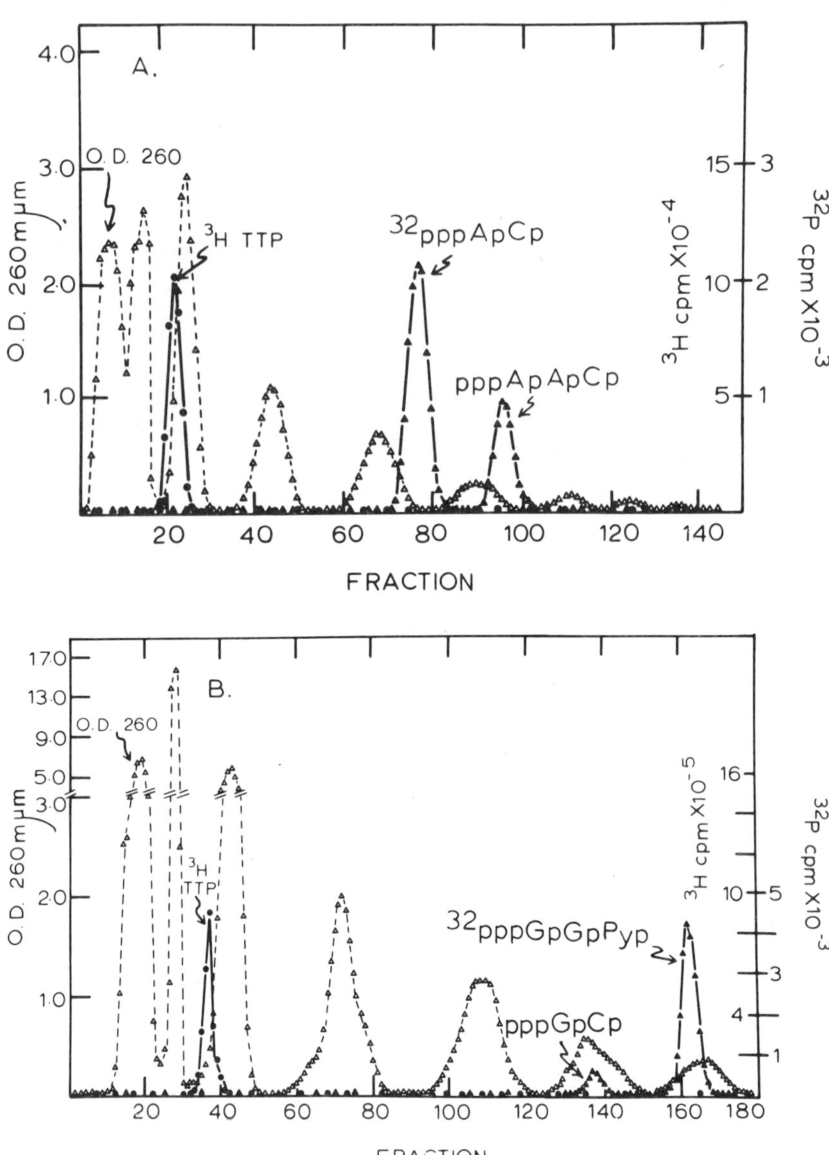

Fig. 5.6 Identification of the [γ-^{32}P]-ATP- or [γ-^{32}P]-GTP-labeled termini in VSV Indiana reaction reaction product RNA. The [γ-^{32}P]-ATP- (A) or [γ-^{32}P]-GTP (B)-labeled products obtained from VSV Indiana transcription reactions (Fig. 5.5), were digested by pancreatic ribonuclease and resolved at pH 5.5 by DEAE cellulose column chromatography. The pppApCp(Gp), pppApApCp (XpGp), pppGpCp. . ., and pppGpGpPyp. . . sequences have been characterized previously ([196, 222] and E. Hefti and D. H. L. Bishop, unpublished results, 1975).

[^{32}P-α]-ATP product as might be expected for mixtures of capped and uncapped product termini. Third, with SAM in the reaction mixture, alkaline phosphatase released phosphate and a second nucleotide fragment which eluted from DEAE cellulose columns with a −4 charge and, from nearest neighbor analyses was found to correspond to a capped nucleotide of sequence m^{7}GpppAmpA (Fig. 5.7). In this case, the distribution of label between inorganic phosphate and the second capped nucleotide was about 1 : 1 for the [^{32}P-α]-GTP product and 2 : 1 for the [^{32}P-α]-ATP product (again as might be expected for mixtures of both capped and uncapped product termini).

The conclusions to be drawn from these results are the following. First, because termini can be labeled by [^{32}P-γ]-purine nucleotides, product RNA synthesis is initiated by purine nucleoside triphosphates and among the product RNA species there are several sequences present.

Second, there are both capped and uncapped product termini.

Third, those sequences which are capped become methylated when SAM is present. This capping is probably selective and may be rate limiting.

The in vitro transcription product RNA sequences analyzed by Banerjee and associates have been found to contain the capped and methylated sequence m^{7}G$^{5'}$ppp$^{5'}$AmpApCpXpGp. . . , where m^{7}G is the capping nucleoside which is methylated at the 7N position, and Amp is a 2′-0-methyl, 3′-phosphate derivative of adenosine ([229] and A. Banerjee, personal communication, 1976). In contrast to our results they have not detected other product termini. The reason is not clear. Hefti and Bishop have confirmed the presence and identity of the m^{7}G$^{5'}$AmpApCpXpGp . . . sequence among the in vitro product RNA sequences initiated by VSV Indiana transcriptase. To date we have no evidence of capping or methylation of the pppApCpGp sequence. Whether the sequences initiated with guanosine nucleotides also become capped and methylated is under investigation.

The proportion of [^{32}P-γ]-ATP- to [^{32}P-γ]-GTP-labeled initiation sequences is dependent upon the length of time that an in vitro reaction mixture is incubated [197]. In long-term reaction mixtures, it has been shown that the ratio of [^{32}P-γ]-purine nucleotide-labeled sequences is 4.5 for pppApCpGp. . . to 3.5 for pppGpGpPyp. . ., to 1.5 for pppApApCpXpGp. . ., to 1.0 for pppGpCp. If one considers that some of the original pppApApCpXpGp sequences were converted into m^{7}G$^{5'}$ppp$^{5'}$ApApCpXpGp sequences and lost their γ-^{32}P label (since the capping guanosine nucleotide supplies its α and β phosphates in lieu of the γ and β phosphates of the initiating adenosine nucleotide triphosphate [224]), therefore there must have been more original initiation sequences starting with pppApApCpXpGp than the above ratio indicates. The data described in Fig. 5.7 with [^{32}P-α]-ATP (minus SAM) support this view.

Banerjee and associates have obtained evidence (224, 225] and A. Banerjee, personal communication, 1976) which suggests that all types of in vivo and in vitro

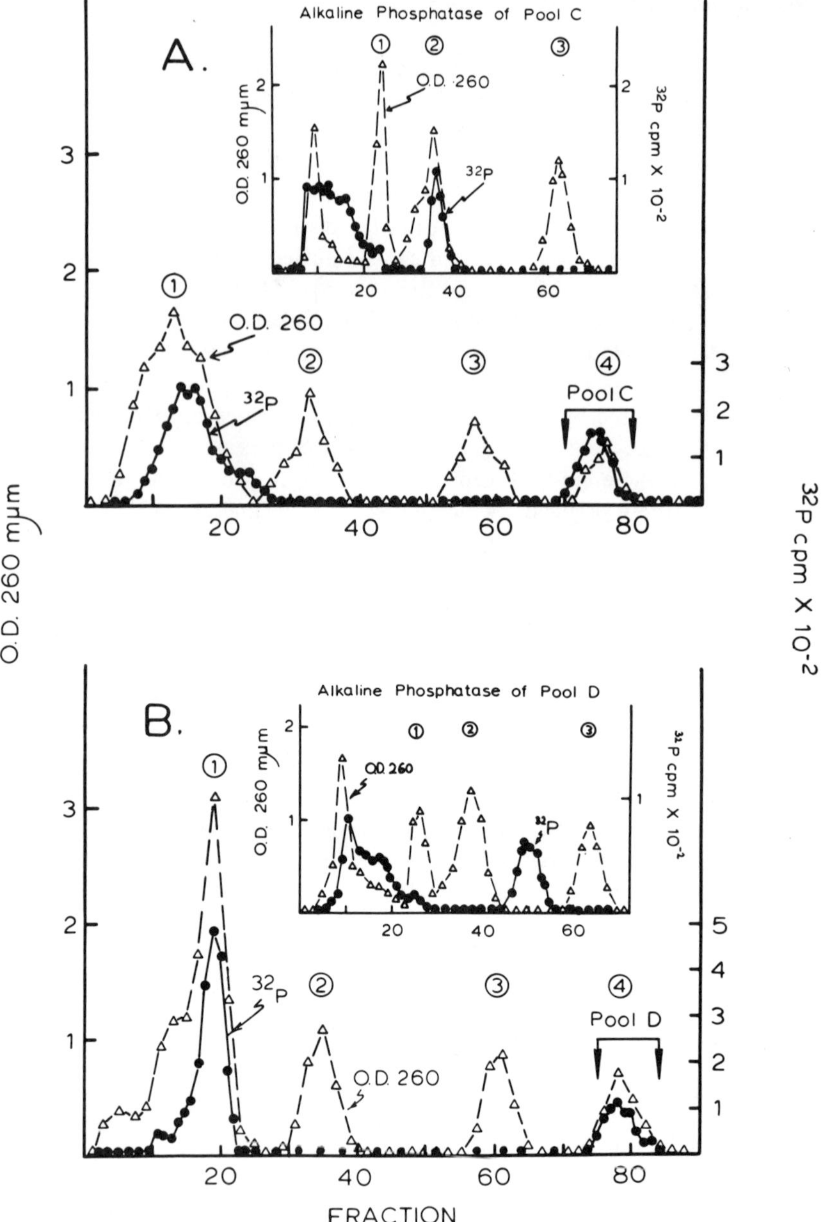

Fig. 5.7 Analysis of VSV Indiana transcription product labeled by (A, B) [α-^{32}P]-ATP or (C,D) [α-^{32}P]-GTP. VSV Indiana transcription product obtained from reaction mixtures with (B,D), or without (A,C) S-adenosyl-L-methionine (SAM), were treated with 2 cycles of alkali digestion [126], chromatographed with suitable oligonucleotide markers (Fig. 5.5), the tetranucleotides were recovered, treated with alkaline phosphatase and rechromatographed (respective inserts). (Unpublished data of E. Hefti and D. H. L. Bishop, 1975).

^{32}P-α-GTP VSV product RNA ⊖ SAM

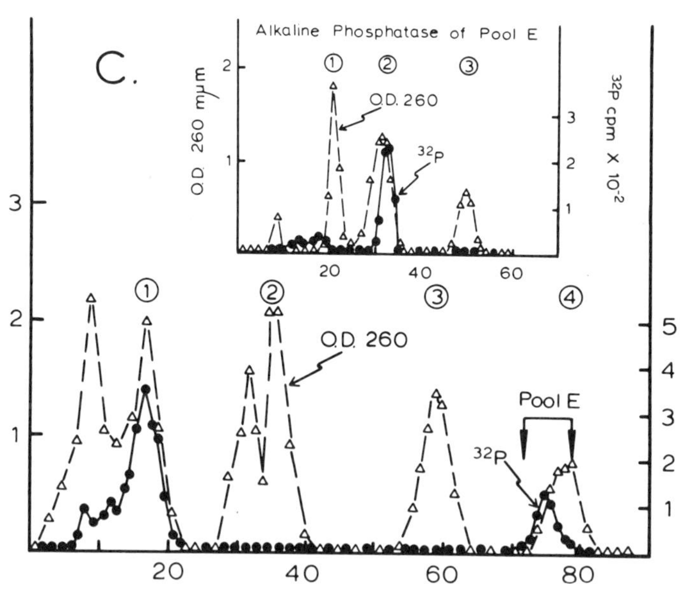

O.D. 260 mμm

^{32}P-α-GTP VSV product RNA ⊕ SAM

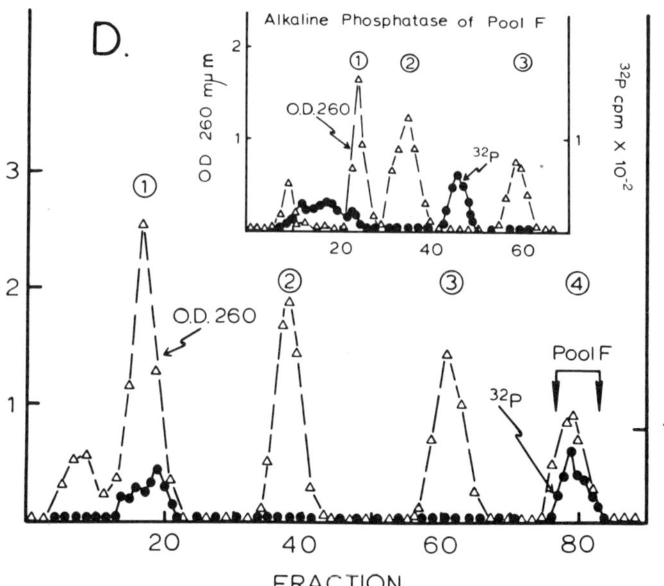

mRNA species possess similar 5′ sequences (m^7GpppAmpApCpXpGp). Evidence obtained by Rose and associates indicates that for each in vivo mRNA there is variation in the terminal sequences as well as in the capping and methylation of some of those sequences.

The relevance of other observed in vitro initiation sequences to the transcription process producing mRNA (as opposed to replication initiation or incorrect in vitro initiations) is not known. We also do not know which proteins constitute the capping or methylation enzyme activities or if they are host- or viral-instructed functions. It is conceivable that both activities may be captured by progeny virions upon their egress from a cell, although whether that is a necessity is not known, since the cell that is to be infected can presumably also supply those functions.

Evidence obtained from both Shatkin's and Banerjee's laboratories indicates that for reovirus and VSV, nonmethylated mRNA species need to be methylated before they can function in an in vitro protein-synthesizing system [226]. Clearly, therefore, if this pertains to all viral mRNA species, posttranscriptional modification is an important function in mRNA processing.

In summary, then, it appears that in vitro transcripts can be initiated by purine nucleotides. Of the various sequences initiated in vitro, at least 1 type is subsequently capped and methylated, another does not appear to be capped, and whether that sequence or other initiated sequences represent replication attempts is not known. All available evidence points to the conclusion that each type of mRNA is separately initiated and therefore that there are particular initiation sites upon the template RNA for the enzyme to recognize and commence transcription. From both the in vivo and in vitro mRNA studies one can conclude that there are probably 5 initiation sites on the VSV genome for mRNA synthesis.

b. Transcription Termination, Polyadenylation. Are there particular stop sites for transcription termination? The answer to this question is not known but is probably in the affirmative. What is clear is that polyadenylation of the mRNA product can occur at the 3′ end of the RNA producing poly(A) stretches of up to 200 nucleotides [185, 186]. This is also true for the VSV in vivo mRNA species [227-231]. Whether polyadenylation occurs on most or all product RNA strands is not certain. In our laboratories we have observed less polyadenylation of VSV transcription products than that reported by others (D. H. L. Bishop, unpublished observations, 1974). This may be caused by our reaction conditions or by the virus or cell strains employed. There is no decisive evidence to indicate whether polyadenylation is an attribute of the virion transcriptase or another enzyme present in minor amounts; however, since the viral RNA-N complex lacking NS and L proteins has neither poly(A) nor transcriptase enzyme activities and both activities are restored on the return of NS and L, this can be taken as suggestive evidence that the poly(A) activity may be an attribute of the virion polymerase (S. U. Emerson, personal communication, 1976). No long stretches of poly(U) have been identified in the viral RNA sequences [227, 228] so presumably the poly(A) elongation proceeds as a nontemplated function.

The poly(A) stretches are heterogeneous in size (50-200 nucleotides), and this may account for the observed heterogeneity of the majority of the transcription product RNAs (12-18 S). It is conceivable that poly(A) synthesis is initiated from a short stretch of viral poly(U) (say, 5-6 nucleotides) and that through slippage or nontemplated activity polyadenylation is continued.

c. Transcriptional Control as Evidenced by In Vitro Analyses. The possibility that there may be some form of transcriptional control operative during in vitro analyses stems from the observation that the in vitro product RNA does not equally represent all parts of the viral genome [198]. The mRNA obtained in vivo also is disproportionate [197]. Despite the obvious problem that the in vitro analyses may not represent in vivo conditions it is worthwhile speculating about potential control mechanisms which are suggested by these analyses.

One possible control point could be the transcription initiation sequences. If, as all evidence currently suggests, the individual messages are transcribed in a 5′ to 3′ direction involving separate 5′ initiations, then the preference of the transcriptase to pick 1 initiation point over another would result in selective transcription. Other than template sequence considerations, how could 1 initiation point be preferred over another? One possibility would involve a role for the NS phosphoprotein, whereby at particular locations on the RNA-N protein complex, the NS protein promotes initiation at greater efficiency than at other locations. The relative efficiencies may reflect the number of NS molecules per location or the degree of NS phosphorylation, etc. In parenthesis, it has been shown that there is a protein kinase activity associated with VSV cores and that the endogenous substrate for this activity is the NS protein [137, 189, 190]. We do not know if all NS proteins are equally phosphorylated, or if some are selectively phosphorylated during transcription. Again, as in the case of the nucleoside triphosphatase, the phosphotransferase, the poly(A) synthesis and the methylation activities, which protein is responsible for the protein kinase activity is not known.

Second, transcription control mechanisms could reflect differences in the rate of initiation as opposed to the rate of RNA elongation. Alternatively there could be differences in the elongation rate of different transcripts whereby longer transcripts are transcribed at a slower rate because of the intrinsic sequences involved, or by requiring greater or more sustained unwinding of the RNA-N protein complex.

Third, although stop or prevention signals in transcription have no supportive evidence at this time, it appears reasonable from the available evidence to suggest such signals exist. If so, then they could function in controlling transcription by holding up the completion or release of some, but not all transcripts. Again this could be a possible role for the NS protein or involve differences in the poly(A) fabrication machinery.

As a fourth possibility for transcriptional control, a sequential synthesis mechanism can be envisaged, whereby there is an interdependence of transcript syntheses; transcript A being synthesized before transcript B synthesis is triggered, with the efficiency of triggering B being less than 100%, etc. Such a procedure may be more plausible when one considers the initial primary transcription process in the infected cells if, after uncoating, an unwinding of the infecting virion RNA-N protein complex is involved.

d. Identification of the Transcriptase Proteins, Reconstruction of In Vitro Enzyme Activity, and Infectivity. Experiments have been undertaken to identify the protein species involved in RNA transcription for VSV Indiana. Subviral "dextran" cores have been obtained by polyethylene glycol-dextran phase separation of detergent- and salt-dissociated preparations of VSV [142]. These cores lack all the viral G and M proteins but possess all the initial virion transcriptase activity. Such cores are also infectious [167, 169].

Dissociation of VSV Indiana virus preparations into RNA-N protein complexes (lacking the G, M, NS, and L proteins) have given structures which are noninfectious and which did not exhibit transcriptase activity [170, 173]. Restoration of purified L protein to these RNA-N protein complexes, when they contain a small amount of NS, restored some transcriptase activity [232]. Restoration of unfractionated, solubilized viral proteins (containing G, M, L, and NS proteins) to the RNA-N protein complex restored both all the initial virion transcriptase activity *as well as some infectivity* [170]. Attempts at using naked RNA as a transcription template together with unfractionated, solubilized viral proteins have so far been unsuccessful, even when the same proteins will efficiently utilize the RNA-N protein complex (D. H. L. Bishop, unpublished observations, 1974).

Similar attempts to use other DNA, RNAs, or synthetic polynucleotides as substrates for the transcription process have also been negative, suggesting that the enzyme is specific for its own type of template [200]. Recent experiments by Emerson and Yu have shown that both L and NS are required for the in vitro restoration of transcriptase activity to RNA-N complexes [174]. The requirement for NS was also shown using an anti-NS serum [233]. How do NS and L interact in transcription? We do not know how they interact, nor do we know if 1 is the polymerase and the other is used to open up the RNA-N template for the enzyme.

Can other rhabdoviral RNA-N complexes be used by the VSV Indiana transcriptase enzyme? Reconstruction of enzyme activity and infectivity for homologous dissociated components of other rhabdoviruses [170] has been achieved (e.g., VSV New Jersey RNA-N complexes plus VSV New Jersey solubilized enzyme). Successful homologous reconstructions were obtained for VSV New Jersey, Cocal, and Chandipura viruses in addition to VSV Indiana. Attempts at reconstituting enzyme activity and infectivity from heterologous dissociated components of these rhabdoviruses have been successful for reciprocal combinations of VSV Indiana and Cocal viruses but unsuccessful for combinations of VSV Indiana, VSV New Jersey, and Chandipura viruses [170]. This parallels in part the weak heterologous complementation that can be obtained between VSV Indiana and Cocal temperature-sensitive mutants but not with temperature-sensitive mutants of VSV New Jersey or Chandipura viruses [234-236]. These results also parallel the slight nucleotide sequence homology between VSV Indiana and Cocal viruses and the relative lack of genome homology between the other members of this subgroup (see Table 5.4).

e. Translation of In Vitro Synthesized mRNA. It has been shown that VSV messenger RNA isolated from infected cells consists of a viral-complementary 28 S and an 11-18 S mixture of species [229, 237]. The 28 S species corresponds to the expected RNA size for, and is the message progenitor of, the viral L protein [238]. Likewise the 11-18 S mRNA species correspond to and are the message progenitors of [238] the viral N, NS, M, and G proteins (see p. 256). The fact that the in vitro (or in vivo) 11-12 S product RNA makes NS and M proteins during in vitro translation, the 14-15 S RNA makes N protein and the 17-18 S RNA makes G protein substantiates these findings [239, 240]. These results also suggest that the true polypeptide size of NS protein may be of the order of 20,000-30,000 daltons.

3. *In Vivo Studies on the Primary and Secondary Transcription of Viral mRNA*

a. Primary Transcription. Within an infected cell, the initial synthesis of viral messenger RNA occurs by transcription from the parental genome (primary transcription). This synthesis can be monitored using either cycloheximide or puromycin to inhibit cellular and viral protein synthesis, thereby preventing the appearance and activity of new transcriptase or replicase enzymes [144, 241-243].

The number of adsorbed viruses which participate in primary transcription has been estimated from the percentage of virus RNA molecules which become ribonuclease resistant as a result of the infection process [144]. Values of about 1 out of 5 are commonly observed for VSV, although occasional values of 1 out of 2 have been obtained [197]. As far as can be ascertained, the inactive particles remain inactive throughout an infection cycle even during high-multiplicity infections; they do not appear capable of being "rescued" by active particles [197, 215]. It does not seem likely (Fig. 5.3) that the inactive virions represent viruses which have not penetrated (since most of the adsorbed viruses become uncoated). Probably their defect is in some structural or enzymatic attribute.

By hybridization experiments, the rate, completeness, and extent of VSV Indiana primary transcription have been measured [144, 215]. The methodology employed involved infecting cells at a low multiplicity of infection with [^{3}H]-nucleoside-labeled virus of high specific activity. The infection was carried out in the presence of cycloheximide but in the absence of any other drug or radioisotope. The infected cell nucleic acids were extracted at various times postinfection and annealed to purified [^{3}H]-viral RNA and the content of viral-complementary RNA determined from the increase in [^{3}H]-ribonuclease resistance. The results of such experiments indicated that primary transcription can occur at a linear rate for up to 6 hr postinfection (Fig. 5.8). The fact that the transcripts represent the entire viral genome was demonstrated [144] by self-annealing the infected cell nucleic acids and rendering the [^{3}H]-RNA therein completely ribonuclease resistant (Fig. 5.9). Flamand and Bishop [144] showed that primary transcription proceeded at a greater rate at 38°C than at 31°C (or 40°C). From the number of active viruses per cell and the rate of viral-complementary RNA synthesis, they calculated that 1 genome mass equivalent of viral-complementary RNA was synthesized every 90 sec [144, 215]. If the 28 S viral messenger RNA represents half the genome and the other 4 messenger RNA species about 12% each, then this rate corresponds to a synthesis time of 45 sec for the 28 S RNA and about 6-18 sec for the other species.

As far as primary transcription is concerned, it has been shown that neither actinomycin D, cycloheximide, puromycin, interferon, nor poly(rI) : poly(rC) pretreatments affect primary transcription, although the latter 2 totally inhibit secondary transcription, presumably by inhibiting or preventing viral protein synthesis [144, 214, 215]. The rate of primary transcription developed in interferon-treated cells was observed to be comparable to that found in cycloheximide- or interferon-plus-cycloheximide-treated cells [214]. These results did not preclude the possibility that the transcripts were less stable in the interferon-treated cells because of induced nucleases or lack of proper processing or other factors. However, they did indicate that the transcriptase itself was not directly affected by interferon-induced products.

b. The Intracellular Secondary Transcription Process of VSV-Indiana. It has been shown that from 45 min to 1 hr postinfection, the rate of viral-complementary RNA synthesis dramatically increased (Fig. 5.10). This

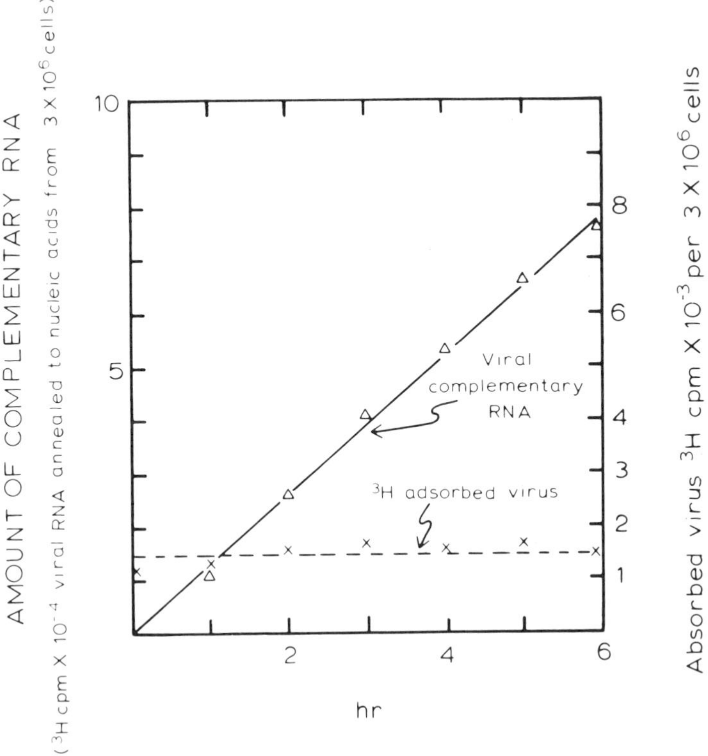

Fig. 5.8 Primary transcription of VSV Indiana in BHK 21 cells. [^{3}H]-Nucleoside-labeled virus, 2 X 10^5 cpm [144], was allowed to adsorb to monolayers of 3 X 10^6 BHK 21 cells for 30 min at 4°C. The monolayers were washed, then incubated at 37°C for up to hr hr in Eagle's medium containing 5% fetal calf serum and 100 μg cycloheximide/ml. The cell nucleic acids were extracted [144] and the content of [^{3}H]-label was determined. The amount of viral-complementary RNA was determined by hybridizing the cell nucleic acids to excess of [^{3}H]-viral RNA obtained from the unused inoculum [215]. The increase in [^{3}H]-resistance represents the accumulated content of viral-complementary RNA [215]. Reprinted from Ref. [215], p. 35, by courtesy of Academic Press.

increase was not affected by treating the cells with actinomycin D but was inhibited by cycloheximide, puromycin, interferon, or poly(rI) : poly(rc) treatments [144, 214, 215]. Neither secondary transcription [215] nor viral-like RNA has been observed in cells infected with group IV or group I temperature-sensitive mutants grown at nonpermissive temperatures for virus production [197]. The conclusion to be reached from those results is that secondary

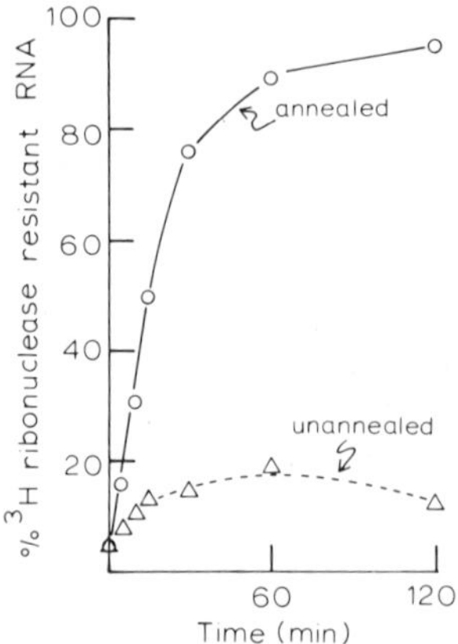

Fig. 5.9 The completeness of in vivo transcription. The increase in [^{3}H]-ribonuclease resistance before or after self-annealing infected cell nucleic acids (obtained as described in Fig. 5.8) was determined. (Unpublished data of A. Flamand and D. H. L. Bishop, 1974).

transcription and the synthesis of viral RNA require the synthesis and expression of new viral gene products.

In an experiment which involved adding cycloheximide at various times postinfection and measuring the linear rates of RNA transcription established thereafter (by comparison with that obtained during primary transcription), the number of intracellular transcriptive intermediates has been estimated (Fig. 5.10). Such estimates have given values of around 450 transcriptive intermediates per infected BHK-21 cells by 3-6 hr postinfection [215]. No increase in the number of intermediates was observed subsequent to 3 hr postinfection, although progeny virus continued to be liberated.

The total amount of viral complementary RNA produced per infected BHK cell has been claculated [215]. Values of 2×10^4 genome mass equivalents per cell (i.e., about 1×10^5 individual mRNA species) have been obtained.

Similar experiments involving a 30-fold higher multiplicity of infection indicated that by 3-6 hr postinfection, equivalent amounts of viral complementary RNA were obtained to that observed for the lower multiplicity of infection, even though the initial rate of RNA synthesis was 30-fold higher [215]. These results suggested that there was some *finite limit to the amount of viral-complementary RNA which could be synthesized in an infected cell* and also some form of host-modulated regulation of transcription.

Comparative experiments employing the same batch of virus and different cell types (e.g., BKH-21, mouse L-cells, and secondary chick embryo fibroblasts have indicated that both the yield of infectious virus per cell as well as the accumulated amount of viral-complementary RNA are dependent on the host cell type [144, 214, 215]. For VSV, the BHK cell line is 1 of the most permissive host cell types we have so far investigated. Which host cell factors influence this permissiveness are not known.

It is noteworthy that secondary transcription induced by VSV in BHK cells starts at a time when the first progeny virus and newly synthesized viral-like RNA can be detected (see Fig. 5.10). These observations strongly suggest that the templates for secondary transcription are the newly synthesized viral-like RNA species. These observations also suggest that once formed, even at such an early stage of the infection cycle (viz., the first rounds of replication), the progeny viral-like RNA species have a possibility of becoming progenitors of infectious virus particles or templates for secondary transcription.

At late stages of the infection cycle when the number of transcriptive intermediates reaches a plateau, a majority of the viral-like RNA species give rise to progeny virus particles [215]. Why? What interrelationships exist between primary transcription and replication, and between replication and secondary transcription? Since the template for primary transcription must be the template for replication (there is no other, unless one proposes stitching together the messenger RNA species), what form of conversion from transcription to replication can be envisaged?

c. Evidence for the Intracellular Control of Transcription. Evidence suggesting that some form of transcriptional control is operative within the infected cell comes from 2 sources. First, it has been shown that there is a disproportionate representation (on a molar basis) of the various VSV messenger RNA species [215]. Which species are involved is not known although it is probable that there are fewer 28 S mRNA species than other species. The reason for this judgement came from hybridization experiments using defective T-particle RNA (VSV-111) and viral RNA to determine the relative molar representation of the different mRNA sequences. On a molar basis, less than 1/3 of the total mRNA annealed to the T-particle RNA by comparison to that which annealed to the viral RNA [215]. It has been recently shown that the T-particle RNA is probably equivalent to part of the 28 S messenger RNA [130].

Second, examination of the relative proportions of messenger RNA species obtained by labeling experiments have also indicated that there is always less 28 S RNA than the other mRNA species [143, 237, 244]. It is not known if some of the other mRNA species in the 11-18 S size range are represented unequally.

In addition to the suggestions discussed previously of how transcription might be controlled, the possibility of an interrelationship between

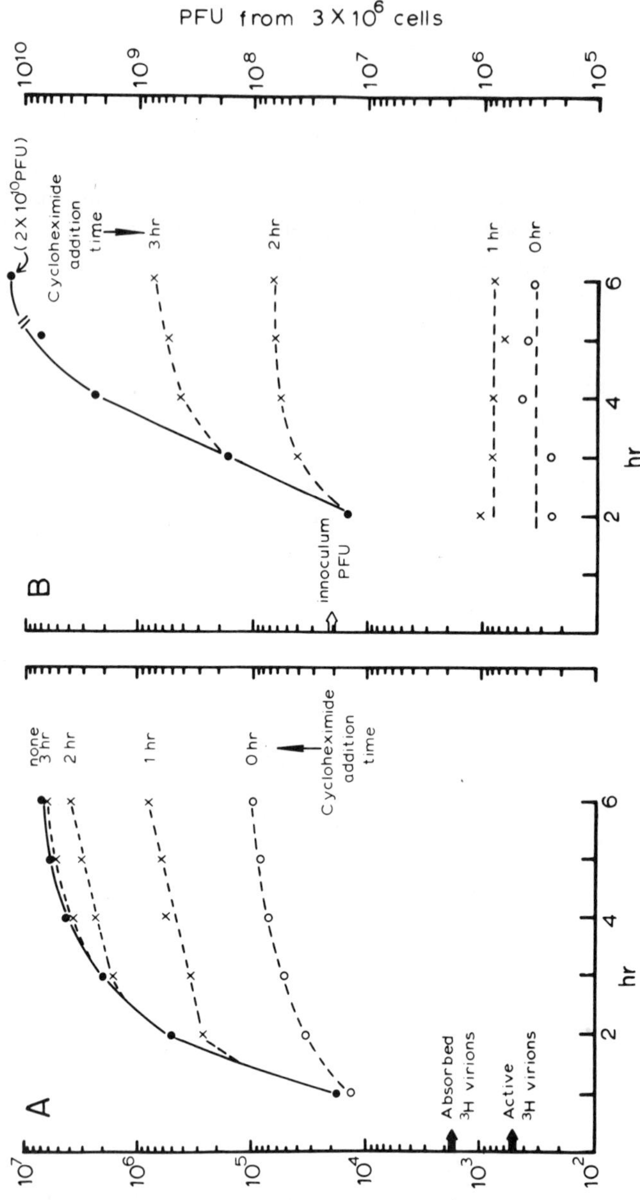
A
B
AMOUNT OF COMPLEMENTARY RNA
(cpm 3H viral RNA annealed to RNA from 3 X 10^6 infected cells)
PFU from 3 X 10^6 cells
hr
none
3 hr
2 hr
1 hr
0 hr
Cycloheximide addition time
Absorbed 3H virions
Active 3H virions
innoculum PFU
(2 X 10^10 PFU)

Fig. 5.10 Effect of addition of cycloheximide at different times upon the production of virus and intracellular accumulation of viral complementary RNA in BHK 21 cells.

Monolayers of BHK 21 cells were infected in the presence of DEAE-dextran with 3×10^4 cpm (2×10^7 PFU) of [^{3}H]-labeled virus and incubated at 38°C. For the monolayers not treated with cycloheximide (continuous lines), the infected cell nucleic acids were extracted and adsorbed [^{3}H]-label ($19 \pm 3.5 \times 10^2$ cpm), its RNAase resistance (20 ± 4%) and the content of viral-complementary RNA were determined (A). For each of these monolayers the medium was changed 1 hr postinfection to remove desorbed virus and fresh warm medium was added. At the time of harvest, the content of released PFU in the supernatant fluids was determined (B), continuous line).

One series of infected monolayers was treated immediately postinfection with 100 μg cycloheximide/ml (broken line, 0 hr) and similarly incubated at 38°C. The released PFU (B) or cell-associated [^{3}H]-label, its RNAase resistance, and the intracellular viral complementary RNA were determined (A). For this series, as well as the next 3 series, the cell-associated [^{3}H]-label and its RNAase resistance were the same as in the untreated monolayers.

For another series of untreated monolayers, after 1 hr of incubation at 38°C, the supernatant fluids were removed and replaced with medium containing per ml 100 μg cycloheximide (broken line, 1 hr) and the subsequent release of PFU determined (B). Monolayers were removed from this series at hourly intervals to determine the [^{3}H]-label, its RNAase resistance, and the intracellular content of viral-complementary RNA (Fig. 5.8). Similar series of monolayers were treated 2 hr and 3 hr postinfection and analyzed as described above.

The hourly rates of viral comlementary RNA synthesis were determined for each set of treated monolayers. A mean value (± S.D.) equivalent to $1.7 \pm 0.3 \times 10^4$ ^{3}H cpm viral-complementary RNA per hour was obtained for the zero time, cycloheximide-treated, sample. For the 1 hr-treated samples, the mean rate of RNA synthesis from 2 to 6 hr postinfection was equivalent to $1.2 \pm 0.9 \times 10^5$ ^{3}H cpm of viral-complementary RNA per hour. For the 2 hr-treated samples, the mean rate of RNA synthesis from 3 to 6 hr postinfection was equivalent to $8.2 \pm 0.8 \times 10^5$ ^{3}H cpm viral-complementary RNA per hour and for the 3 hr drug-treated samples, a mean rate equivalent to 11 ± 10^5 ^{3}H cpm viral-complementary RNA was obtained between 4 and 6 hr postinfection. Reprinted from Ref. [215], p. 39, by courtesy of Academic Press.

transcription and translation should not be overlooked. Are RNA transcripts picked by the translational machinery with an equal probability despite sequence, size, and conformational differences? If there is a selection (or selective processing procedure), then there could also be a feedback mechanism. Until these questions can be answered, we will not fully understand the transcriptional control mechanisms or their ramifications.

d. Characterization of the Viral mRNA Species Formed In Vivo. The viral mRNA species for VSV Indiana have been identified as a 28 S RNA which translates to give the viral L protein, a 17-18 S RNA which can be translated into G protein, a 14-15 S RNA which makes N protein, and a 11-12 S RNA which codes for both NS and M proteins [238-240]. These mRNA species possess 3′-polyadenylated sequences which facilitates their isolation through the use of oligo-dT cellulose columns. It is not known if all VSV mRNA species possess 3′-polyadenylated sequences.

In recent work from the laboratory of J. Rose, it appears that most VSV Indiana mRNA species are capped and methylated, although some uncapped sequences were also found (J. Rose, personal communication, 1976). In Banerjee's laboratory only 1 capped sequence was identified [225]. Rose found that each size class of VSV Indiana messenger RNA contained a variety of 5′ sequences. The majority (65-70%) were either $m^7G^{5'}ppp^{5'}AmpAp$ . . . or $m^7G^{5'}ppp^{5'}mAmpAp$ . . . (where mA is a base methylated adenosine). Of the other 5′ termini, 20% were in $m^7G^{5'}AmpmAmpCp$ . . . and $m^7G^{5'}ppp^{5'}$-mAmpmAmpCp . . . sequences. The remaining 10-15% of the terminal sequences possessed pppAp . . . or pppGp nucleotides and were on mRNA size transcripts not associated with ribosomes. Clearly from these results it is evident that both ribose and base methylations are variable functions and that both may be required before selection for translation by ribosomes. That adenosine and guanosine triphosphates were also found is an interesting observation and may tie in with some of the previously reported in vitro results.

4. *A Summary and Model of Primary Transcription*

What is known about the in vivo and in vitro RNA transcription process is summarized below.

1. An infecting virion possesses several RNA-instructed RNA polymerase enzymes which are responsible for the primary transcription of viral-complementary mRNA. The direction of RNA transcription is 5′ to 3′.

2. Each mRNA is individually initiated with a purine nucleoside triphosphate. The 5′ end of these RNA species can be capped (which may be a rate-limiting process) and variously methylated, while their 3′ ends possess a polyadenosine nucleotide sequence of variable length.
3. The viral RNA-N protein complex is the template for the transcriptase (L + NS). At best, transcription involves only a temporal removal of N from the RNA.
4. The transcription process is highly repetitive, somewhat selective with regard to which mRNA species are transcribed most frequently, and possibly sequential.

The events involved in primary transcription are summarized in diagrammatic form in Fig. 5.11. We suggest that the process possesses the following attributes.

1. There are a limited number of specific transcription initiation sites on the template complex.
2. There are specific transcription stop regions, e.g., short poly(U) sequences which trigger slippage, release, and polyadenylation of the 3′ end of the mRNA.
3. The N proteins are contiguous. They are hydrogen bonded to each other and to the phosphate backbone of the viral RNA. Such N protein associations protect both the 3′ and 5′ ends (as well as internal regions of the RNA) from nuclease digestion. This sequestering function of N protein is in addition to its role in maintaining the helical structure of the nucleocapsid in the virion, as well as its function in transcription and replication.
4. The function of the NS phosphoprotein is to lift the N protein off the RNA during product RNA synthesis by substituting its phosphate for the phosphate of the template RNA. This creates a bubble in the N protein framework and allows access of the transcriptase to the RNA. It is not necessary for this function that the NS molecules are integral parts of the transcriptase (as shown in Fig. 5.11); they could just initiate a bubble by attaching to N protein at particular sites.

   Either way, we propose this as *the bubble hypothesis* of RNA transcription.
5. Ribosomes can attach to transcripts after capping and methylation. This may occur either prior (as shown in Fig. 5.11) or subsequent to the completion of RNA transcription, depending upon the efficiency of the capping and methylation processes.

As an alternative to the bubble hypothesis we postulate a *groove hypothesis*. In this case we suggest that the attributes of the RNA-N complex are

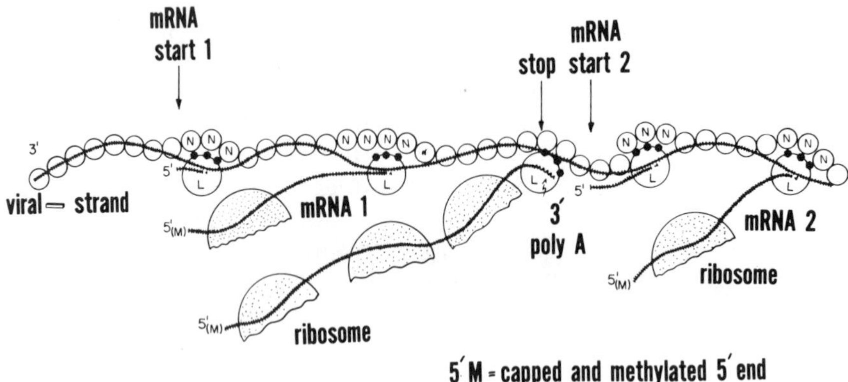

Fig. 5.11 A model of primary transcription.

functionally the same as those described above, except that the bases of the RNA are exposed in a groove and accessible to the transcriptase. During transcription there would be no liftoff of the N protein. We suggest that structural variations of the groove or enzyme-NS interactions with it would designate transcription initiation or termination sites.

5. *In Vivo Studies on the Replication of Viral RNA*

Without discounting the value of the published work there are no reports which in our opinion have clearly identified the in vivo replication apparatus of VSV Indiana viral RNA as opposed to the intracellular transcription apparatus. The reason for this opinion will be clear later. There have been reports of infected cell extracts capable of synthesizing RNA for a limited period, but all the extracts so far described and analyzed [245-247] appear to make viral-complementary RNA (i.e., represent transcriptive complexes). Likewise double- or multistranded RNA species isolated from infected cells are probably, for the most part, transcriptive intermediates [121, 248-253]. The studies by Huang and colleagues using defective-interfering T particles and identifying the T RNA replicative species induced in cells coinfected with B and T particles [242, 254] are certainly more conclusive with regard to characterizing the replicative procedure.

Why has it been so difficult to identify the replicative machinery? The reason is evident from the experiments presented in Fig. 5.12, in which it is shown that in infected cells, *viral RNA synthesis is always less than viral-complementary (VC) RNA synthesis.* In these experiments, the synthesis and amount of intracellular VC RNA (mRNA, plus strands) have been related not

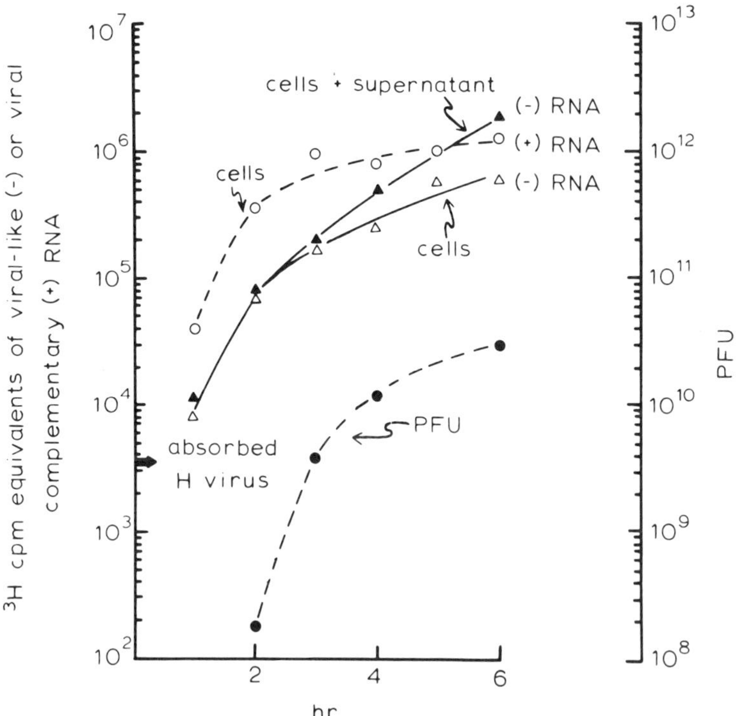

Fig. 5.12 Viral and viral-complementary RNA syntheses at 38°C in 3 X 10^6 BHK 21 cells infected by wild type VSV Indiana. The accumulated synthesis of viral-complementary (+) RNA in BHK 21 cells infected with wild-type VSV Indiana was determined as described in Fig. 5.10. The release of PFU from the infected 3 X 10^6 cells was also determined (Fig. 5.10). The intracellular and extracellular accumulation of viral-like (–) RNA was determined as described in the text.

only to an infection timecourse, but also to the production of viral-like RNA (minus strands). By extracting both cells and the released progeny virus, the intracellular and extracellular accumulation of viral-like RNA has also been enumerated (A. Flamand and D. H. L. Bishop, unpublished data, 1975). The protocol of the experiment was as follows. High specific activity [^{3}H]-nucleoside-labeled virus was used to infect monolayers of 3 X 10^6 BHK cells. After 30 min of adsorption at 4°C, the cells were washed and incubated at 37°C. Cell monolayers were removed at intervals, the released virus PFU was determined in the supernatant fluids, and both the cells and supernatant fluids were extracted for RNA. The average ^{3}H label associated with the monolayers was determined (adsorbed virus ^{3}H cpm) and the amount of VC

RNA determined by annealing aliquots of the infected cell nucleic acids to excesses of [^{3}H]-viral RNA (obtained from the unused inoculum virus). It was observed that the increase in ribonuclease-resistant [^{3}H]-RNA equated to the accumulated amount of viral-complementary RNA [144, 215]. These values are plotted in Fig. 5.12 and enumerated as the plus strand RNA. Other aliquots of the cell nucleic acids were self-annealed and the percent [^{3}H]-ribonuclease resistance was determined (see Fig. 5.13 for equivalent experiments conducted at 31, 38, or 40°C). The percentage of [^{3}H]-RNA which was rendered ribonuclease resistant was found to increase to 100% and then decrease as a function of the infection timecourse and temperature of incubation. Since the percentage of [^{3}H]-ribonuclease resistance stayed at 100% when replication was inhibited (e.g., in the presence of cycloheximide [144], or for group I and group IV temperature-sensitive mutants when held at nonpermissive temperatures [144, 215], it is evident that the observed decrease in [^{3}H]-ribonuclease resistance was caused by competing, unlabeled,

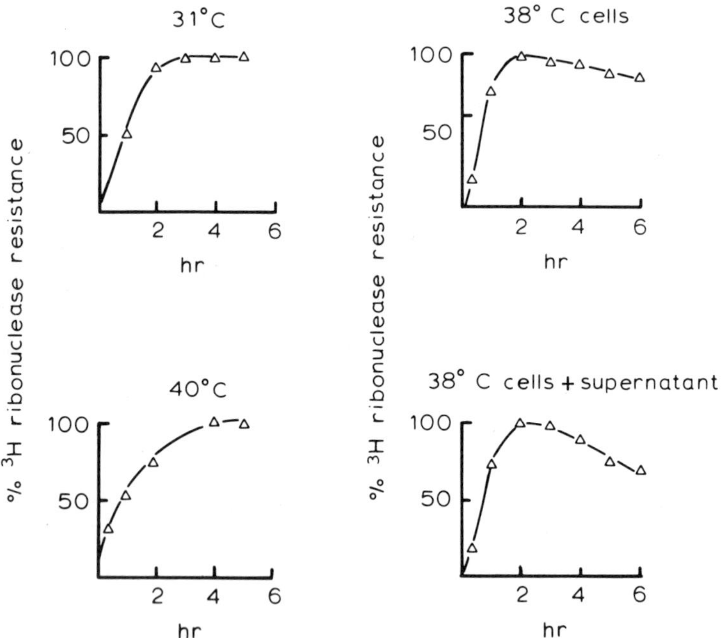

Fig. 5.13 An enumeration of viral-like RNA synthesis in infected cells. BHK 21 cell monolayers were infected with [^{3}H]-labeled virus and grown at 31°C, 38°C, or 40°C. RNA was extracted from the cells and supernatant fluids. After the infected cell nucleic acids were self-annealed, the [^{3}H]-ribonuclease resistance was determined. The decrease in ^{3}H resistance (38°C extract ± supernatant extract) was used to calculate the amount of competing unlabeled viral-like RNA (see Fig. 5.12 and text).

newly synthesized viral-like RNA. That this is true is evident from the fact that the decrease in percent [^{3}H]-ribonuclease resistance was greater when the RNA from the supernatant fluids were included (Fig. 5.13). The decrease in [^{3}H]-ribonuclease resistance was used to calculate the unlabeled competing viral-like RNA. From these observed decreases, and the amount of vc RNA determined as described above, both the intracellular accumulation and amount of released viral-like RNA were calculated (Fig. 5.12).

The results of these and other experiments [144, 215] show that:

1. Viral complementary RNA synthesis continued throughout the productive infection timecourse. Initially the rate of VC RNA synthesis was linear (primary transcription), and then exponential (secondary transcription), and finally, when the majority of progeny virions were being released, it was again linear.
2. Viral RNA replication occurred 45 min postinfection as evidenced by the onset of secondary transcription *and* the presence of intracellular new viral RNA.
3. Viral RNA synthesis was exponential from 45 min postinfection but tailed off later in the infection timecourse.
4. After 3 hr postinfection, the accumulation of viral-like RNA in released progeny virus was greater than that remaining in the cell.
5. The rate of synthesis of viral RNA was never greater than that of viral-complementary RNA at any stage of the infection cycle.

The last fact indicated that *even under optimal conditions there was more RNA transcription than replication* (note the amount of virus produced from 3 $\times$ 10^6 BHK cells by 6 hr postinfection when all cells were already showing cytopathic effects).

Consequently it is not surprising that most extracts made from infected cells represent transcription activities. It should also be noted that the production of viral RNA (even more than transcription) is quite temperature dependent; within the cell at 31 or 40°C, it is evident from the results presented in Fig. 5.13 that more transcription products were present than replication products.

A recent finding by Soria et al. [255] of intracellular 42 S plus strand RNA in an N-RNA complex in the first tangible evidence of an RNA intermediate which might be involved in replication. We have found similar structures in infected cells and this has led us, like Huang, to believe that replication involves an RNA-N protein intermediate which contains a 42 S strand RNA.

An intriguing question concerning the replication process is its interrelationship with transcription. It is clear that the synthesis of viral-like RNA depends on viral gene products becoming available through translation of the primary mRNA transcripts. This is shown by the fact that temperature-sensitive mutants of group I and group IV do not synthesize viral-like RNA at nonpermissive temperatures [197, 289]. Mutants of group III and group V, in contrast, do synthesize viral-like RNA at high temperatures [197]. Also, no viral RNA is synthesized by wild type virus when protein synthesis is inhibited by cycloheximide or puromycin.

We envisage replication in 2 steps. The step 1 phase of replication is the synthesis of a complete 42 S viral complementary, plus strand RNA [255]. The step 2 phase is the synthesis of 42 S viral minus strand RNA from the 42 S plus strand template. The first step is the intriguing one. How does it relate to the ongoing transcription process? In order to provide working hypotheses we suggest 2 models for the step 1 replication.

6. *Models for the Interrelation of Transcription and Step 1 Replication*

The first model we propose is a *simultaneous model for transcription and replication.* This is exemplified in Fig. 5.14. The essential features of this model are as follows.

1. Transcription and replication can occur on the same template *without restriction* of each other.
2. Replication starts at the 3′ end of the viral RNA; transcription may also initiate there but possibly does not (as indicated in the scheme, Fig. 5.14).

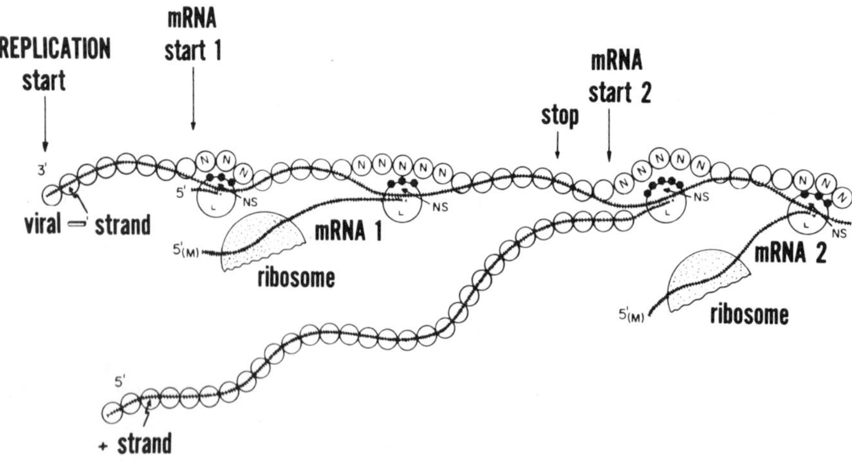

Fig. 5.14 A scheme of simultaneous transcription and replication (step 1).

3. The replication process is envisaged as employing the same bubble or groove methods of synthesis that are postulated to be used in primary transcription.
4. We propose that the 42 S plus strand becomes coated with N protein as it is synthesized. Whether 42 S replication is held up until N protein or a replicase is available is not critical to the model. Either possibility may in fact regulate the process. If the transcriptase is the replicase, then we postulate that naked 42 S plus strand RNA may be made by the transcriptase, perhaps at relatively low frequency, until N protein is available. Alternatively, if the replicase is a modified transcriptase (such as 1 possessing more NS molecules, as indicated in the scheme, or super- or dephosphorylated NS species or even an additional, newly synthesized, gene product), then we suggest that replication is held up until they become available. Similarly, if the replicase is not related to the transcriptase then we should expect replication to await the availability of the replicase. This last possibility, wherein the replicase is not the transcriptase, we find hard to believe in view of the fact that L, G, N, NS, and M account for almost all the viral genetic information.

How in this model does the replication process relate to the mRNA initiation sites? Clearly for an RNA in progress of being synthesized, any internal initiation site would pose little problem to the ability of an enzyme to read through that site. Replication initiation at internal mRNA sites may be a means by which deletion RNA species are generated (replicating to give defective progeny genomes).

What about the postulated stop signals? One would imagine that if the replicase was the transcriptase, then stop signals (if they existed) would be a hazard. Although such a hazard may be a chance the replication process takes, it is conceivable that the ongrowing 42 S plus strand RNA-N complex might be better equipped to ignore a stop signal because of its distinctive arrangement (like a car driver with no brakes). Alternatively if the replicase is a modified transcriptase (as suggested above), the modification may make the enzyme less sensitive to the stop signals.

How and why, in the proposed model, could the 42 S plus strands become coated with N protein when the mRNA plus strands do not? If there is a difference in 5′ initiation sequence of 42 S plus strand RNA (e.g., pppApCpGp. . .) as opposed to most mRNA species (e.g., pppApApCpXpGp. . .), then it is conceivable that because of their sequence the latter are capped and methylated while the former are not and that this allows, on the one hand, ribosome binding to mRNA and, on the other hand, N protein attachment to the nascent 42 S plus strand RNA. Presumably the continuous addition of N protein would give a contiguous N protein complex on the 42 S plus strand RNA. This we suggest would then prevent ribosome binding at any internal site along the RNA. It is also conceivable that as far as the 42 S RNA species is concerned, ribosome binding or translation may not even be detrimental to subsequent N protein addition.

This model of simultaneous transcription and replication makes certain predictions which can be investigated experimentally.

The second alternative model we propose is a *nonsimultaneous transcription and replication process,* whereby either the viral genome template is used in transcription (e.g., primary transcription), or it is used in replication. A possible scheme is shown in Fig. 5.15. In this case we suggest that internal mRNA initiation sites and stop signals are blocked. As discussed above and as exemplified in the scheme presented in Fig. 5.15, the virion transcriptase could function as a replicase (with or without prior modification) or the replicase could be a new gene product. Blocking in this scheme could be achieved by NS proteins complexing with the N protein at mRNA initiation or stop sites, causing permanent liftoff or changes in the groove structure. Other site blocking mechanisms can also be envisaged.

7. *Model for Step 2 Replication*

Step 2 replication is defined as the synthesis of viral minus strand RNA from the 42 S plus strand RNA template. We propose, as has Huang [255], that

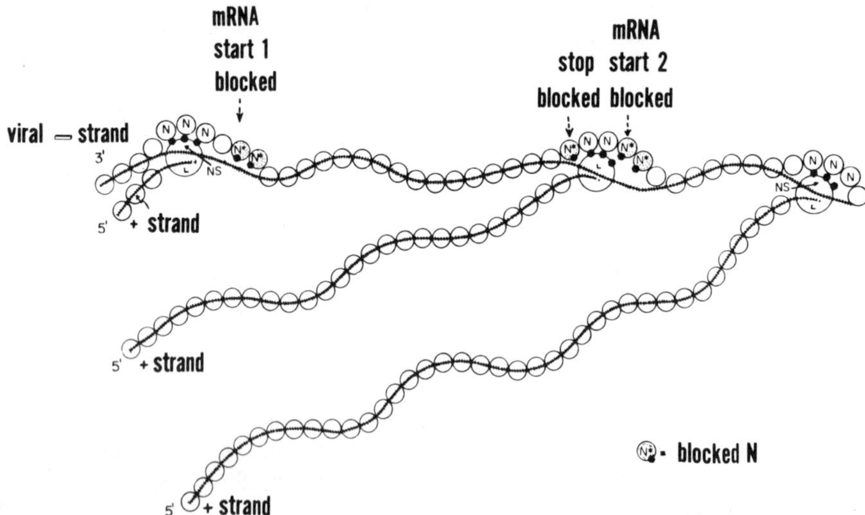

Fig. 5.15 A scheme of nonsimultaneous transcription and replication (step 1).

the template for this phase of the replication process is a 42 S plus strand RNA-N protein complex (Fig. 5.16) and not a double-stranded or naked RNA replicative intermediate. Without internal initiation sites present on the plus strand RNA, the virion transcriptase could presumably function as the polymerase. If synthesis of viral RNA leads directly into progeny virion packaging then the enzyme may even become enclosed within the virus particle. Alternatively, a modified transcriptase might function as a replicase (as in the simultaneous transcription-replication scheme described above).

This step 2 replication process would again employ either the bubble or the groove mechanisms of enzyme-template interaction. The essential distinctive feature of this process, like that of the simultaneous step 1 replication scheme, is that N protein covers both the template plus strand and the product minus strands. It would be in keeping with this scheme of events if N protein availability controlled this stage of RNA replication.

Other mechanisms of RNA synthesis could be and have been proposed; however, we believe that the simultaneous transcription-replication and step 2 replication procedures outlined above are more plausible mechanisms. It is hoped it will be possible in the not too distant future to obtain evidence which will lead to a more definitive model for the mechanism of RNA synthesis.

In conclusion, 2 observations concerning the VSV infection process should be mentioned. The first is that there is more NS protein in the infected cell than is ever recovered in virus particles. This abundance of NS protein occurs early in the growth cycle as shown by Kang and Prevec [260].

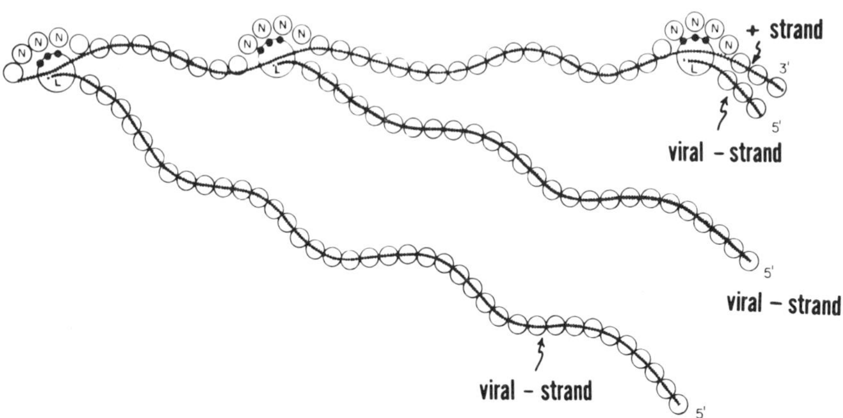

Fig. 5.16 A scheme of replication (step 2).

Second, after pulse-labeling experiments of an infected cell late in the infection cycle, the observed RNA species for wild type virus consist of 42 S viral and 28 S and 11-18 S messenger RNA species. If cycloheximide is added before the pulse labeling, only messenger RNA is obtained; no 42 S viral RNA is produced [244].

One interpretation of both these results is that whatever protein controls the 42 S viral RNA production, it is required to be synthesized continuously. This could be true for NS protein or a replicase or a replicase component, or for N protein or some other structural entity required for processing the viral RNA. If this protein function is part of the replicase then it must have a short half-life, as, for instance, in the case of NS phosphorylation whereby the amount of phosphorylation could control its ability to function as a replicase subunit.

In addition to the above hypotheses, the possibility that host proteins contribute to the replication process should not be overlooked. None of the preceding arguments precludes host proteins being involved. If they are, then they must be present in a wide variety of cells–insect, fish, avian, and mammalian–since VSV can productively infect many different cell types.

It would greatly aid an unraveling of the interrelationships between primary transcription and replication if we could identify the replicative machinery in the infected cell, study it in vitro, and determine its protein composition and *modus operandi,* etc. Until that is achieved, we can only speculate on what may be involved.

C. Protein Synthesis and the Topography of the Infection Process

In consideration of the role of protein synthesis in an infection developed by a rhabdovirus, one has to take into account both the sites at which each event occurs and the timeliness and extent of the various processes. VSV Indiana, Mokola, and Klamath can productively infect enucleated cells [114, 256]. This indicates that all of the processes involved in an infection can occur in the cytoplasm. This is possibly not true for rabies virus infections [114].

Heine and Schnaitman showed that, following VSV adsorption, the viral G and M proteins remained at the cell plasma membrane while the nucleocapsid (RNA-N, L, and NS) penetrated into the cytoplasm [163]. Exactly where primary transcription occurs is not known; however, for VSV Indiana, all mRNA species are synthesized within a few minutes postinfection [144]. The start of viral RNA replication occurs by 45 min of incubation at 38°C and the first progeny viruses are released 15 min later [215]. Within this hour, therefore, all the necessary steps for a productive infection must have occurred.

Although it has been shown that an infection by VSV substantially inhibits the host-determined syntheses of protein and RNA, it is not completely certain that the synthesis of all host proteins is equally inhibited [120, 257-259].

There is no convincing evidence that there are more than 5 virus-specified proteins in infected cells, although the possibility remains that there is another.

Mudd and Summers showed that there were no detectable large polypeptide precursors for viral proteins in infected cells [120, 259], an observation which is in keeping with later findings of individual viral messenger RNA species coding for particular viral proteins. Morrison and co-workers were the first to show that VSV mRNA could be translated in vitro to give specific viral polypeptides [238]. These results have been amplified by Knipe et al. [239], as well as by Both et al. [240], who have shown that distinct in vivo or in vitro viral complementary mRNA species code for particular viral proteins. Their results indicate that the smallest mRNAs (11-12 S) [143], which apparently are a mixture of RNA species each about 0.3×10^6 MW, code for the viral NS and M proteins. The next large mRNA (13-14 S, $\sim 0.5\text{-}0.6 \times 10^6$ MW) codes for the N protein while the 17-18 S mRNA species ($\sim 0.7 \times 10^6$ MW) codes for G protein. It has been postulated that the 28 S RNA codes for L protein, although the evidence obtained for this is less convincing than that obtained for the other RNA species [238]. By summing the molecular weights of the different progenitor mRNAs (or the proteins they code for: L, G, N, NS, and M), it is evident that most, if not all, of the genetic information of the viral RNA resides in these 5 virion proteins [143].

It has been found that the in vitro synthesized G polypeptide migrates in polyacrylamide SDS-gels slightly faster than the virion G protein, as expected for a nonglycosylated version of G. Similar polypeptides have been identified by Kang and Prevec in infected cells [260].

Wagner and associates have shown that the cell plasma membrane is the location at which most of the VSV G protein is recovered [261]. It has been suggested that the site of G protein synthesis, and the occurrence of the 17-18 S mRNA which codes for it, is distinct by comparison to the sites at which other proteins are synthesized, although how this occurs is a moot point [240]. The interrelationships of G protein synthesis and its subsequent glycosylation, association with, and insertion through the plasma membrane are not understood. Once it is inserted through the cell membrane, how G protein proceeds to the site on the plasma membrane where virus budding occurs is also not known.

Kang and Prevec [260], as well as Wagner [259], have studied the time-course of the syntheses of the various viral proteins. Of interest in their findings is the relative paucity of L protein throughout an infection, the early occurrence of NS and N proteins, and later buildup of G and M proteins. In part this parallels the low recovery of 28 S mRNA (the progenitor of L protein) indicating that there is transcriptional control of 28 S mRNA synthesis. Whether the observed variations in amounts of the other proteins reflect transcriptional or translational control is not known.

The majority of the viral N protein in infected cells is recovered in association with the intracellular viral RNA; relatively little is recovered free [259]. This suggests that the affinity of N protein for viral RNA is rather high. What determines the specificity of N for its own viral RNA is not known, nor is it known if N protein regulates or is rate limiting for viral RNA synthesis.

The intracellular NS protein is a phosphoprotein which occurs in soluble form as well as in association with nucleocapsid [229, 259, 262]. The cellular content of NS is rather large by comparison to the NS content of virions [259, 260]. It is believed that the cellular and virion NS proteins are identical [189, 233]. Although both the soluble and nucleocapsid-associated NS appear to be phosphorylated to equivalent extents [137, 189], it is not known if the same amino acid residues are phosphorylated. Neither do we know which enzymes are responsible for the phosphorylation of the NS protein. We have observed that the amount of NS associated with intracellular nucleocapsids appears to be greater than that present in the virion nucleocapsids (D. H. L. Bishop, unpublished observations, 1975). This suggests that there is an intracellular nucleocapsid NS function which is not needed by the virus particle. Could this function relate to packaging, transcription, and/or replication? Again we do not know.

Most of the M protein present in infected cells is found in association with cellular membranes [259]. The rapid transferral to membranes has been demonstrated by pulse-chase experiments, by kinetic analyses of M protein synthesis, and by determining the site where M is located [261, 263, 264]. It has been suggested from these experiments that M protein proceeds rapidly to cell membranes and that its synthesis is a rate-limiting step in the assembly of virus particles [262].

How do virus particles mature? All the evidence so far obtained indicates that virus particles mature by budding from the cell surface or from intracellular membranes (especially for rabies). If we assume that viral nucleocapsids of VSV Indiana (RNA-N, L, and NS) assemble prior to budding and that NS is involved solely in that assembly process, then the question resolves itself into how M, G, and the nucleocapsids recognize each other and come together within the maze of a cell. We have no notion at present of how this occurs; however, in order to provide a working hypothesis the following scheme is suggested.

1. Viral glycoprotein is synthesized and is glycosylated by cellular enzymes at some intracellular site(s) by a process which is determined by the G protein sequence and secondary structure. Because of the glycosylation, the G protein is inserted through the plasma membrane

and, once through, it is free to roam on the cell surface with its hydrophobic tail embedded in the cell lipid membrane.

2. The viral M protein is synthesized on polysomes associated with the endoplasmic reticulum. M protein inserts itself into the plasma membrane by reason of its hydrophobic tail. The hydrophilic portion of M associates with the nucleocapsid in a mechanism involving N (but possibly NS transiently). This initial association leads into conglomerates of M being arranged around the nucleocapsid. This causes the nucleocapsid to coil, which prevents the associated transcriptase from transcribing RNA.
3. As part and parcel of this buildup, hydrophobic G protein tails associate with M protein tails in the lipid membrane and these G-M bridges facilitate a local change in the membrane structure (both internal and external) which causes membrane distortion, budding, and eventual release of the virus particle.

Although this scheme is hypothetical, it focuses attention on 3 fundamental traits of a rhabdovirus architecture. These are the RNA-N nucleocapsid structure, the M protein(s), and the G protein. It seems reasonable to suggest that there are M-G and M-N relationships in the virion which are necessary for virion synthesis and structure, although discovering exactly what these relationships are must await further experimentation.

D. Genetic Analyses of Rhabdovirus Development

In order to progress further in the study of the molecular biology of rhabdoviruses, conditional lethal mutants, such as those which are temperature sensitive (*ts*), host restricted (*hr*), complementation dependent (*cd*), and conditionally temperature sensitive (*td*, e.g., mutants which are *ts* in 1 but not another host [313]), potentially provide a set of tools which could be useful in unraveling the intracacies of the viral growth cycle. Over the last 6 years considerable effort has gone into the isolation and characterization of *ts* mutants of various members of the VSV subgroup of viruses. Mutants of other rhabdoviruses have not yet been as fully characterized.

1. VSV Indiana Genetic Analyses; Isolation and Characterization of ts Mutants

From the initial studies of Flamand [265, 266], Pringle [267, 268], or Holloway and his associates [269], over 400 VSV Indiana *ts* mutants have been characterized into 5 groups by complementation analyses [235]. Since these mutants were isolated by different procedures, it has been deemed desirable to maintain the source of origin of the various *ts* mutants in the nomenclature. Consequently, Flamand's mutants are designated as the Orsay (*O*) mutants, Pringle's mutants as the Glasgow (*G*) mutants, and Holloway's as the Winnepeg (*W*) mutants. By following this convention, Baltimore's recently described mutants [270] are described as the Massachusetts (*M*) mutants.

Flamand, using chick embryo cells, isolated 71 spontaneous *ts* mutants; they occurred at a frequency of 2.3% [266]. She obtained higher frequencies (up to 8%) by limited serial passage of her wild type strain at 39.3°C (up to 9 passages [271]; thereafter the rate reduced to 2-3%). Holloway and associates with a heat-resistant (HR) strain of VSV Indiana and mouse L-cells, obtained by mutagenesis 25 *ts* mutants (frequency ~1%). They employed either 5-fluorouracil, nitrous acid, proflavine, or ethyl methane sulfonate as mutagenic agents [269]. Pringle has isolated more than 200 mutants of VSV Indiana using principally 5-fluorouracil and BKH cells [235].

Five complementation groups of mutants have been established, and these are categorized by roman numerals (I-V). The functions represented by some of these groups are under debate; however, Lafay has provided convincing evidence that group V mutants involve G protein alterations, while group III mutants involve M protein changes [272-274]. There is some evidence obtained by both Printz and his associates and Buller in Glasgow that the group II mutants have a defect in their N protein (A. Combard and C. R. Pringle, personal communication, 1976). Groups I and IV (and some group II mutants) exhibit reduced RNA synthesis at nonpermissive temperatures in vivo; however, elucidation of their gene products has given somewhat ambiguous results, basically, as will be seen, because the mutants of either group are pleiotropic in behavior and this has complicated interpretation of either the in vivo or the in vitro studies.

In vitro studies at different temperatures have been undertaken by Hunt and Wagner [275] to determine the ability of solubilized components of several group I *ts* mutant viruses to restore transcriptase activity to template fractions of wild type virus (and the reverse combinations). From these studies they proposed that group I mutants possessed a defect in their transcriptase but not in their template (RNA-N). Similar results have been obtained by Ngan et al. [276]. Szilágyi and Pringle, in earlier studies on the lability of the virion transcriptases of various *ts* mutants, provided evidence that some group I *ts* mutants (such as *ts* G I-114) possess labile transcriptase functions [218]. Since both NS and L are involved in transcription, none of the foregoing results tells us which protein is the group I gene product.

It has been noted in an earlier part of this review that the in vitro transcriptase activity of wild type virus is not as active or sustained at 39.5°C as it is at 31°C. There are, in fact, no reports of continued linear in vitro synthesis of RNA for long periods at high temperatures, at rates comparable to the 31°C in vitro rate, let alone commensurate with the fact that in vivo primary transcription is faster at 39.5°C than that at 31°C [144]. Consequently, these problems complicate the interpretations of the in vitro studies. A second complication of poor virion enzyme specific activity for group I mutants was demonstrated by Cairns et al. [277] and confirmed by Hunt and Wagner [275] as well as by Bishop and Flamand [278]. These last investigators demonstrated that when assayed at 28, 31, or 39°C the in vitro specific enzyme activity of whole virions of group I temperature-sensitive mutants was less, often considerably less, than that of the wild-type virus. This was also shown to be true for some, but not all, group IV mutants. By way of contrast, a group III mutant gave a specific enzyme activity approximately the same as the wild type virus [278]. Bishop and Flamand further demonstrated that the in vitro transcriptase specific activity *increased* (!) at 31 and/or 39.5°C for a group I or a group IV mutant when subviral dextran cores were prepared (RNA-N, NS, and L), but not for the group III or wild type virus [278]. Since in preparing subviral cores G, M, and some L are removed, this result poses the possibility that the lower transcriptase specific activities of the group I and group IV whole virion enzymes are caused by some soluble, inhibitory component. In addition, it was found that when group I or group IV whole virion preparations were mixed with a group III (or wild type) whole virion preparation, the recovered enzyme specific activity (at 31 or 39.5°C) was *less* than expected, a result which also argues for inhibitory substances being present in the group I or IV virus preparations. Mixtures of dextran cores of group I or group IV viruses with those of a group III virus gave the expected transcriptase values. Could these inhibitors be defunct transcriptase components capable of binding to the template and inhibiting functional enzymes?

Assays of virus mixtures of group I and group IV mutants were shown by Cairns and associates [277] to give *more* enzyme activity than expected; this result was confirmed by Bishop and Flamand [278]. However, when dextran cores of the 2 viruses were mixed, they gave the expected values [278]. It was suggested that despite the presence of inhibitory substances, one could obtain an effective in vitro complementation between these viruses although what was involved was not clear [277, 278].

It can be argued that if group I or group IV viruses possess soluble inhibitory substances, then the observed lack of reconstitution of transcriptase activity using soluble components of these viruses with templates of a wild type virus is meaningless, even if the inhibitor represents defunct transcriptase components. The results of Hunt and Wagner [275] rightly were interpreted to indicate that group I gene product is not N protein since the RNA-N complexes of group I mutants could be used by wild type solubilized enzyme at high or low temperatures. Clearly, therefore, the N protein of group I mutants is not defective. They interpreted their results to indicate that the group I gene product was L or NS.

Ngan and associates [276], using an RNA-N template derived from a group IV mutant with wild type enzyme, obtained no reconstitution of enzyme activity. However, the solubilized components of their group IV viruses did give activity with wild type virus template. They suggested that the group IV defect was in N protein, although they recognized the possibility that the template may have contained some NS and that the NS may have been inhibitory.

In order to attempt to clarify the in vitro results, Flamand and Bishop investigated the in vivo primary transcription rates of certain *ts* I, *ts* II, *ts* IV, and *ts* V Orsay mutants [144]. In both published and unpublished experiments, it was found that for *ts O* I-5, *ts O* I-80, *ts O* I-53, and *ts O* I-78 virus preparations the in vivo primary transcription rates at 39.5 or 40°C were approximately the same as those observed at 31°C [144, 197, 215]. These results indicated that for those 4 group I mutants, the transcriptase, when synthesized at permissive temperatures (31°C), was able to function at permissive and at nonpermissive temperatures. Repik and Bishop (unpublished observations, 1975), have obtained evidence that one group I mutant, *ts G* I-114, possesses a thermolabile transcriptase when assayed in vivo for primary transcription at 39 and 40°C by comparison to its activity at 31°C. These results agreed with those of Szilágyi and Pringle who has previously demonstrated that *ts G* I-114 possessed little transcriptase activity when assayed in vitro at high temperautres [218]. The results for the in vivo *ts O* I-80 and *ts G* I-114 experiments are presented in Fig. 5.17. It is clear from these results that the *ts G* I-114 behaves quite differently from the wild type virus or from the *ts O* I-80 mutant. The conclusion to be drawn from these results, and the in vitro experiments, is that when some group I mutant viruses are grown at permissive temperatures, the enzyme may be capable of functioning either at permissive temperatures or at nonpermissive temperatures in vivo or in vitro. Some group I mutants grown at permissive temperatures, however, may not be capable of transcriptase activity at nonpermissive temperatures. Group I mutants are therefore pleiotropic in behavior. One way of interpreting such differences in behavior is that during translation at 31°C (permissive temperature), the *ts G* I-114 enzyme folds to a functional form (A) which can be changed to a nonfunctional form (B) when that enzyme is put at a 39°C. Upon translation at nonpermissive temperatures only form B is obtained. For the other type of group I mutant (e.g., *ts O* I-80), at 31°C a functional form A is obtained which remains in form A confirguration when that enzyme is shifted to 39°C. However, if a *ts O* I-80-infected cell is making enzyme at 39°C, a nonfunctional form B is obtained.

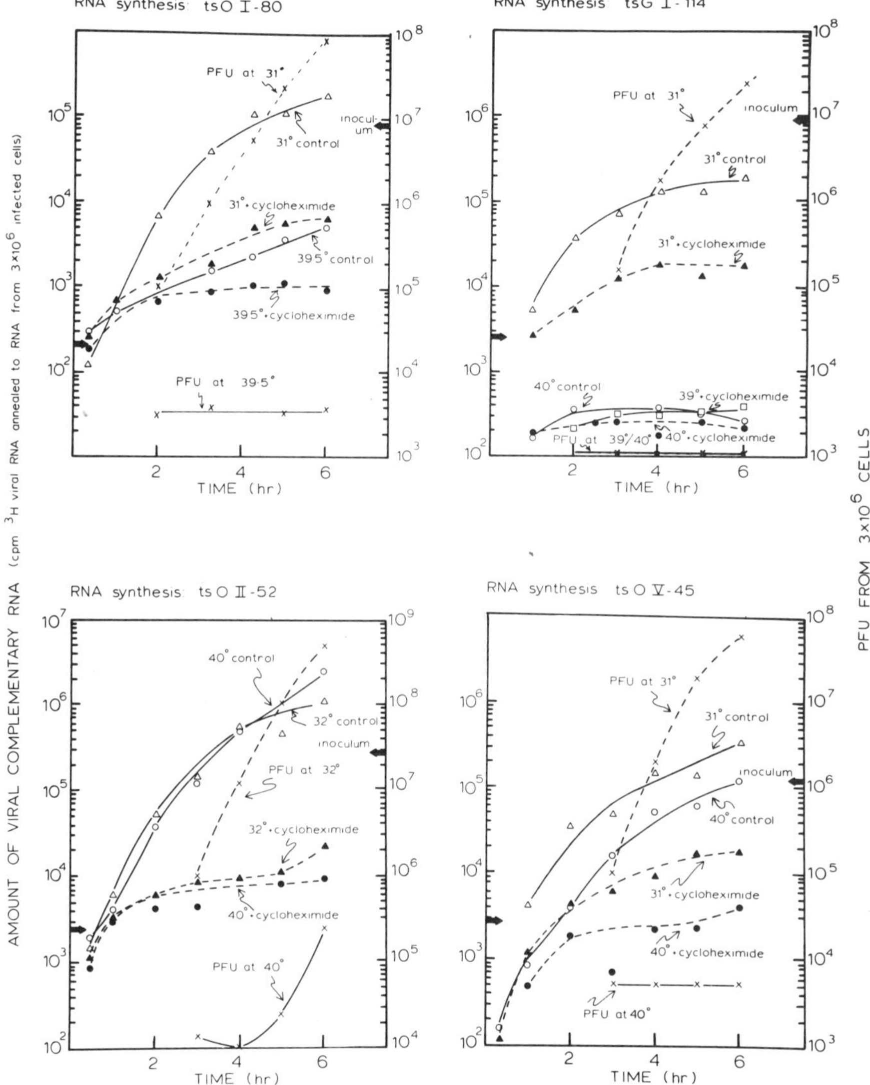

Fig. 5.17 Viral complementary RNA syntheses and release of PFU in cells infected with *ts O* I-80, *ts G* I-114, *ts O* II-52, or *ts O* V-45. The methods employed for enumerating released PFU and viral-complementary RNA syntheses at permissive or nonpermissive temperatures (± cycloheximide) have been described [215]. Arrows at the left ordinates are the adsorbed ^3H cpm. Arrows at the right ordinates are the infecting virus PFU.

None of the results so far reported indicates whether the group I gene product represents NS or L protein. In recent unpublished work (S. U. Emerson and M. Hunt, personal communication, 1976), using purified NS or L proteins from certain group I mutants, it appears that only the L protein is thermolabile. This suggests that the group I gene product is L protein.

What about the group IV gene product? If every other decision on gene product assignment were correct, then one could be easily persuaded that the group IV gene product was the NS protein. There are as yet no results which unambiguously support this contention. As in the case of certain group I mutants, primary transcription (but not secondary transcription) has been demonstrated [144, 197, 215] for all group IV *ts* mutants so far investigated at nonpermissive temperatures (e.g., *ts O* IV-100, *ts O* IV-194). It has been shown that in vitro some group IV *ts* mutants possess a low transcriptase specific activity at 31°C and even less activity at 39.5°C when the assays were conducted with whole virion preparations [278]. However, when dextran cores preparations were made, the enzyme specific activity increased at both temperatures and approached that of the wild type virus (or a group III mutant virus). This observation has yet to be satisfactorily explained. As mentioned previously, studies by Ngan and collaborators using dissociated components of group IV mutants plus template fractions from the VSV HR wild type virus have suggested that the group IV defect is in the viral N or possibly NS protein [276]. It is quite possible that group IV mutants are also pleiotropic; it is also possible that defective NS proteins bind to RNA-N complexes and inhibit transcription by viable enzyme.

There is one further possibility which should be considered and that is that another complementation group exists. This has been suggested by Rettenmier and collaborators [270]; however, their results do not exclude completely the possibility that the putative group VI mutants may be exhibiting intracistronic complementation with 1 of the other groups, a suggestion made previously by Flamand [266] for mutants with low complementation indices. In support of 6 complementation groups is the claim from Pringle's studies of 6 complementation groups for VSV New Jersey.

Other VSV Indiana mutants which have been obtained and preliminarily characterized include host-range (*hr*) mutants [279, 280], some of which are also *ts*, including some which are *ts* in one host but not in the other (the *td* mutants described by Szilágyi and Pringle [313]). Also nontemperature-sensitive, complementation-dependent (*cd*) mutants have been obtained [270]. The nature of the *cd, hr,* or *td* mutants is not known and, although it would be useful if they were amber-ocher-opal types of mutants, such a possibility seems unlikely. It has been suggested that the *td* chick embryo-restricted mutants have an associated defect in their virion polymerase and that this relates to host factors which may be involved in normal polymerase function [313]. Since already several enzymatic activities have been identified in virus preparations, some of which are almost certainly host originated, this does not seem to be too unlikely a proposition.

2. *Use of VSV Indiana ts Mutants to Characterize the Viral Infection Process*

a. RNA Synthesis. All the available information indicates that group III, group V, and some group II mutants synthesize RNA at both permissive and nonpermissive temperatures for virus production [144, 197, 215, 269, 272, 281-284]. The mutant *ts G* II-21 is classed as RNA positive, while *ts G* II-22 is classified as RNA negative [235]. Flamand and Lafay showed that the amount of RNA synthesis induced by *ts O* II-S 52 is dependent on the multiplicity of infection; the meaning of this observation is not clear [284]. The results presented in Fig. 5.17 also document that for either the *ts O* II-52 or the *ts O* V-45 mutants viral complementary RNA is synthesized at 31°C and 39.5 or 40°C. The syntheses of both viral-like and viral-complementary RNA at 31 or 40°C for *ts O* III-23 has been documented previously [197]. As discussed previously, all group IV mutants so far examined, as well as certain group I mutants (but not *ts G* I-114), perform primary transcription and synthesize viral-complementary RNA at nonpermissive temperatures [144, 197, 215]. No viral-like RNA synthesis and no secondary transcription has been observed for any group I or group IV mutant-infected cell held at nonpermissive temperatures from the time of adsorption [197].

These results basically agree with results published by all other investigators using group I and IV mutant-infected cells [244, 281-283, 285-289], and indicate that at nonpermissive temperatures no gene products are made which are capable of replicating the viral RNA. Because no progeny viral RNA is made, no secondary transcription occurs.

Combard and associates have shown that at nonpermissive temperatures, different *ts O* IV mutants exhibited various levels of primary transcription, depending on the mutant [285]. They performed shiftup experiments whereby cells infected with a group IV mutant and incubated at 30°C for 3 hr were shifted up to 39.2°C. The RNA labeled at the highest tempeature was then analyzed by sucrose gradient centrifugation. In these experiments they showed that for *ts O* IV-111 transcription continued after the shiftup, while for *ts O* IV-100 both transcription and replication continued after the shiftup. For cells infected with either mutant, the transcription at 39.2°C declined without any specific alteration in the various RNA species. When shiftup experiments were conducted with the addition of cycloheximide at the time of changing the temperature, only transcription was observed, and this transcription was greater than that obtained without cycloheximide. As noted previously, wild type virus-infected cells, treated with cycloheximide, also abruptly stop viral RNA synthesis but not transcription [244]. Combard and associates [285] have also analyzed the intracellular proteins synthesized under permissive, nonpermissive, or shiftup conditions and found that in both *ts O* IV-100- and *ts O* IV-111-infected cells, L, G, N, NS, and M proteins were synthesized (although giving lesser amounts of N protein for the *ts O* IV-111 mutant).

The pleiotropic behavior of group IV mutants with regard to their capability to synthesize viral RNA at nonpermissive temperature parallels to some extent the pleiotropism of group I mutants *vis à vis* their in vivo primary transcription capability. If group IV mutants possess a defective replicase component when translated at nonpermissive temperatures (giving a nonfunctional form B secondary structure), and it is quite possible that this same replicase component synthesized at permissive temperatures (and giving a functional form A secondary structure) may then be capable of functioning at nonpermissive temperatures since it does not refold to give form B. Other mutants may behave differently.

Studies on wild type and group I and group IV mutant-infected cells have been performed in Huang's laboratories and have led her to hypothesize that a relationship exists between replication and transcription [244, 288]. Using *ts G* I-114, she and her colleagues found that this mutant was unable to perform either primary or secondary transcription at nonpermissive temperatures. Although the nonpermissive temperature chosen was low (37°C), since the original mutant was selected using a nonpermissive temperature of 39°C, it was shown that in shiftup experiments (31-37°C), viral RNA synthesis continued and transcription was somewhat inhibited but not abolished. The latter observation may have been caused to some extent by the low "nonpermissive" temperature employed. Shiftup experiments in the presence of cycloheximide led to a total inhibition of viral RNA synthesis but elevated mRNA transcription rates giving amounts comparable to those found in control cells held at 31°C; this transcription ability decayed the longer the cells were held at 37°C. Again transcription by *ts G* I-114 that they obtained at nonpermissive temperatures could have been caused by the low nonpermissive temperature employed. What do these results mean? Huang has suggested that there is an interdependence between transcription and replication [244, 288], a suggestion derived from her observations that cycloheximide inhibits wild type RNA replication and promotes greater amounts of RNA transcription. Likewise, Printz and associates [289, 311] have suggested that a common subunit links both transcription and replication. Although studies with group I or, for that matter, group IV mutants do not unambiguously prove that there is such an interdependence of RNA transcription and replication is a plausible hypothesis.

b. Protein Synthesis and Utilization. Identifying viral-induced proteins in infected cells using radioactive amino acid precursors always involves a problem of detecting the viral protein species over a background of cellular protein synthesis. For group I, or RNA-negative group II, or group IV mutant-infected cells, in which there is no secondary transcription of mRNA, little or no viral protein synthesis has been detected at nonpermissive temperatures by comparison to that obtained at 31°C [290, 291]. For group III, or group V, or the RNA-positive group II mutant-infected cells, viral protein synthesis is obtained at nonpermissive temperatures [272, 273, 285, 290, 291]. Although some group III mutants may synthesize little M protein at high temperatures, others apparently synthesize normal amounts [273]. Lafay has shown that upon labeling at nonpermissive temperatures group III mutants synthesize a defective M protein which is the only protein which fails to be incorporated into virions when cells are shifted down to permissive temperatures [273]. Lafay has recently shown [273] that group V mutants synthesize a defective G protein when grown at high temperatures; the nature of the observed defect appears to involve an inability of G protein to participate in virion maturation at high temperatures (perhaps by not being inserted in the plasma membrane). This inability is reversed when infected cells are shifted down, which suggests that the defect is reversible. Some, but not all, group V virions are thermolabile [273, 292], which also argues for group V gene protein being the G protein.

3. *Recombination Between VSV Indiana ts Mutants*

At present there is little hard evidence that recombination occurs among VSV Indiana mutants of the same complementation group or of different complementation groups. While the possibility still exists, unambiguous evidence has not been obtained [235].

E. Phenotypic Mixing and the Procedure of Pseudotypes

The ability of cells infected with 2 (or more) enveloped viruses to produce particles comprised of components derived from each virus was initially demonstrated by Compans and Choppin using VSV Indiana and the paramyxovirus, SV5 [149, 293]. Similar phenotypically mixed particles (pseudotypes) have been obtained from mixed virus infections involving VSV Indiana and the myxovirus, fowl plague virus [294], or oncornaviruses, such as avian myeloblastosis virus, AMV [295], or the murine leukemia viruses [296, 297] or Rous sarcoma virus [298].

Composition analyses of the progeny derived from mixed virus infections of VSV Indiana and SV5 showed that, in addition to virions of the 2 parental types, 2 forms of phenotypically mixed, bullet-shaped particles were present which possessed VSV genomes [293]. One form (~1% of the progeny) contained SV5 glycoproteins, while the other form (9-45%) contained viral glycoproteins of both parents [293]. The existence of both forms of phenotypically mixed particles was demonstrated by VSV or SV5 antibody neutralization tests, electron microscopy involving ferritin conjugated anti-SV5 antibody, hemagglutination tests, and glucosamine labeling followed by protein gel electrophoresis [149, 293]. The results obtained indicated that only the glycoproteins of the 2 viruses could be interchanged. This suggests that as far as VSV maturation is concerned, there is not an absolute requirement for its own glycoprotein; glycoproteins of other viruses can be substituted. However, there does appear to be a requirement for the viral N and M proteins. It is not known how absolute these requirements are in dual virus infections involving 2 rhabdoviruses (e.g., 2 members of the VSV subgroup).

VSV pseudotypes possessing glycoprotein(s) derived from a virus which is limited in its host range (such as the avian or murine leukosis viruses) have been found to exhibit host-range specificities of the leukosis virus type [295-298].

VSV grown in BALB/c murine cells (which produce N-tropic MuLV) have given VSV pseudotypes with MuLV coats capable of growing in murine cells resistant to N-tropic and/or B-tropic viruses [297]. This result indicates that the N, B, or NB tropisms of murine leukemia viruses are probably not functions related to their glycoproteins.

Závada has obtained VSV pseudotypes by growing VSV in human mammary carcinoma cells [299]. These pseudotypes, which could not be neutralized by anti-VSV antibody, were capable of infecting human but not chick or other cell types [299]. Závada has also demonstrated that AMV can complement VSV *ts O* V-45 infections at nonpermissive temperatures but not *ts* mutants belonging to complementation groups I or III [295]. The virus particles produced dual virus infections of AMV and VSV *ts O* V-45 (at permissive or nonpermissive temperatures) were more thermostable than the parental *ts O* V-45 mutant.

The ability of VSV Indiana glycoprotein extracts to restore infectivity in vitro to spikeless particles of VSV Indiana or VSV New Jersey has also been demonstrated [150]. These glycoprotein extracts (obtained by Triton N101 extraction of virus followed by removal of Triton by absorption to Amberlite XAD-2 resin), contain viral lipids and probably consist of liposomes. These liposomes can be fused with spikeless particles (obtained by protease digestion of whole virus preparations of the same or another virus) and, upon fusion, infectivity is restored because of the presence of the glycoprotein [150]. Such in vitro pseudotypes have been obtained for a variety of rhabdoviruses (e.g., rabies, KCV, PFR, Chandipura) and in each case it has been found that the infectivity of the heterologous particles can only be inhibited by antibody raised againt the virus whose glycoprotein was used and not by antibody raised against virus from which the spikeless particles were obtained (P. Repik and D. H. L. Bishop, unpublished observations, 1975).

F. Defective-Interfering Particles, Autointerference

Since the original demonstration by Cooper and Bellett [300] of transmissible agents in VSV preparations capable of interfering with B particle replication, various defective interfering (DI) particles have been detected for a variety of rhabdoviruses [2, 115, 116, 122, 128-130, 132, 203-205, 258, 301]. Such particles have been shown to be capable of reducing the productivity of infections induced by the homologous B-type virus [115, 205, 206, 242, 250, 258, 302, 303]. As discussed previously, these DI particles represent deletion derivatives of the whole virus genome and occur in the form of abbreviated or "truncated" virions (T particles) [304, 305]. Defective particles possessing various RNA sizes [119, 130], as well as either plus or minus strand forms [129], have been described. One of the more common T particles possesses an RNA minus strand genome equivalent to approximately 1/3 of the B particle genome [118, 121, 132, 200, 248, 306]. Such T-particle preparations do not exhibit an endogenously templated transcriptase activity [173, 198, 200], although a functional transcriptase can be extracted from them and used to transcribe B-particle RNA-N protein templates [173]. Other T particles do appear to possess an endogenously templated transcriptase [129, 200].

Stocks of defective particles are commonly obtained by serial passages at high multiplicity of infection [206, 307] and, although such procedures do not guarantee a unique clone of defective particles, frequently unique particle types are obtained.

As discussed previously, some, but not all, defective particles possess a higher propensity for homotypic interference than for heterotypic interference, although this depends on the defective particle stock [204]. Many intriguing questions are raised about what is involved in heterotypic interference, such as whether the DI particles contribute gene products, but as yet these questions remain unanswered.

How do defective particles originate? The fact that different defective particles represent different parts of the viral genome [130] suggests that there is not one unique site of origin for all T particles. Presumably incorrect replication initiation as well as incorrect replication termination could contribute to the evolution of defective genomes, although what circumstances govern their origin and evolution are not known. Do defective genomes originate during plus or minus strand replication or even by mRNA replication? What are the prerequisites for replicating defective genomes? These are questions for which we have, as yet, no answers.

It has been postulated that defective particles play a role in establishing chronically infected cells, presumably by their ability to inhibit the cytolytic infection process of complete particles [307]. Similarly it has been claimed that T particles can protect mice against infection by homologous B virions [308].

How do defective particles interfere with the infection process of B virions? Again the answer to this question is not known, although it seems probable that the interference occurs at the level of removing B-particle gene products (transcriptase, N, M, and G proteins) from intracellular pools and thereby depleting their availability for B-particle replication and maturation [206]. Since T particles possess curtailed genomes, they may also possess a selective replication advantage as demonstrated for the Qβ deltion RNA species [309].

It has been shown that in B- and T-coinfected cells, certain RNA species can be recovered by phenol extraction which are not present in B-particle-infected cells [250]. These new species include both single-stranded (23 S) and double- or multistranded RNA structures (13-19 S). Since these RNA species are not found in B-particle-infected cells, it has been concluded that they relate to the replication process of the T particles. The interrelationships of the various RNA species to the replication process of T RNA have yet to be unraveled but possibly may provide a needed key to understanding the process involved in viral replication.

VII. Postscript

A new rhabdovirus isolate has recently been isolated by Dr. D. I. H. Simpson from a pool of *Mansonia uniformis* collected at Kampong, Tijirak, Sarawak in 1969. This isolate (S 1643) appears to be serological unrelated to other known rhabdoviruses (S. Buckley and D. I. H. Simpson, personal communication, 1976).

Acknowledgments

We are indebted to the following colleagues for graciously allowing us to use their data and for constructive help in the preparation of this review: Dr. F. Brown, Dr. H. F. Clark, Dr. A. Flamand, Ms. E. Hefti, Dr. F. A. Murphy, Dr.

J. F. Obijeski, Ms. P. Repik, Dr. P. Roy, Dr. L. G. Schneider, Dr. R. E. Shope, and Dr. T. J. Wiktor. We are also indebted to Dr. A. Banerjee, Dr. S. U. Emerson, Dr. J. J. McSharry, Dr. J. Rose, and Dr. R. R. Wagner for personal communications of their findings. This study was supported by grant AI-13402 from the National Institute of Allergy and Infectious Diseases.

References

1. R. I. B. Francki, Adv. Virus Res., 18, 257 (1973).
2. R. R. Wagner, in Comprehensive Virology, Vol. 4, Reproduction, Large RNA Viruses (H. Fraenkel-Conrat and R. R. Wagner, eds.), Plenum, New York, 1975, pp. 1-93.
3. F. A. Murphy, in Viruses, Evolution and Cancer (E. Kurstak and K. Maramorosch, eds.), Academic, New York, 1974, pp. 699-722.
4. D. L. Knudson, J. Gen. Virol., 20, 105 (1973).
5. K. Hummeler, in Comparative Virology (K. Maramorosch and E. Kurstak, eds.), Academic, New York, 1971, pp. 361-386.
6. A. F. Howatson, Adv. Virus Res., 16, 195 (1970).
7. S. Matsumoto, Adv. Virus Res., 16, 257 (1970).
8. R. P. Hanson, Bacteriol. Rev., 16, 179 (1952).
9. J. D. Almeida, A. P. Waterson, and D. I. H. Simpson, in Marburg Virus Disease (G. A. Martini and R. Siegert, eds.), Springer-Verlag, Berlin, 1971, pp. 84-97.
10. S. Theiler, Dtsch. Tieraerztl. Wochschr., 9, 131 (1901).
11. W. E. Cotton, Vet. Med., 22, 169 (1927).
12. A. J. Garcia Pirazzia, C. H. Caggiano, and A. Alonso Fernandez, Publication Tecnica No. 2, CANEFA, Ministry of Agriculture, Buenos Aires, Argentina, 1963.
13. K. B. Federer, R. Burrows, and J. B. Brooksby, Res. Vet. Sci., 8, 103 (1967).
14. A. H. Jonkers, R. E. Shope, T. H. G. Aitken, and L. Spence, Am. J. Vet. Res., 25, 236 (1964).
15. G. H. Bergold and K. Munz, Arch. Gesamte Virusforsch., 31, 152 (1970).
16. P. N. Bhatt and F. M. Rodrigues, Indian J. Med. Res., 55, 1295 (1967).
17. H. N. Johnson, Calif. Health, 23, 35 (1965).
18. C. Sanmartin, as cited in Am. J. Trop. Med. Hyg., 20, 1035 (1971).
19. R. A. de Haas, A. H. Jonkers, and D. W. Heinemann, Am. J. Trop. Med. Hyg., 15, 954 (1966).
20. N. Karabatsos, M. B. Lipman, M. S. Garrison, and C. A. Mongillo, J. Gen. Virol., 21, 429 (1973).
21. R. H. Kokernot, B. M. McIntosh, C. B. Worth, and J. de Sousa, Am. J. Trop. Med. Hyg., 11, 683 (1962).
22. G. W. Burgess, Vet. Bull., 41, 887 (1971).
23. B. Van der Westhuizen, Onderstepoort J. Vet. Res., 34, 29 (1967).

24. Y. Inaba, Y. Tanaka, K. Sato, H. Ito, T. Omori, and M. Matumoto, Jap. J. Microbiol., 12, 253 (1968).
25. N. Saski, K. Kodama, I. Iwamoto, A. Izumida, and T. Matsubara, Jap. J. Microbiol., 12, 251 (1968).
26. R. L. Doherty, H. A. Standfast, and I. A. Clark, Aust. J. Sci., 31, 365 (1969).
27. W. A. Snowdon, Aust. Vet. J., 46, 258 (1969).
28. P. B. Spradbrow and J. Francis, Aust. Vet. J., 45, 525 (1969).
29. D. Metselaar, M. C. Williams, D. I. H. Simpson, R. West, and F. A. Mutere, Arch. Gesamte Virusforsch., 26, 183 (1969).
30. T. O. Berge (Ed.), International Catalogue of Arboviruses, 2nd Ed., U.S. Public Health Service, Washington, D.C., 1975.
31. E. Whitney, Am. J. Trop. Med. Hyg., 13, 123 (1964).
32. F. A. Murphy, P. H. Coleman, and S. G. Whitfield, Virology, 30, 314 (1966).
33. P. L'Héritier, Adv. Virus Res., 5, 195 (1958).
34. A. Berkaloff, J. C. Bregliano, and A. Ohanessian, Compt. Rend. Acad. Sci. (Paris), 260, 5956 (1965).
35. P. L'Héritier, Evol. Biol., 4, 185 (1970).
36. P. L'Héritier, Ann. Inst. Past., 102, 511 (1962).
37. P. De Kinkelin, B. Galimard, and R. Bootsma, Nature, 241, 465 (1973).
38. P. A. Bachmann and W. Ahne, Nature, 244, 235 (1973).
39. N. Fijan, Z. Petrinec, D. Sulimanovic, and L. O. Zwillenberg, Veter. Arhiv., 41, 125 (1971).
40. P. A. Bachmann and W. Ahne, Arch. Gesamte Virusforsch., 44, 261 (1974).
41. W. Schäperclaus, Fischkrankheiten, 3rd Ed., Berlin, 1954.
42. M. H. Jensen, Ann. N.Y. Acad. Sci., 126, 423 (1965).
43. P. E. V. Jørgensen, Symp. Zool. Soc. London, 30, 333 (1972).
44. R. R. Rucker, W. J. Whipple, J. R. Parvin, and C. A. Evans, U.S. Fish Wildl. Serv. Fish Bull., 76(54), 35 (1953).
45. A. J. Ross, J. Pelnar, and R. R. Rucker, Trans. Am. Fish. Soc., 89, 160 (1960).
46. E. S. Tierkel, in Working Conference on Rabies, Proceedings (Y. Nagano and F. M. Davenport, eds.), Baltimore Press, Baltimore, 1971.
47. J. S. Porterfield, D. H. Hill, and A. D. Morris, Brit. Vet. J., 114, 1 (1958).
48. V. Kubes and F. Gallia, Bol. Inst. Invest. Vet., 1, 1 (1942).
49. C. D. Meredith, A. P. Rossouw, and H. Van Praag Koch, S. Afr. Med. J., 45, 767 (1972).
50. L. R. Boulger and J. S. Porterfield, Trans. Roy. Soc. Trop. Med. Hyg., 52, 421 (1958).
51. R. E. Shope, F. A. Murphy, A. K. Harrison, O. R. Causey, G. E. Kemp, D. I. H. Simpson, and D. L. Moore, J. Virol., 6, 690 (1970).
52. G. E. Kemp, O. R. Causey, D. L. Moore, A. Odelola, and A. Fabiyi, Am. J. Trop. Med. Hyg., 21, 356 (1972).

53. J. B. Familusi and D. L. Moore, Afr. J. Med. Sci., 3, 93 (1972).
54. G. E. Kemp, V. H. Lee, D. L. Moore, R. E. Shope, O. R. Causey, and F. A. Murphy, Am. J. Epidemiol., 98, 43 (1973).
55. J. R. Schmidt, M. C. Williams, M. Lule, A. Mivule, and E. Mujomba, E. Afr. Virus Res. Inst. Rep., 15, 24 (1966).
56. J. R. Mohler, U.S. Bur. Anim. Ind. Circ., 51, 000 (1904).
57. D. H. Ferris, R. P. Hanson, R. J. Diche, and R. H. Roberts, J. Infect. Dis., 96, 184 (1955).
58. M. Mussgay and D. Suárez, Virology, 17, 202 (1962).
59. R. B. Tesh, B. N. Chaniotis, and K. M. Johnson, Am. J. Epidemiol., 93, 491 (1971).
60. S. M. Buckley, Proc. Sco. Exp. Biol. Med., 131, 625 (1969).
61. Y. J. Yang, D. B. Stoltz, and L. Prevec, J. Gen. Virol., 5, 473 (1969).
62. H. Artsob and L. Sepence, Acta Virol., 18, 331 (1974).
63. A. H. Jonkers, Am. J. Epidemiol., 86, 286 (1967).
64. H. N. Johnson, in Viral and Rickettsial Infections of Man (F. L. Horsfall and I. Tamm, eds.), J. B. Lippincott Co., Philadelphia, 1965, pp. 814-840.
65. R. E. Dierks, F. A. Murphy, and A. K. Harrison, Am. J. Pathol., 54, 251 (1969).
66. H. F. Clark and T. J. Wiktor, in Strains of Human Viruses (J. Majer and A. Plotkin, eds.(, Krager, Basel, 1972, pp. 177-182.
67. G. M. Baer, T. R. Shanthaveerappa, and G. H. Bourne, Bull. WHO, 33, 783 (1965).
68. H. R. Fischman and M. Schaeffer, Ann. N.Y. Acad. Sci., 177, 78 (1971).
69. R. A. Goldwasser, R. E. Kissling, R. T. Carski, and T. S. Hosty, Bull. WHO, 20, 579 (1959).
70. R. T. Johnson, J. Neuropathol. Exp. Neurol., 24, 662 (1965).
71. F. A. Murphy, A. K. Harrison, W. C. Winn, and S. P. Bauer, Lab. Invest., 29, 1 (1973).
72. F. A. Murphy, S. P. Bauer, A. K. Harrison, and W. C. Winn, Lab. Invest., 28, 361 (1973).
73. L. G. Schneider, Zentralbl. Bakteriol., 211, 281 (1969).
74. L. G. Schneider, Zentralbl. Bakteriol., 212, 1 (1969).
75. L. G. Schneider and I. Hamann, Zentralbl. Bakteriol., 212, 13 (1969).
76. V. Hronovský and R. Benda, Acta Virol. Prague, Engl. Ed., 13, 198 (1968).
77. V. Hronovský, Acta Virol. Prague, Engl. Ed., 15, 58 (1971).
78. D. G. Constantine, Publ. No. 1617, Public Health Service Washington, D.C., 1967.
79. W. G. Winkler, Bull. Wildl. Dis. Ass., 4, 37 (1968).
80. D. G. Constantine, R. W. Emmons, and J. D. Woodie, Science, 175, 1255 (1972).
81. S. E. Sulkin and R. Allen, Monographs in Virology, Vol. 8, Karger, Basel, 1974, pp. 10-20.

82. E. P. Correa-Giron, R. Allen, and S. E. Sulkin, Am. J. Epidemiol., 91, 203 (1970).
83. H. R. Fischman and F. E. Ward, Am. J. Epidemiol., 88, 132 (1968).
84. G. M. Baer, M. K. Abelseth, and J. G. Debbie, Am. J. Epidemiol., 93, 487 (1970).
85. J. G. Debbie, in Progress in Mecical Virology, Vol. 18, (J. L. Melnick, ed.), Karger, Basel, 1974, pp. 241-256.
86. Y. Iwasaki, S. Ohtani, and H. F. Clark, J. Virol., 15, 1020 (1975).
87. Y. Iwasaki and H. F. Clark, Lab. Invest., in press, 1975.
88. G. E. Kemp, D. L. Moore, T. T. Isoun, and A. Fabiyi, Arch. Gesamte Virusforsch., 43, 242 (1973).
89. G. H. Tignor, R. E. Shope, P. N. Bhatt, and D. H. Percy, J. Infect. Dis., 128, 471 (1973).
90. F. A. Murphy, W. P. Taylor, C. A. Mims, and C. G. Whitfield, Arch. Gesamte Virusforsch., 38, 234 (1972).
91. F. A. Murphy, H. N. Johnson, A. K. Harrison, and R. E. Shope, Arch. Gesamte Virusforsch., 37, 323 (1972).
92. F. A. Murphy, R. E. Shope, D. Metselaar, and D. I. H. Simpson, Virology, 40, 288 (1970).
93. F. A. Murphy and B. N. Fields, Virology, 33, 625 (1967).
94. A. B. Jenson, E. R. Rabin, R. D. Wende, and J. L. Melnick, Exp. Mol. Pathol., 7, 1 (1967).
95. W. P. Heuschele, Arch. Gesamte Virusforsch., 30, 195 (1970).
96. I. H. Holmes and R. L. Doherty, J. Virol., 5, 91 (1970).
97. D. Teninges, Arch. Gesamte Virusforsch., 23, 378 (1968).
98. F. Bussereau. Ann. Inst. Pasteur, 118, 367, and 626 (1970).
99. P. Printz, Adv. Virus Res., 18, 143 (1973).
100. P. Printz, Compr. Rend. Acad. Sci. (Paris), 265, 169 (1967).
101. F. Bussereau. Ann. Inst. Pasteur, 121, 223 (1971).
102. C. J. Rasmussen, Ann. N.Y. Acad. Sci., 126, 427 (1965).
103. K. Wolf, Adv. Virus Res., 12, 35 (1966).
104. P. Ghittino, Ann. N.Y. Acad. Sci., 126, 468 (1965).
105. L. O. Zwillenberg, M. H. Jensen, and H. H. L. Zwillenberg, Arch. Gesamte Virusforsch., 17, 1 (1965).
106. B. B. McCain, J. L. Fryer, and K. S. Pilcher, Proc. Soc. Exp. Biol. Med., 137, 1042 (1971).
107. D. F. Amend and V. C. Chambers, J. Fish. Res. Bd. Can., 27, 1285 (1970).
108. D. F. Amend and L. Smith, Infect. Immun., 11, 171 (1975).
109. H. F. Clark and E. Z. Soriano, Infect. Immun., 10, 180 (1974).
110. R. W. Darlington, R. Trafford, and K. Wolf, Arch. Gesamte Virusforsch., 39, 257 (1972).
111. B. Kocylowski, Ann. N.Y. Acad. Sci., 126, 616 (1965).
112. R. Bootsma and C. J. A. H. V. Van Vorstenbosch, Netherlands J. Vet. Sci., 98, 86 (1973).
113. T. J. Parisot, W. T. Yusutake, and G. W. Klontz, Ann. N.Y. Acad. Sci., 126, 502 (1965).

114. T. J. Wiktor and H. Koprowski, J. Virol., 14, 300 (1974).
115. A. S. Huang and R. R. Wagner, Virology, 30, 173 (1966).
116. M. E. Reichmann, C. R. Pringle, and E. A. C. Follett, J. Virol., 8, 154 (1971).
117. T. L. Chow, F. H. Chow, and R. P. Hanson, J. Bacteriol., 68, 724 (1954).
118. P. Repik and D. H. L. Bishop, J. Virol., 12, 969 (1973),
119. D. H. L. Bishop and P. Roy, J. Mol. Biol., 57, 513 (1971).
120. J. A. Mudd and D. F. Summers, Virology, 42, 328 (1970).
121. M. P. Kiley and R. R. Wagner, J. Virol., 10, 244 (1972).
122. D. H. L. Bishop, H. G. Aaslestad, H. F. Clark, A. Flamand, J. F. Obijeski, P. Repik, and P. Roy, in Negative Strand Viruses (R. D. Barry and B. W. J. Mahy, eds.), in press, 1975.
123. R. Roy, H. F. Clark, H. P. Madore, and D. H. L. Bishop, J. Virol., 15, 338 (1975).
124. F. L. Schaffer, M. E. Soergel, G. Tegtmeier, and I. L. Shechmeister, Virology, 47, 236 (1972).
125. D. Baltimore, Bacteriol. Rev., 35, 235 (1971).
126. E. Hefti and D. H. L. Bishop, J. Virol., 15, 90 (1975).
127. H. G. Aaslestad and C. Urbano, J. Virol., 8, 922 (1971).
128. J. Crick and F. Brown, J. Gen. Virol., 22, 147 (1974).
129. P. Roy, P. Repik, E. Hefti, and D. H. L. Bishop, J. Virol., 11, 915 (1973).
130. R. N. Leamnson and M. E. Reichmann, J. Mol. Biol., 85, 551 (1974).
131. P. Repik, A. Flamand, H. F. Clark, J. F. Obijeski, P. Roy, and D. H. L. Bishop, J. Virol., 13, 250 (1974).
132. A. S. Huang and R. R. Wagner, J. Mol. Biol. 22, 381 (1966).
133. R. L. Erikson, in Fundamental Techniques in Virology (K. Habel and N. P. Salzman, eds.), Academic, New York, 1969, pp. 451-459.
134. R. R. Wagner, L. Prevec, F. Brown, D. F. Summers, F. Sokol, and R. MacLeod, J. Virol., 10, 1228 (1972).
135. J. F. Obijeski, A. T. Marchenko, D. H. L. Bishop, B. W. Cann, and F. A. Murphy, J. Gen. Virol., 22, 21 (1974).
136. W. H. Wunner and C. R. Pringle, J. Gen. Virol., 16, 1 (1972).
137. F. Sokol and H. F. Clark, Virology, 52, 246 (1973).
138. F. Sokol, H. F. Clark, T. J. Wiktor, M. L. McFalls, D. H. L. Bishop, and J. F. Obijeski, J. Gen. Virol., 24, 433 (1974).
139. P. E. McAllister and R. R. Wagner, J. Virol., 15, 733 (1975).
140. F. Sokol, D. Stanček, and H. Koprowski, J. Virol., 7, 241 (1971).
141. F. Sokol and H. Koprowski, Proc. Natl. Acad. Sci. U.S., 72, 933 (1975).
142. D. H. L. Bishop and P. Roy, J. Virol., 10, 234 (1972).
143. J. K. Rose and D. Knipe, J. Virol., 15, 994 (1975).
144. A. Flamand and D. H. L. Bishop, J. Virol., 12, 1238 (1973).
145. B. Cartwright, C. J. Smale, F. Brown, and R. Hull, J. Virol., 10, 256 (1972).

146. R. R. Wagner, T. C. Schnaitman, and R. M. Synder, J. Virol., 3, 395 (1969).
147. J. A. Mudd, Virology, 62, 573 (1974).
148. B. Cartwright, C. J. Smale, and F. Brown, J. Gen. Virol., 5, 1 (1969).
149. J. J. McSharry, R. W. Compans, and P. W. Choppin, J. Virol., 8, 722 (1971).
150. D. H. L. Bishop, P. Repik, J. F. Obijeski, N. Moore, and R. R. Wagner, J. Virol., 16, 75 (1975).
151. R. H. Schloemer and R. R. Wagner, J. Virol., 14, 270 (1974).
152. R. H. Schloemer and R. R. Wagner, J. Virol., 15, 882 (1975).
153. B. W. Burge and A. S. Huang, J. Virol., 6, 176 (1971).
154. J. R. Etchison and J. J. Holland, Virology, 60, 217 (1974).
155. M. J. Grubman, E. Ehrenfeld, and D. F. Summers, J. Virol., 14, 560 (1974).
156. J. J. McSharry and R. R. Wagner, J. Virol., 7, 412 (1971).
157. R. H. Schloemer and R. R. Wagner, J. Virol., 15, 1029 (1975).
158. P. E. Halonen, F. A. Murphy, B. N. Fields, and D. R. Reese, Proc. Soc. Exp. Biol. Med., 127, 1037 (1968).
159. J. M. Kelley, S. U. Emerson, and R. R. Wagner, J. Virol., 10, 1231 (1972).
160. B. Dietzschold, L. G. Schneider, and J. H. Cox, J. Virol., 14, 1 (1974).
161. T. J. Wiktor, E. György, H. D. Schlumberger, F. Sokol, and H. Koprowski, J. Immunol., 110, 269 (1973).
162. R. W. Simpson, R. E. Hauser, and S. Dales, Virology, 37, 285 (1969).
163. J. W. Heine and C. A. Schnaitman, J. Virol., 3, 619 (1969).
164. A. R. Neurath, S. K. Vernon, M. D. Dobkin, and B. A. Rubin, J. Gen. Virol., 14, 33 (1972).
165. G. Walter and J. A. Mudd, Virology, 52, 574 (1973).
166. N. F. Moore, J. M. Kelley, and R. R. Wagner, Virology, 61, 292 (1974).
167. B. Cartwright, C. J. Smale, and F. Brown, J. Gen. Virol., 7, 19 (1970).
168. B. Cartwright, P. Talbot, and F. Brown, J. Gen. Virol., 7, 267 (1970).
169. J. F. Szilágyi and L. Uryvayev, J. Virol., 11, 279 (1973).
170. D. H. L. Bishop, S. U. Emerson, and A. Flamand, J. Virol., 14, 139 (1974).
171. B. Cartwright and F. Brown, J. Gen. Virol., 16, 391 (1972).
172. M. Stampfer and D. Baltimore, J. Virol., 11, 520 (1973).
173. S. U. Emerson and R. R. Wagner, J. Virol., 10, 297 (1972).
174. S. U. Emerson and Y.-H. Yu, J. Virol., 15, 1348 (1975).
175. J. J. McSharry and R. R. Wagner, J. Virol., 7, 59 (1971).
176. H. R. Schlesinger, H. J. Wells, and K. Hummeler, J. Virol., 12, 1028 (1973).
177. S. R. Bates and G. H. Rothblat, J. Virol., 9, 883 (1972).
178. K.-H. Tsai and J. Lenard, Nature, 253, 554 (1975).
179. B. Cartwright and C. A. Pearce, J. Gen. Virol., 2, 207 (1968).
180. B. Cartwright and F. Brown, J. Gen. Virol., 15, 243 (1972).
181. H.-D. Klenk and P. W. Choppin, J. Virol., 7, 416 (1971).

182. T. T. Hecht and D. F. Summers, J. Virol., 10, 578 (1972).
183. P. Roy and D. H. L. Bishop, Biochem. Biophys. Acta, 235, 191 (1971).
184. M. Strand and J. T. August, Nature, New Biol., 233, 137 (1971).
185. L. P. Villarreal and J. J. Holland, Nature, New Biol., 246, 17 (1973).
186. A. K. Banerjee and D. P. Rhodes, Proc. Natl. Acad. Sci. U.S., 70, 3566 (1973).
187. D. P. Rhodes, S. A. Moyer, and A. K. Banerjee, Cell, 3, 327 (1974).
188. E. Hefti, P. Roy, and D. H. L. Bishop, in Negative Strand Viruses (B. W. J. Mahy and R. D. Barry, eds.), Academic, London, 1975, pp. 307-326.
189. R. L. Imblum and R. R. Wagner, J. Virol., 13, 113 (1974).
190. S. A. Moyer and D. F. Summers, J. Virol., 13, 455 (1974).
191. Y. Furuichi, Nucl. Acid Res., 1, 809 (1974).
192. A. J. Shatkin, Proc. Natl. Acad. Sci. U.S., 71, 3205 (1974).
193. C. M. Wei and B. Moss, Proc. Natl. Acad. Sci. U.S., 71, 3014 (1974).
194. D. Baltimore, A. S. Huang, and M. Stampfer, Proc. Natl. Acad. Sci., U.S., 66, 572 (1970).
195. H. G. Aaslestad, H. F. Clark, D. H. L. Bishop, and H. Koprowski, J. Virol., 7, 726 (1971).
196. S. H. Chang, E. Hefti, J. F. Obijeski, and D. H. L. Bishop, J. Virol., 13, 652 (1974).
197. D. H. L. Bishop and A. Flamand, in Soc. Gen. Micro. Symp. 25, Cambridge University Press, Cambridge, England, 1975, pp. 95-152.
198. P. Roy and D. H. L. Bishop, J. Virol., 9, 946 (1972).
199. J. Perrault and J. J. Holland, Virology, 50, 159 (1972).
200. D. H. L. Bishop and P. Roy, J. Mol. Biol., 58, 799 (1971).
201. F. Brown, B. Cartwright, and C. J. Smale, J. Virol., 1, 368 (1967).
202. G. Baer, Rabies, Academic, New York, in press, 1975.
203. J. C. Crick and F. Brown, J. Gen. Virol., 18, 79 (1973).
204. L. Prevec and C. Y. Kang, Nature, 228, 25 (1970).
205. J. Crick, B. Cartwright, and F. Brown, Arch. Gesamte Virusforsch, 27, 221 (1969).
206. A. S. Huang, Ann. Rev. Micro., 27, 101 (1973).
207. L. G. Schneider, R. E. Dierks, W. Matthaeus, P.-J. Enzmann, and K. Strohmaier, J. Virol., 11, 748 (1973).
208. G. H. Tignor and R. E. Shope, J. Infect. Dis., 125, 322 (1972).
209. P. de Kinkelin, M. LeBerre, and G. Lenoir, Ann. Microbiol. Inst. Pasteur, 125A, 93 (1974).
210. B. J. Hill, B. O. Underwood, C. J. Smale, and F. Brown, J. Gen. Virol., in press, 1975.
211. P. E. McAllister, J. L. Fryer, and K. S. Pilcher, J. Wildl. Dis., 10, 101 (1974).
212. J. Crick and F. Brown, in The Biology of Large RNA Viruses (R. D. Barry and B. W. J. Mahy, eds.), Academic, London, 1970, pp. 133-140.
213. R. B. Tesh, B. N. Chaniotis, and K. M. Johnson, Science, 175, 1477 (1972).

214. P. Repik, A. Flamand, and D. H. L. Bishop, J. Virol., 14, 1169 (1974).
215. A. Flamand and D. H. L. Bishop, J. Mol. Biol., 87, 31 (1974).
216. D. H. L. Bishop, J. F. Obijeski, and R. W. Simpson, J. Virol., 8, 66 (1971).
217. A. S. Huang, D. Baltimore, and M. A. Bratt, J. Virol., 7, 389 (1971).
218. J. F. Szilágyi and C. R. Pringle, J. Mol. Biol., 71, 281 (1972).
219. D. H. L. Bishop, R. Ruprecht, R. W. Simpson, and S. Spiegelman, J. Virol., 8, 730 (1971).
220. P. Roy and D. H. L. Bishop, Fed. Proc., 30, 1153 (1971).
221. D. H. L. Bishop, J. Virol., 7, 486 (1971).
222. P. Roy and D. H. L. Bishop, J. Virol., 11, 487 (1973).
223. Y. Furuichi, M. Morgan, S. Muthukrishnan, and A. J. Shatkin, Proc. Natl. Acad. Sci. U.S., 72, 362 (1975).
224. G. Abraham, D. P. Rhodes, and A. K. Banerjee, Cell, 5, 51 (1975).
225. S. A. Moyer, G. Abraham, R. Adler, and A. K. Banerjee, Cell, 5, 59 (1975).
226. G. W. Both, A. K. Banerjee, and A. J. Shatkin, Proc. Natl. Acad. Sci. U.S., 72, 1189 (1975).
227. E. Ehrenfeld, J. Virol., 13, 1055 (1974).
228. E. Ehrenfeld and D. F. Summers, J. Virol., 10, 683 (1972).
229. J. A. Mudd and D. F. Summers, Virology, 42, 958 (1970).
230. M. Soria and A. S. Huang, J. Mol. Biol., 77, 449 (1973).
231. H. Galet and L. Prevec, Nature, New Biol., 243, 200 (1973).
232. S. U. Emerson and R. R. Wagner, J. Virol., 12, 1325 (1973).
233. R. L. Imblum and R. R. Wagner, J. Virol., 15, 1357 (1975).
234. C. R. Pringle and W. H. Wunner, J. Virol., 12, 677 (1973).
235. C. R. Pringle, Curr. Top. Microbiol. Immunol., 69, 85 (1975).
236. C. R. Pringle, I. B. Duncan, and M. Stevenson, J. Virol., 8, 836 (1971).
237. A. S. Huang, D. Baltimore, and M. Stampfer, Virology, 42, 946 (1970).
238. T. Morrison, M. Stampfer, D. Baltimore, and H. F. Lodish, J. Virol., 13, 62 (1974).
239. D. Knipe, J. K. Rose, and H. F. Lodish, J. Virol., 15, 1004 (1975).
240. G. W. Both, S. A. Moyer, and A. K. Banerjee, J. Virol., 15, 1012 (1975).
241. P. I. Marcus, D. L. Engelhardt, J. M. Hunt, and M. J. Sekellick, Science, 174, 593 (1971).
242. A. S. Huang and E. Manders, J. Virol., 9, 909 (1972).
243. G. W. Wertz and M. Levine, J. Virol., 12, 253 (1973).
244. S. M. Perlman and A. S. Huang, J. Virol., 12, 1395 (1973).
245. H. Galet, J. G. Shedlarski, Jr., and L. Prevec, Can. J. Biochem., 51, 721 (1973).
246. L. P. Villarreal and J. J. Holland, J. Virol., 14, 441 (1974).
247. R. G. Wilson and J. P. Bader, Biochem. Biophys. Acta, 103, 549 (1965).
248. F. L. Schaffer, A. J. Hackett, and M. E. Soergel, Biochem. Biophys. Res. Comm., 31, 685 (1968).
249. J. F. E. Newman and F. Brown, J. Gen. Virol., 5, 305 (1969).
250. M. Stampfer, D. Baltimore, and A. S. Huang, J. Virol., 4, 154 (1969).

251. A. L. Schincarial and A. F. Howatson, Virology, 42, 732 (1970).
252. A. L. Schincarial and A. F. Howatson, Virology, 49, 766 (1972).
253. T. F. Wild, J. Gen. Virol., 13, 295 (1971).
254. E. L. Palma, S. M. Perlman, and A. S. Huang, J. Mol. Biol., 85, 127 (1974).
255. M. Soria, S. P. Little, and A. S. Huang, Virology, 61, 270 (1974).
256. E. A. C. Follett, C. R. Pringle, W. H. Wunner, and J. J. Skehel, J. Virol., 13, 394 (1974).
257. A. S. Huang and R. R. Wagner, Proc. Natl. Acad. Sci. U.S., 54, 1579 (1965).
258. M. Petric and L. Prevec, Virology, 41, 615 (1970).
259. R. R. Wagner, R. M. Snyder, and S. Yamazaki, J. Virol., 5, 548 (1970).
260. C. Y. Kang and L. Prevec, Virology, 46, 678 (1971).
261. R. R. Wagner, M. P. Kiley, R. M. Snyder, and C. A. Schnaitman, J. Virol., 9, 672 (1972).
262. B. Cartwright, J. Gen. Virol., 21, 407 (1973).
263. G. H. Cohen, P. H. Atkinson, and D. F. Summers, Nature, New Biol., 231, 121 (1971).
264. A. E. David, J. Mol. Biol., 76, 135 (1973).
265. A. Flamand, Compt. Rend. Acad. Sci. (Paris), 268, 2305 (1969).
266. A. Flamand, J. Gen. Virol., 8, 187 (1970).
267. C. R. Pringle, in The Biology of Large RNA Viruses (R. D. Barry and B. W. J. Mahy, eds.), Academic, London, 1970, pp. 567-582.
268. C. R. Pringle, J. Virol., 5, 559 (1970).
269. A. F. Holloway, P. K. Y. Wong, and D. V. Cormack, Virology, 42, 917 (1970).
270. C. Rettenmier, R. Dumont, and D. Baltimore, J. Virol., 15, 41 (1975).
271. A. Flamand, Mutation Res., 17, 177 (1973).
272. F. Lafay, J. Gen. Virol., 13, 449 (1971).
273. F. Lafay, J. Gen. Virol., 14, 1220 (1974).
274. F. Lafay and A. Berkaloff, Conpt. Rend. Acad. Sci. (Paris) Ser. D., 269, 1031 (1969).
275. D. M. Hunt and R. R. Wagner, J. Virol., 13, 28 (1974).
276. J. S. C. Ngan, A. F. Holloway, and D. V. Cormack, J. Virol., 14, 765 (1974).
277. J. E. Cairns, A. F. Holloway, and D. V. Cormack, J. Virol., 10, 1130 (1972).
278. D. H. L. Bishop and A. Flamand, in Negative Strand Viruses (R. D. Barry and B. W. J. Mahy, eds.), Academic, London, 1975, p. 327.
279. R. W. Simpson and J. F. Obijeski, Virology, 57, 357 (1974).
280. J. F. Obijeski and R. W. Simpson, Virology, 57, 369 (1974).
281. C. R. Pringle and I. B. Duncan, J. Virol., 8, 56 (1971).
282. P. K. Y. Wong, A. F. Holloway, and D. V. Cormack, Virology, 50, 829 (1972).
283. J. T. Unger and M. E. Reichmann, J. Virol., 12, 570 (1973).
284. A. Flamand and F. Lafay, Ann. Microbiol. Inst. Pasteur, 124, 261 (1973).

285. A. Combard, C. Martinet, C. Printz-Ané, A. Friedman, and P. Printz, J. Virol., 13, 922 (1974).
286. V. Deutsch, Compt. Rend. Acad. Sci. (Paris), 271, 273 (1970).
287. C. Martinet and C. Printz-Ané, Ann. Inst. Pasteur (Paris), 119, 411 (1970).
288. S. M. Perlman and A. S. Huang, Intervirology, 2, 312 (1974).
289. C. Printz-Ané, A. Combard, and C. Martinet, J. Virol., 10, 889 (1972).
290. W. H. Wunner and C. R. Pringle, Virology, 48, 104 (1972).
291. P. Printz and R. R. Wagner, J. Virol., 7, 651 (1971).
292. V. Deutsch and A. Berkaloff, Ann. Inst. Pasteur, 121, 101 (1971).
293. P. W. Choppin and R. W. Compans, J. Virol., 5, 609 (1970).
294. J. Závada and M. Rosenbergová, Acta Virol., 16, 103 (1972).
295. J. Závada, Nature, New Biol., 240, 122 (1972).
296. J. Závada, J. Gen. Virol., 15, 183 (1972).
297. A. S. Huang, P. Besmer, L. Chu, and D. Baltimore, J. Virol., 12, 659 (1973).
298. D. N. Love and R. A. Weiss, Virology, 57, 271 (1974).
299. J. Závada, Z. Závadova, A. Malíř, and A. Kocént, Nature, New Biol., 240, 124 (1972).
300. P. D. Cooper and A. J. D. Bellett, J. Gen. Microbiol., 21, 485 (1959).
301. F. Brown, S. J. Martin, B. Cartwright, and J. Crick, J. Gen. Virol., 1, 479 (1967).
302. J. Crick, B. Cartwright, and F. Brown, Nature, 211, 1204 (1966).
303. A. J. Hackett, F. L. Schaffer, and S. H. Madin, Virology, 31, 114 (1967).
304. E. Reczko, Arch. Ges. Virusforsch., 10, 588 (1960).
305. A. J. Hackett, Virology, 24, 51 (1964).
306. F. L. Schaffer and M. E. Soergel, Arch. Ges. Virusforsch., 39, 203 (1972).
307. E. L. Palma and A. S. Huang, J. Infect. Dis., 129, 402 (1974).
308. M. Doyle and J. J. Holland, Proc. Natl. Acad. Sci. U.S., 70, 2105 (1973).
309. S. Spiegelman, N. R. Pace, D. R. Mills, R. Levisohn, T. C. Eikhom, M. M. Taylor, R. L. Peterson, and D. H. L. Bishop, Cold Spring Harbor Symp. Quant. Biol., 33, 101 (1968).
310. J. Bernard and A. M. Petitjean. Compt. Rend. Acad. Sci. (Paris), 280, 2269 (1975).
311. P. Printz, in Negative Strand Viruses (R. D. Barry and P. W. J. Mahy, eds.), Academic, London, 1975.
312. E. A. C. Follett, C. R. Pringle, and Pennington. J. Gen. Virol. 26, 183 (1975).
313. J. F. Szilágyi and C. R. Pringle. J. Virol. 16, 927 (1975).
314. E. Hefti and D. H. L. Bishop. Biochem. Biophys. Res. Comm. 66, 785 (1975).
315. E. Hefti and D. H. L. Bishop. Biochem. Biophys. Res. Comm. 67 (176).

Chapter 6

The Biology of Myxoviruses

Debi Prosad Nayak

Department of Microbiology and Immunology
School of Medicine
University of California at Los Angeles
Los Angeles, California

I. Introduction

This chapter presents the biology and chemistry of myxoviruses. Several specialized reviews dealing in depth with specific aspects of influenza viruses have been published recently, including a comprehensive monograph [1] and treatises on epidemiology and immunology [2-4], structure [5], replication [6], viral proteins [7, 8], viral envelope [9], RNA [10, 11], and genetics [12]. In addition, Barry and Mahy [13] have edited a large collection of papers on negative-strand viruses and a comprehensive book edited by Kilbourne has been published recently [231]. The objective of this review is to present what I consider the central themes in studying the biology of myxoviruses: their structure, the function of structural components, the processes involved in replication, transcription, translation, processing, and assembly of subviral components into virions; as well as some unique properties of these viruses. The data presented in this chapter, unless otherwise mentioned, come primarily from the studies on influenza A viruses. However, it is likely that many of the properties of type B and type C viruses which have not been studied so extensively are similar to those of type A viruses. For additional information and in-depth analysis, readers are directed to the above-mentioned reviews and other papers cited in the text.

II. The Nature of Myxoviruses

A. Properties

Originally, myxoviruses were described as a group of viruses that possess a strong affinity to certain mucoproteins and an enzyme that removes neuraminic acid from the mucoproteins. Both of these properties are common to other viruses, particularly the paramyxoviruses [14]. Recently, however, the term "myxovirus" (or orthomyxovirus) has been restricted to the influenza viruses of man and animals [15]. Table 6.1 lists some of the characteristics that are common to all influenza viruses and that differentiate them from other viruses of similar morphology, such as paramyxoviruses. Among the most distinguishing features are the segmented genome, the nature of the ribonucleoprotein, and 2 distinct surface glycoproteins–hemagglutinin and neuraminidase.

B. Classification

The present system of classification and strain designation for myxoviruses was recommended by the World Health Organization Expert Committee on Influenza Viruses [16]. This is primarily based on (a) antigenic type of ribonucleoprotein, that is, type A, B, or C; (b) host of origin if isolated from nonhuman host; (c) geographical origin; (d) strain number; and (e) year of isolation. For influenza A viruses, antigenic designation follows strain designation and includes hemagglutinin (HA) subtype (e.g., human: H0, H1, H2; equine: Heq1, Heq2; avian: Hav1, Hav2) and neuraminidase (NA) subtype (e.g., human: N1, N2; equine: Neq1, Neq2; avian: Nav1, Nav2). A full designation would include all of the above information, e.g., A/Hong Kong/1/68 (H3 N2). Some common prototype isolates with antigenic designation are shown in Table 6.2. An examination of these strains shows that isolates from different species do not contain any common HA but do possess some common NA, as, for example, N1 and N2 being present among the isolates from both swine and avian species.

Designation of recombinant or hybrid viruses produced in the laboratory from 2 known strains should indicate the inherited antigenic characteristics of each parent virus strains, e.g., A/BEL/42 (HO)-A/Singapore/1/57 (N2), or abbreviated as A/BEL (HO)-Sing (N2).

Type B and C viruses are found only in humans. Although there is strain variation among B and C viruses, there is not enough information about HA and NA subtypes for antigenic designation. The description of B and C viruses is therefore limited to strain designation, e.g., B/England/5/66, C/Paris/1/67.

Table 6.1 Properties of Myxoviruses

Morphology

Spherical; enveloped; possess spikes on surface projections; 80-110 nm in diameter; pleomorphic, some are long rods up to 4 μm in length.

Structural Components

Virion contains 5 separate polypeptides: hemagglutinin (HA), neuraminidase (NA), membrane protein (M), nucleoprotein (NP), and transcriptase (P). HA may be cleaved into HA_1 and HA_2. Both HA and NA are glycoproteins.

RNP complex consists of RNA and NP and P proteins. RNP is segmented and heterogenous. Multiple RNP molecules representing multiple RNA segments are present in virions.

Viral genome consists of single-strand RNA, segmented, 6-8 RNA segments per virion. Total molecular weight of RNA is $4\text{-}6 \times 10^6$.

Replication

RNA is transcribed as complementary RNA (CRNA) which serves as messenger as well as template for VRNA replication.

The nucleus is involved in virus replication; enucleated cells or actinomycin D- or UV-treated cells cannot support viral infection.

Mature virus buds from plasma membrane but contains little or no host proteins.

Biological Properties

Antigenic shift: Major change in envelope proteins, starting new pandemic strains of influenza virus every 10-15 years.

Pleomorphism or heterogeneity in viral shape and size is common.

Von Magnus virus or multiplicity-dependent noninfectious virus formation is extremely common among influenza viruses; these incomplete or defective viruses lack one or more RNA segment.

Formation of hybrid or recombinant viruses occurs at a higher frequency than observed among other single-strand RNA viruses.

Marker's rescue, cross-reactivations, multiplicity-dependent reactivation are also common among influenza viruses.

Table 6.2 Strain Description and Antigenic Designation of Some Common Influenza Viruses

A Viruses	B Viruses	C Viruses
A/PR/8/34 (HONI)	B/Lee/1940	C/1233/1947[a]
A/Weiss/43 (HONI)	B/Bon/Australia/1943	C/JJ/1950[a]
A/FM/1/47 (HINI)	B/Great Lakes/1954	
A/England/51 (HINI)	B/Johanesburg/1958	
A/Denver/57 (HINI)	B/Taiwan/1962	
A/Tokyo/3/67 (H2N2)		
A/Hong Kong/1/68 (H3N2)[b]		
A/Hong Kong/107/71 (H3N2)[b]		
A/England/42/72 (H3N2)[b]		
A/Port Chalmers/1/73 (H3N2)[b]		
A/Swine/Wisconsin/15/30 (Hsw1 N1)		
A/equine/Prague/1/56 (Heq1 Neq1)		
A/equine/Miami/1/63 (Heq2 Neq2)		
A/FPV/Dutch/27 (Hav1 Neq1)		
A/chicken/Germany "N"/49 (Hav2 N1)		
A/duck/England/56 (Hav3 Nav1)		
A/duck/Czechoslovakia/56 (Hav4 Nav1)		
A/tern/S. Africa/61 (Hav5 Nav2)		
A/Turkey/England/63 (Hav1 Nav3)		
A/Turkey/Massachusetts/65 (Hav6 N2)		
A/duck/Ukraine/1/63 (Hav 7 Neq2)		
A/Turkey/Ontario/6118/68 (Hav8 Nav4)		

[a]Antigenically identical.

[b]Although all of these isolates subsequent to Hong Kong strain are antigenically designated as H3N2, they vary considerably among themselves and also from A/Hong Kong/1/68. However, these variations are not good enough to warrant a new antigenic designation.

C. Pathogenesis

Influenza is still one of the major uncontrolled viral diseases of man and animals. It is caused by an elusive virus that by changing its character (and its coat) escapes effective prophylactic control. In humans, influenza is characterized by either dreadful pandemics or seasonal epidemics. The advent of a new strain of virus with a major antigenic shift is the cause of new pandemics, which occur every 10-15 years. Jet age mobility spreads the disease faster, and in a relatively short period of time humanity is in the grip of flu: in a major pandemic almost the entire world population (up to 90%) is affected. In spite of the remarkable progress of modern medicine, influenza inflicts high morbidity and personal misery and suffering. Furthermore, an influenza pandemic has a tremendous socioeconomic impact at a global level; at its peak, it can cause up to a 90% reduction in goods and services produced by a factory or a community. The economic impact of influenza pandemics in terms of monetary loss demands further careful study in the future.

Seasonal epidemics are less severe and usually affect a population group, particularly in winter in temperate regions. The disease "influenza" is highly contagious and has a short incubation period. It starts abruptly with catarrh in the respiratory tract, high fever, and severe fatigue. The acute phase in an uncomplicated case runs for 2-4 days, followed by a prolonged (2-4 weeks) convalescence accompanied by fatigue. Secondary bacterial invasion (e.g., *Hemophilus influenza, Klebsiella pneumoniae, Staphylococcus aureus,* or beta hemolytic group A streptococci) produces many of the bronchopulmonary complications. The detailed clinical aspect of influenza has been discussed in *Clinical Virology* by Debre and Cellers [17].

Influenza affects other animals, particularly swine, equines, and a variety of avian species (fowl, turkey, ducks). Epidemics, however, are species specific, although there is some evidence of cross-species infection in nature. In all animals, including humans, the virus primarily affects the epithelium of the respiratory tract, although occasionally it was been implicated in cardiac or neurological involvement. The 1957 pandemic virus has been particularly implicated in some neurological complications, such as encephalitis and meningitis. In chickens, fowl plague virus causes a more generalized infection, including viremia, cardiac and neurologic lesions, and high mortality.

Experimentally, influenza can be produced in human volunteers [18], ferrets, mice, pigs, turkeys, and other susceptible animals by intranasal inoculation with specific viruses. Virulence of a virus strain for a host species is a multigenic factor that can be either transmitted by recombination with a virulent virus or acquired by a selection process through repeated passage in a specific host. In an experimental infection via the intranasal route the influenza virus produces a descending infection; i.e., the cells of the upper respiratory tract (nasal mucosa and trachea) become infected first, and subsequently the virus spreads to the bronchial and alveolar epithelia [19-21]. Influenza virus has been also reported experimentally as a teratogen in pregnant monkeys [22].

D. Antigenic Variation

Although there are antigenic subtypes among other DNA and RNA viruses, these strain variants are rather stable in nature and there is no evidence of such constant modification of viral genome and viral proteins as occurs among influenza viruses. Persistent change in the coat proteins hemagglutinin (HA) and neuraminidase (NA) is rather unique and restricted to influenza viruses. Two types of antigenic variation are observed among these viruses, antigenic drift and antigenic shift. As shown in Table 6.2, the major antigenic variation occurs primarily among influenza A viruses.

1. Antigenic Drift

Antigenic drift is a gradual change observed in HA and NA proteins among the virus isolates. In recent years 2 major pandemic strains have emerged in a span of 10 years: Asian virus in 1957–A/Singapore/1/57 (H2N2)–and Hong Kong virus in 1968–A/Hong Kong/1/68 (H3N2). From 1957 to 1968 gradual variations in Asian virus HA and NA were observed: the virus was relatively stable from 1958 to 1962; then, minor changes were observed among the isolates from 1962 to 1966; and finally, the virus became rather unstable from 1966 to 1968, i.e., just before the Hong Kong strain emerged. During this last period, heterogeneity was more common and diverse among the virus isolates, although they were still related to the Asian virus designated as H2N2. Immediately before the advent of the 1957 pandemic Asian strain an increased variation among virus isolates had likewise been observed [23].

An essentially similar pattern is being observed in the course of the current pandemic strain (Hong Kong 1968). The later strains (Hong Kong 1971, England 1972, and Port Chalmers 1973) (Table 6.2) are variants of the Hong Kong 1968 strain [2]. Recent isolates are more diverse, and an increased frequency of antigenic variation among the isolates is now being observed [24]. It remains to be seen whether this diversity among the current isolates is an omen of a new pandemic strain with a major antigenic shift that will emerge in the next few years.

The antigenic drift in HA and NA is caused by mutational changes in the viral genome and selective pressure from immunity in human populations.

The drift is usually slow and limited after a new pandemic strain appears but becomes more extensive and frequent toward the end of a pandemic era and before the beginning of a new pandemic strain [25, 26].

In laboratory experiments such variation has been observed among viruses grown in the presence of subneutralizing antibodies in ovo, as well as in partially immune animals in vivo. These laboratory-grown variants contained changes in peptide maps of the HA subunit [27].

2. *Antigenic Shift*

Antigenic shift causes the advent of new pandemic viruses that are very different from the strains prevalent before or during the period the new virus emerged. Thus, major changes are apparent in 1947 (H1N1), 1957 (H2N2), and 1968 (H3N2) viruses [25]. These changes take place invariably in HA and sometimes also in NA antigen (1947 and 1957 strains). The mechanism of the antigenic shift in nature is not clearly understood. Two hypotheses have been proposed.

The *direct mutation theory* maintains that antigenic shift is an extension of antigenic drift in which mutation and selection pressure eventually produce a new strain different enough to cause a pandemic and to become prevalent worldwide. This view is supported by the observation that pandemics are followed by slow antigenic drift, which toward the end of a pandemic era grows more frequent and diverse until the new strain emerges. Such a period of unstability was observed before the emergence of the 1957 and 1968 strains; recent studies indicate that diversity among recent isolates is becoming more frequent. Furthermore, pandemic strains appear at rather regular intervals (1947, 1957, 1968). During the 10-15 year periods the population becomes saturated with antibodies to previous strains. The major objection to the direct mutation theory is the finding of a large number of differences in the amino acid sequences of the HA polypeptides of the Asian (1957) and Hong Kong (1968) strains [25, 28]. It is difficult to visualize how so many mutations could have taken place all at once.

The *genetic recombination theory* proposes alternatively that antigenic shift is caused by wide variation in the HA polypeptide sequence and cannot be explained by mutation alone. These changes would therefore be caused by gentic exchange among influenza A viruses prevalent in nature. In support of this theory, its proponents claim that influenza A viruses are widespread among many species of birds and mammals [29]; that, although most of the viruses are species specific in pathogenesis, transpecies infection without apparent disease can take place; and that such recombinants or hybrid viruses are produced by mixed infection in cell culture, in eggs as well as in animals [4, 30-33]; and finally that many of the viruses from different species share common antigens (Table 6.2).

There is as yet no definitive evidence that such an exchange does take place in nature and is the cause of the advent of new pandemic viruses with a major antigenic shift. The recombination theory cannot explain why major antigenic shifts take place regularly every 10-15 years. Furthermore, an examination of the virus isolates listed in Table 6.2 shows there is no common antigenic determinant of HA among different animal and human virus strains. However, the common HA antigenic determinant represents only a minor number of amino acid sequences compared to the amino acid sequence of the entire HA molecule, and minor modifications caused by mutation of the antigenic site may take place either before or after the exchange of HA and the formation of a hybrid virus. This may explain why there is no common HA determinant among different influenza A viruses isolated from man and animals. Homology sequence analysis by peptide mapping shows that a portion of the light chain of HA/Hong Kong/1/68 (H3N2) is identical to that of A/equine/Miami/1/63 (Heq2 Neg2) and A/duck/Ukraine/1/63 (Heq7 Neq2) but is completely different from A2 Asian strains isolated before 1968 [31]. This evidence and the research on genetic exchange among influenza viruses in cell culture, chicken eggs, and animals [4, 12, 30-33] suggest that new pandemic strans may originate because of genetic recombination from mixed infection of a host by more than 1 virus.

III. Virion Structure

A. Morphology

Influenza virions appear as spherical particles containing multiple helical ribonucleoprotein molecules enclosed in a protein membrane which is further surrounded by a lipoprotein envelope containing "spikes" (Fig. 6.1). Although virus particles are usually spherical with an average diameter of 80-110 nm, some in the population may be filamentous, up to 4 μm in length (Fig. 6.2) [34-38]. In spite of morphological differences, spheres and filaments are equally infectious and possess essentially the same structural proteins. Compared to any other known viruses, pleomorphism or heterogeneity in size and shape is much more common among the influenza viruses (Fig. 6.2B). The basic mechanism for pleomorphism is far from being understood. At least 4 variables could be responsible for the observed heterogeneity. First, it may be a technical artifact produced by handling, storage, and preparation of viruses during electron microscopy [5]. Second, it may be a host-induced variation; for example, freshly isolated human influenza virus strains are filamentous [39] and gradually become spherical during several passages in embryonic chicken eggs. Third, virion morphology can be modified by surface-acting agents (e.g.,

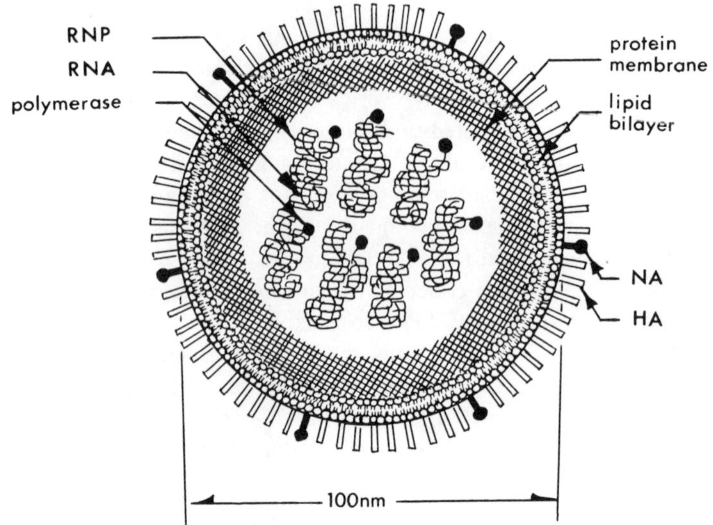

Fig. 6.1 Diagramatic illustration of influenza virus.

vitamin A and lipids) in the cellular environment [40]; these agents alter membrane fluidity and thus modify the virion structure. Fourth, shape and size of virion structure are also genetically determined [12, 36, 41]. Thus the morphology of virions may be the result of a number of factors influenced by host, viral genome, and environment.

All influenza virions, whether spherical or filamentous, have surface projections called "spikes," a lipid envelope, a protein coat (the membrane or matrix), and an internal helical core (ribonucleoprotein, RNP).

1. Spikes

The surface projections or "spikes", around 10 nm in length, are easily discernible by negative staining (Figs. 6.1 and 6.2); their structure and properties have been reviewed in detail by Laver and Webster [3, 42]. There are approximately 500 spikes on an average virion 100 nm in diameter [43]. The spikes contain 2 distinct proteins, hemagglutinin (HA) and neuraminidase (NA). Isolated hemagglutinin "spikes" are triangular and rod shaped and contain 2 or 3 protein molecules. They are monovalent and attach to the receptors by 1 end only, the hydrophylic end; the other end is hydrophobic and forms lipoprotein bonds to the virion's lipid bilayer. The functional neuraminidase spike is a tetramer in the form of a rod with a fibrous hydrophobic tail that interacts with the lipid

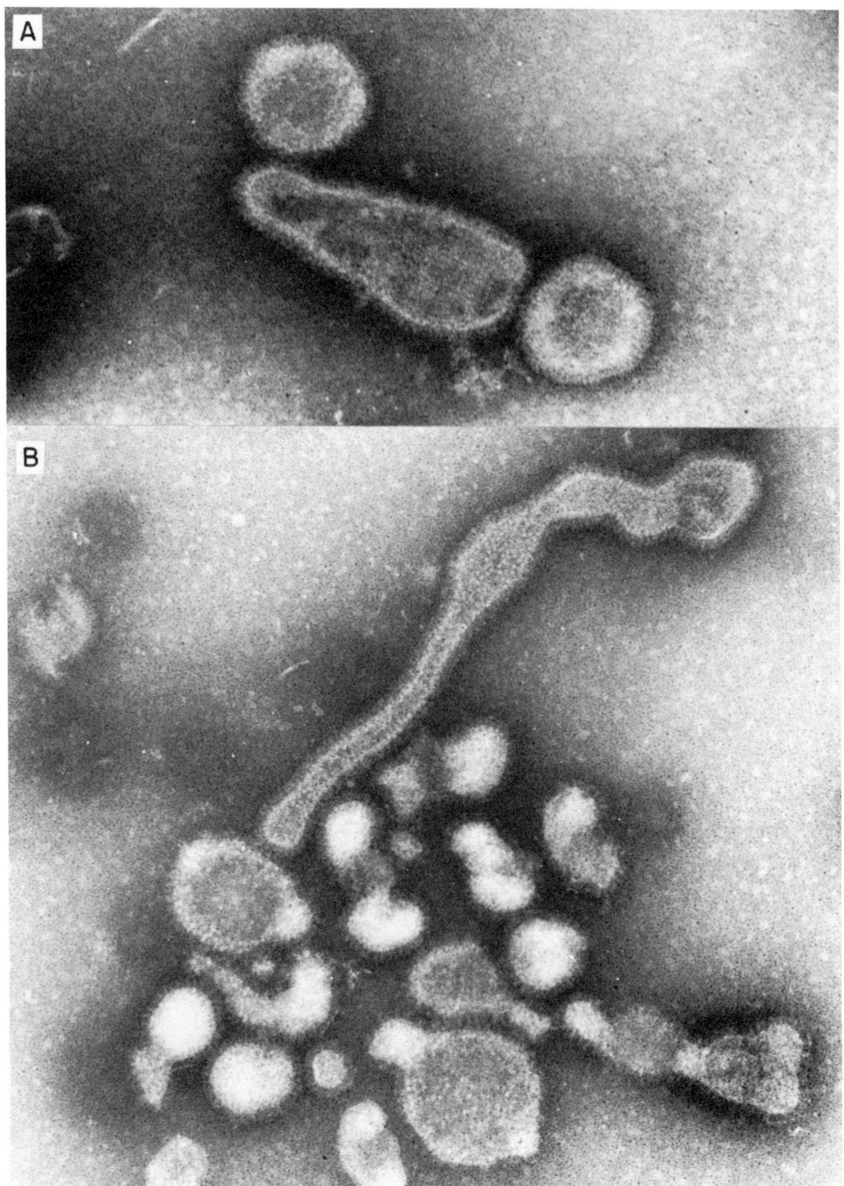

Fig. 6.2 (A) Electron micrograph of influenza virus by negative staining; × 151,250. (B) Electron micrograph of influenza virus particles by negative staining; × 151,250.

layer. Of the total spikes present on the virion surface, HA spikes represent more than 90% and NA spikes about 10% [44-48].

2. *Lipid Bilayer*

Next to the spikes is the lipid bilayer, which comprises about 20 to 25% of the virion mass (Fig. 6.1) and renders the virion sensitive to ether and other lipid solvents. Over 95% of virion lipids are phospholipids and cholesterol; only a small amount (1-2%) are glycolipids. Most of the virion lipids are host specified. Although there is a selection in the quality of lipid composition, the virion incorporates prelabeled host lipids in its envelope. Moreover, virions produced in different cells contain lipids that reflect qualitatively the lipid composition of the host cell membrane. However, the mechanism of lipid selection during the virion budding and assembly processes is not yet understood. The work of Choppin et al. [49] suggests that the metabolic alteration produced by viral infection induces changes in the plasma membrane that reflect the selection of viral lipids during budding and eventually in the viral envelope. Blough and Merlie [50], however, suggest that viral lipids are selected by the nature of the viral proteins; these authors have found that viral lipids in incomplete virus vary from those present in complete virions, a difference they claim is from the variation of arrangement of structural proteins. Blough [37] also found that lipids made after infection were incorporated into virions; as lipids were synthesized at the endoplasmic reticulum and transported to the cell surface membrane, he suggested that viral membrane biogenesis took place at the hot spots caused by the interaction of viral lipids and viral polypeptides. The observation by Holland and Keihn [51] that host proteins are selectively displaced by viral proteins at the budding area suggests selection and reassembly of lipid layers before budding.

The arrangement of lipids in the viral envelope has not been clearly resolved. One model proposed by Tiffany and Blough [43] claims that the viral lipids form a micellum and not a complete layer; the outer spike proteins would interact through these micelles with membrane proteins, suggesting that spike proteins are held by protein-protein interaction. The alternate model, proposed by Schulze [5, 52, 53] suggests that the lipid layer is complete and that there is no direct interaction between the spike proteins and the membrane proteins. This latter model is supported by the observation that extensive treatment by proteolytic enzymes does not degrade nonglycosylated M (membrane) protein, suggesting that the lipid bilayer is a barrier between spike proteins and membrane proteins. Inability to iodinate M protein by lactoperoxidase following extensive protease treatment also supports the existence of the lipid barrier between M proteins and spikes [54]. Electron spin resonance studies by Landsberger et al. [55, 56] also support the formation of a lipid bilayer rather than of a micellar structure in the virion envelope. Thus, the lipid bilayer model proposed by Schulze [5] has stronger experimental support and is more likely to depict the actual orientation of lipids in the viral envelope.

3. Lipid-Protein Interaction

If spike proteins do not interact with membrane proteins because of the existence of a lipid bilayer, they must form a stable bond with lipids. The nature of this noncovalent lipoprotein bond is not clear. Both hemagglutinin and neuraminidase spikes possess hydrophobic ends which would bind to the fatty acid chains of viral lipids [57, 58]. The light polypeptide chain of hemagglutinin is strongly hydrophobic and likely to interact with lipids; further evidence that this hydrophobic end of the light chain interacts with lipids came from the fact that bromalein-released light chain is more hydrophylic than the intact light chain. Bromalein-released light chain is also shorter and has lost about 3000 daltons in molecular weight [59]; this indicates that a small segment of light chain (about 25 amino acids) contains the hydrophobic property which possibly interacts with lipids. This hydrophobic peptide is at the C terminal end of the light chain and contains about 50% serine residues [235].

4. Membrane or Matrix

Underneath the lipid layer is an electron-dense layer that encompasses the ribonucleoprotein core [60-62]. This layer is a protein coat consisting of a single-protein component, "M" protein, which is the most abundant protein in virion structure. When chymotrypsin-derived cores are stained with uranyl acetate, a layer of protein (6 nm) becomes clearly visible beneath the lipid layer. The membrane protein layer, along with the lipid bilayer, protects RNA from nuclease and gives structural integrity to virions, for virions retain their spherical form even after their spikes are removed [53, 63, 64].

5. Ribonucleoprotein

Ribonucleoprotein (RNP) is the internal or core component of virions and has 3 structural components: 2 proteins (NP and P) and RNA. It contains transcriptase [65] and appears to be responsible for the biological activity of virions [66]. The number of RNP components per virion is not definitely determined. Each virion RNA segment is present as a separate RNP, and therefore the number of RNP segments will vary with the RNA content of the virions. Thus, complete virions must contain at least 8 RNP molecules whereas incomplete virions (or von Magnus particles) contain less RNA and fewer RNP molecules. RNP becomes visible either in thin sections or following virion disruption after detergent treatment. RNP strands possess striations giving the appearance of a helical symmetry [60, 67]. In addition, a superstructure of a double helix with the strand folded back on itself has been suggested for RNP; this model was based on the electron microscopic observation of alternate deep and shallow grooves and a loop at 1 end [68, 69]. It is still unclear, however, whether this superstructure exists in all RNA molecules, which vary considerably in length.

The average diameter of RNP is about 6 nm [236], 10 nm [69], or 15 nm [70]; the difference in diameter could result from the condition of RNP isolation and consequently from the presence and absence of coiled superstructure. The RNP length appears variable and depends primarily on the size of RNA; 3 classes based on size (90-110 nm, 60-90 nm, and 30-35 nm) have been observed [69]. Also RNP up to 2 μm have been reported [236]. RNP isolated from purified virions or infected cells has a density of 1.26 g/ml. It possesses 10-12% RNA, variable in length and S value [71]. Although virion RNP contains only VRNA, intracellular RNP contains about 10% CRNA an 90% VRNA [72]. It is therefore not clear whether all of the intracellular RNA molecules contain the same structural components as RNP isolated from virions, or whether some of the intracellular RNP molecules possess entirely different structural components and arrangement but appear as RNP because of the association of RNA molecules with proteins. Furthermore, although labeled CRNA was measured by hybridization with VRNA, the amount of labeled VRNA associated with RNP was not assayed directly.

Unlike the helical nucleocapsids of other viruses, the structure of virion RNP is also different. For example, RNA present in influenza RNP can be degraded by RNase treatment, whereas RNA in helical nucleocapsids either naked (e.g., tobaco mossaic virus) or isolated from other enveloped viruses (e.g., paramyxoviruses) is resistant to RNase treatment.

Virtually nothing is known about the nature of binding and the specificity of interaction between the protein components and RNA molecules. The NP protein, which is the major structural component of the RNP, binds to the negatively charged phosphate groups of RNA. Polyvinylsulfate, a negatively charged molecule, can displace RNA quantitatively without any major structural alteration in RNP [73]. Both VRNA and CRNA will bind to NP [74]. The role of the P proteins (transcriptase?), which is also associated with RNP and may be involved in maintaining the RNP structure, is unknown. As there are less than 40 P molecules, compared to 1000 NP molecules per virion, P proteins are unlikely to play any major structural role, although their location at strategic places (e.g., at the 3′OH end of RNA), providing structural stability, cannot be ruled out. Furthermore, the interaction of RNP to M protein, which encloses RNP completely and provides protection from external nucleases and proteases, has not been resolved.

IV. Virion Composition

Influenza virions contain approximately 1-2% RNA, 20-25% lipid, 5-8% carbohydrates, and 68-70% protein. Virion density is 1.21 g/ml in sucrose gradients.

A. Virion Proteins

Influenza virions contain 5-7 structural proteins, depending on whether HA is cleaved into HA_1 and HA_2 and whether both P_1 and P_2 are resolved by polyacrylamide gel electrophoresis (PAGE) analysis (Fig. 6.3, Table 6.3). Of these, 2 or 3 are glycosylated and the rest are nonglycosylated proteins.

1. Hemagglutinin (HA)

Hirst [75] and McClelland and Hare [76] first demonstrated that influenza virus can agglutinate chick red blood cells. This observation was soon confirmed and extended to the agglutination of red cells of other species. Since then, HA has been used extensively in the quantification of virus particles (approximately 1-2 $\times$ 10^6 virus particles per HA unit) and viral antigens. Hemagglutinin, along with neuramindase, confers the type specificity of an isolate. The antibodies against HA are the major components of neutralizing (and therefore protecting) antibodies. Thus, HA spikes are the main viral receptor for viral adsorption to host cells.

Laver, Webster, and their colleagues [8, 42, 58] are mainly responsible for the isolation, characterization, and elucidation of structural properties of the HA antigen. They isolated the spikes containing HA by detergent treatment (sodium deoxycholate, Tween-40, or sodium dodecyl sulfate) and freed HA from neuraminidase by adsorption to red cells. Purified HA spikes are rod shaped (14 nm $\times$ 4 nm) and monovalent; i.e., they adsorb to red cells and interfere with the agglutination of red cells by virus particles, although they do not cause hemagglutination. When sodium dodecyl sulfate is removed, the HA spikes form a rosette, become multivalent, and can cause agglutination of red cells. By polyacrylamide gel electrophoresis (PAGE) analysis, these purified HA spikes were shown to contain a single glycoprotein (MW 77,000) which would dissociate further into 2 smaller glycoproteins (HA_1, MW 55,000; HA_2, MW 25,000) after treatment with reducing agents. This showed that the larger polypeptide (HA) is composed of 2 small

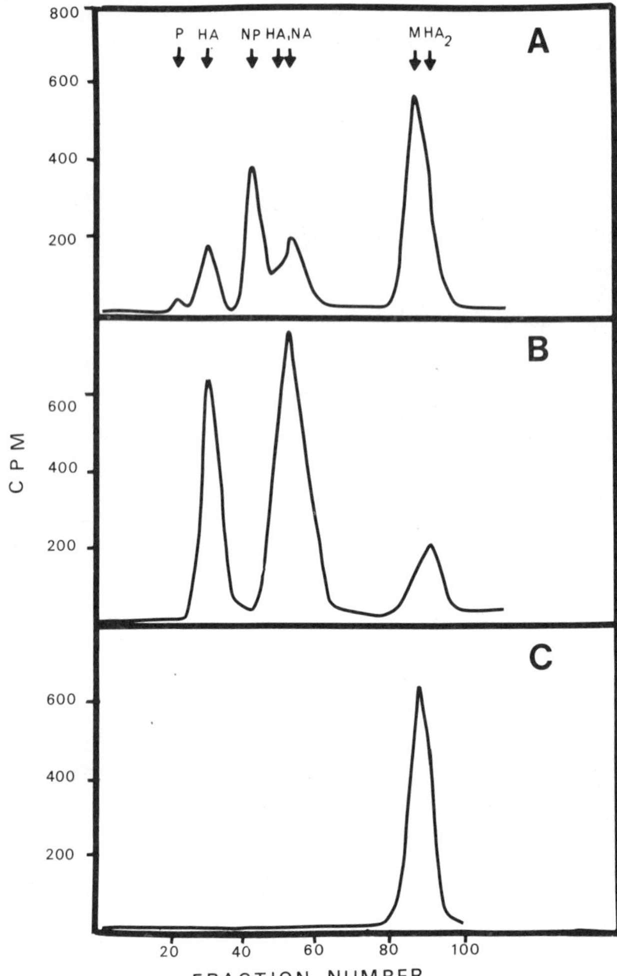

Fig. 6.3 Polyacrylamide gel electrophoresis (PAGE) analysis of the proteins of WSN virus labeled with [^{3}H]-amino acids and [^{14}C]-glucosamine. (A) [^{3}H]-labeled whole viral proteins. (B) [^{14}C]-labeled viral glycoproteins. (C) [^{3}H]-counts of the virus material extracted into acidic chloroform-methanol (M protein). From A. Gregoriades [234] with the permission of Academic Press.

Table 6.3 Proteins of Influenza Viruses[a]

Designation	Molecular weight $\times 10^{-3}$	Percentage of total protein	No. of molecules per virion	Presence of carbohydrates	Location	Function
P1	89-96	0.2	12	–	A component of internal ribonucleoprotein	Unknown P1, P2, or both may be involved with a RNA-dependent RNA polymerase activity
P2	80-83	0.9	35	–	Same as P1	
HA	75-80	0-24[b]	0-950[b]	+	Outer envelope, a component of spike	Agglutination of red cells and adsorption to cell receptors
NP	53-60	17-31	500-1170	–	Internal core	The main component of internal ribonucleoprotein
NA	45-70	2.5-6.9	100-240[c]	+	Outer envelope spikes	Neuraminidase; splits a glycosidic bond containing sialic acids
HA_1	49-60	0-21[b]	0-950[b]	+	Spikes of outer envelope	Hemagglutination; the heavy-chain peptide component following cleavage of HA
HA_2	23-30	0-12[b]	0-950[b]	+	Same as HA_1	Hemagglutination; the light chain or the smaller peptide component following the cleavage of HA; possibly binds HA spikes to the viral lipid bilayer
M	21-27	30-40	2500-3200	–	Associated with inner surface of lipid layer	"Matrix" or membrane protein of the virus core

[a]This table was prepared from data presented by Compans et al. [62], Schulze [52], Skehel [96], Lazarowitz et al. [79], Laver [8], and Compans and Choppin [6].

[b]HA may be present in virion as the uncleaved molecule or may be cleaved into HA_1 and HA_2. Two such molecules held by disulfide bonds constitute spikes with hemagglutinin activity.

[c]Neuraminidase spikes are tetramers, i.e., 25-60 NA spikes are present on the virion envelope.

polypeptides (HA_1 and HA_2) held together by disulfide bonds. From molecular weight estimates for spikes it appears that each spike contains 2 or 3 molecules of HA, or 2 or 3 each of HA_1 and HA_2. Although many strains of influenza virus contain HA_1 and HA_2, some viruses (i.e., Ao/BEL) grown in calf kidney cells were shown to contain instead the large glycoprotein (HA, MW 77,000), which was not dissociated into the smaller glycopeptides by reducing agents [77]. Thus, analyses of viruses containing only 1 HA protein and, later, pulse-chase experiments in infected cells demonstrated conclusively that hemagglutinin is synthesized as a single protein molecule (HA), glycosylated, and later cleaved into 2 smaller glycoproteins (HA_1 and HA_2). The order of biosynthesis is NH_2 HA_1-HA_2 COOH [235]. This proteolytic cleavage of HA into HA_1 and HA_2 is a host-dependent function and is not required for virus assembly and hemagglutinating activity [78-80] but is necessary for infectivity of virions [239, 240].

By weight, HA peptide contains 17-20% carbohydrate, covalently linked to both HA_1 and HA_2. Glycosylation of HA protein is mainly a host-controlled function that involves a considerable number of sugar transferase enzymes. The length and distribution of the carbohydrate chains in HA remain uncertain. The total molecular weight of the carbohydrate is 12,000 and is possibly distributed in 6-12 chains with an average molecular weight of 1000-2000. Although unlikely, it is possible that the entire carbohydrate moiety is present as a giant molecule of 12,000-15,000 daltons [81]. The carbohydrate residue contains N-acetylglucosomine, mannose, galactose, and fucose (6 : 6 : 1 : 1 molar ratio) and is possibly attached to the peptide molecule via amino acids, such as aspartic acid, glutamic acid, threonine, and serine. The carbohydrate moiety is added in a stepwise manner on the protein backbone. As sugar transferases add labeled monosaccharide introduced during infection, preformed carbohydrates polymerized into an oligosaccharide chain before infection are not likely to be incorporated into viral proteins.

2. *Neuraminidase (NA)*

Neuraminidase (sialidase or mucopolysaccharide N-acetyl neuraminylhydrolase, E.C. 3.2.1.18) is a glycoprotein enzyme (MW 45,000-70,000) that splits off glycosidically bound neuraminic acid residues from glycoproteins and gangliosides. It is located on the viral envelope as a component of spike protein and is functional as a tetramer (or dimer) [58, 44-47, 82]. Studies involving genetic recombination as well as temperature-sensitive (*ts*) mutation have shown conclusively that neuraminidase is a virally coded protein [45, 58, 83, 84]. Palese et al. [83] have identified temperature-sensitive mutants (*ts*3, *ts*11) with a defect in neuraminidase. Although the exact molecular nature of this defect (i.e., substitution of a specific amino acid) in the neuraminidase component has not been determined, it has been shown that the entire molecules is made at the restrictive temperature and is also present in the virion but is nonfunctional because of its increased thermal lability at higher temperatures.

Neuraminidase removes all neuraminic acid residues from virions and causes elution of virus from red cells. It is also believed that neuraminidase enables the virus to elute from inhibitory mucoproteins present in the respiratory mucosa until it finds its way to infect epithelial cells. Although much is understood about the function of this enzyme, the role of neuraminidase in the biology of the infectious process is not clear. Obviously, neuraminidase does not aid directly in the adsorption of virions. More likely, it may prevent adsorption by removing receptors containing neuraminic acid, which also serves as the receptor for hemagglutinin—the antigen believed to be primarily responsible for virus adsorption. Following penetration, neuraminidase does not affect directly the intercellular processes (including budding) involved in virus replication; in fact, this enzyme has little effect in the single-step growth cycle of influenza viruses. It is clear, however, that neuraminidase plays a significant role in the multicycle replication of virus, both in animals and in cell culture, by facilitating cell to cell infection. Using recombinant viruses, Kilbourne and his colleagues [85-87] have shown that antineuraminidase antibodies help in prophylaxis by restricting multicycle replication and thus self-limiting the infection in animals. In comparison with controls, animals possessing antineuraminidase antibodies exhibit milder symptoms and produce lesser virus titer. Essentially the same happens in embryonated chicken eggs or cells in culture: when the culture medium contains antineuraminidase antibodies or inhibitors of neuraminidase (e.g., 2-deoxy-2,3-dehydro-N-trifluoroacetyl neuraminic acid), the plaque size is smaller and the virus yeild in multiple-cycle replication is less than in the controls [86, 88-90]. Similar results have been obtained at the restrictive temperature with temperature-sensitive mutants (*ts*3, *ts*11) in which the *ts* lesion was located in the neuraminidase [83].

Three hypotheses have been proposed to explain the role of neuraminidase in multiple-cycle infection: (a) Neuraminidase facilitates release of virus from the cell surface glycoproteins by removing all neuraminic acid residues from virion and cell membrane; virions can therefore infect neighboring cells [85, 89, 90]. (b) Neuraminidase removes sialic acid residues from the viral envelope and thereby prevents agglutination of virus particles into clumps [83, 91]. (c) Neuraminidase may be involved in the assembly and maturation of virions at the cell membrane [89, 92]. The first 2 hypotheses are not mutually exclusive, as both virus agglutination and increase in cell-associated virus clumping have been observed in absence of neuraminidase activity. There is little evidence for the third hypothesis at present. One must be cautious, however, in interpreting the results of the continued presence of antibodies and inhibitors in culture medium as these agents may not only prevent virus release but may also interact with released virions and interfere with viral adsorption. Antibodies against neuraminidase because their bivalent nature cause clumping of virus particles and thus reduce the infectivity of virus particles. The strongest evidence for the role of neuraminidase came from the study of temperature-sensitive mutants [83]. Although NA function is important in the biology of viral infection, only a minimal amount of NA is required. Palese and Schulman [93] did not find any significant difference in plaque size or virus yield between 2 recombinants, MPN^+ and MPN^-, although MPN^+ combined 8 times more NA than MPN^-.

Being a glycoprotein, NA also contains carbohydrate residues [46, 47]. No precise estimate is available about the amount of carbohydrate per NA molecule or the number and nature of the side chains. There are about 5.7 glucasamine residues per NA molecule [46]. Information about other carbohydrate molecules is not available. It is not known whether the sugar composition is the same as in hemagglutinin. Much of the carbohydrate is distributed toward the base of the NA spikes, suggesting the carbohydrate moiety of NA may be involved in the interaction with glycolipids of the lipid bilayer [47, 94, 95].

3. *M Protein*

M protein is the most abundant protein in the virion (33% of the total protein mass). It is a nonglycosylated protein with a molecular weight of 20,000-25,000. The existence of this major protein component in virion became evident only after PAGE analysis; it had been missed completely by serological studies using antibodies prepared against intact virion, isolated RNP, or infected cells. In retrospect, none of these procedures would produce significant amounts of antibodies against M protein because of the structural location of the latter in the virion and of the lack of abundant free M protein molecule in infected cells. In fact, compared with other virion proteins, only a few M protein molecules are present in infected cells, and synthesis of M protein appears to be the step-limiting component in virion maturation [237].

The following evidence suggests that M protein constitutes the membrane layer underneath the lipid envelope: (a) Deoxycholate treatment of virion releases M protein from RNP. (b) Proteolytic enzyme treatment that releases HA and NA does not degrade or release M protein. (c) "Cores" produced by proteolytic treatment contain M protein along with P and NP [60, 62-64]. (d) Iodination by lactoperoxidase of intact virion or cores produced by proteolytic enzyme does not iodinate M protein. (e) Phospholipase C treatment renders M protein accessible to proteolytic digestion, indicating the lipid bilayer forms a barrier against the proteolytic degradation of M protein. (f) If cores are treated with glutaraldehyde and subsequently treated with NP 40 to remove lipids, smaller cores containing M proteins are formed.

4. *P Protein*

The largest protein molecule (MW 90,000) found in the virion, P protein, is nonglycosylated and is associated with the internal virion component. While it is believed to be the transcriptase molecule in the virion, there is no direct evidence as yet. A reconstitution experiment in vitro (dissociating P from RNP and restoring transcriptase activity by adding P) has yet to be done. Furthermore, it also remains to be seen whether the antibodies against purified P will inhibit transcriptase function as well. Although only a few molecules of P (<50) are present in the virion, they may be located in the strategic RNA position to give structural stability, to link RNP segments, or to produce a recognition site for polymerase molecules. Skehel [63, 96] has reported 2 separate P proteins (P1, P2). The significance of 2 proteins instead of 1 is unknown at present; it is not clear whether both of these proteins are involved in transcription, or 1 in transcription and the other in replication, or in some other viral function. Clearly, more work is needed in defining the functions of P proteins in viral replication.

5. *NP Protein*

NP is the major component (>95% of the protein) of the RNP complex. It is a homogenous, nonglycosylated protein with a molecular weight of 50,000-65,000. Approximately 500-1000 NP molecules are present within a virion. The relative amount of virion NP varies with its RNA content.

NP is the major antigen responsible for virus type specificity (type A, B, or C). Amino acid analysis of NPs from type A and type B viruses shows that they are different. Within a type, NP confers the common antigenic determinant and is immunologically indistinguishable among the different strains isolated from humans and animals. Antibodies against NP, an internal protein, do not confer any protection against infection and thus do not produce any selective pressure toward mutation or recombination. It still remains to be seen, however, whether immunological identity within a type results from the identity of the primary sequence of amino acids of the entire molecule or from that of the antigenic determinants only.

NP possesses a marked affinity in vitro for influenza viral RNA (both CRNA and VRNA) and the small AMP-rich RNA from normal cells, and therefore possibly binds to the negatively charged phosphate groups of the RNA molecules [73]. NP also binds (although to a lesser extent) to other viral RNAs, e.g., Newcastle disease virus RNA, Sindbis virus RNA [74]. It also imparts structural integrity to the RNA genome, thus facilitating binding and the transcriptase function of the P protein.

B. Virion RNA

Approximately 1% of the total virion mass is RNA [97, 98]. Influenza viral RNA is single stranded and segmented; the precise number of segments and the total complexity of the virion RNA are yet to be determined. The present estimates range from 6 to 8 segments of RNA per virion; the total molecular weight is 4-6 $\times 10^6$ ([99, 101] and Mahy, personal communication). Higher estimates of molecular weight are based on increased number of RNA segments observed in gel electrophoresis with better resolution.

1. *Segmented Nature of RNA*

Unlike other single-strand RNA viruses (e.g., poliomyelitis virus, tobacco mosaic virus), earlier attempts to isolate RNA from influenza virus did not yield a single RNA molecule of the expected size, a failure attributed to technical problems and nuclease contamination. Later on, several laboratories showed independently that purified influenza viral RNA (of any virus strain) when analyzed in sucrose gradients yielded heterogenous RNA molecules with a median S value of 18 S (range 9-21 S) instead of the single large RNA expected for the mass of 2.5-3 $\times 10^6$ daltons. These results were obtained consistently under the mild conditions used for isolating influenza viral RNA, at which other viral RNA or even large RNA molecules (e.g., RNAs of tumor virus or of Newcastle disease virus) could be isolated without degradation in the same preparation [102-104]. The segmentation of RNA was further confirmed by polyacrylamide gel electrophoresis, where at least 5 and (later) 6-8 discreet peaks of RNA were found [99, 101, 105-107] (Fig. 6.4).

It was further concluded that some of the earlier results showing the presence of 35 S or larger RNA in the virion resulted from an aggregation caused by either monovalent or (especially) divalent cations [103, 108]. Further evidence that the segmentation of RNA was not caused by a technical artifact of the isolation procedure was provided by Duesberg [71], and later confirmed by Pons [72] with the finding that RNP is also segmented. The size of the RNP molecules match fairly closely with the RNA they contain, i.e., large RNPs contain large RNA molecules and small RNPs contain smaller RNA molecules [69, 71]. RNA is therefore already segmented in RNP.

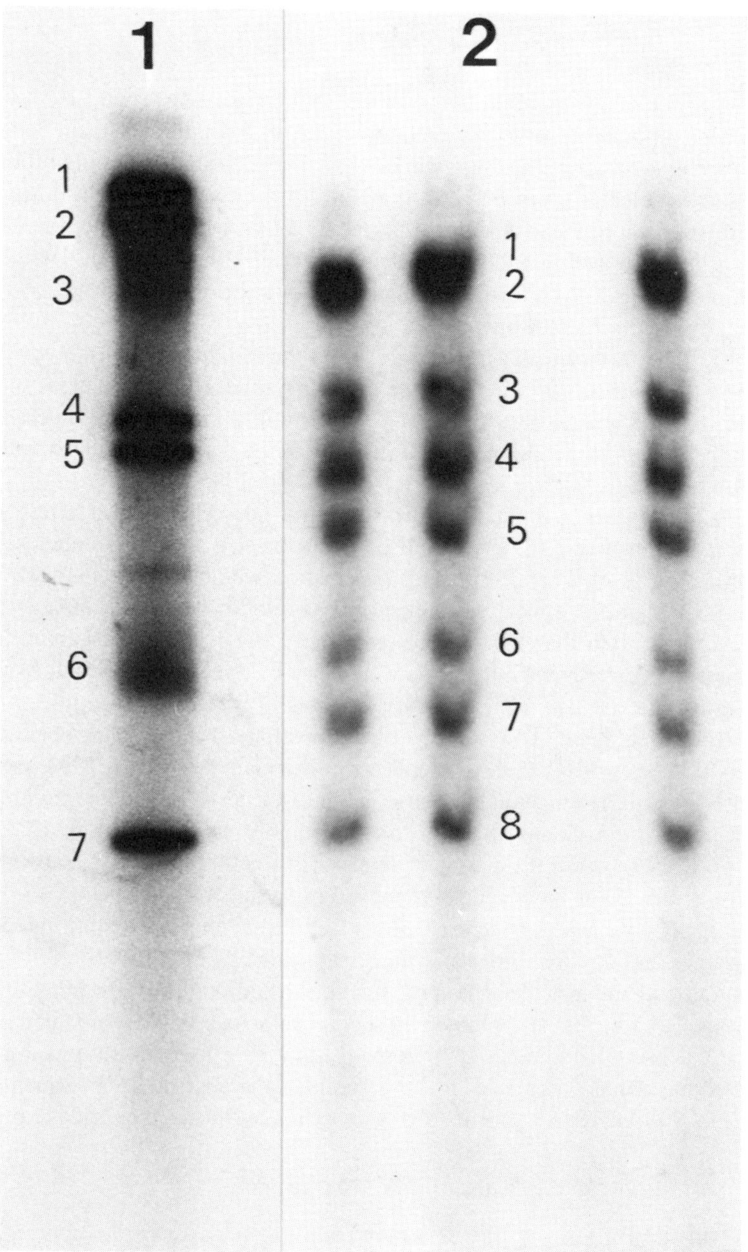

Fig. 6.4 Polyacrylamide gel electrophoresis of [^{32}P]-labeled RNAs extracted from WSN strain influenza virus grown in MDBK cells. Electrophoresis was carried out in (1) urea polyacrylamide gel (2% acrylamide + 0.15% N,N'-methylene bis acrylamide), or in (2) formamide polyacrylamide gel (3.5%). Both gels were run for 16 hr at 1.7 MA/cm at 25°C.

In spite of the fact that heterogenous and discreet RNA and RNP species have been obtained consistently even under the mild condition of isolation, there are 2 possibilities for the origin of RNA molecules: RNA could replicate as a single molecule, then be cleaved at specific vulnerable spots by nuclease to give rise to a number of discreet segments; or each RNA segment could replicate independently and not be the product of the cleavage of 1 large RNA molecule. The following evidence clearly indicates that the second alternative is the correct one.

1. The intracellular virus-specific RNA comigrates with virion RNA. There is no evidence for the presence of a large intracellular precursor of virion RNA. The large RNA molecules often found inside the cell are apparently caused by complex formation and disassociate into smaller virion RNAs following denaturation [109].

2. Duplexes or replicative forms found inside cells are also heterogenous and segmented and correspond fairly to the size and the number of virion RNAs [110, 111, 232]. This observation would support the view that each RNA segment replicates independently on its own duplex; duplexes, however, are often the end product rather than the active intermediates in the synthesis of viral RNA. Furthermore, the size of duplexes generated from replicative intermediate (RI) by RNase vary, depending on the condition of the RNase treatment. It has not yet been possible to isolate and separate different species of RI (i.e., the replicative intermediate of each RNA segment). Each RI is also heterogenous because of the presence of a variable number of single-strand RNA chains attached to the double-strand backbone [112-115]. The presence of different classes of duplex is therefore supportive but not conclusive evidence for independent replication of each RNA segment.

3. The strongest evidence that the RNA segments arise from independent replication and not from nuclease processing came from the work of Lewandowski and his colleagues [100], who showed that the 5′ terminus of each size class RNA molecules obtained by sucrose velocity gradient contained a 5′-ppp-A-p and a 3′-U-OH terminal. If RNA segments were caused by RNA cleavage by RNase, one would find a 5′OH and 3′-monophosphate in the RNA segments.

4. Specific RNA segments segregate by reassortment in hybride viruses [241].

2. *Nucleotide Sequences of RNA Segments*

The evidence for RNA segments arising from independent replication per se does not show whether these RNA segments are different or contain overlapping sequences. For example, one can envision 6-8 classes of 21 S RNA molecules (each 1 being different and coding for different proteins), and that the smaller RNA molecules (less than 21 S) result from partial (or defective) replication, because of either internal initiation or termination. Such RNA segments would then contain a 5′-triphosphate and a 3′OH end and would behave as separate classes of RNA in gels and in gradients. These smaller RNA molecules, however, would not represent separate genetic sequences but would contain a portion of the same sequences present in the larger RNA molecules.

Attempts to answer these questions by cross-hybridization using 3 classes of single-strand and corresponding double-strand RNA isolated by PAGE analysis have produced rather ambiguous results. Although some sequences present in the smallest size class were absent in the largest size class, there was extensive cross-hybridization among the 3 classes of RNA [116]. Similarly, some distinct oligonucleotide sequences also identified in the smallest size RNA class are absent in the largest size class [117]. While these experiments support the idea that each RNA segment contains different nucleic acid sequences, they do not provide the conclusive evidence of different nucleic acid sequences for each RNA segment. The major problem (the separation of each or at least the major RNA segments relatively free from contamination by each other) is created by the heterogeneity of the RNA segments and the similarity of their molecular weights. Although different virion proteins (structural and nonstructural) closely approximate the size of different VRNA segments, until recently there was no evidence for any specific VRNA (or CRNA) coding for a specific protein. The search for such a direct proof is further complicated by the fact that CRNA (not VRNA) serves as messenger, and therefore VRNA cannot be used directly in an in vitro protein-synthesizing system. Recently, Palese and Schulman (personal communication) have mapped the specific RNA segment coding for HA and NA of PR8 and Hong Kong viruses.

Genetic studies have provided supportive evidence that each RNA segment observed by PAGE analysis codes for a separate functional protein. The work by Sugiura and his colleagues [221] has shown the presence of 7 complementation recombination groups among the *ts* mutants. These authors have further demonstrated that the number of mutants within each group varies from 7 to 1. Expecting that mutation is random among the nucleotides, the number of mutants in a group is indicative of its relative size. However, the number of mutants of influenza viruses within a group cannot be used to estimate the relative size of a gene because of the specific selective pressure employed to obtain these mutants and small number of mutants within a group. For example, the number of mutants within a group, in addition to the size of a particular gene, may depend on the tendency toward reversibility and leakiness. Mutants defective in late function tend to be leaky. There are only a few *ts* mutants containing a defect in HA protein, although the gene for HA is not the smallest RNA segment (Palese, personal communication). Thus, other criteria such as the size of the peptide rather than the number of mutants in a group should be considered in estimating the size of a gene (see Recent Developments).

3. *Complexity of the Viral Genome*

The complexity of the viral genome can be defined as its total coding capacity (often determined as the molecular weight estimate, assuming that repetitive sequences, if any, are few within the genome). The total molecular weight of influenza viral RNA has been estimated as 2.5-6 $\times 10^6$. These estimates were based on either the chemical composition (i.e., RNA comprises 1% of the total virion mass [97, 98], or the size and number of peaks of labeled viral RNA in gel electrophoresis [99-101]). The difficulty with chemical determination is that the precision for the estimation of the mass required may be beyond the limit of such chemical determination, e.g., a variation from 1 to 2% of the total mass will double the coding capacity. Also, the required very high degree of purity from host cell contamination may be difficult to obtain. Furthermore, the presence of random defective viruses [118] as well as specific defective viruses, such as von Magnus virus [119], will reduce the percentage of RNA per virus particle. In contrast, the determination of the molecular weight of virion RNA from gel electrophoresis requires that the relative specific activity of each RNA segment be the same, that all species of RNA be clearly defined by PAGE analysis, and also that each RNA peak contain a different nucleic acid sequence.

The earlier estimate of the molecular weight of viral RNA (2.5 $\times 10^6$) was found insufficient for coding all of the structural as well as the nonstructural viral proteins. These proteins would require a minimum coding capacity of approximately 5 $\times 10^6$ daltons of RNA [120], provided all of these structural and nonstructural proteins contain different amino acid sequences. As explained above, however, the present methods of determining RNA molecular weight are insufficient for estimating the coding capacity of the influenza viral genome. Further work is clearly needed in this area.

4. *Number of Segments per Virion*

Although an average of 6-8 RNA segments per virion has been estimated, there are no hard data as to the precise number of RNA segments contained in a single virus particle. The number, as well as the nature of the segments, would depend on whether the processing of RNP molecules is random or selective during budding, which would in turn be determined by the presence or absence of any specific linkage between VRNA segments, either inside the cell or in the virion. The possible presence of labile linkage has been suggested by Li and Seto [121] and Almeida and Brand [236]. There is no definitive evidence to support or disprove random selection versus specific selection of VRNA segments. The observation of randomly defective viruses that complement each other [118] would support random assortment of RNP or RNA during budding. The formation of infectious viruses, however, may indeed require selective processing, for random processing would need a larger number of RNA segments to assure the presence of the entire genome in an infectious virus particle [62]. Neither random nor selective processing can explain the formation of both randomly defective and fully infectious viruses at the same time without some additional constraints in the hypotheses.

5. *Biological Implications of Segmented RNA*

Influenza viruses exhibit some unique properties not shared by other single-strand RNA viruses in general or by the paramyxoviruses in particular, which otherwise resemble influenza viruses in many respects, such as morphology and morphogenesis (e.g., budding), the possession of similar structural components, and many biological properties. Characteristics unique to influenza viruses include a high incidence of pleomorphism [40, 122]; a very high rate of genetic recombination or hybrid virus formation [123]; muliplicity reactivation [124, 125]; cross-reactivation [126]; stepwise inactivation of viral functions by chemical agents, ultraviolet light, or undiluted passage [127, 128]; as well as a high incidence of multiplicity-dependent von Magnus virus formation [119]. In fact, before the chemical nature of viral genome was established, Hirst [123] proposed the segmented nature of viral genome to explain some of these properties, particularly the high frequency of genetic recombination observed earlier in crosses among related strains [129].

Finally, it has been proposed that the major antigenic shifts or the evolution of pandemic influenza strains arise from genetic exchange coupled with mutation in different strains of influenza A viruses prevalent in nature among different animal species and humans [25-32]. Thus, both the chemical and biologic data strongly support the segmented nature of influenza viral genome, although the possibility of labile linkage among the RNA segments cannot be completely ruled out.

C. Carbohydrates

In addition to lipids, proteins, and RNA, influenza virions contain carbohydrates: 5-8% of the virion mass is carbohydrate and consists primarily of galactose, mannose, glucosamine, and fucose [130]. As discussed earlier, most of the carbohydrate is covalently attached to hemagglutinin and neuraminidase polypeptides: approximately 17% of these glycoproteins are carbohydrate.

The carbohydrate moieties are attached in a stepwise manner during and after synthesis of these virion polypeptides (HA and NA). As sugar transferases involved in this glycosylation process are host specified, the nature and the amount of the carbohydrate in virus preparations depend on the host. In addition, the host antigen found in purified virus preparations is also carbohydrate in nature [81, 131]. These host antigens are also covalently linked to HA. There is no evidence of a free carbohydrate moiety in virions. Small amounts of carbohydrate are also present as glycolipids [57].

V. Host-Virus Interaction

A. Host Range

Although influenza A viruses infect and produce disease in many animal species (e.g., avian, swine, equine, human), each virus isolate is relatively species specific and cross-species infection by influenza virus is rather rare in nature. In the laboratory, most of these virus isolates can be adapted to grow in embryonated chicken eggs or in chorioallantoic cells. In a majority of cases, inability to infect cells of other species is not for lack of the capacity to be adsorbed, as the viral receptors (i.e., hemagglutinin spikes) and cell receptors (neuraminic acid) are present in most viruses and cells, respectively. It is known, however, that some differences do exist among different host and different viral receptors, as shown by the receptor gradient among viruses [132]. The reason for receptor gradient is unclear at present. The mechanism of host specificity may thus be an intracellular event involving multigenic factors not yet understood. Genetic crosses have shown that the ability of a virus to grow in a specific cell resides in the viral genome and can be transferred to a hybrid virus [133].

In addition to the viral genome, host genes play a major role in replication and therefore in the outcome of viral infections. For example, WSN virus infects chick embryo fibroblast and produces defective as well as infectious particles, depending on the multiplicity of infection, whereas HeLa cells, L-cells, and T24 cells (a human bladder tumor derived) favor the production of noninfectious particles. MDBK cells, in contrast, favor the production of infectious virus [134]. Any of these host systems can be forced to produce either infectious or noninfectious virus by selecting the nature of the input virus, i.e., the ratio of noninfectious to infectious virus and the multiplicity of infection.

B. Effect of Virus Infection on Host Macromolecule Synthesis

Influenza virus causes essentially a lytic infection in which the host cell is eventually killed. Indeed, cell death from virus infection can be demonstrated both in intact animals (i.e., necrosis and exfoliation of respiratory mucosa) as well as in tissue culture (plaques, cytopathic effect). Influenza viruses differ from other cytopathic RNA viruses in that some host nuclear function is required after infection for virus replication and in that sometimes cells may continue to release viral particles for days after infection.

In tissue culture, infected cells become round, refractile, detach from adhering surface, and become more agglutinable by lectins [95]. These infected cells show hemadsorption and produce clumping because of the presence of viral HA on the cell membrane.

Under the electron microscope, nucleoli appear enlarged and spotty in appearance [135]. In immunofluorescent assay, NP becomes first visible in the nucleus, while helical structures similar in morphology to viral RNP appear in the cytoplasm [136]. Elongated and striated structures resembling viral RNP have also been found in infected nuclei [137, 138], although it has never been shown that these structures are truly viral RNP.

The inhibitory effect of virus infection on cellular RNA and protein synthesis is rather slow. In fact, the overall RNA synthesis is stimulated after infection and begins to decrease only 3 hr afterward [138]. The initial increase appears to be caused by nucleoplasmic DNA-dependent RNA polymerase (form II), which is stimulated early in infection [139]. The nucleolar (form I, ribosomal) polymerase is inhibited. Viral replication is also inhibited by α-amanitin, which suppresses form II polymerase, suggesting the functional implication of form II polymerase in viral replication. Inhibition of form I polymerase is also evident from the reduction of ribosomal RNA synthesis and the alteration of nucleolar structure [140]. The viral functions involved in inducing these changes have not been defined. The most likely candidate is the NS protein, which becomes bound to nucleoli, although there is as yet no direct evidence that NS is involved in suppressing rRNA synthesis.

The shutdown of host protein synthesis is not as drastic as in picornaviruses. There is a gradual reduction of host protein synthesis, and by 4 hr postinfection most of the protein synthesis is directed toward viral protein. The inhibitory effect on host protein synthesis is also variable; it is more pronounced in some cell lines (HeLa, MDCK) than in others (MDBK, CEF). Inhibition of host protein synthesis is not caused by lack of host mRNA (for mRNA may be present) but takes place at the translational level. In fact, mRNA synthesis (form II polymerase) is stimulated. It is not clear whether, as in other RNA viruses, the inhibiting activity is caused by a viral protein or by double-strand RNA. It is likewise not clear whether the NS peptide is also involved in inhibiting host protein synthesis [78, 141]. Experiments by Long and Burke [142] suggest that a virally coded protein is involved in suppressing host protein synthesis.

The effect of virus infection on DNA synthesis and on carbohydrate and lipid metabolism is possibly indirect, through its effect on host RNA and protein synthesis.

VI. Infectious Cycle

The molecular events of an infectious cycle of influenza virus are schematically presented in Fig. 6.5.

A. Single-Step Growth Cycle

Determining the single-step growth cycle of influenza virus is somewhat difficult, as the high multiplicity of infection necessary for a synchronous infection will often favor the production of noninfectious viruses. Low multiplicity of infection, in contrast, favors the production of infectious viruses, but cells are not infected synchronously. Some cells, such as MDBK, can be infected synchronously with WSN virus at a higher multiplicity and still produce infectious viruses [134]. The total span of the infectious cycle with influenza viruses is somewhat variable, depending on the virus strain and the host cells. The latent period can vary from 3 to 6 hr, and peak infectious virus production can be reached as early as 8 hr postinfection and may continue for days in some systems. Figure 6.6 shows the single-step growth cycle of a fowl plague virus in chick embryo fibroblasts [143]. Cell-associated virus production precedes the virus released in the supernate by 2-3 hr. Supernate virus can be detected at 3 hr with maximum production around 8 hr.

B. Attachment, Penetration, and Uncoating

Initial interaction between the virus and the host cell is mediated by attachment of viral hemagglutinin to the cell receptors containing neuraminic acid [144]. Although viral NA also reacts to neuraminic acid, it has little or no effect on the initial attachment [86] because, compared to HA, only a few NA molecules (50 NA spikes vs. 500 HA spikes) are present in the virion

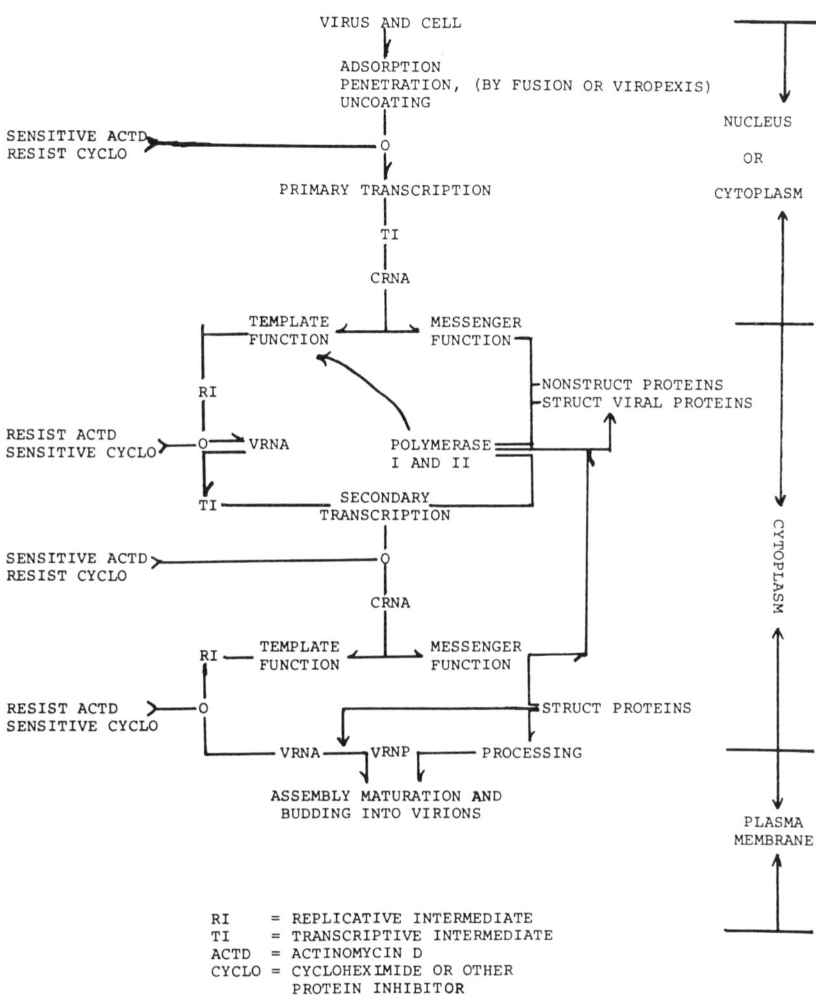

Fig. 6.5 Molecular events of influenza virus infectious cycle.

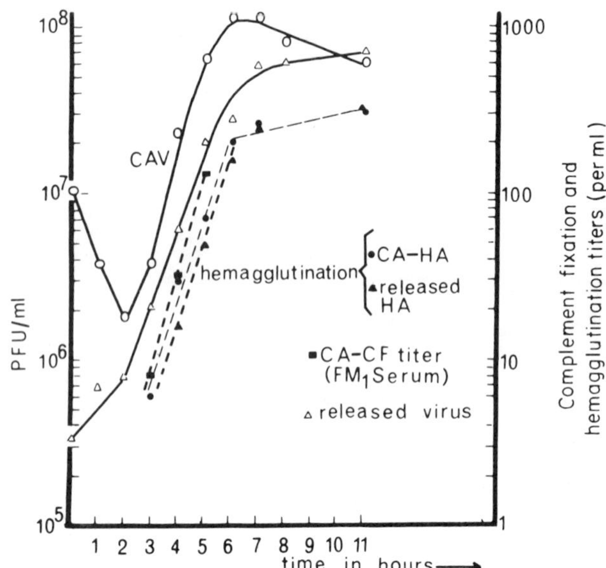

Fig. 6.6 Single-step growth cycle of fowl plague virus in chick embryo cells. From Franklin and Henry [143] with the permission of Academic Press.

envelope. Virus adsorption follows first-order kinetics when the host cell receptors are in excess. Under such conditions, 70-90% [145, 146] of input virus become cell associated within 2 hr. Approximately 1000-2000 receptor sites per cell are available in cells of chick embryo chorioallantoic membrane [145]. The studies on adsorption kinetics should be repeated because in the early work virus particles were assayed by 50% egg infectivity (EID50) or by HA units, methods less precise than the plaque assay now available.

Penetration and uncoating of virus particles follow adsorption. Viruses penetrate the cell membrane either by phagocytosis in vacuoles (also called viropexis) [147-149], or by a process in which viral envelope and cell membrane fuse, releasing viral genome into the cytoplasm [150]. If virus enters by viropexis, uncoating would take place in vacuoles, possibly aided by lysosomal enzymes. If fusion is the main penetration process, however, uncoating would occur at the cell membrane. Dourmashkin and Tyrrell [149] observed virus particles in vacuoles and noted that uncoating took place in the cytoplasm without the aid of lysosomal enzyme. An understanding of the mechanism of penetration and uncoating of influenza virus has been further complicated by the recent observation of Stephenson and Dimmock [151] that penetration and uncoating can occur at 4°C, a temperature at which neither fusion nor viropexis should take place. Which, if any, of these processes is involved in influenza virus penetration is not yet settled.

In influenza viruses, uncoating can be further divided into 2 processes: removal of viral envelope and membrane protein (thereby releasing RNP), and activation of transcriptase. The first process can be followed by RNase sensitivity of the viral genome, and the second can be monitored by the synthesis of complementary RNA, as well as by the formation of RNase-resistant double-strand RNA. Both of these processes, i.e., release of RNP and activation of transcriptase, may take place simultaneously, or activation of transcriptase may follow RNP release.

Host cell nuclei play a significant role in the replication of influenza viruses. Enucleated cells or cells in which the nucleus is inactivated by ultraviolet light, actinomycin D, and mytomycin C [152-159] cannot support the growth of influenza virus. The role of the host nuclei in influenza replication is still unclear. Apparently, DNA-dependent RNA transcription of host genome is required for viral replication (see V.B). This nuclear function is crucial at the early stages of replication. Recent work by Stephenson and Dommock [151] indicates that host nuclei may also be involved in uncoating; they found that at 4°C 50% of the virions become nucleus associated within 15 min, and 85% at 3 hr, whereas at 37°C possible migration to the nucleus happens faster. Subsequently, virions exit from the nucleus into the cytoplasm. At 37°C, 70-80% of the virions migrate back to the cytoplasm, and the rest (20-30%) remain in the nucleus. Neither cycloheximide nor actinomycin D prevented either the entry or the exit of the virions from the nucleus, although both of these drugs retarded somewhat the rate of reentry into the cytoplasm from the nucleus. Two points remain unclear from these experiments: (a) Is nuclear association caused by entry into nucleoplasm or by association with perinuclear membrane? The authors used homogenization rather than detergent treatment in isolating and fractionating nuclei from cytoplasm; it is difficult to remove perinuclear membrane without detergent treatment. (b) Is all of the cell-associated virion RNA present in the form of nucleoprotein? If so, this would imply that fusion of viral and cellular membrane at the cell surface is the likely mechanism of penetration. The role of host nuclei and DNA-dependent RNA synthesis in influenza replication is further discussed below.

C. RNA Replication

1. *Virion-Associated Transcriptase*

The presence of virion-associated enzyme was reported first by Chow and Simpson [65] and confirmed by others [160-162]. Virion-associated transcriptase has the following characteristics.

1. The enzyme is located at the internal RNP complex of the virion, which consists of P, NP, and RNA [65, 160-162].

2. The enzyme requires the presence of all 4 ribotriphosphates (0.2-0.8 mM); divalent cations (8 mM Mg^{2+}, 1-2 mM Mn^{2+}); monovalent cations (0.1-0.2 M Na^+, Li^+, K^+, NH_4^+ ions); and nonionic detergent (1-10 mg/ml Triton N101). Influenza viral transcriptases do not have an obligatory requirement of dithiothreitol, mercaptoethanol, or other sulfhydryl reagents [163].

3. The optimal temperature for the reaction in vitro is 31°C and not 37°C [163], although there is variation among different strains [228].

4. The reaction is linear up to at least 7 hr [163].

5. The product is only complementary RNA, which remains hydrogen bonded to the template even after 2 hr or more of incubation, suggesting that the product is not released from the template and that in vitro transcription continues for only 1 cycle without reinitiation [164].

6. After 2 hr of reaction, only 7% of the template RNA becomes RNase resistant, demonstrating that in vitro transcription involves only a minority of the templates. PAGE analysis, however, showed that labeled RNA was associated with all virion segments, indicating there was no selection or specificity in the transcription of the segments. The product after denaturation was smaller than the template because of either the presence of nicks in the product, internal termination before readout of the entire template, or internal initiation [164]. The dinucleosides, such as ApG and GpG markedly stimulate CRNA synthesis in vitro and the size of the product approaches the size of the template (R. Krug, personal communication).

7. The enzyme uses endogenous template only and cannot use exogenous templates because either a structural conformation of the template (as in RNP) or the cooperation of a number of proteins (NP, P, etc.) is required for polymerization. The specific function of the proteins of the RNP complex in transcription is not yet resolved.

8. Transcription in both virion-associated and cell-associated polymerase continues in vitro in the presence of actinomycin D, although transcription in vivo is blocked by this drug [65, 160-170]. While the specific reason for this difference between in vivo and in vitro reactions is not clear, it can be proposed that slow polymerization kinetics, inability of products to be dissociated from the template, lack of reinitiation, nonselectivity of segments in transcription, and involvement of a small percentage of templates in polymerization indicate that the synthesis in vitro may not represent polymerization in vivo. The suboptimal in vitro reaction may be caused by a number of missing factors, as yet undetermined. The present evidence, however, supports the view that 1 of these factors is identical to the actinomycin D-sensitive factor required for primary as well as secondary transcription in vivo. This actinomycin D-sensitive factor may be an RNA (but not a protein, because primary transcription is sensitive to actinomycin D only, and not to cycloheximide)

with a very short half-life that requires it to be transcribed continuously from host DNA. It would therefore have properties similar to those of short-lived heterogenous nuclear RNA. It is not clear whether such a host-coded RNA factor is also transcribed in uninfected cells or induced only after infection. This unstable, actinomycin D-sensitive factor may be responsible for the releases of CRNA and the reinitiation that does not take place in vitro but is required in vivo for the infectious cycle. However, even in the absence of this factor, RNA synthesis in vitro can be detected because of an increased sensitivity of the assay (limited only by the amount of radioisotopes in the reaction mixture) and the high (nonphysiological) concentration of triphosphates and cations (both divalent and monovalent) in the reaction mixture. These artificial conditions may force polymerization in vitro that would not take place in the physiological environment inside the cell. Whether actinomycin D prevents RNA synthesis by blocking the release of the CRNA in vivo remains to be seen.

The main function of the virion-associated transcriptase appears to be the transcription of the infecting viral genome (primary transcription). The kinetics of virion-specific RNA synthesis show a biphasic curve: primary transcription peaks at 1 hr after infection, and secondary transcription reaches maximum around 3 hr postinfection [138, 168, 171]. Evidence for the primary transcription being caused by the infecting virion-associated enzyme is provided by the finding that the primary transcription does take place in the presence of cycloheximide, an inhibitor of protein synthesis [172, 173].

2. *Cell-Associated Virion-Specific Polymerase*

Before virion-associated polymerase was discovered, intracellular virion-specific polymerase was predicted and found in infected cells by a number of workers [165-171]. From the nature of virion-specific RNA synthesized in infected cells, 2 polymerases could be expected: a replicase for VRNA, and a transcriptase for CRNA synthesis. Alternatively, the same polymerase, with modifying factors for template specificity, may be involved in both transcription and replication. In HeLa cells infected with Ao/NWS, 2 RNA polymerases have been reported [165]. One of these intracellular polymerases had the identical divalent (Mn^{2+}) and monovalent (K^{+}) cation requirement as, and kinetics similar to, the virion-associated polymerase; the other polymerase had different cation requirements. The products were not analyzed, however, and therefore it could not be ascertained whether the latter polymerase was involved in VRNA synthesis.

Location of the cell-associated polymerase has been studied either by assaying polymerase in vitro of subcellular fractions, isolated from infected cells, or by assaying the nature of labeled RNA by hybridization present in subcellular fractions after short pulses of [^{3}H]-uridine. Both nuclear and cytoplasmic fractions contained RNA-dependent RNA polymerase activity by in vitro assay. Hastie and Mahy [168] found that the polymerase in cell nuclei could be detected 1 hr postinfection and reached a maximum at 3-4 hr after infection. Although cytoplasmic contamination of nuclei cannot be ruled out entirely, evidence for the presence of a genuine intranuclear polymerase is much strengthened by the recent observation that influenza viruses migrate into nuclei immediately after infection [151].

Cytoplasmic polymerase is primarily located in the microsomal fraction, detectable at 2 hr and peaking at 5-6 hr after infection. Again, as in virion-associated polymerase, most of the product is CRNA and remains associated as a double-strand RNA [166-171]. Only under certain conditions is in vitro synthesized CRNA released as single-strand RNA [171]. A single report [169] that VRNA was actually synthesized has yet to be confirmed. Cytoplasmic RNP complexes containing NP, P, and RNA contain the bulk of the cytoplasmic transcriptase.

Because of the accumulation of a relatively large amount of RNP protein containing polymerase in the cytoplasm, and because of the occurrence of virion polymerase in nuclei, the question has been raised as to the location of the synthesis of VRNA and CRNA in infected cells. Several alternative hypotheses have been presented. One suggestion is that both CRNA and VRNA are synthesized in cell nuclei, and that CRNA synthesis observed in vitro is caused by the activation of transcriptase present in inactive RNP complex stored for processing and assembly into virions. The other possibility is that VRNA that has not yet been synthesized in vitro is replicated only in the nucleus, and CRNA in the cytoplasm. The third possibility is that the nuclear phase is not significant in viral replication and that synthesis of both VRNA and CRNA occurs in the cytoplasm. Hastie and Mahy [168] suggested that VRNA synthesis takes place in the nucleus because they found only 40% of the RNA synthesized with nuclear extract to be CRNA; the nature of the remaining 60% of the product claimed to be VRNA remains unresolved.

Using short-pulse labeling of 5 min or less, Nayak [174] found CRNA associated with rough endoplasmic reticulum. These experiments suggest that active CRNA synthesis takes place in cytoplasm 4 hr after infection. Nayak further found that, unlike the case of polio virus, transcription of influenza viral RNA and translation of viral messengers take place in the same subcellular component, i.e., the rough endoplasmic reticulum. It is unlikely that both in vitro and in vivo transcription and translation in the rough endoplasmic reticulum are cuased merely by the sedimentation properties and contamination of rough endoplasmic reticulum by RNP complexes. Nayak and his colleagues have recently suggested the possible presence of a transcriptase-translation complex that may explain the association of transcription to the rough endoplasmic reticulum membrane (see Section 6.VI.C.5).

3. *Classes of Intracellular Virus-Specific RNA*

In many RNA viruses (e.g., poliomyelitis virus) one can study the synthesis of virus-specific RNA under conditions that totally inhibit cellular RNA synthesis (e.g., actinomycin D, 5 μg/ml). Unlike these RNA viruses, the analysis of virus-specific RNA in influenza virus-infected cells is rather difficult because of the lack of a specific inhibitor able to suppress cellular RNA synthesis without affecting viral replication. Thus, the information about influenza viral RNA synthesis in infected cells has become available either by using inhibitors, such as actinomycin D (which, if added 2 hr postinfection inhibits CRNA synthesis but allows VRNA synthesis), or by hybridization of pulse-labeled intracellular RNA using unlabeled VRNA or CRNA as probes. Studies [103, 109, 110, 112-115] have suggested the presence of 5 classes of virus-specific RNA in infected cells: VRNA, CRNA (both single-strand RNAs), transcriptive (TI) and replicative (RI) intermediates (partially single- and partially double-strand RNA), and RNA duplexes (or replicative form, totally double-strand RNA).

Early studies by Nayak and Baluda [109] and Duesberg and Robinson [103] showed that virus-specific RNA could be demonstrated in infected cells 4 hr postinfection by actinomycin D treatment. The RNA profile was heterogenous and similar to that of viral RNA in sucrose gradients [109] and by gel electrophoresis [106]. Nayak and his colleagues [109, 113-115] further demonstrated that infected cells contained single-strand CRNA, implicating involvement of CRNA in functions other than the template activity for VRNA replication. Replicative intermediates with partial double and single strandedness were also demonstrated by Nayak and Baluda [112, 113] in the presence of actinomycin D. RNase-resistant RNA arising either from inactive RNA duplex [115] or from the double-strand core of replicative intermediates [106, 110, 112] was demonstrated in cells with or without actinomycin D treatment. RNase-resistant RNA has been resolved by gel electrophoresis [106-110] into 5-8 separate species that corresponded fairly in size to single-strand viral RNA segments [105-107]. Electronmicroscopic measurement has yielded similar results [232].

4. *Kinetics of RNA Synthesis*

a. VRNA and CRNA Synthesis. The kinetics of VRNA and CRNA synthesis in infected cells was followed by Scholtissek and his colleagues [171, 175-178] by labeling infected cells at different times after infection and annealing them to an excess of unlabeled C- or V-RNA. CRNA was prepared from the microsomal fraction of infected cells by RNA polymerization in vitro. RNA released in the supernate was mostly CRNA, and any small amount of VRNA was removed by preannealing, thus permitting only CRNA to hybridize with the VRNA of infected cells. In these experiments it was found that in chick embryo fibroblast (CEF) cells infected with fowl plague virus almost equal amounts of C- and V-RNA were present up to 2 hr after infection, when the rate of CRNA synthesis reached a peak, and were followed by the rate of VRNA maximum at 3 hr postinfection. At later times (3-7 hr after infection), both V- and C-RNA synthesis declined, although the rate of VRNA synthesis remained twice that of CRNA synthesis from 3 hr postinfection. Small but significant amounts of both V- and C-RNA syntheses were demonstrated 1 hr after infection.

Bean and Simpson [173], using [^{32}P]-RNA into the RNase-resistant form, also obtained the first evidence of transcription about 40-60 min after infection. They further dissected the transcription of VRNA into 2 phases. The rather slow primary transcription was detected around 40-60 min postinfection; it was independent of protein synthesis. Secondary transcription, and enhanced rate of CRNA synthesis, was evident from 90 to 120 min after infection and was dependent on protein synthesis. In the presence of cycloheximide, an inhibitor of protein synthesis, primary transcription continues linearly up to 6 hr postinfection, while an enhanced rate for CRNA synthesis is not evident; under this condition 20% of VRNA becomes double stranded.

Essentially similar results were also obtained by Stephenson and Dimmock [151]. Using CEF infected with a recombinant (FP/BEL) virus, these authors found that, in the presence of cycloheximide, CRNA transcription was evident 1 hr after infection and continued linearly, whereas in untreated cells an accelerated CRNA transcription was evident 1.5 hr postinfection. In infected cells without cycloheximide, a maximum of 40% VRNA became RNase resistant after annealing, whereas in cells treated with cycloheximide only 15% of input VRNA became double stranded, indicating decreased transcription in the presence of cycloheximide. They further analyzed the synthesis of CRNA in cytoplasmic and nuclear fraction and found that both primary and secondary transcription occurred in nucleus as well as in cytoplasm. The rate of CRNA synthesis remained constant in nucleus but increased in cytoplasm up to 4 hr postinfection.

Virtually nothing is known about the temporal control of transcription. In some experiments only about 15-20% of input RNA became RNase-resistant in infected cells [172, 173]. However, these experiments cannot differentiate transcription and hybridization of a specific segment of VRNA from a fraction of the total VRNA population. Thus, definitive evidence of the temporal control of viral transcription is lacking.

b. Replicative (RI) and Transcriptive (TI) Intermediates and Duplexes (DS). Influenza viral RNA replicates via replicative intermediates (RI). RI with partially double- and single-strand structure was identified and characterized in infected cells by Nayak and his colleagues [112-114]. They found that

RI was heterogenous, approximately sedimented at 14 S, and had the properties of partial single and double strandedness. In the presence of actinomycin D, RI becomes preferentially labeled in shorter pulses. With longer pulses a higher amount of label appears in single-strand RNA and the percentage of RNase resistance decreases. In similar kinetic experiments, as well as in quasi-pulse-chase experiments (the large pool of nucleosides and nucleotides in animal cells makes complete pulse-change experiments not feasible), they concluded that RI was an active intermediate in the synthesis of single-stranded RNA (Fig. 6.7). The percentages of single and double strandedness in RI are somewhat variable and may depend on the isolation procedures. Nayak and Baluda [112] obtained 11 S double-strand RNA from RI by mild RNase treatment but only a 7 S duplex by rigorous RNase treatment; they concluded that single-stranded regions in a double-strand backbone were susceptible to rigorous RNase treatment. The amount of double-stranded regions in RI may be variable and

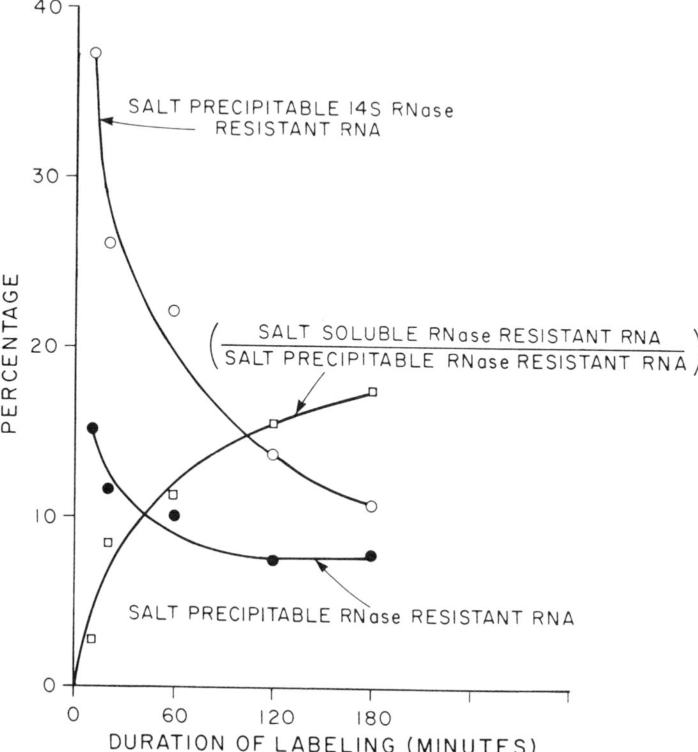

Fig. 6.7 Kinetics of the synthesis of different influenza viral RNA species: single-strand RNA, replicative intermediates, and double-strand RNA. From Nayak and Baluda [112].

may depend on the isolation procedures. They also isolated an RNA duplex of approximately 12 S from infected cell RNA by 2 M NaCl fractionation and found that these RNA duplexes are inactive and possibly arise as the RI end product.

Pons [110] and Duesberg [106] have isolated, after RNase treatment, RNase-resistant RNA from infected cells. They further found that these RNA duplexes could be separated by PAGE analysis into 5-8 sizes that correspond fairly well with the size of the expected duplexes of influenza viral RNA. Similar results were obtained from the electronmicroscopic studies by Chang and Seto [232]. Identification of these RNA duplexes implicates independent replication of each viral RNA segment.

The findings of single-stranded CRNA in infected cells [109, 113-115] and transcriptase in virions [65, 160-162] predict the existence of transcriptive intermediates (similar to RI) for the synthesis of CRNA. Although the nature of these transcriptive intermediates (TI) must be similar to RI, they have been neither isolated nor characterized. Most of the partial RNase-resistant RNAs studied by Nayak [113-115] under the condition of actinomycin D treatment were RI, as actinomycin D suppresses the synthesis of CRNA. The presence of transcriptase in virion should make is possible now to isolate and study the properties of TI in vitro.

5. *Messenger RNA*

Two approaches have been used to characterize the messenger RNA of influenza virus: Analysis of the polysome-associated RNA in infected cells, and synthesis of proteins made in vitro by using either virion RNA or complementary RNA. Although some uncertainty remains as to the precise nature of viral messenger RNA, recent experiments by Nayak [114, 244], Pons [179, 180], and Etkind and Krug [181] support the view that the CRNAs are the sole messengers in infected cells.

Nayak and Baluda [109] first demonstrated the existence of single-stranded complementary RNA not associated as template of RI in infected cells. Later, Nayak [114] further analyzed RNA associated with polysomes by uridine labeling in the presence of actinomycin D at 3.5-6.5 hr postinfection. RNA found in the polysomes of infected cells had a profile similar to that of viral RNA and contained both virion and complementary RNA. These data provided the first evidence that CRNA may be involved as messenger, although it was not clear whether CRNA was the only messenger RNA. The evidence that virus contains transcriptase [65, 160-162] provided further support to the view that the primary translation at the beginning of infection must be carried out by the CRNA. Later on, Pons [179, 180] found further evidence that CRNA is the major functional messenger in infected cells, although 30-50% of the total RNA associated with polysomes was still not accounted for. Recently, Etkind and Krug [181] used cordicepin to reduce host mRNA synthesis and freed mRNA from other RNA by polysome dissociation. They also found that under these conditions polysome-released mRNA consisted mainly of CRNA. In none of these experiments, however, have the investigators used any direct assay for measuring VRNA, although CRNA was measured directly and therefore the small amounts of VRNAs serving as messengers might have been overlooked. Recently, Etkind and Krug [238] and Palese et al. (unpublished) have isolated CRNA by oligo dT-cellular chromatography and synthesized in vitro proteins which comigrated with P,NP,MP, and possibly NS.

Nayak and his colleagues [244] have recently analyzed carefully the viral messengers and used a direct assay method for measuring both CRNA and VRNA by hybridizing with excess VRNA and CRNA, respectively. Unlabeled CRNA for this probe was prepared by treating infected cells with cycloheximide, which preferentially inhibits VRNA. The cellular RNAs so treated were self-annealed to remove any VRNA, and thus CRNA only was available for hybridization. Using these probes they analyzed by hybridization both polysome-associated and polysome-released mRNA in infected cells without any inhibitors, and in cells treated with various concentrations of actinomycin D. They found that, when infected cells without any inhibitor were labeled with [^{3}H]-uridine for 30 min, 30-40% of the label in polysome-associated RNA of infected cells was in CRNA and none in the VRNA. The rest (approximately 60%) of the label must therefore be in the cellular RNA, although this was not shown directly. When free mRNA was isolated following polysome dissociation, over 70% of the free messengers were composed of CRNA, and again no VRNA was present in free mRNA. Most of the label associated with host RNA in polysomes was found in the low molecular weight RNA (transfer RNA), although some was found in the ribosomal RNA. Only 30% of the total label in free mRNA could result from host messengers.

When a high concentration of actinomycin D was used and the label was added 30 min later, there was little or no measurable radioactivity in the polysome, confirming observations by Pons [179] that CRNA synthesis is selectively inhibited by actinomycin D, which also inhibits incorporation of labeled RNA in polysome. However, when a lower concentration (0.2 μg/ml) of actinomycin D was used and the label ([^{3}H]-uridine) was added soon after or along with actinomycin D, the polysome-associated RNA contained an equal amount of CRNA and VRNA (1 : 1 ratio). Free messengers, isolated by dissociation of polysomes of actinomycin D-treated cells, contained only CRNA and no VRNA, however, Furthermore, it was found that VRNA in the polysomes did not result from contamination by RNP or by replicative complexes. These experiments therefore confirm the early observation that both labeled CRNA and VRNA can be detected in polysomes under certain conditions of actinomycin D treatment and labeling [114].

To explain these results, 3 alternative models are shown: Figure 6.8A shows single-stranded CRNA as the functional messenger; however, this would not explain the results obtained in the presence of actinomycin D. The model in Fig. 6.8B, showing some segments of VRNA and some of CRNA, would serve as mRNA, but is even more untenable because there is no evidence of single-stranded VRNAs as messengers. The model shown in Fig. 6.8C seems to explain all of the results obtained so far. It has the following characteristics: CRNA is the only functional messenger; the newly synthesized CRNA, while associated with VRNA in transcriptive complex, is associated with polysomes and is being translated—this complex would be a transcription-translation complex. However, the transcription-translation complex cannot be the only protein-synthesizing system in infected cells. In addition, single-stranded CRNAs (Fig. 6.8A) must also serve as functional messengers. This model would explain that under certain conditions of labeling (such as short pulses in absence of actinomycin D) there should be no appreciable amount of labeled VRNA in

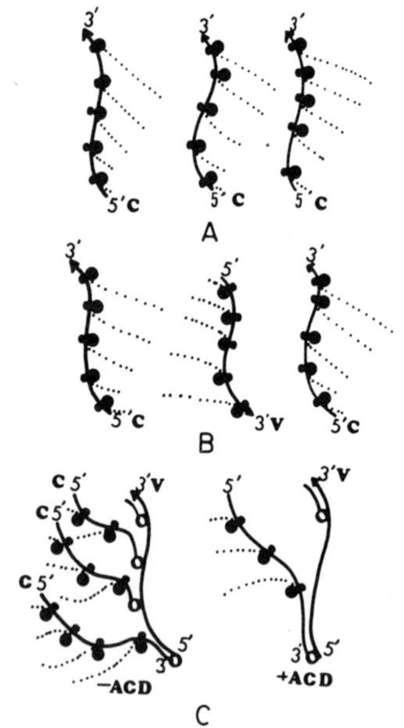

Fig. 6.8 Models proposed for influenza viral messengers. (A) Single-stranded CRNA as functional messenger. (B) Some CRNA and some VRNA segments as messengers. (C) Transcription-translation complex in the presence or absence of actinomycin D, CRNA serving as messenger.

in polysomes, that at high concentrations of actinomycin D there should be no label associated with polysomes, that only with partial inhibition of CRNA synthesis are both VRNA and CRNA found in the polysomes, and that under all conditions the free mRNA is solely CRNA. However, it should be made clear that the transcription-translation complex is only a suggestive model without any direct evidence to support it.

Further evidence that CRNA is the only functional messenger was provided by Nayak [174] and Etkind and Krug [181] in the finding that only CRNA contained poly(A) sequences at the 3′ OH, and VRNA did not, and that CRNA (and not VRNA) coded for virus-specific proteins in vitro [238] (Palese, unpublished). Kingsbury and Webster [182] found that the total RNA from an infected cell produced a virus-specific protein in in vitro synthesis, while VRNA did not. This would further support the messenger function of CRNA present in infected cells. Furthermore, although the sizes of VRNA (and possibly CRNA) segments correspond fairly to the size of virion polypeptides, direct information is not available as to which RNA segment codes for which polypeptide. Recent studies by Palese and his co-worker [241] show that HA and NA are coded by specific viral RNA segments. This area will be the subject of intensive research in the next few years.

In all of these studies involving replication, transcription, or translation, the role of VRNA or CRNA has been discussed. However, these RNA molecules in intracellular milieu are not naked but are usually associated with protein molecules. NP protein has been shown to be associated with both VRNA and CRNA during replication, transcription [166, 167], and translation [233]. These NP molecules in RNA-protein complexes may provide both structural stability of these functional complexes and/or a regulatory role. However, it is not known whether these RNA nucleoprotein complexes involved in replication, transcription, or translation possess identical morphological structure and biophysical properties. Furthermore, it is not clear if all of these complexes are identical to viral RNP molecules found in matured virion. It is possible that the role of the NP molecules vary depending on the nature and the function of RNP complex and a quantitative difference in the ratio of protein to RNA may provide stability as well as flexibility required for different functions.

VII. Virus-Specific Protein Synthesis in Infected Cells

Since 1950, numerous investigators have studied the kinetics of the synthesis of viral antigens. In early studies, the development of intracellular HA, NA, and RNP was followed and quantified by hemagglutination, assay of neuraminidase activity, and complement-fixation assay of nucleoprotein (CF-NP), respectively. Later on, immunofluorescent techniques were used to follow the kinetics of development, localization, migration, and processing of specific antigens in virions. Finally, during the last 6-8 years the kinetics of viral protein synthesis have been studied intensively by using analytical methods of protein chemistry. Sensitive methods of protein analysis, such as PAGE (polyacrylamide gel electrophoresis) analysis, radioisotope labeling, together with pulse-chase experiments and cell fractionation procedures, have been used to quantitate the rate of synthesis, the processing and the fate of individual viral proteins, both structural and nonstructural.

A. Localization and Development of Viral Antigens

The kinetics of the development of 3 major viral antigens (HA, NA, and RNP) have been studied by using immunofluorescent techniques in both cultured cells and experimental animals. The development of these antigens was followed using antibodies against V antigens (which contain both HA and NA) and g antigens (S or NP) in the early studies [183-185], and monospecific antibodies against HA, NA, or NP in the later studies [186]; the results were essentially the same.

Among the viral antigens, NP was detected first (usually 3 hr after infection), at least an hour before HA. That NP is among the first viral antigens detected has been confirmed by both serological tests and by PAGE analysis [187]. Using immunofluorescence, NP was first detected 3 hr postinfection. Fluorescence intensity increased on the nuclear membrane and eventually, by 4 hr, NP antigens began to spread out into the cytoplasm. Subsequently, the intensity of nuclear fluorescence decreased with the increase of NP antigen at the cell surface. Intranuclear migration of NP has also been demonstrated by PAGE analysis [188, 189]. It is not clear, however, whether intranuclear migration is an obligatory step in processing, antigenic configuration, and assembly of NP into RNP.

Both HA and NA, in contrast, appear first in cytoplasm about 4 hr after infection, remain localized in cytoplasm, and finally become concentrated on the plasma membrane. These studies show that different viral antigens are processed through different subcellular organelles and eventually accumulate on the cell surface, where virion budding takes place. With the availability of monotypic antisera it should now be possible to study the development of viral antigens by more sensitive immunological techniques, such as radioimmune assay, and to correlate the development of antigens to the synthesis of specific polypeptides and their processing.

B. Virus-Specific Polypeptides in Infected Cells

The kinetics of the synthesis of virus-specific polypeptides is variable with the specific host-virus system and may depend on a number of factors, such as inhibition of host protein synthesis and replacement by viral protein synthesis. The degree of inhibition of host protein synthesis is variable, and does not reflect the ability of the host to produce infecitous virus. For example, in cells such as HeLa, BHK, and MDCK, virus infection shuts off host protein synthesis quite rapidly, and most of the viral proteins synthesized after 4 hr of infection are virus-specific, whereas in MDBK and CEF cells the degree of inhibition is minimal until late in the infectious cycle.

In spite of the variability in the kinetics of total viral proteins, the temporal relationship of the synthesis of different viral proteins is fairly stable in different cell-virus systems. In CEF cells infected with fowl plague virus, viral polypeptides become detectable at 2 hr and reach a peak at 5 hr after infection. Thereafter, a steady state of synthesis is maintained for the next 3 hr and gradually declines later [96].

The control and regulation of individual proteins is not clearly understood. From the early studies of the quantitation of viral antigens by serology and later from more sophisticated studies with PAGE analysis, it became clear that the relative amounts of viral proteins in infected cells vary greatly and do not correspond to their relative abundance in virion structures. It has been known for a long time that NP antigen appears first and is the most abundant protein in infected cells. Recent studies by Skehel [187] have shown that P, NP, and NS are the earliest peptides synthesized. Using cycloheximide to inhibit secondary transcription and releasing the block to determine the kinetics of labeling, Skehel showed that these 3 peptides are indicative of the product of primary transcription. If this were correct it would also mean that primary transcription is selective regarding which specific RNA segments are transcribed. Protein M, which is abundant in the virion, is hardly detectable in infected cells. Only a small amount of P protein is present, both in virions and in infected cells. Although NP is the most abundant protein in infected cells, the number of NP molecules in virions is smaller than that of M and HA. Thus, there are at least 2 levels of control in virion formation: protein molecule synthesis, and selection during virion formation.

The role of transcriptional and translational control in the synthesis of viral proteins is not yet clear. The ratios of RNA segments in the virion as well as the RNA in the infected cell in presence of actinomycin D are essentially the same. It is not clear, however, if that transcriptional control plays a major role in the regulation of viral protein synthesis. Critical analysis of the rates of mRNA synthesis and viral protein synthesis, to determine whether the control at the transcription level or at the translation level plays the major role in viral protein synthesis has not yet been done.

Many of the proteins, as described below, go through a selective transport phase, in association with specific membranes before they are incorporated into virions. It is not yet clear how the final selection and processing of protein into the virion takes place. Obviously, it does not depend solely on the intracellular concentration of specific proteins, whereby the relative intracellular concentration of virion proteins would reflect their relative amounts in virions. The kinetics and processing of individual proteins are discussed below.

1. Hemagglutinin (HA)

Pulse-chase experiments [174] show that HA is synthesized in rough endoplasmic reticulum and transported soon into smooth membrane, where it accumulates (Fig. 6.9). The half-time ($t_{1/2}$) for this process is about 10 min [80, 190]. Hemagglutinin is glycosylated and cleaved in some cell-virus systems, although cleavage is not necessary for hemagglutination activity but required for infectivity [78-80, 239, 240]. Glycosylation and cleavage have been studied by short pulse followed by chase, as well as by using deoxyglucose and high concentrations of D-glucosamine, which are effective glycosylation inhibitors [191-200]. Pulse-chase experiments gave direct evidence that HA is synthesized as a single molecule and undergoes glycosylation either during synthesis on polysomes or immediately thereafter. Cleavage of HA into HA_1 and HA_2 takes place primarily at the plasma membrane and is a host-controlled phenomenon. Cleavage at the cell membrane is also regulated by the presence of proteolytic enzymes, such as serum plasminogen, in the medium.

Klenk and his colleagues [193-196] used inhibitors to block the synthesis of HA and NA. They blocked the glycosylation of HA with D-glucosamine and 2-deoxy-D-glucose and found that nonglycosylated precursors continued to accumulate. Cleavage takes place only after glycosylation and occurs at the cell membrane [194, 198]. Glycosylation appears to proceed in a stepwise manner in which glucosamine is added during or right after translation, while the peptide is still attached to the rough endoplasmic reticulum, and fucose is added later in smooth membranes. The order of biosynthesis has been determined by Skehel and Waterfield [235]. They found HA1 at the N terminal and HA2 at the C terminal end. Although the synthesis of NA was blocked by the inhibitors of glycosylation, the nonglycosylated form of NA peptide has not been demonstrated.

Numerous investigators [80, 174, 191-201] have observed that hemagglutinin, synthesized at the rough ER membrane, is transported into smooth membrane; the significance of this observation is unknown. It is postulated that this step is an obligatory intermediate in the sequential processing toward

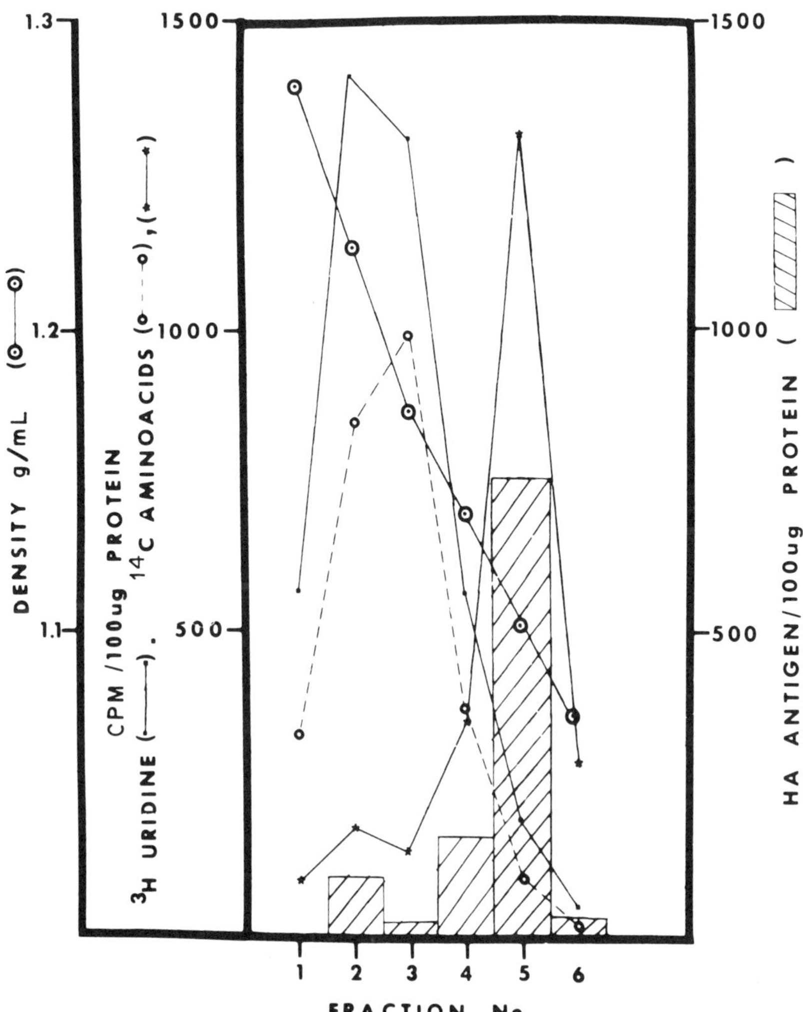

Fig. 6.9 Pulse-chase experiments demonstrating the migration of labeled HA from rough endoplasmic reticulum to smooth membrane. Cells at 4 hr postinfection were labeled with [H^3]-amino acids for 5 min and harvested immediately, or chased for 2 hr in unlabeled medium after 5 min pulse labeling before harvest. Cellular membranes were isolated in a tight-fitting steel Dounce homogenizer. Membrane fractions were isolated by equilibrium centrifugation; 6 fractions were collected. Sucrose densities in each fraction were calculated from the refractive index. Membrane in each fraction was diluted 4 times and pelleted by centrifugation. Each pellet was analyzed for hemagglutination activity and TCA precipitable radioactivity. In a parallel experiment, each pellet was analyzed by thin section in electron microscopy. Fractions 1, 2, and 3 were mostly rough ER; fractions 4, 5, and 6 were smooth membranes. Radioactivity after a short pulse cosediments with the rough ER; after 2 hr chase the radioactivity migrates to the smooth membrane and cosediments with the bulk HA. Data taken from Nayak [174].

incorporation of HA into plasma membrane and finally into virions. An alternate explanation would be that most of the HA is not incorporated into virions but is stored in smooth membrane, and that the HA that is incorporated into virions is processed via different pathways. Although there are as yet no hard data to support or reject either possibility, the favored hypothesis is:

$$\text{HA} \twoheadrightarrow \text{rough ER} \twoheadrightarrow \text{smooth membrane} \twoheadrightarrow \text{plasma membrane} \twoheadrightarrow (\text{HA}_1 + \text{HA}_2) \twoheadrightarrow \text{virion}$$

2. *Neuraminidase (NA)*

By both immunofluorescent observation and PAGE analysis, NA, the other glycosylated virion protein, is synthesized possibly at the same time as HA. Although the kinetics of these processes have not been studied extensively, synthesis, glycosylation, and transport of NA may follow in general the same pattern as in the case of HA. Technical difficulties are mainly caused by the synthesis of fewer NA molecules in cells. Furthermore, the presence of a large excess of NP often obscures the presence of a small amount of NA because NP and NA often migrate to a similar position in PAGE. More refined techniques, such as PAGE analysis after immune precipitation using monospecific antisera, will be required to study further details of the kinetics of synthesis and the sequential steps involved in the processing and assembly of NA into virions. The rate of transfer and processing of NA appears to be 2-3 times slower than that of HA [202].

3. *M Protein*

Virtually nothing is known about the kinetics of the synthesis and assembly of M protein. Indeed, M protein appears to be the rate-limiting component in virion production, being present in virions in large quantities and hardly detectable in infected cells [201, 237]. As there is little or no pool of M protein in infected cells, the steps (not yet fully resolved) involved in processing M protein from synthesis to virion assembly must be fairly rapid. Klenk and colleagues [194] report that M protein is present in smooth membrane after pulse-chase experiments. The presence of M protein is also obscured by the large amount of NS protein in infected cells.

4. *P Protein*

Only a few molecules of P protein (which may function as transcriptase) are found in virions. Like NP, it is not associated with the membranous fraction but is found mostly in the postmicrosomal supernate. The factors that regulate the transcription of RNP by P and its assembly into RNP are not known.

5. *Nucleoprotein (NP)*

This is the most abundant of the structural viral proteins in infected cells and is not associated with any specific cellular membrane. Over 80% of NP is present in the postmicrosomal supernate (100,000Xg supernate fraction) [191]. After synthesis, NP appears to accumulate in the nucleus and eventually migrates from the nucleus into cytoplasm [188]. However, it is not clear whether the nuclear phase of NP is obligatory for viral replication. It is possible that the intranuclear phase of NP has no significance in viral biology. It is also possible that this intranuclear phase is very important, particularly in relation to the host-nuclear involvement in influenza infection. Studies using actinomycin D, enucleated cells, and fused cells (fusing actinomycin D-treated BHK cells with chick red cells) show that host cell nucleus is required for the early stages of viral replication [203]. Is the migration of NP (or RNP) into nucleus involved in this important process? Furthermore, how and where does the assembly of NP, P, and RNA into RNP take place? Is it cytoplasmic or nuclear? What controls NP synthesis? Is there a free pool of NP, or are all NP molecules present as RNP? It is known that NP, along with P, constitutes an important part of the transcription complex, where NP functions either as a stabilizer of the template or as a part of the enzyme complex itself. The answers to these important questions are unknown at present. Finally, the manner in which RNP molecules are assembled into virions is virtually unknown.

6. *NS Protein*

One of the first viral proteins detected (along with NP and P), NS is the most abundant polypeptide found in infected cells, although none is present in virions [190]. The function of NS in the infectious process is unknown. Labeled NS molecules are detected in polysomes but rapidly migrate into nucleus; the $t_{1/2}$ is less than 5 min [188]. NS is rich in arginine and accumulates in the nucleolus [204]. It continues to be synthesized at a high rate up to 6 hr postinfection, and afterwards its rate of synthesis gradually decreases [204]. As mentioned earlier, NS is reported to be involved in controlling the host ribosomal RNA synthesis and is also implicated in the inhibition of host protein synthesis after infection.

VIII. Processing and Budding

It was early recognized that the assembly of viral components into virions takes place only at the cell membrane by a process called budding. Although all of the

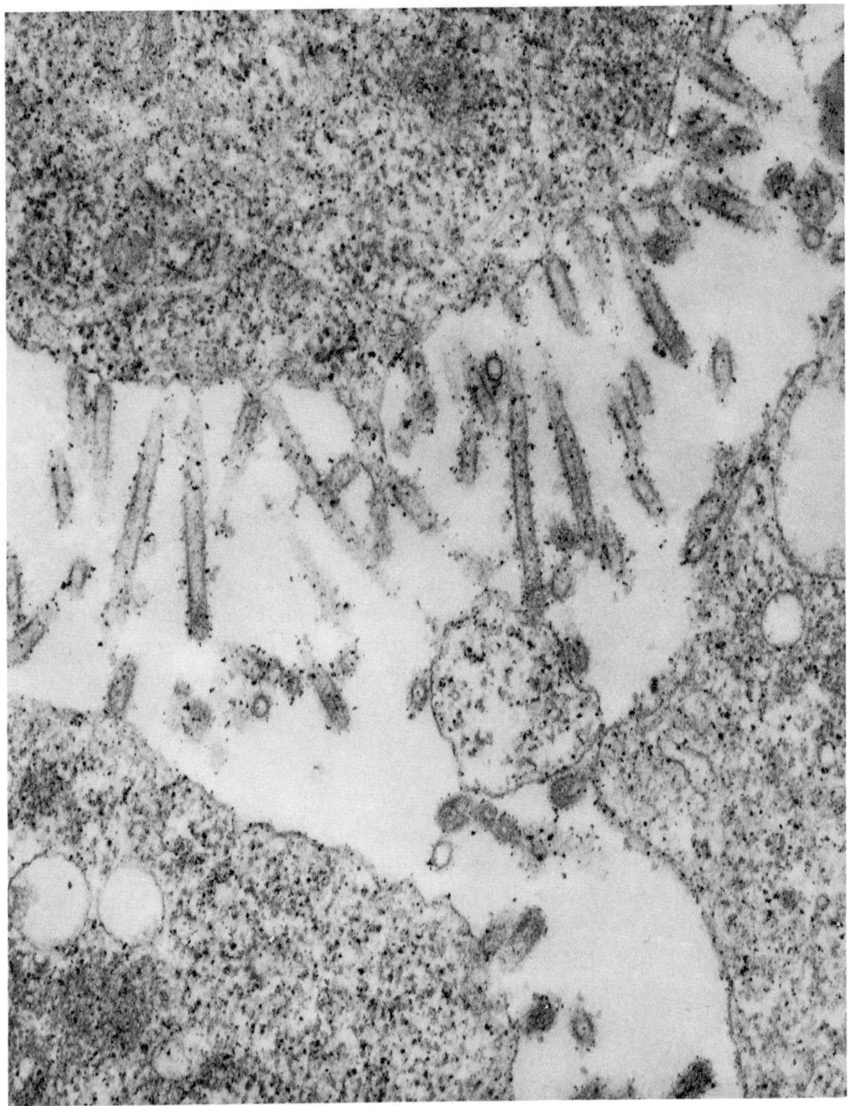

Fig. 6.10 Thin section of influenza virus-infected cells demonstrating the filamentous processes that lead to the production of mature virus particles.

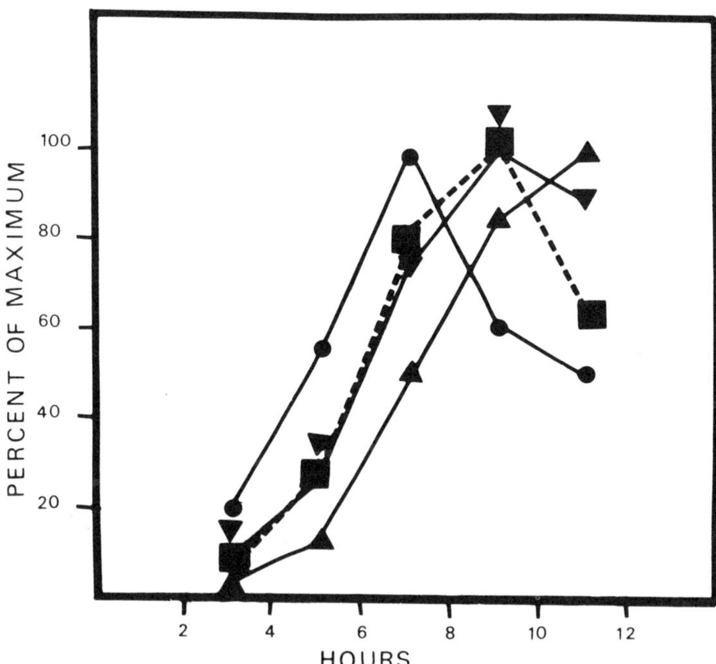

Fig. 6.11 Rates of synthesis and incorporation into WSN virions of polypeptides P, NP, M, and the total glycoproteins. The virions were grown in MDBK cells and labeled with [^{3}H]-leucine (10 μc/ml) from 3 to 4 hr, 5-6 hr, 7-8 hr, 9-10 hr, and 11-12 hr after infection. Following the labeling period, cells were washed and incubated further in unlabeled medium until virus was harvested and purified at 24 hr after infection. The virion polypeptides were analyzed by polyacrylamide gel electrophoresis. The total protein applied to each gel was determined, and the total radioactivity in each peak was calculated and normalized to equivalent amounts of protein. (●—●) P, (■--■) NP, (▼—▼) total glycoprotein, (▲—▲) M. From H. Meier-Ewert and R. W. Compans [201] with the permission of the American Society of Microbiologists.

viral components, including RNP, have been found inside infected cells, neither complete virions nor budding virions have been observed in the cell interior. Many events are still unresolved in the jigsaw puzzle of processing, assembly, budding, and release of viral components into virions. Electron microscopic (Fig. 6.10) and biochemical studies (Fig. 6.11) have revealed some of the sequences of events that eventually lead to the formation of virions.

A. Modification of Cell Membrane

The observation that viral envelopes contain little or no host protein suggested extensive modification of cell membrane before budding. Essentially, this process demands complete replacement of host proteins by viral antigens. Modification of cell membrane was suggested early from the observation that infected cell membrane adsorbs to red cells at specific areas that still appear normal by electron microscopy. Later, in the course of infection, more defined structures such as spikes become apparent befure budding. Electron-dense membrane proteins become associated within specific spots, where RNP eventually aggregates and the budding process continues.

Holland and Keihn [51] and later Lazarowitz et al. [190] and Hay [202] investigated the kinetics of assembly of viral antigens in host membrane by PAGE analysis and found that viral proteins become associated with the plasma, nuclear, mitochondrial, and microsomal membranes of infected cells. A surprising observation was that viral antigens were not only added in the specific spots of cell membranes, but that they completely displaced all preexisting cellular proteins from discreet areas that are precursors of viral envelopes. Further changes in the budding area were reported by Klenk [9], who observed that budding spots did not contain any neuraminc acid residue, which was present in adjacent areas of cellular membrane.

The mechanism of the displacement of host protein by viral proteins is unknown. What does initiate nucleation of viral proteins? How are host proteins displaced? Is there an assembly of lipoprotein molecules before their assembly into membrane? Holland and Keihn [51] suggested that nucleation of viral proteins may take place at random or in weak spots of plasma membrane; these localized spots would grow by aggregation of viral proteins and eventually contain only viral proteins. Virion budding would continue to take place from the same spots. However, the observed storage of viral glycoproteins (HA and NA) in smooth membranes [120, 174, 191-202, 204, 205] may suggest that these smooth membrane-associated protein compartments are precursors of viral envelope and that a conveyor belt-type sequence from smooth membrane to plasma membrane takes place in virion assembly. Another alternative pathway would be the concept of a fluid membrane structure in which all proteins including viral proteins are freely mobile. Such a condition would allow viral RNP molecules directly under the plasma membrane to attract and interact with viral glycoprotein molecules and thus exclude cellular proteins during budding. Neither of these suggestions has solid experimental support. Although the mechanism of insertion of HA and NA into viral membrane is not clear, Hay [202] finds that HA is inserted into discrete regions of plasma membrane, whereas NA has a broader distribution. The membrane protein is synthesized in the proximity of plasma membrane and is directly incorporated into the plasma membrane at sites containing HA. The incorporation of M protein occurs directly at the site of assembly and completes the formation of viral envelope during the final stage of budding.

B. Processing of RNP

The how of RNP molecule selection and processing into budding virions is even more unclear than the processing of viral envelope proteins. Two processes have been postulated: a selective processing of RNP molecules by linkage via RNA or protein linkers, or random processing from the pool of intracellular RNP without any selection. Although there is no solid evidence in favor of against either of these postulates, some experimental support has been obtained for both. The occasional finding of larger RNA and RNP molecules in both cells and virions suggests the possibility of a linkage. If it exists at all, such a linkage would be extremely labile and would dissociate by mild denaturation or by the procedures used for the isolation of RNA or RNP, including centrifugation. The linkers would be RNA molecules, either small (approximately 4 S or smaller) joining ends of VRNA segments, or VDNA segments may join end to end by base pairing. Small RNA molecules have been found associated with virions.* The amount of base pairing (fewer than 15 nucleotides) would form a rather unstable bond but would be sufficient to help in the selection of viral RNPs. Large RNA molecules, possibly formed by unstable linkers, have been found by electron microscopy of isolated virion RNA [121] and RNP [236]. In addition to possible RNA linkers, protein linkers have also been suggested to play a role in the selection of RNA molecules [101].

The alternate idea that RNP molecules are selected and packaged at random into virions [62] is supported by the finding that a high frequency of recombination among the progeny viruses is observed when the same cell is infected by 2 viruses [123]; the expected percentages of hybrid viruses often follow the expected random assortment of segments. In addition, randomly defective viruses that can complement each other have been observed [118]; these viruses apparently lack 1 or more segments of viral RNA at random. These observations by themselves cannot fully justify the notion of random assortment being the dominant process involved in assembly, because it would be very inefficient and a much larger number of RNA segments would be needed to insure at least 1 each of individual RNP molecules per infectious virion. Also, the main bulk of the virus particles follows the principle that 1 virus particle is able to initiate the infectious process and, therefore, may not

*D. P. Nayak, unpublished observations.

be randomly defective. However, these experiments cannot distinguish if 1% or more of virus particles are fully infectious and the rest are defective because of the lack of RNA segment at random or are noninfectious for other reasons (e.g., structural defect, thermal inactivation). Whether processes involved in producing randomly defective virus are the main assembly line of virus production remain unresolved. Thus, neither of the above postulates has any firm basis, and more information is required before the actual mechanism of RNP selection is known.

C. Budding

Again, the biochemical explanation of the mechanical process involved in budding is a complete mystery. Why are RNP molecules attracted toward the selective spots containing viral antigens as opposed to normal host membranes? Why are some viral buds on cell membrane filamentous, while others are spherical? Are the majority of spherical viruses produced from the processing of filaments? Are the filaments produced by the fusion of spheroidal particles? Answers to these and other questions must be known before the process of budding is resolved. Finally, the eventual release of the virion from the cell membrane appears to be caused by virion-associated neuraminidase (see above), which helps virions to disseminate from their parent host, infect other hosts, and continue the infectious cycle.

IX. von Magnus Virus

Incomplete virus formation was first observed in 1951 by von Magnus [119], who noted that when influenza viruses were passaged successively at high multiplicity [e.g., 0.2-0.5 ml of undiluted infected chorioallantoic (CA) fluid per egg], the infectivity of progeny virus was reduced more than its hemagglutinin content. The ratio of infectivity versus hemagglutinin becomes lower on successive passages, with the resultant production of more noninfectious particles. This property is called the von Magnus phenomenon or multiplicity-dependent incomplete virus production. The resulting particles are referred to as von Magnus, incomplete, defective (DI), or noninfectious viruses.

DI particles have been found for almost all viruses that have been extensively studied. Thus, passing viruses at high multiplicity can generate defective particles from rhabdoviruses (T particles), polyoma and SV-40 viruses, poliomyelitis viruses, Newcastle disease virus, Sendai virus, SV-5 virus, Rift Valley fever virus, lymphocytic choriomeningitis virus, and others [206].

All of the DI viruses, including von Magnus influenza virus, have the following properties in common: (a) They possess most of the structural proteins and antigens of the infectious virus, although there may be some quantitative differences between the realtive amounts of structural proteins of infectious and DI viruses [207] ; these virus particles, therefore, can be adsorbed to host cells with the same efficiency. (b) DI viruses are not infectious by themselves and need the help of infectious viruses for replication; in other words, the same cell has to be infected by at least 1 complete and 1 or more DI viruses for replication. (c) DI viruses interfere with the synthesis of infectious particles and thus produce defective particles; this is best demonstrated by the interactions between B (infectious) and T (defective) particles of rhabdoviruses [206]. (d) All of the DI viruses appear as deletion mutants; e.g., von Magnus influenza virus lacks the largest RNA molecule (21 S) [105, 106]. In other viruses, up to 2/3 of the viral genome may be missing in incomplete particles [208] ; the specific amount and extent of deletion are yet unresolved in influenza viruses. (e) von Magnus influenza virus is different from the so-called "non-plague-forming particles," which possibly lack some segment of the viral genome at random [118].

The mechanism of interference with the replication of complete viruses by DI influenza viruses is unknown. Both host and virus are involved. The role of host factor in virus replication is demonstrated by the fact that L-cells or HeLa cells produce mostly incomplete particles (abortive infection) [209], whereas MDBK cells favor the production of complete virus particles [105, 134]. In chick embryo cells one can produce either infectious or noninfectious particles, depending on the nature of input virus and the multiplicity of infection [127]. In addition to host regulation, the presence of noninfectious particles in the input virus inoculum also controls the nature of progeny virus. For example, if the input virus contains fewer incomplete viruses, the progeny virus will be mostly infectious, whereas if the input virus contains large numbers of noninfectious virus, the progeny virus will be mostly noninfectious. Thus, the same cell infected by 2 virus particles, 1 infectious and the other noninfectious, will favor the production of von Magnus virus, the relative amount of which will depend on the cell type.

Although von Magnus viruses are biologically noninfectious (or at least less infectious) and biochemically identifiable because of the loss of the largest (21 S) RNA segment, the molecular mechanism involved in interference by von Magnus virus is not yet known. Von Magnus viruses arise mostly through the interference of the replication of infectious particles by DI particles, although occasionally a few DI particles may be produced by spontaneous deletion of a part of the complete viral genome [206]. Because the influenza viral genome is segmented, and each RNA molecule has to be transcribed and replicated independently and finally assembled into the virion, 3 general mechanisms of interference can be postulated: (a) defective synthesis (either transcription or replication) of viral RNA molecules may form incomplete virus particles, (b) all of the viral RNA molecules are synthesized as in normal infectious cycle but a defective translation, or (c) assembly produces incomplete virus particles, and a combination of both these processes.

A defect in the assembly process may indeed by involved in von Magnus virus formation. It may be the major cause of abortive infection in L-cells and HeLa cells, where all species of viral RNA molecules are synthesized but no infectious virus is produced [209]. However, in chick embryo cells that can yield either von Magnus or infectious virus particles, depending on the condition of infection, preliminary evidence [62, 105, 210] suggests that a shift in intracellular viral RNA synthesis results in the production of von Magnus viruses. Although it is suggested that smaller RNA molecules replicate more rapidly and thus interfere with the replication of large RNA molecules, the precise mechanism of the interference has not been determined [105]. It is not known whether the transcriptive or the replicative phase of VRNA is involved. Another possibility is a defect in translation, either independent of RNA synthesis or caused by aberrant RNA synthesis. Compared to infectious viruses the relative amount of structural proteins of DI viruses vary greatly [207]. Furthermore, the specific deletion of largest RNA alone may not account for the loss of infectivity in DI viruses [242].

X. Genetics

In 1951, Burnet and Lind [211] injected 2 influenza virus strains into mouse brain and isolated (in addition to the parent viruses) a hybrid virus with some properties from both parent viruses. This was the first demonstration of the exchange of phenotypic markers among animal viruses and of the formation of a recombinant virus; it marks the beginning of animal virus genetics. Burnet and Lind also demonstrated the formation of hybrid virus following a mixed infection of chick embryo chorioallantois [212]. These early studies in genetics were soon followed by a number of workers using more sophisticated methods and better markers. Simpson and Hirst [126, 213] developed the plaque assay for influenza viruses and used cloned viruses to study genetic recombination; they also investigated the formation of recombinant virus following mixed infection in cell culture. Tumova and Pereira [30] demonstrated genetic exchange between human and animal viruses by using ultraviolet-treated fowl plague and infectious human and other animal viruses.

Several investigators [4, 12, 32, 214-218] have analyzed recombination among antigenically distinguishable viruses and have demonstrated the exchange of antigenic markers such as HA and NA among the parent viruses. Subsequently, *ts* mutants were isolated and classified according to complementation and recombination groups by a number of workers [219-225]. Webster and his group [31, 32] have carried out extensive recombination experiments in turkeys using 2 antigenically distinct influenza viruses and have demonstrated the emergence of hybrid viruses following mixed infection in animals. These recombinant viruses were gentically stable; some were highly virulent and infectious and others were not.

The main results of these extensive genetic studies can be summarized as follows:

1. The formation of stable hybrids among influenza viruses is caused by genetic exchange. Compared to other RNA viruses, this phenomenon is much more common among influenza viruses following mixed infection of a single cell by 2 or more viruses.

2. The closer the parent viruses to each other in growth characteristics the higher the frequency of formation of hybrid virus. Antigenic and gentic relatedness among influenza A viruses do not play a major role in recombination frequency. However, if a cell is infected with 2 completely unrelated viruses (e.g., types B and C), the occurrence of a stable hybrid virus has not been demonstrated.

3. Temperature-sensitive (*ts*) mutants can be isolated from influenza viruses. Five to 8 complementation and recombination groups have been demonstrated among these mutants. Stable wild viruses are formed after mixed infection of *ts* mutants belonging to different complementation groups, showing that true genetic exchange can take place among influenza viruses.

4. Complementation and recombination grouping of *ts* mutants are most likely to be caused by a lesion in a single functional peptide [222]. Recently, Palese et al. [83] have shown that 2 mutants (*ts*3 and *ts*11) of the same recombination group are neuraminidase defective, whereas *ts*104 (with a possible defect in HA) belongs to a different group (Ueda and Kilbourne, unpublished). Using the same group of mutants, Sugiura et al. [221] and Krug et al. [222] have reported that, of the 7 complementation groups of *ts* mutants, 4 are RNA^- and 3 are RNA^+. All of the RNA^- mutants are incapable of replicating virion RNA. Some of the RNA^- may be a secondary effect rather than from a primary lesion in the RNA synthesizing system. In addition, some *ts* mutants may have a defect in the synthesis of CRNA, while others may be defective in VRNA. It is likely that the 4 RNA^- mutants may have mutation in genes coding for P1, P2, NP, and NS proteins which are presumed to be involved in RNA synthesis. Of the 3 RNA^+ mutants, 2 are assigned to NA and HA. The third one makes all of the known viral proteins except M protein but not complete virus particles. This group may have a mutation in M protein or some other factor required in assembly of virus particles. If so, the mutants containing a defect in M protein have not been isolated yet.

5. Animal experiments show that genetic exchange of influenza virus is not restricted to eggs or cells in culture but can take place in infected animals

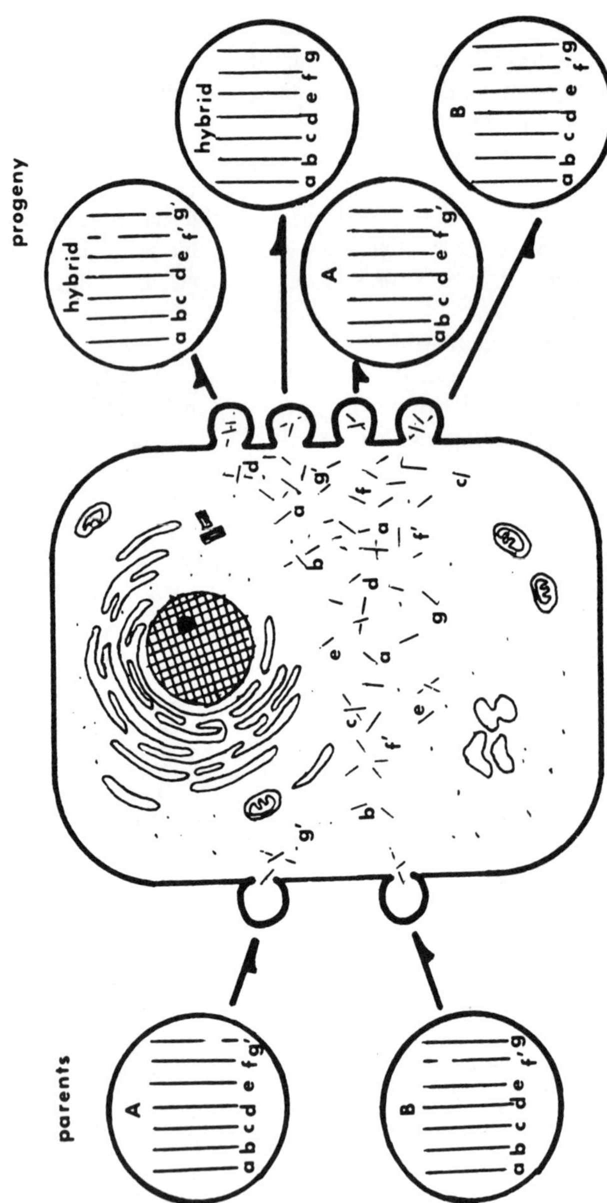

Fig. 6.12 Schematic presentation of hybrid virus formation following infection of a single cell by 2 viruses.

as well as in contact animals. Hybrid viruses are formed between different animal viruses and between animal and human viruses. Webster et al. [226] isolated hybrid viruses from turkeys infected with 2 viruses, T/Ont and T/Wis. They also isolated hybrids from turkeys that were exposed (contact) to turkeys infected with both viruses. The hybrid virus, T/Ont(H)-T/Wis(N), always had the HA of the more virulent virus. In pigs infected with Hong Kong virus and swine virus, both kinds of stable hybrids, HK(H)-SW(N) and SW(H)-HK(N), were isolated.

A. Mechanism of Genetic Exchange

The high frequency of hybrid virus formation following mixed infection is a puzzle not completely understood. Classical genetic exchange occurring among DNA viruses is caused by recombination, which involves crossing over, excision, and ligation of double-strand DNA molecules. The occurrence of such recombinational events among the single-strand RNA viruses is rare and does not take place with any measurable frequency among NDV, and other RNA viruses, although complementation among *ts* mutants can be demonstrated. The phenomenon of increased frequency of recombination among influenza viruses led Hirst [123] to propose the multisegmented genome model of influenza viruses, which has been upheld by subsequent biochemical and biological experiments. It would then be expected that, following infection of a single cell by 2 viruses, the nucleic acid components of each virus will replicate independently. Subsequently, during processing and assembly of viral RNP into virions, by reason of either random or selective RNP assortment, hybrid viruses will be formed along with parent viruses (Fig. 6.12). This sequence of events would predict that recombination events follow 2-hit kinetics. However, Hirst and Pons [118, 227] have recently shown that the recombination event may be more frequent than predicted by 2-hit kinetics because of the clumping or aggregate formation among the input virus particles. Whether the heterozygotes among the progeny viruses play a major role in recombination is not known.

Recent Developments

Recently important advances have been made towards mapping the influenza viral RNA segments coding for specific proteins. In one approach (Table 6.4, III), Palese and his colleagues [241, 243] have correlated differential migration behavior of viral RNA segments to the migration behavior or serological specificity of different viral proteins from a number of viral strains and their recombinants. Using this correlation, they have assigned specific VRNA segments to specific viral proteins. On the other hand, Suguira and his colleagues have determined the relative target size of viral RNA segments from the complementation data of

TABLE 6.4 Assignment of Viral RNA Segments to Viral Coded Proteins

	I	II		III	
				Mapping VRNA segments according to their migration behavior (b)	
Viral proteins	Molecular weight of viral proteins $\times 10^{-3}$	Relative target size of *ts* groups (a)	Known *ts* function	Assigned proteins	Molecular weight of VRNA $\times 10^{-6}$
P1	89-96	I	CRNA	P3	1.15
P2	80-83	VI	HA	P1	0.92
HA	75-80	III	CRNA	P2	0.89
NP	53-60	II	VRNA	HA	0.73
NA	45-70	VIII	?	NP	0.64
M	21-27	V	VRNA	NA	0.62
NS1	23-30	IV	NA	M	0.40
NS2	23-30	VII	?	NS	0.39

[a]The target size of VRNA segments was studied by K. Nakajima and A. Suguira (unpublished).
[b]From Ritchey, Palese, and Schulman [243].

ultraviolet inactivated wild type virus with temperature-sensitive mutants containing known temperature-sensitive functions (Table 6.4, II). As seen in Table 6.4, there is a discrepancy in the assignment of RNA segment for HA and NA proteins. These coding assignments of viral RNA segments must await direct proof that the CRNA of a specific segment of VRNA codes for the specific protein.

Acknowledgments

Part of the work in the author's laboratory was supported by a grant from the National Institute of Allergy and Infectious Diseases (AI 12749). Electron micrographs were taken by Mr. Zane Price. The author also gratefully acknowledges the assistance of Alan Davis, Kamales Som, Abantika Nayak, Roel Sanchez, and Bharathi Vayuvegula in the preparation of this manuscript. Helpful comments from Drs. J. Seto, A. Sugiura, J. Holland, and Peter Palese are also greatly appreciated.

References

1. L. Hoyle, Virology Monographs, Vol. 4, The Influenza Viruses, Springer-Verlag, New York, 1968.
2. W. R. Dowdle, M. T. Coleman, and M. B. Gregg, Prog. Med. Virol., 17, 91 (1974).
3. R. G. Webster and W. G. Laver, Prog. Med. Virol., 13, 271 (1971).
4. R. G. Webster, Curr. Top. Microbiol. Immunol., 59, 75 (1972).
5. I. T. Schulze, Adv. Virus Res., 18, 1 (1973).
6. R. W. Compans and P. W. Choppin, in Comprehensive Virology, Vol. 4 (H. Frankel-Conrat and R. R. Wagner, eds.), Plenum, New York, 1975, p. 179.
7. D. O. White, Curr. Top. Microbiol. Immunol., 63, 1 (1974).
8. W. G. Laver, Adv. Virus Res., 18 , 57 (1973).
9. H. D. Klenk, Curr. Top. Microbiol. Immunol., 68, 29 (1974).
10. M. W. Pons, Curr. Top. Microbiol. Immunol., 52, 142 (1970).
11. A. J. Shatkin, Bacteriol. Rev., 35, 250 (1971).
12. E. D. Kilbourne, Prog. Med. Virol., 5, 79 (1963).
13. R. D. Barry and B. W. J. Mahy, Negative Strand Viruses, Academic, New York, 1975.
14. C. H. Andrews, F. B. Bang, and F. M. Burnet, Virology, 1, 176 (1955).
15. J. L. Melnick, Prog. Med. Virol., 17, 290 (1974).
16. R. M. Chanock et al., WHO Bull., 45 (1971).
17. R. Debre and J. Cellers, Clinical Virology, Saunders, Philadelphia, 1970.
18. J. A. Kasel, R. H. Alford, and V. Knight, Nature, 206, 41 (1965).
19. D. P. Nayak, G. W. Kelley, and N. R. Underdahl, Am. J. Vet. Res., 26, 984 (1965).

20. D. P. Nayak, M. J. Twiehaus, G. W. Kelley, and N. R. Underdahl, Am. J. Vet. Res., 26, 1271 (1965).
21. D. P. Nayak, G. W. Kelley, G. A. Young, and N. R. Underdahl, Proc. Soc. Exp. Biol. Med., 116, 200 (1964).
22. W. T. London, D. A. Fuccillo, J. A. Sever, and S. G. Kent, Nature, 255, 483 (1975).
23. F. M. Davenport, E. Minuse, A. V. Hennessy, and T. Francis, Jr., Bull. WHO, 41, 453 (1969).
24. G. C. Schild, J. S. Oxford, W. R. Dowdle, M. Coleman, N. S. Pereira, and P. Chakraverty, Bull. WHO, 51, 1 (1975).
25. R. G. Webster and W. G. Laver, Virology, 48, 433 (1972).
26. W. G. Laver, J. C. Downie, and R. G. Webster, Virology, 59, 230 (1974).
27. W. G. Laver and R. G. Webster, Virology, 34, 193 (1968).
28. W. G. Laver and R. G. Webster, Virology, 48, 445 (1972).
29. R. G. Webster and H. G. Pereira, J. Gen. Virol., 3, 201 (1968).
30. B. Tumova and H. G. Pereira, Virology, 27, 253 (1965).
31. W. G. Laver and R. G. Webster, Virology, 51, 383 (1973).
32. R. G. Webster and C. H. Campbell, Virology, 62, 404 (1974).
33. E. D. Kilbourne, Science, 160, 74 (1968).
34. R. W. Horne, A. P. Waterson, P. Wildy, and A. E. Farnham, Virology, 11, 79 (1960).
35. L. Hoyle, R. W. Horne, and A. P. Waterson, Virology, 13, 448 (1961).
36. P. W. Choppin, Virology, 21, 342 (1963).
37. H. A. Blough, Nature, 251, 333 (1974).
38. J. G. Cruickshank, Ciba Found. Symp. Cell. Biol. Myxovir., 5 (1964).
39. F. M. Brunet and P. E. Lind, Arch. Ges. Virsuforsch., 7, 413 (1957).
40. H. A. Blough, Ciba Found. Symp. Cell. Biol. Myxovir. Infect., 120 (1964).
41. J. S. Murphy and F. B. Bang, J. Exp. Med., 95, 259 (1952).
42. W. G. Laver, J. Mol. Biol., 9, 109 (1964).
43. J. M. Tiffany and H. A. Blough, Proc. Natl. Acad. Sci. U.S., 65, 1105 (1970).
44. D. J. Bucher and E. D. Kilbourne, J. Virol., 10, 60 (1972).
45. W. G. Laver and E. D. Kilbourne, Virology, 30, 493 (1966).
46. A. P. Kendal and E. A. Eckert, Biochim. Biophys. Acta, 258, 484 (1972).
47. I. Lazdins, E. A. Haslam, and D. O. White, Virology, 49, 758 (1972).
48. R. Drzeniek, H. Frank, and R. Rott, Virology, 36, 703 (1968).
49. P. W. Choppin, H. D. Klenk, R. W. Compans, and L. A. Caliguiri, Perspect. Virol., 7, 127(1971).
50. H. A. Blough and J. Merlie, Virology, 40, 685 (1970).
51. J. J. Holland and E. D. Keihn, Science, 167, 202 (1970).
52. I. T. Schulze, Virology, 42, 890 (1970).
53. I. T. Schulze, Virology, 47, 181 (1972).
54. P. Stanley and E. A. Haslam, Virology, 46, 764 (1971).
55. F. R. Landsberger, J. Lenard, J. Paxton, and R. W. Compans, Proc. Natl. Acad. Sci. U.S., 68, 2579 (1971).

56. F. R. Landsberger, R. W. Compans, P. W. Choppin, and J. L. Lenard, Biochemistry, 12, 4498 (1973).
57. H. D. Klenk, R. Rott, and H. Becht, Virology, 47, 579 (1972).
58. W. G. Laver and R. C. Valentine, Virology, 38, 105 (1969).
59. C. M. Brand and J. J. Skehel, Nature, New Biol., 238, 145 (1972).
60. K. Apostolov, T. H. Flewett, and A. P. Kendal, in The Biology of Large RNA Viruses (R. D. Barry and B. W. J. Mahy, eds.), Academic, New York, 1970, p. 3.
61. T. Bächi, W. Gerhard, J. Lindenmann, and K. Mühlethaler, J. Virol., 4, 769 (1969).
62. R. W. Compans, N. J. Dimmock, and H. Meier-Ewert, in The Biology of Large RNA Viruses (R. D. Barry and B. W. J. Mahy, eds.), Academic, New York, 1970, p. 87.
63. J. J. Skehel, Virology, 44, 409 (1971).
64. M. V. Nermut, J. Gen. Virol., 17, 317 (1972).
65. N. L. Chow and R. W. Simpson, Proc. Natl. Acad. Sci. U.S., 68, 752 (1971).
66. G. K. Hirst and M. W. Pons, Virology, 47, 546 (1972).
67. J. D. Almeida, and A. P. Waterson, in The Biology of Large RNA Viruses (R. D. Barry and B. W. J. Mahy, eds.), Academic, New York, 1970, p. 27.
68. I. T. Schulze, M. W. Pons, and G. K. Hirst, in The Biology of Large RNA Viruses (R. D. Barry and B. W. J. Mahy, eds.), Academic, New York, 1970, p. 324.
69. R. W. Compans, J. Content, and P. H. Duesberg, J. Virol., 10, 795 (1972).
70. M. W. Pons, I. T. Schulze, and G. K. Hirst, Virology, 39, 250 (1969).
71. P. H. Duesberg, J. Mol. Biol., 42, 485 (1969).
72. M. W. Pons, Virology, 46, 149 (1971).
73. E. A. Goldstein and M. W. Pons, Virology, 41, 382 (1970).
74. C. Scholtissek and H. Becht, J. Gen. Virol., 10, 11 (1971).
75. G. K. Hirst, Science, 94, 22 (1941).
76. L. McClelland and R. Hare, Can. Publ. Hlth. J. 32, 530 (1941).
77. D. O. White, J. M. Taylor, E. A. Haslam, and A. W. Hampson, in The Biology of Large RNA Viruses (R. D. Barry and B. W. J. Mahy, eds.), Academic, New York, 1970, p. 602.
78. S. G. Lazarowitz, R. W. Compans, and P. W. Choppin, Virology, 52, 199 (1973).
79. S. G. Lazarowitz, A. R. Goldberg, and P. W. Choppin, Virology, 56, 172 (1973).
80. P. Stanley, S. S. Gandhi, and D. O. White, Virology, 53, 92 (1973).
81. W. G. Laver and R. G. Webster, Virology, 30, 104 (1966).
82. N. G. Wrigley, J. J. Skehel, P. A. Charlwood, and C. M. Brand, Virology, 51, 525 (1973).
83. P. Palese, K. Tobita, M. Ueda, and R. W. Compans, Virology, 61, 397 (1974).
84. A. B. Germanov, L. Ya. Zakstelskaya, E. V. Molibog, N. A. Evstigneeva, N. A. Parasiuk, and G. V. Kornilaeva, Intervirology (Moscow), 2, 273 (1973-1974).

85. E. D. Kilbourne, F. S. Lief, J. L. Schulman, R. I. Jahel, and W. G. Laver, Perspect. Virol., 5, 87 (1968).
86. E. D. Kilbourne, W. G. Laver, J. L. Schulman, and R. G. Webster, J. Virol., 2, 281 (1968).
87. J. L. Schulman, M. Khakpour, and E. D. Kilbourne, J. Virol., 2, 778 (1968).
88. J. T. Seto and R. Rott, Virology, 30, 731 (1966).
89. P. Palese, J. L. Schulman, G. Bodo, and P. Meindl, Virology, 59, 490 (1970).
90. R. G. Webster and W. G. Laver, J. Immunol., 99, 49 (1967).
91. P. Palese and R. W. Compans, unpublished data, 1975.
92. H. Becht, U. Hämmerling, and R. Rott, Virology, 46, 337 (1971).
93. P. Palese and J. Schulman, Virology, 57, 227 (1974).
94. H. Becht, R. Rott, and H. D. Klenk, Z. Med. Mikrobiol. Immunol., 156, 305 (1971).
95. H. Becht, R. Rott, and H. D. Klenk, J. Gen. Virol., 14, 1 (1972).
96. J. J. Skehel, Virology, 49, 23 (1972).
97. G. L. Ada and B. T. Perry, J. Exp. Biol. Med. Sci., 32, 453 (1954).
98. W. Frisch-Niggemeyer and L. Hoyle, J. Hyg., 54, 201 (1956).
99. D. H. L. Bishop, J. F. Obijeski, and R. W. Simpson, J. Virol., 8, 74 (1971).
100. L. J. Lewandowski, J. Content, and S. H. Leppla, J. Virol., 8, 701 (1971).
101. J. J. Skehel, J. Gen. Virol., 11, 103 (1971).
102. P. Davies and R. D. Barry, Nature, 221, 384 (1966).
103. P. H. Duesberg and W. S. Robinson, J. Mol. Biol., 25, 383 (1967).
104. D. P. Nayak and M. A. Baluda, J. Virol., 1, 1217 (1967).
105. P. W. Choppin and M. W. Pons, Virology, 42, 603 (1970).
106. P. Duesberg, Proc. Natl. Acad. Sci. U.S., 59, 930 (1968).
107. M. W. Pons and G. K. Hirst, Virology, 34, 385 (1968).
108. R. D. Barry and P. Davies, J. Gen. Virol., 2, 59 (1968).
109. D. P. Nayak and M. A. Baluda, J. Virol., 2, 99 (1968).
110. M. W. Pons, Virology, 69, 789 (1976).
111. R. J. Young and J. Content, Nature, New Biol., 230, 140 (1971).
112. D. P. Nayak and M. A. Baluda, Proc. Natl. Acad. Sci. U.S., 59, 184 (1968).
113. D. P. Nayak, Fed. Proc., 28, 1858 (1969).
114. D. P. Nayak, in The Biology of Large RNA Viruses (R. D. Barry and B. W. J. Mahy, eds.), Academic, New York, 1970, p. 371.
115. D. P. Nayak and M. A. Baluda, J. Virol., 3, 318 (1969).
116. J. Content and P. H. Duesberg, J. Mol. Biol., 62, 273 (1971).
117. J. Horst, J. Content, S. Mandeles, H. Frankel-Conrat, and P. Duesberg, J. Mol. Biol., 69, 209 (1972).
118. G. K. Hirst and M. W. Pons, Virology, 56, 620 (1973).
119. P. von Magnus, Adv. Virus Res., 2, 59 (1954).
120. N. J. Dimmock and D. H. Watson, J. Gen. Virol., 5, 499 (1969).

121. K. K. Li and J. T. Seto, J. Virol., 7, 524 (1971).
122. F. B. Bang, Ann. Rev. Microbiol., 9, 21 (1955).
123. G. K. Hirst, Cold Spring Harbor Symp. Quant. Biol., 27, 303 (1962).
124. W. Henle and O. C. Liu, J. Exp. Med., 94, 305 (1951).
125. R. D. Barry, Virology, 14, 398 (1961).
126. R. W. Simpson and G. K. Hirst, Virology, 15, 436 (1961).
127. C. Scholtissek, R. Drzeniek, and R. Rott, Virology, 30, 313 (1966).
128. C. Scholtissek and R. Rott, Virology, 22, 169 (1964).
129. F. M. Burnet, Science, 123, 1101 (1956).
130. L. H. Frommhagen, C. A. Knight, and N. K. Freeman, Virology, 8, 176 (1959).
131. A. Harboe, Acta Pathol. Microbiol. Scand., 57, 488 (1963).
132. F. M. Burnet, J. F. McCrea, and J. D. Stone, J. Exp. Pathol., 27, 228 (1946).
133. E. D. Kilbourne and J. S. Murphy, J. Exp. Med., 111, 387 (1960).
134. P. W. Choppin, Virology, 39, 130 (1969).
135. R. W. Compans and N. J. Dimmock, Virology, 39, 499 (1969).
136. A. Porebska, H. G. Pereira, and J. A. Armstrong, J. Med. Microbiol., 1, 145 (1968).
137. J. Archetti, E. Bereczky, F. Rosati-Valente, and D. Steve-Bocciarelli, Arch. Ges. Virusforsch., 29, 275 (1970).
138. B. W. J. Mahy, in The Biology of Large RNA Viruses (R. D. Barry and B. W. J. Mahy, eds.), Academic, New York, 1970, p. 392.
139. B. W. J. Mahy, N. D. Hastie, and S. J. Armstrong, Proc. Natl. Acad. Sci. U.S., 69, 1421 (1972).
140. J. R. Stephenson and N. J. Dimmock, in Negative Stranded Viruses (R. D. Barry and B. W. J. Mahy, eds.), Academic, New York, 1975, p. 485.
141. R. M. Krug and P. R. Etkind, Virology, 56, 334 (1973).
142. W. F. Long and D. C. Burke, J. Gen. Virol., 6, 1 (1970).
143. R. M. Franklin and C. Henry, Virology, 10, 406 (1960).
144. S. Fazekas De St. Groth, Aust. J. Exp. Biol. Med. Sci., 27, 65 (1949).
145. W. Henle, J. Exp. Med., 90, 1 (1949).
146. D. O. White, Virology, 10, 21 (1960).
147. S. Fazekas De St. Groth, Nature, 162, 294 (1948).
148. S. Dales and P. W. Choppin, Virology, 18, 489 (1962).
149. R. R. Dourmashkin and D. A. J. Tyrrell, J. Gen. Virol., 24, 129 (1974).
150. C. Morgan and H. M. Rose, J. Virol., 2, 925 (1968).
151. J. R. Stephenson and N. J. Dimmock, Virology, 65, 77 (1975).
152. R. D. Barry, Ciba Found. Symp. Cell. Biol. Myxovir. Infect., 51 (1964).
153. R. D. Barry, Virology, 24, 563 (1964).
154. D. P. Nayak and A. F. Rasmussen, Virology, 30, 673 (1966).
155. A. Gregoriades, Virology, 42, 905 (1970).
156. R. Rott, S. Saber, and C. Scholtissek, Nature, 205, 1187 (1965).
157. Yu. Ghendon, V. P. Ginsburg, G. Ya. Soloviev, and S. G. Markushin, J. Gen. Virol., 6, 249 (1970).

158. D. C. Kelly, R. J. Avery, and N. J. Dimmock, J. Virol., 13, 1155 (1974).
159. D. C. Kelly and N. J. Dimmock, in Negative Strand Viruses, Vol. 1 (B. W. J. Mahy and R. D. Barry, eds.), Academic, 1975, p. 469.
160. J. J. Skehel, Virology, 45, 793 (1971).
161. E. Penhoet, H. Miller, M. Doyle, and S. Blatti, Proc. Natl. Acad. Sci. U.S., 68, 1369 (1971).
162. R. T. Schwarz and C. Scholtissek, Z. Naturforsch., 28C, 202 (1973).
163. D. H. L. Bishop, J. F. Obijeski, and R. W. Simpson, J. Virol., 8, 66 (1971).
164. D. H. L. Bishop, P. Roy, W. J. Bean, Jr., and R. W. Simpson, J. Virol., 10, 689 (1972).
165. M. Horisberger and L. E. Guskey, J. Virol., 13, 230 (1974).
166. L. A. Caliguiri and R. W. Compans, J. Virol., 14, 191 (1974).
167. R. W. Compans and L. A. Caliguiri, J. Virol., 11, 441 (1973).
168. N. D. Hastie and B. W. J. Mahy, J. Virol., 12, 951 (1973).
169. B. J. Ruck, K. W. Brammer, M. G. Page, and J. D. Coombes, Virology, 39, 31 (1969).
170. C. Scholtissek and R. Rott, Virology, 40, 989 (1970).
171. C. Scholtissek, in The Biology of Large RNA Viruses (R. D. Barry and B. W. J. Mahy, eds.), Academic, New York, 1970, p. 438.
172. R. J. Avery and N. J. Dimmock, Virology, 64, 409 (1975).
173. W. J. Bean, Jr., and R. W. Simpson, Virology, 56, 646 (1973).
174. D. P. Nayak, Intl. Symp. Influenza Vaccines Men Horses, London, 20, 362 (1973).
175. C. Scholtissek, Biochim. Biophys. Acta, 179, 389 (1969).
176. C. Scholtissek and R. Rott, J. Gen. Virol., 4, 125 (1969).
177. C. Scholtissek and R. Rott, Virology, 39, 400 (1969).
178. C. Scholtissek, G. Kaluza, and R. Rott, J. Gen. Virol., 17, 213 (1972).
179. M. W. Pons, Virology, 51, 120 (1973).
180. M. W. Pons, Virology, 47, 823 (1972).
181. P. R. Etkind and R. M. Krug, Virology, 62(1), 38 (1974).
182. D. W. Kingsbury and R. G. Webster, Virology, 56, 654 (1973).
183. C. Liu, J. Exp. Med., 101, 665 (1955).
184. P. M. Breitenfeld and W. Schäfer, Virology, 4, 328 (1957).
185. O. A. Holterman, W. D. Hillis, and M. A. J. Moffat, Acta Pathol. Microbiol. Scand., 50, 398 (1960).
186. K. Maeno and E. D. Kilbourne, J. Virol., 5, 153 (1970).
187. J. J. Skehel, Virology, 56, 394 (1973).
188. J. M. Taylor, A. W. Hampson, J. E. Layton, and D. O. White, Virology, 42, 744 (1970).
189. R. M. Krug and R. Soeiro, Virology, 64, 378 (1975).
190. S. G. Lazarowitz, R. W. Compans, and P. W. Choppin, Virology, 46, 830 (1971).
191. R. W. Compans, Virology, 51, 56 (1973).
192. H. D. Klenk and R. Rott, J. Virol., 11, 823 (1973).
193. H. D. Klenk, C. Scholtissek, and R. Rott, Virology, 49, 723 (1972).
194. H. D. Klenk, W. Wöllert, R. Rott, and C. Scholtissek, Virology, 57, 28 (1974).

195. C. Scholtissek, R. Rott, and H. D. Klenk, Virology, 63, 191 (1975).
196. C. Scholtissek, Adv. Biosci., 11, 193 (1973).
197. J. Etchison, M. Doyle, E. Penhoet, and J. Holland, J. Virol., 7, 155 (1971).
198. S. S. Gandhi, P. Stanley, J. A. Taylor, and D. D. White, Microbios, 5, 41 (1972).
199. R. T. Schwarz and H. D. Klenk, J. Virol., 14(5), 1023 (1974).
200. G. Kaluza, C. Scholtissek, and R. Rott, J. Gen. Virol., 14, 251, 1972.
201. H. Meier-Ewert and R. W. Compans, J. Virol., 14, 1083 (1974).
202. A. J. Hay, Virology, 60(2), 398 (1974).
203. D. C. Kelley and J. Dimmock, in Negative Strand Viruses (R. D. Barry and B. W. J. Mahy, eds.), Academic, New York, 1975, p. 469.
204. H. Becht, J. Virol., 7, 204 (1971).
205. C. Morgan, K. C. Hsu, R. A. Rifkind, A. W. Knox, and H. M. Rose, J. Exp. Med., 114, 825 (1961).
206. A. S. Huang and D. Baltimore, Nature, 226, 325 (1970).
207. J. Lenard and R. W. Compans, Virology, 65, 418 (1975).
208. P. Repik and D. H. L. Bishop, J. Virol., 12, 969 (1973).
209. R. A. Lerner and L. D. Hodge, Proc. Natl. Acad. Sci. U.S., 64, 544 (1969).
210. D. P. Nayak, J. Gen. Virol., 14, 63 (1972).
211. F. M. Burnet and P. E. Lind, J. Gen. Microbiol., 5, 59 (1951).
212. F. M. Burnet and P. E. Lind, Aust. J. Exp. Biol. Med. Sci., 30, 469 (1952).
213. R. W. Simpson, Ciba Found. Symp. Cell. Biol. Myxovir. Infect., 187 (1964).
214. A. Sugiura and E. D. Kilbourne, Virology, 26, 478 (1965).
215. A. Sugiura and M. Ueda, J. Virol., 7, 499 (1971).
216. R. Schwarz, C. Scholtissek, and R. Rott, Med. Microbiol. Immunol., 158, 54 (1972).
217. W. Anschutz, C. Scholtissek, and R. Rott, Med. Microbiol. Immunol., 158, 26, 1972.
218. D. Hobson and C. Scholtissek, J. Gen. Virol., 9, 151 (1970).
219. R. W. Simpson and G. K. Hirst, Virology, 35, 41 (1968).
220. A. Sugiura, K. Tobita, and E. D. Kilbourne, J. Virol., 10, 639 (1972).
221. A. Sugiura, M. Ueda, K. Tobita, and C. Enomoto, Virology, 65, 363 (1975).
222. R. M. Krug, M. Ueda, and P. Palese, J. Virol., 16, 790 (1975).
223. J. S. Mackenzie, J. Gen. Virol., 6, 63 (1970).
224. Y. Z. Ghendon, S. G. Markushin, A. T. Marchenko, B. S. Sitnikov, and V. P. Ginzburg, Virology, 55, 305 (1973).
225. C. Scholtissek, R. Kruczinna, R. Rott, and H. D. Klenk, Virology, 58, 317 (1974).
226. R. G. Webster, C. H. Campbell, and A. Granoff, Virology, 51, 149 (1973).

227. G. K. Hirst, Virology, 55, 81 (1973).
228. W. J. Bean and R. W. Simpson, J. Virol., 16, 516 (1975).
229. J. S. Mackenzie and N. J. Dimmock, J. Gen. Virol., 19, 51 (1973).
230. S. B. Spring, S. R. Nusinoff, E. L. Tierney, D. D. Richman, B. R. Murphy, and R. M. Chanock, Virology, 66, 542 (1975).
231. E. Kilbourne, (ed.), Influenza Viruses, Academic, New York, 1975.
232. S. S. Chang and J. Seto, J. Gen. Virol., 17, 99 (1972).
233. M. W. Pons, Virology, 67, 209 (1975).
234. A. Gregoriodes, Virology, 54, 369 (1973).
235. J. J. Skehel and M. D. Waterfield, PNAS, 72, 93 (1975).
236. J. D. Almeida and C. M. Brand, J. Gen. Virol., 27, 313 (1975).
237. S. G. Lazarowitz, R. W. Compans, and P. W. Choppin, Virology, 46, 830 (1971).
238. P. R. Etkind and R. M. Krug, J. Virol., 16, 1464 (1975).
239. H. Klenk, R. Rott, M. Orlich, and J. Blödorn, Virology, 68, 426 (1975).
240. S. G. Lazarowtiz and P. W. Choppin, Virology, 68, 440 (1975).
241. P. Palese and J. L. Schulman, Proc. Natl. Acad. Sci., U.S., 73, 2142 (1976).
242. W. J. Bean and R. W. Simpson, J. Virol., 18, 365 (1976).
243. M. B. Ritchey, P. Palese, and J. L. Schulman, J. Virol., 20, 307 (1976).
244. D. P. Nayak, Emma D'Andrea, and F. O. Wettstein, J. Virol., 20, 107 (1976).

Chapter 7

Paramyxoviruses

David W. Kingsbury

Laboratories of Virology
St. Jude Children's Research Hospital
Memphis, Tennessee

I. Introduction

Paramyxoviruses are enveloped, single-stranded RNA-containing viruses with helical nucleocapsid symmetry whose site of replication is the cell cytoplasm. Their replication strategy is identical to that of a similar virus group, the rhabdoviruses. The key point about this replication strategy is that the single-stranded RNA in the virion is not message for viral protein. Instead, the viral messenger RNAs are complementary in base sequences to the virion RNA. Viruses exhibiting this life style have been aptly termed "negative-strand" viruses, based on the convention that viral message is designated the positive strand [7, 12].

The negative-strand life style is analogous to the *modus vivendi* of DNA-containing viruses. For these viruses, also, viral genome replication is a process readily distinguishable from the transcription of messenger RNA. However, negative-strand RNA-virus replication stands apart from the classical picture of RNA-virus replication developed from the study of RNA bacteriophages and picornaviruses. For these positive-strand viruses, the genome is the message: Transcription is obviated; genome replication and messenger RNA production are the same event [7].

A requirement for transcription presents a regulatory opportunity: The relative abundances of the various virus polypeptides may be controlled by the relative abundances of their RNA templates. As we shall see, recent evidence suggests that this opportunity has been exploited by paramyxoviruses.

Therefore, at this stage in our understanding of paramyxovirus replication, there are 2 crucial regulatory mechanisms which need to be explained. These are the switch from transcriptional to replicative employment of the negative-strand RNA genome as template for RNA synthesis, and the regulation of polypeptide abundances. Much of this review is devoted to what we know about these processes and what we may hope to learn in the future.

II. Paramyxoviruses in Nature

Paramyxoviruses have been found in a variety of vertebrate species, both wild and domesticated (Table 7.1), and there are no doubt many more agents still be be discovered in wild animals.

Paramyxoviruses known to be pathogenic for humans are measles virus, mumps virus, and 4 antigenically distinguishable parainfluenza types [34, 37, 88, 118, 123]. Respiratory syncytial virus is an important pathogen which resembles paramyxoviruses biologically and structurally [96, 119, 150], but it has a slightly narrower nucleocapsid [91, 120]. Various degrees of morbidity, ranging from mild upper respiratory syndromes to incapacitating generalized

Table 7.1 The Paramyxoviruses

Host	Virus	References
Man	Measles	256
	Mumps	34
	Parainfluenza 1, 2, 3, 4	118
	Respiratory syncytial	119
Monkey	SV41	161
Dog	Distemper	141
	SV5	110
Ungulate	Parainfluenza 3 (bovine)	1
	Rinderpest	189
Mouse	Sendai (HVJ)	74
	Pneumonia (virus) of mice	97
Bird	Newcastle disease	83
	Yucaipa	60
	Enders-Ruckle	248

illness with occasional mortality, chiefly in the pediatric population, are associated with these agents. Although measles and mumps vaccines are widely used, no effective prophylaxis has been developed for the parainfluenza and respiratory syncytial viruses [13, 128].

Readers who wish to explore the clinical and pathological manifestations of paramyxovirus infections in man and animals are directed to Refs. [37], [69], [88], [118], [119], [146], [160], and [166].

There is an aspect of disease involving paramyxoviruses which deserves some comment here, however, because it has recently attracted a lot of interest. Paramyxoviruses are now implicated in certain chronic central nervous system diseases. Measles virus, or a variant thereof, is known to be responsible for the rare disorder, subacute sclerosing panencephalitis (SSPE) [58, 246]. The more common but devastating neurological disease, multiple sclerosis, is still of unknown etiology, but recent evidence suggests an altered host response to viral antigens, those of measles virus in particular [25, 79, 179, 180, 188, 210].

Among other animals, Newcastle disease virus of birds [83], distemper virus of canines [114], and parainfluenza 3 [1, 59, 72, 221, 229] and rinderpest [189, 245] viruses of ungulates are the best known examples, because of their economic importance. Antigenic cross-reactivity among measles, canine distemper, and rinderpest viruses has been reported [113, 183]. This seems to reflect more fundamental relationships, as indicated also by similarities in virus polypeptide composition [245, 254].

Sendai virus, also known as hemagglutinating virus of Japan (HVJ), is a useful laboratory model of the paramyxovirus group. Although antigenically related to human parainfluenza 1 virus and originally thought to be a human pathogen [74], it is probably of murine origin [185]. Another murine virus of interest is pneumonia virus of mice [97]. It has a narrower nucleocapsid than the standard paramyxoviruses do, a feature that it shares with respiratory syncytial virus [84]. If neither of these viruses deserves to be classified as a paramyxovirus, they might be called "metamyxoviruses" [159], a more euphonious term than "paramyxovirus-like."

Simian virus (SV) 5 has been studied extensively by Choppin and colleagues [39, 40, 50, 52]. As indicated by its name, it was first isolated from cultured monkey cells [109], but it has been classified as a canine virus in Table 7.1, reflecting recent epidemiological information [5, 16, 105, 121, 122, 149, 222].

III. Practical Aspects of Biochemical Work with Paramyxoviruses

Despite their medical and economic importance and a long history as tools for investigations of biological phenomena, paramyxoviruses have received less attention from molecular biologists than the single rhabdovirus, vesicular stomatitis virus (VSV). VSV is currently the most popular model of negative-strand viruses [251]. The reasons are simple. It grows abundantly in a short time in a wide variety of cell types, including cells which multiply rapidly in suspension cultures. It generates large quantities of virus-specific macromolecules in infected cells and its virions contain an RNA-dependent RNA polymerase which is much more active than the RNA polymerase in any paramyxovirus [8]. In contrast, none of the paramyxoviruses has been found to replicate well in continuous cell lines and many (especially the medically interesting ones) replicate poorly in any type of cell [104, 153, 230]. Some of the better preparations of measles virus, for example, represent yields of 1 infectious unit per infected cell, and this may be obtained only after a day or more [172, 194], whereas it is routine to obtain at least 10^3 infectious units of VSV per cell in a few hours [249].

We shall see later (Section 7.IV.D.) that extracellular proteolytic cleavage of a surface glycoprotein of Sendai virus is required to make the virus infectious and that cell types differ in their ability to provide the needed enzyme. A mechanism of this kind may be involved in the restricted growth of other paramyxoviruses in cell cultures [117, 165] or in nature [42].

For studying single-cycle replicative events, the best performing paramyxoviruses are Sendai virus, SV5, and NDV. SV5 and Sendai virus appear to be more capable than most paramyxoviruses of growing productively in cells from a variety of species. SV5 grows best in early passage monkey kidney cells [40]. Sendai virus grows readily in embryonated eggs, primary or secondary cultures of avian cells, and a variety of vertebrate tissue cells of the epithelial type [57, 223, 226, 240, 253]. Its growth in continuous cell lines is restricted to a single cycle, for the reason given above. NDV grows well in embryonated eggs, less well in avian cell cultures, and poorly in mammalian cells [87, 162, 261, 262].

Under optimal conditions, the above viruses may be produced in amounts approaching 10^3 particles per infected cell, and many of the biochemical events relevant to virus replication proceed at rates sufficiently rapid to permit their study in monolayer cultures with only a little less convenience than with VSV grown in suspended cell cultures. However, when large amounts of material are needed, working with monolayer cultures is much less convenient.

NDV is the fastest growing paramyxovirus, with an apparent latent period of about 3 hr, the maximum rate of virus synthesis being reached by 6-8 hr and cell death ensuing between 12 and 24 hr [208]. Sendai virus and SV5 have longer latent periods, up to 24 hr, but virus is synthesized for longer periods, sometimes indefinitely, without cell death [17, 40, 57]. Maximum rates of virus production are reached by 48-72 hr after infection. Rates of virus-specific RNA and protein synthesis parallel the rates of virus synthesis in each system [17, 21].

IV. The Virion

We will discuss the virion from the inside out, starting with the RNA, proceeding through the nucleocapsid, and ending with the viral envelope. Sendai virus will be the principal model (Figs. 7.1 and 7.2).

A. The Paramyxovirus Genome

The paramyxovirus genome is represented by a single species of rapidly sedimenting single-stranded RNA. Sedimentation coefficients ranging from about 40 to 60 S have been obtained with various paramyxoviruses by different workers [11, 17, 52, 61, 63, 82, 130, 173, 216, 231]. The sedimentation coefficient of a single-stranded RNA molecule depends on its environment, however, and whenever the RNAs from 2 different viruses have been examined under the same conditions in the same laboratory, their sedimentation coefficients have been identical, or nearly so. Similar sedimentation coefficients

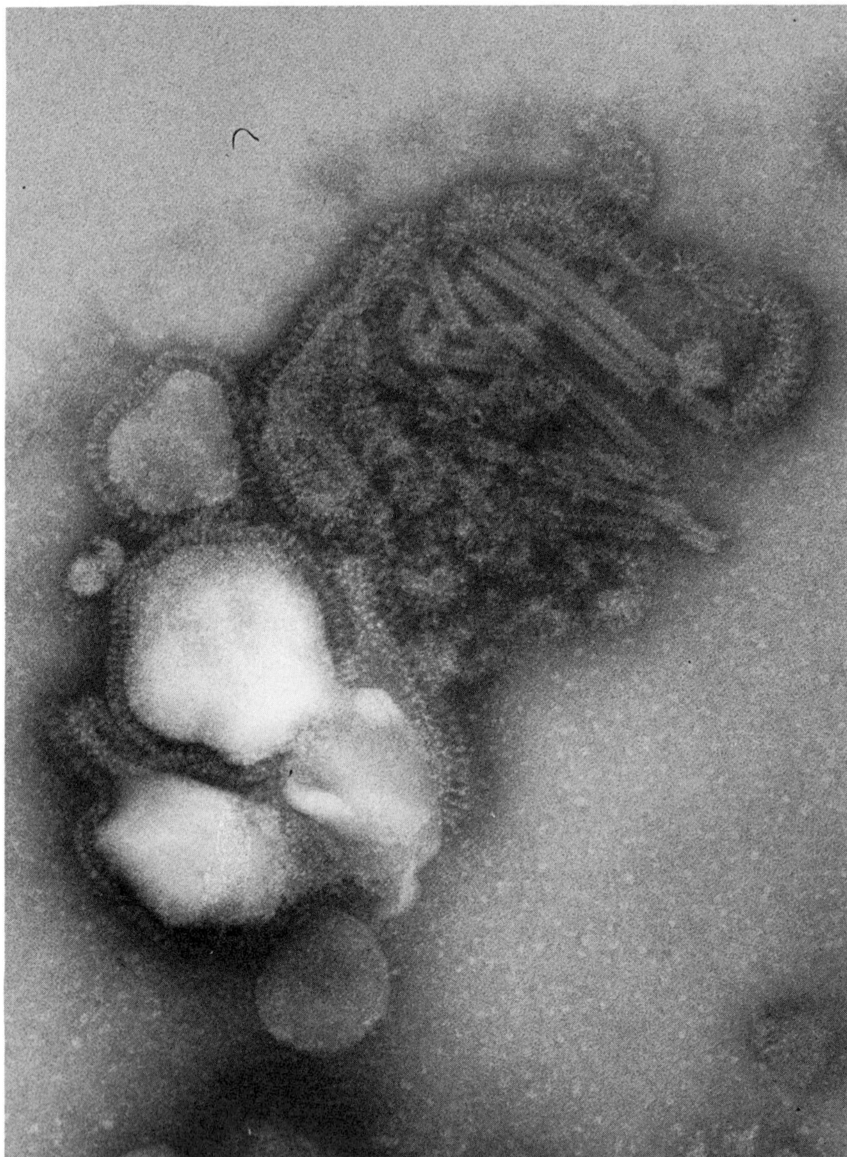

Fig. 7.1 Electron micrograph (X300,000) of several Sendai virus particles stained with phosphotungstate. One of the virus particles is disrupted, revealing broken segments of the nuclecapsid. (Courtesy of the Virus Laboratory, University of California, Berkeley.)

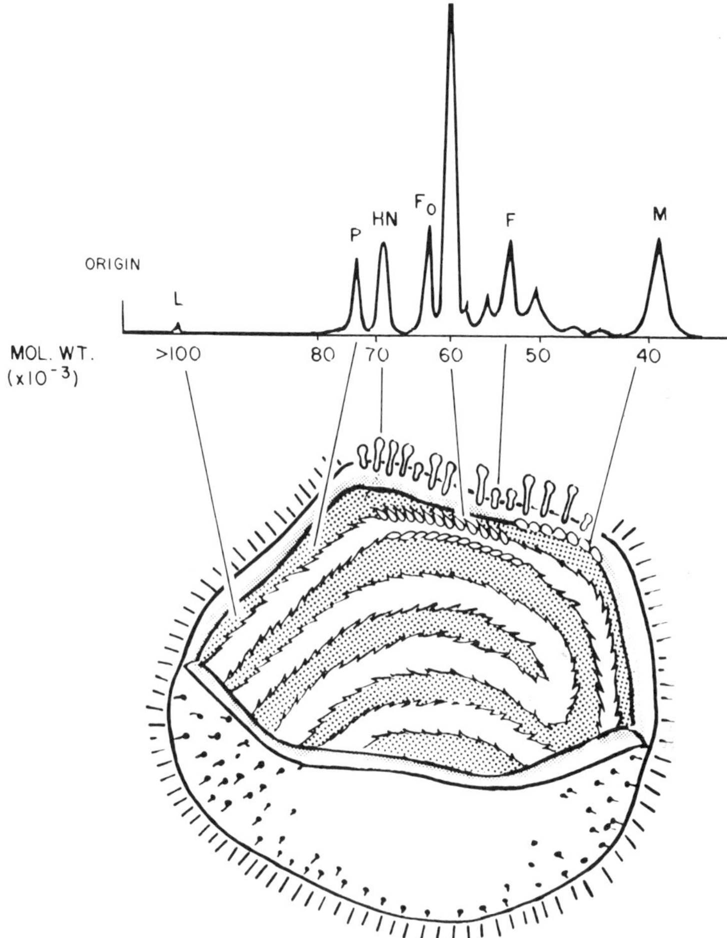

Fig. 7.2 Relationship between Sendai virion polypeptides and virion components. The upper diagram represents the pattern of polypeptides resolved by polyacrylamide gel electrophoresis in a discontinuous, SDS-containing system [148, 191]. The designations are according to Scheid and Choppin [215]. The lower diagram represents a virus particle from which a segment of envelope has been removed, giving a cross-sectional view of the envelope and revealing the nucleocapsid within (not to scale).

suggest similar molecular weights, so all paramyxoviruses probably have about the same amount of genetic information. The best available current estimate of the molecular weight of a paramyxovirus genomic RNA, that of Sendai virus, is 5×10^6, based on electron microscopic contour length measurements [143, 144].

The contour length method is especially useful, because paramyxovirus genomic RNA appears to have considerably more intramolecular base pairing (secondary structure) in solution than the "average" RNA molecule. This hampers deduction of its molecular weight from transport data. Thus, the RNA of Sendai virus sediments more rapidly in aqueous salt solutions than expected from its molecular weight but more slowly than expected in organic solvents [142, 143]; its electrophoretic behavior in formamide is also anomalous [62]. As will be seen in Section 7.V, this potential for intramolecular base pairing is probably never realized at any point in the life cycle of the virus because of the structure of the viral nucleocapsid.

A single-stranded RNA molecule with a molecular weight of 5×10^6 can specify about 5×10^5 daltons of protein. In the case of Sendai virus, about 80% of this coding potential can be accounted for in the structural proteins of the virion (Fig. 7.2). Lamb and Mahy [148] have observed several new polypeptides in cells infected by Sendai virus. These may be nonstructural viral polypeptides. Their aggregate molecular weights would account for the remainder of the genetic information in the virus [148].

B. Positive and Negative RNA Molecules in Virus Particles

An unusual feature of paramyxoviruses is the frequent occurrence of full length (50 S) *positive-strand* RNA in virus particles. The existence of such material was first indicated when certain preparations of NDV or Sendai virus RNA were self-annealed at high concentrations. Marked conversion of the RNA to ribonuclease-resistant (double-stranded) form was observed [192, 199]. Careful work was required to distinguish whether this self-hybridization was intramolecular, between self-complementary portions of a single RNA molecule, or intermolecular, between different RNA molecules. Kolakofsky et al. [144] were able to demonstrate that the hybridization was intermolecular, establishing the existence of genome sized positive strands in Sendai virus particles. The reannealing kinetics of an equimolar mixture of positive and negative 50 S RNA species revealed no anomalies; a single smooth curve was obtained that had properties consistent with the 5×10^6 molecular weight estimate [144]. Thus, it appears that there is no significant redundancy in the Sendai virus genome, and that each 50 S positive strand represents the complete base sequence of the viral genome in complementary form.

The genome sized RNA produced in infected cells is likewise a mixture of positive and negative strands [19, 145]. The positive strands are evidently assembled into virions because they are encapsidated and there is little, if any, discrimination between positive and negative nucleocapsids in virus assembly [145]. It is not possible to separate virus particles containing positive-strand RNA from those which contain the negative-strand genomes, so there is no way of testing whether positive strands are capable of initiating any event in infection by themselves. A priori, it may be argued that they are inert, since they apparently do not function as messenger RNA [134]. Attempts at physical separation of the 2 kinds of 50 S RNA have failed [200], but they can be separated by other means. Advantage can be taken of the fact that positive strands are always in the minority, usually about 30% or less of the total [144, 145, 192, 199]. Self-annealing the mixture at high concentration will convert all of the minority positive strands to double-stranded complexes with an equivalent number of the majority negative strands. The remaining free negative strands can be separated from the complexes by cellulose chromatography [73]. Isolation of pure positive strands is more complicated, but it can be approached using the procedures devised by Roy and Bishop [207] for isolating the complementary (in this case, negative) RNA strand occurring in poliovirus replicative forms. Purified positive and negative strands should be useful hybridization probes for exploring the structure of the paramyxovirus genome, messenger RNA synthesis, and viral RNA replication.

C. The Nucleocapsid

1. Basic Properties

The nucleocapsid is the ribonucleoprotein core of the virus, comprising a single piece of 50 S RNA associated with many protein subunits in a rod-shaped structure with helical symmetry. The architecture of the paramyxovirus nucleocapsid is similar to that of the rod-shaped plant viruses, of which tobacco mosaic virus (TMV) is the best studied example. There are, however, important differences in dimensions and protein composition. Properties of both particles are compared in Table 7.2 The most obvious differences between TMV and paramyxovirus nucleocapsids are in the lengths of the particles (0.3 vs. 1.0 μm), in keeping with the molecular weights of the viral RNAs within them (2×10^6 vs. 5×10^6) and in the molecular weights of the protein subunits (17,400 vs. 60,000). The dimensions, conformations and chemical reactivities of the protein subunits are the principal determinants of the physical properties of the particles [36].

A particularly important property is helix flexibility. TMV is a rigid rod, the adjacent helical turns of which are tightly held together [36, 140]. The

Table 7.2 Properties of Tobacco Mosaic Virus (TMV) and of Paramyxovirus Nucleocapsids

Property	Quantity		
	TMV	Paramyxovirus	References[a]
RNA molecular weight	2×10^6	5×10^6	143
Percent RNA	5.1	4-5	32, 50, 98, 132
Particle molecular weight	39×10^6	$100\text{-}125 \times 10^6$	b
Length of particle	300 nm	1000 nm	51, 70
Diameter (mean) of particle	15 nm	17 nm	95
Diameter of hollow core	4 nm	5-12 nm	4, 70
Periodicity of helix	2.3 nm	5 nm	4, 95
Number of helical turns	130	200-220	99, 172
Hand of the helix	Right	Left	54, 178
Protein subunit molecular weight	17,400	60,000	68, 85, 168
Number of subunits per particle	2130	1600-2000[c]	b
Number of subunits per helical turn	16.3	9-13[c]	70, b
Number of RNA nucleotides per particle	6400	16,000	b
Number of nucleotides per protein subunit	3	8-10	b

[a] All data for TMV from Ref. [140], except the hand of the helix [71].
[b] Values calculated by D. W. Kingsbury.
[c] Finch and Gibbs [70] give 2400 or 2800 subunits per particle, based on 11 or 13 subunits per helical turn.

attractive forces between adjacent turns of paramyxovirus nucleocapsids seem to be considerably weaker. Finch and Gibbs [70] have remarked on the need for flexibility in paramyxovirus nucleocapsids because these 1 μm rods are fitted into virus particles almost an order of magnitude smaller in diameter. These authors have also deduced that the radial axes of the protein subunits are not perpendicular to the long axis of the rod; they all form an angle of about 60° to the long axis, an arrangement which may facilitate the sliding of adjacent helical turns beneath each other in the process of bending, like scales on a fish. This orientation of the protein subunits confers an obvious polarity on the nucleocapsid. Finch and Gibbs compared the electron microscopic appearance to "a series of interlocking arrow-heads pointing to one end of the nucleocapsid" [70]. No doubt the RNA within has a defined polarity, but no information is available on whether it is the 5′ or 3′ end that is to be found in the head of the "arrow."

Another manifestation of looseness between helical turns is the frequent appearance of stretched nucleocapsids in electron micrographs [39, 172, 247]. We shall see later that this stretchability of paramyxovirus nucleocapsids may permit the RNA to have a template function without being detached from the proteins of the helix.

2. *Functions of the Polypeptides*

TMV contains a single species of protein subunit and no enzyme activities. An RNA-dependent RNA polymerase is not needed, since TMV is a positive-strand virus [64, 198]. Paramyxovirus nucleocapsids are more complex, however, containing, in addition to the RNA and the major structural protein (NP), at least 1 enzyme, an RNA-dependent RNA polymerase (transcriptase) and probably an additional (nonenzymatic) polypeptide. The role of the transcriptase in virus replication is treated in section 7.V.A.1. Current information and speculations about the functions of nucleocapsid polypeptides are dealt with here.

The major role of NP has already been discussed: It is the matrix of the nucleocapsid. Little more is known about its properties beyond the following. If nucleocapsids are treated with trypsin, NP can be cleaved almost quantitatively to a 40,000 molecular weight fragment which remains associated with the RNA; the remaining 20,000 daltons of NP is reduced to low molecular weight fragments [170]. The latter are thought to represent the external surface of the nucleocapsid and to contribute a hydrophobic character to its surface, since intact nucleocapsids tend to aggregate and hydrophobic interactions may be involved in the recognition of the nascent virus envelope by nucleocapsid in virus maturation [170]. What remains of the nucleocapsid after trypsin treatment is a more rigid, more hydrophilic rod with less tendency to aggregate [98, 170, 255]. Thus, there is a topographical distribution of internal hydrophilic and external hydrophobic domains in each NP polypeptide which is reflected in the nucleocapsid overall.

A second polypeptide, 75,000 in molecular weight, has been found associated with Sendai virus nucleocapsids isolated either from virus particles or from infected cells [156, 236, 265]. Nucleocapsids from cells contain about 1/5 as many molecules of this polypeptide as of NP, but virion nucleocapsids contain fewer, about half as many. No explanation for this difference is readily apparent, reflecting our ignorance of the function of the 75,000 molecular weight polypeptide. It has, however, been designated P (for polymerase) because nucleocapsids which contain it are active in RNA synthesis and they become inactive when it is removed [215, 234, 235].

A polypeptide like P has not been identified in nucleocapsids isolated from paramyxoviruses other than Sendai virus but, generally, sodium deoxycholate or concentrated cesium chloride has been used in the isolation of these nucleocapsids [15, 80, 202], and these reagents remove P from Sendai virus nucleocapsids [55, 168, 234, 236]. Nevertheless, when more gentle procedures have been used, a 75,000 molecular weight polypeptide is not apparent in polyacrylamide gel analyses of nucleocapsids from Newcastle disease virus or SV5. Another polypeptide of lower molecular weight, migrating between NP and the 40,000 molecular weight polypeptide M, can be seen, though, and this might represent an analog of Sendai virus polypeptide P [213, 214].

Recently, improved techniques of polyacrylamide gel electrophoresis have revealed a very high molecular weight polypeptide in nucleocapsids from Sendai virus and Newcastle disease virus (NDV). It is present in minute amounts and it migrates in gels with an apparent molecular weight of more than 100,000, so it has been designated L for large [48, 89, 148]. Its high molecular weight and low abundance make it a more attractive candidate for the RNA polymerase than the superabundant P [236]. Therefore, L may turn out to be a fortunate designation, since this is the label which has been applied to a comparably sized minor protein in VSV nucleocapsids which is presently the best candidate for the RNA polymerase of that virus [65, 66].

It may be instructive to push the analogy between Sendai virus and VSV even further. The so-called NS polypeptide of VSV, about 40,000 in molecular weight, might be an analog of P. It has been shown recently that NS is needed for in vitro VSV RNA transcription by encapsidated virion RNA, together with polypeptide L [66, 115]. There is a large pool of NS in infected cells and it can be removed from VSV nucleocapsids with relative ease, suggesting a superficial location in the nucleocapsids [66, 250]. Likewise, there appears to be an excess of the Sendai virus P in infected cells [236] and its removal by deoxycholate without disturbing the rest of the nucleocapsid suggests a similar surface localization. Moreover, A. Portner has recently learned that newly synthesized P exchanges with P already attached to nucleocapsids within infected cells, whereas NP forms a stable association with the viral RNA [191]. Additionally, the Sendai virus P and the NS polypeptide of VSV are phosphopolypeptides [147, 232]. All of this suggests the possibility that P and NS have regulatory roles in RNA synthesis by the respective viruses [66].

D. The Envelope

The paramyxovirus envelope is a modified segment of cell surface membrane comprising 2 major classes of chemical constituents: lipids and proteins. The composition of the virion lipids is essentially identical to that of surface membranes from uninfected cells [19, 136, 137, 242], but the proteins are

acid should cause aggregation of influenza viruses and certain paramyxoviruses since other paramyxoviruses and enveloped viruses, such as rhabdoviruses and arboviruses, manage well without neuraminidase and have no tendency to aggregate despite the retention of sialic acid in their glycoproteins [31, 67, 238].

2. *Glycopolypeptide F*

A second prominent glycopolypeptide in envelopes of SV5, NDV, and Sendai virus has been designated F for "fusion factor." It is smaller than HN, migrating in polyacrylamide gels with an apparent molecular weight of about 50,000. There is a strong case for its participation in virus entry into host cells, as well as in virus-mediated cell fusion and hemolysis, which are somewhat artificial manifestations of the same activity [3, 6, 22, 23, 45, 77, 78, 103, 127, 163, 164, 176, 182, 252]. This interesting story begins with the observation that Sendai virus grown in L-cells or other mammalian cell types in culture was noninfectious for these same mammalian cells, despite respectable hemagglutination titers, but it was infectious for embryonated eggs [116, 158]. Homma [92] reasoned that a product of mammalian cells masked a virus function required for infectivity. This cellular function appeared to be protein, because low concentrations of trypsin conferred infectivity as well as fusing and hemolytic activities on the virus preparations [92, 94]. Later work showed that the mammalian cells were not adding a proteinaceous blocking substance but were deficient in a proteolytic enzyme function necessary to make active virus. Polyacrylamide gel electrophoresis revealed that glycopolypeptide F was absent in Sendai virus grown in mammalian cells. Instead, these virus particles contained another, larger glycopolypeptide, migrating between HN and NP, with an apparent molecular weight of about 65,000 [93, 215]. Trypsin treatment converted this glycopolypeptide to F at the same time that it conferred infectivity on the virus particles. In recognition of this precursor-product relationship, the larger glycopolypeptide has been designated F_0 [215]. The embryonated egg apparently supplies the proteolytic function required for conversion of F_0 to F, accounting for the absence of F_0 in egg-grown virions and for the activation of infectivity when virus grown in mammalian cells is inoculated into eggs. Choppin et al. [42] have generated some F_0 cleavage mutants of Sendai virus, selected for ability to form plaques in mammalian cells in the absence of trypsin but in the presence of proteolytic enzymes with different specificities. Mutants which require chymotrypsin for conversion of F_0 to F will not grow in embryonated eggs unless small amounts of chymotrypsin are provided, and polyacrylamide gel patterns reveal that polypeptide F_0 of these mutants is cleaved to F by chymotrypsin but not by trypsin [42].

It is important to note that the cleavage of F_0 is not an intracellular event; it occurs only after F_0 has been inserted into the viral envelope. Thus, the full activation of virus infectivity depends on an environmental agency, a protease with trypsin-like activity. This peculiar mechanism no doubt has survival value, although precisely how is not immediately apparent.

Less information is available on protease-dependent processing of a surface glycoprotein analogous to F among other paramyxoviruses, but this type of phenomenon is probably the basis for the enhancement of infectivity of human parainfluenza virus types 1 and 4 by trypsin [117, 165], and kinetic studies of NDV polypeptide synthesis in avian cells indicate that a similar processing event takes place in that system [79, 211, 212].

Aside from the cleavage involved in the formation of F, and contrasting with what is obtained for many other animal viruses, there is no evidence for any other precursor-product relationship in the formation of paramyxovirus proteins [89, 211, 212]. This is consistent with the idea that each polypeptide is specified by a separate monocistronic messenger RNA [46, 134].

Although F is an essential ingredient in the fusion capability of Sendai virus, it is not sufficient to function as a fusion factor unaided. Hosaka and co-workers, in a series of ingenious biochemical reconstruction experiments, have shown that fusion requires membrane phospholipids and polypeptide HN in addition to F [100-102, 224]. Both of the requirements can be rationalized: phospholipids, in the form of a lipid bilayer, constitute a matrix which can dissolve into the lipids of the target cell membrane [81, 86, 152], and HN is required for initial attachment of the fusing complex to the target [102, 220]. In the same light, the nonpermissive particles of Sendai virus *ts* mutant 271 (Section 7.IV.D.1.) contain F, but they are incapable of infecting, fusing or hemolyzing cells, because HN, which mediates virus-cell attachment is absent [196].

3. *Polypeptide M*

Another major constituent of paramyxovirus envelopes is the nonglycosylated polypeptide M, the membrane or "matrix" protein. This has a molecular weight of 40,000. It is released from the envelope by nonionic detergents, as the glycopolypeptide-containing spikes are, but it is not soluble in aqueous solution unless high concentrations of salt are present [213, 214]. Several lines of evidence indicate that it is internal to the envelope and that it may mediate recognition between viral nuclecapsids and nascent envelope during virus assembly [38, 170].

V. Virus Replication

The major events in paramyxovirus replication are indicated diagrammatically in Fig. 7.3. They all seem to occur in the cytoplasm, independent of any concurrent nuclear functions [9, 10, 21, 41, 129, 131, 204, 227, 259], although there are some contrary opinions [26, 27, 35, 217]. In summary, these events are:

1. *Entry.* Release of the viral nucleocapsid into the cell cytoplasm as the virus envelope fuses with the cell surface membrane.
2. *Transcription.* A viral RNA-dependent RNA polymerase, brought in as part of the nucleocapsid, synthesizes viral messenger RNA, using the virion RNA as template.
3. *Translation.* Viral messenger RNA specifies viral polypeptides in polyribosomes.
4. *Replication.* Viral proteins direct the infecting viral genome to function as template in the production of genome replicas.

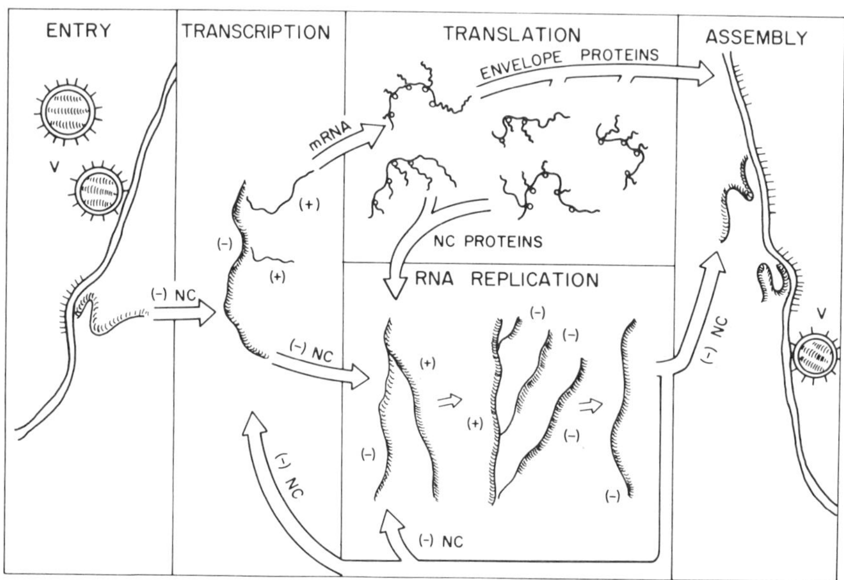

Fig. 7.3 A scheme of paramyxovirus replication. V, virus particle; NC, nucleocapsid; (−), negative strand RNA (the viral genome); (+), RNA complementary to the viral genome, either smaller and unencapsidated (the RNA made by transcription), or genome sized and encapsidated (the intermediate template in RNA replication). For further details, see the text.

5. *Secondary rounds of transcription, translation, and genome replication.* The genomes produced in step 4 are an amplification of viral genetic information in the infected cell. They recruit more of the cell's resources to participate in the sequence of steps 2, 3, and 4. Successive rounds of amplification continue until checked by cellular or viral controls.
6. *Assembly.* Viral macromolecules enter progeny virions as soon as sufficient concentrations have accumulated. Virus production continues as long as the cell provides the needed resources.

Two kinds of macromolecules must be produced under the direction of the viral genome to make progeny paramyxovirus particles: viral polypeptides and genomic RNA. The rates of production of these macromolecules are regulated: they are made in proportions which resemble their representation in virus particles [148, 236]. The following discussion will deal first with the production of virus polypeptides and regulation of their abundances, with emphasis on transcriptional events which generate the messenger RNAs specifying these polypeptides. We will then review what little is known about events in viral genome replication and try to find clues as to how genome replication is controlled.

A. Transcription, Messenger RNA and Viral Polypeptide Synthesis

1. *Virion Transcriptase and Primary Transcription*

The initial biosynthetic event which must take place upon entry of the paramyxovirus genome into the infected cell is transcription. Because the base sequence of the genomic RNA is of the opposite polarity to viral messenger RNA, it cannot be translated into a meaningful polypeptide; it can only be used as a template which specifies messenger RNA molecules of complementary base sequence by the action of an RNA-dependent RNA polymerase (transcriptase). So, it is easy to understand why naked paramyxovirion RNA has never been shown to be infectious [130]: The transcriptase must also be supplied. Since the cytoplasm of a eukaryotic cell evidently contains no such enzyme, it must be supplied in fully functional form by the infecting virus. Paramyxovirus particles do contain such an enzyme, which can be demonstrated in vitro by disrupting the virus envelope with nonionic detergents and supplying suitable concentrations of a divalent cation, monovalent ions, and nucleoside triphosphates [14, 107, 111, 202, 235]. So far, NDV has been found to contain the most active virion-associated transcriptase, but even this is far less potent than the transcriptase found in VSV, so that in early experiments the small amount of NDV activity was overlooked [8, 107]. Nevertheless, sufficient products can be made in vitro to permit further analysis and, under optimal conditions, paramyxovirion transcriptase produces messenger RNAs of a size and relative abundance similar to those which appear in infected cells [258]. Furthermore, the in vitro transcripts contain polyadenylate and blocked methylated 5′ termini [47, 258], as viral and cellular messenger RNAs do [2, 157, 171, 197, 257]. These findings are encouraging, because they indicate that virus particles contain all the ingredients necessary to manifest controlled transcription of biologically meaningful messenger RNA, which may simplify the task of isolating these ingredients and analyzing the mechanisms involved in their functioning.

2. *Lack of a Temporal Transcriptional Program*

Another simplification is indicated by evidence that there is no temporal program of viral genome expression. The polypeptides made early in infection and the polypeptides made late, after viral genome replication is well under way, are identical in kind and relative proportion [89, 148]. Of course, quantitatively more viral polypeptides are made late, but they are not qualitatively different in kinds or relative abundances from those which first become detectable. Likewise, the messenger RNAs which are transcribed from the infecting genomes are made in the same proportions relative to each other as the viral messengers made later, presumably from progeny genomes (secondary transcription), although again, the messenger RNAs made late are far more abundant overall [46]. Thus, anything learned about virus-specific regulation of transcription or translation at any time after infection should be relevant to the entire infectious process.

3. *The Nucleocapsid as Transcription Template*

The template for transcription at all times appears to be encapsidated 50 S RNA and not naked RNA free of its protein coat. The nucleocapsid has been the irreducible minimum structure obtainable from virus particles that has transcriptase activity [48, 156]; transcriptase activity from infected cells likewise resides only in nucleocapsids [28, 30, 155, 218, 236].

Isolated paramyxovirus nucleocapsids are insensitive to ribonuclease; the RNA within them is protected from attack [132]. How then, can encapsidated RNA act as a template for RNA polymerase? The answer probably lies in a feature of helical nucleocapsid symmetry mentioned earlier, in the discussion of the forces between adjacent helical turns (Section 7.IV.C.1.). An analog of a helical nucleocapsid is a spring, and it is fundamental to a spring that it can be stretched or compressed. As usually visualized in the electron microscope, paramyxovirus nucleocapsids are in a compressed state, with adjacent helical turns closely juxtaposed. However, the stretched forms which are occasionally seen provide evidence that the forces holding adjacent turns together are readily overcome. Assuming that the RNA helix is situated in relation to the protein helix as the RNA of TMV is, between helical turns, stretching of the helix will expose the RNA to the environment. However, the phosphate residues of the RNA will remain bound to protein subunits which provide counterbalancing positive charges. This means that a stretched nucleocapsid helix will become a sinuous ribbon of protein with a ribbon of RNA attached (Fig. 7.4). We can conceive that the RNA bases will be exposed and accessible for recognition by polymerase when the helix is in this extended form. Viewed in this light, encapsidation of the template might be advantageous. It is possible that the helical nucleocapsid structure constrains the template to a conformation which is compatible with recognition by the transcriptase, by eliminating the possibility of intramolecular base pairing.

The polarity of paramyxovirus nucleocapsids which derives from the orientation of the NP subunits has been pointed out (Section 7.IV.C.1.); another manifestation of polarity to be anticipated is the association of the RNA with only one face of each and every NP subunit along the helix axis (Fig. 7.4).

Helix stretching need not occur throughout the entire nucleocapsid; it might be local, in the vicinity of the advancing polymerase, and the helix might snap back to the closed conformation downstream from the advancing polymerase. The energy which drives the helix apart might come from the polymerization of the RNA.

The above model provides a mechanism for utilization of the template RNA without covalent bond breakage or removal of nucleocapsid proteins, relying solely on conformational alterations in the nucleocapsid. A different view has been expounded by Bukrinskaya and co-workers [29, 30, 264]. They propose that nucleocapsid template activity depends on conformational changes secondary to proteolytic removal of portions of the proteins. This is reminiscent of the Rockefeller group's findings that nucleocapsid protein is cleaved by added protease [170]. There is a discrepancy, however, because the American workers find that nucleocapsids with cleaved proteins are more rigid and compact, whereas the Russian workers have observed that cleavage relaxes the helix [30].

4. *Messenger RNA Species*

Transcription produces a spectrum of messenger RNA molecules which are complementary in base sequences to the genomic RNA template. Early

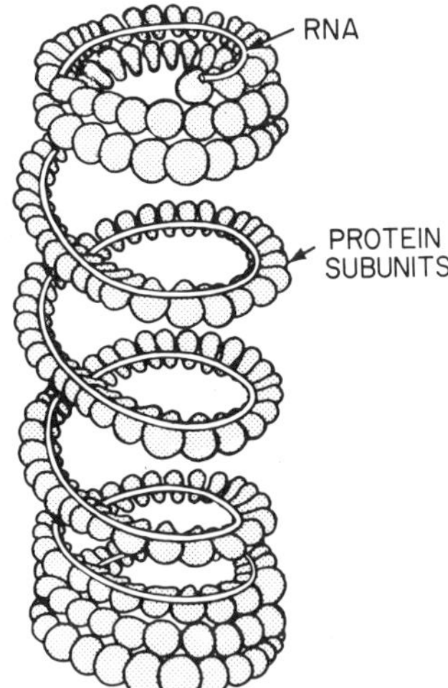

Fig. 7.4 Local stretching of a helical nucleocapsid exposes the RNA within. (Drawing by Roger Burns.)

studies indicated that there were 3 sedimenting classes of NDV- or Sendai virus-specific RNA molecules which were single stranded and complementary to the viral genome and therefore candidate messenger RNAs [11, 17, 21]. The most abundant class, representing about 60% of the genome, includes several species which sediment closely together as a group with a peak distribution at about 18 S [21, 46]. RNA molecules in this group from Sendai virus-infected cells have been translated in vitro, yielding P, NP, and small amounts of material which may represent M [134]. It is not clear whether any glycopolypeptides were made in vitro; neither HN nor F_0 was detected, but there was some material migrating in the position of F. The latter may have been a non-glycosylated polypeptide; polyacrylamide gel profiles of Sendai virion polypeptides indicate more than 1 component in this region of the gel (Fig. 7.2). The identifiable polypeptides made in vitro were synthesized in proportions resembling the proportions made in infected cells, indicating that the mechanisms which regulate polypeptide abundances are faithfully duplicated in vitro [134]. The major contributing factor appears to be the relative abundances of the messenger RNAs for each polypeptide species: when 18 S RNA from NDV-infected cells is separated by polyacrylamide gel electrophoresis several RNA species are resolved and these differ in relative abundance in a pattern that is analogous to the relative abundances of the polypeptides they are presumed to specify [46, 126]. Direct tests of this idea are currently underway. Each of the RNA species separated by polyacrylamide gel electrophoresis is being tested individually in a cell-free protein-synthesizing system to determine which polypeptide it codes for [24].

The 2 other less abundant classes of complementary RNA which have been described are those observed at 22 S and 35 S [17, 21]. Hybridization tests indicate that the 22 S class contains base sequences already represented in the 18 S class; the 22 S RNA appears to represent a conformational variant of 18 S RNA [206]. In contrast, the 35 S RNA may represent a distinct messenger RNA species, perhaps the one which codes for the very large nucleocapsid-associated minor polypeptide, L. The 35 S RNA hybridizes with base sequences in the Sendai virus genome which are distinct from those complementary to the 18 S (and 22 S) RNA populations [24, 206]. Since the sequences complementary to 35 S RNA represent about 40% of the Sendai virus genome, it appears that the coding potential of essentially all of the genome has been accounted for [24, 206].

If the above preliminary indications are correct, another parallel between the paramyxoviruses and rhabdoviruses would lie in the organization of their genomes. VSV produces a major class of messenger RNAs sedimenting at about 13-15 S; these code for all of the viral polypeptides except the putative polymerase, L, which is specified by a distinct 28 S RNA species [20, 141, 167]. For paramyxoviruses, the 18-22 S RNA may represent messages for all proteins except L, with the messenger for the latter being the 35 S RNA.

5. *Regulation of Polypeptide Abundances*

The evidence discussed above indicates that paramyxovirus messenger RNA abundances are a major determinant of viral polypeptide abundances. There are 2 ways by which the relative abundances of the viral messenger RNAs themselves might be controlled. Each messenger RNA might be independently initiated and terminated by transcriptase, differences in the abundance of each RNA being determined by differential affinities of the transcriptase for each promoter (initiation) sequence. Alternatively, the whole genome might be transcribed as a single complementary strand (polytranscript), followed by cleavages and selective degradations of those messenger RNAs destined to be less abundant. Although the latter model has not been formally ruled out, it is unattractive for several reasons. First, as will be discussed below (Section 7.V.B.), although genome sized complementary (positive) strands do appear in infected cells and in virions, they exist only in an encapsidated form. The polytranscript model requires a mechanism for discriminating between genome sized positive strands destined for encapsidation (and a template function in RNA replication) and those which must be cleaved and selectively degraded. Second, transcription in vitro with virion RNA polymerase supplies complementary RNA molecules which are messenger size [258]. There is no evidence of a large intermediate or of a nuclease activity in virus particles which might cleave it [234, 237]. Third, short pulses of [^{3}H]-uridine labeling in infected cells always yield 18 S RNA as the first detectable product and 50 S RNA (or even 35 S RNA) requires longer labeling periods [191]. Fourth, in the presence of cycloheximide, 50 S RNA synthesis declines much more rapidly than 18 S RNA synthesis [63, 193, 201].

The existence of a transcriptional regulation of viral polypeptide abundances does not preclude regulatory modulation at the translational level. Just as a single transcriptase might have different affinities for the promoter of each message, the frequency with which the message for each polypeptide is translated can be a function of the affinity of that message for the factors involved in polypeptide chain initiation [151].

For example, if Sendai virus 35 S RNA is the template for the L polypeptide, its translation would seem to be less efficient than that of other viral messages, since its abundance, compared to the other messenger RNA species, is greater than the abundance of L, compared to the other polypeptides [17, 21, 148]. Translational modulation might be seen as a means for "fine tuning" the production of viral polypeptides superimposed on a "coarse control" at the transcription level.

B. RNA Replication

Ater the infecting viral RNA has been transcribed and the viral messages have been translated into viral polypeptides, the next major event is viral genome replication. This sequence, genome replication following transcription and translation, is indicated by a requirement for new protein to be made if genome replication is to proceed [63, 124, 193, 201]. In contrast, transcription of the incoming viral genome occurs despite protein synthesis inhibition, because it is mediated by the transcriptase which enters with the infecting RNA [44, 201].

It is likely that at least 1 protein required for viral genome replication is virus specified. A mutant of Sendai virus, *ts*105, when shifted up to nonpermissive temperature, exhibits rapidly declining genome replication in the face of continued transcription, in a manner analogous to what is seen when inhibitors of protein synthesis are administered [191, 195]. Thus, the infecting viral genome is used initially for transcriptive template functions, until sufficient amounts of 1 or more viral proteins needed for RNA replication are synthesized. These gene products then interact with the genome, switching it from transcriptive to replicative functions. The required protein might be a new and different RNA polymerase, or "replicase," distinct from the transcriptase in structure and in its utilization of the viral genome template.

Alternatively, if the original transcriptase is the only kind of RNA polymerase involved in paramyxovirus RNA synthesis, the new protein may be a factor which modifies its activity, instructing it to perform the replicative function [186]. In either case, the first step in viral genome replication involves the production of a single, complete transcript of the genome, complementary in base sequence to it. The new or modified RNA polymerase must ignore the initiation and termination signals in the viral genome which specify individual messenger RNA molecules and proceed from end to end of the template without interruption. The product, a genome sized transcript (positive strand) provides the template on which replicas of the viral genome are produced [135].

It has been remarked earlier that the RNA template for transcription is encapsidated (Section 7.V.A.2.) and this also seems to be the case for all of the template RNA species involved in genome replication. By definition, all these RNA molecules are genome sized, and Blair and Robinson have shown that genome sized RNAs are never free but are always encapsidated in the infected cells [18, 201].

It appears that the paramyxoviruses and the rhabdoviruses have evolved a mode of RNA synthesis utilizing an encapsidated RNA template for both transcription and replication [135, 233]. The encapsidation process and the production of the protein required for its accomplishment [125] are thus revealed to be essential participants in viral RNA replication. If the replicase and transcriptase were identical and if NP, the structural protein, were the agent which modifies the action of the enzyme, an ingredient essential for the function of RNA replicas would be the signal which turns on their production. If it becomes possible to reconstitute paramyxovirus nucleocapsids from isolated proteins and RNA, this hypothesis can be tested. Nucleocapsids involved in transcription (the only RNA snythetic activity so far domonstrable in vitro) might synthesize virion-type RNA or at least the 50 S positive strand when capsid proteins are provided to the in vitro system under conditions leading to encapsidation of the product RNA [135].

The final product of paramyxovirus genome replication is an encapsidated 50 S negative strand. Such nucleocapsids have 3 possible fates: they may travel to the cell surface membrane where viral envelope proteins have been inserted and be incorporated into progeny virus particles; they may be utilized as templates for transcription; or they may be utilized as templates in replication of the genome RNA (Fig. 7.3). The availability of modified surface membrane sites probably controls the apportionment of nucleocapsids into virus particles [49]. The availability of capsid proteins may well determine which fraction of the remainder is involved in transcription or replication, according to the model advanced above.

VI. Concluding Remarks

The foregoing was intended to provide a conceptual framework for the reader who desires to be acquainted with the molecular biology of paramyxoviruses. Additional information and opinions can be found in some other recent reviews [12, 43, 53, 133].

Among the topics which I have not discussed, 2 are of special interest on their own and in relation to disease processes. These are defective-interfering virus particles and persistent infections. References [106] and [108] will be of help with the former. As to persistent infections, the literature is large and mainly descriptive; Refs. [40], [181], [203], [239], [241], and [263] are recommended as an introduction. One mechanism for producing a persistently infected state is for the genetic information of the virus to be integrated into the cellular genome in the form of DNA. This unorthodox idea has recently received some startling experimental support [75, 228, 266]. If these results can be verified, a new era of paramyxovirus molecular biology will have begun.

Recent Developments

Since June, 1975, when this chapter was written, there have been numerous important contributions to the literature on paramyxoviruses, but space limitations permit mention of only a few points.

The classification of the paramyxoviruses is under review by the Paramyxovirus Study Group of the International Committee on Taxonomy of Viruses (ICTV). The Study Group plans to endorse a recommendation of the ICTV Executive Committee, Madrid, 1975, to divide the family *Paramyxoviridae* into three genera: *Paramyxovirus* (agents that contain neuraminidase, including the parainfluenza viruses, NDV, SV 5, and Sendai virus), *morbillivirus* (measles, canine distemper and rinderpest viruses, and other agents with typical paramyxovirus morphology, but without neuraminidase), and *pneumovirus* (the morphologically distinct respiratory syncytial virus and pneumonia virus of mice).

Evidence for a precursor of the HN polypeptide of NDV has been obtained [267]. Designated HN_0, the precursor has an apparent molecular weight of 82,000 and is converted by mild proteolysis to the 74,000 dalton HN. Virulence of NDV strains appears to depend on how readily HN_0 of a strain is cleaved in the course of infection [267].

Transcriptional regulation of messenger RNA abundances does indeed seem to be the major determinant of virus polypeptide abundances. What's more, this regulation is accomplished in an elegant manner. The way ultraviolet (UV) radiation inactivates transcription can reveal how the transcriptive complex is organized [268, 269]. When applied to Sendai virus in infected cells, UV light inactivated the templates for each of the virus genes in an additive manner, indicating that the entire virus genome behaves as a single transcription unit, with only one promoter at its 3′ end [270]. With the aid of several assumptions, the data suggested the following order of genes in the Sendai virus genome: 3′-NP-F_0-M-P-HN-L-5′, corresponding to the order of rates of synthesis of the virus polypeptides [270, 271]. VSV behaves the same way [272, 273]. Although this discovery fits a polytranscript model of negative strand virus transcription [272-274], the contrary arguments in Section 7.V.A.5. still look cogent to me. Perhaps there are really several separate issues here. All transcription might start at one point, but it does not necessarily have to end at a single point at the other end of the template. Decreasing efficiencies of transcription could provide decreasing abundances of the various messenger RNA species as a function of distance from the single promoter. To produce individual monocistronic messages, nucleolytic cleavage may or may not be required, depending on whether the transcriptase links adjacent messages as it crosses the intercistronic divide [270].

Acknowledgment

This work was supported by Public Health Service Research Grant AI-05343 and by ALSAC.

References

1. F. R. Abinanti, R. M. Chanock, M. K. Cook, D. Wong, and M. Warfield, Proc. Soc. Exp. Biol. Med., 106, 466 (1961).
2. J. M. Adams and S. Cory, Nature, 255, 28 (1975).
3. W. R. Adams and A. M. Prince, J. Exp. Med., 106, 617 (1957).
4. Y. Amano, T. Takano, K. Takahashi, and N. Ishida, Jap. J. Microbiol., 15, 549 (1971).
5. T. Atoynatan and G. D. Hsiung, Am. J. Epidemiol. 89, 472 (1969).
6. T. Bächi, M. Aguet, and C. Howe, J. Virol., 11, 1004 (1973).
7. D. Baltimore, Bacteriol. Rev., 35, 235 (1971).
8. D. Baltimore, A. S. Huang, and M. Stampfer, Proc. Natl. Acad. Sci. U.S., 66, 572 (1970).
9. R. D. Barry, D. R. Ives, and J. G. Cruickshank, Nature, 194, 1139 (1962).
10. R. D. Barry, Virology, 24, 563 (1964).
11. R. D. Barry and A. G. Bukrinskaya, J. Gen. Virol., 2, 71 (1968).
12. B. W. J. Mahy and R. D. Barry, in Negative Strand Viruses (B. W. J. Mahy and R. D. Barry, eds.), Academic, New York, 1975, pp. xi-xii.

13. A. J. Beale, Proc. Roy. Soc. Med., 67, 1116 (1974).
14. J. P. Bernard and R. L. Northrop, J. Virol., 14, 183 (1974).
15. I. Bikel and P. H. Duesberg, J. Virol., 4, 388 (1969).
16. L. N. Binn, G. Eddy, E. C. Lazar, J. Holmes, and T. Murnane, Proc. Soc. Exp. Biol. Med., 126, 140 (1967).
17. C. D. Blair and W. S. Robinson, Virology, 35, 537 (1968).
18. C. D. Blair and W. S. Robinson, J. Virol., 5, 639 (1970).
19. H. A. Blough and D. E. M. Lawson, Virology, 36, 286 (1968).
20. G. W. Both, S. A. Moyer, and A. K. Banerjee, Proc. Natl. Acad. Sci. U.S., 72, 274 (1975).
21. M. A. Bratt and W. S. Robinson, J. Mol. Biol., 23, 1 (1967).
22. M. A. Bratt and W. R. Gallaher, Proc. Natl. Acad. Sci. U.S., 64, 536 (1969).
23. M. A. Bratt and L. A. Clavell, Appl. Microbiol., 23, 454 (1972).
24. T. G. Morrison, S. Weiss, L. Hightower, B. Spanier Collins, and M. A. Bratt, in In Vitro Transcription and Translation of Viral Genomes (A.-L. Haenni and G. Beaud, eds.), Editions Inserm, Paris, 1975, pp. 281-290.
25. P. Brown, F. Cathala, and D. C. Gajdusek, Proc. Soc. Exp. Biol. Med., 143, 828 (1973).
26. A. G. Bukrinskaya, O. Burducea, and G. K. Vorkunova, Proc. Soc. Exp. Biol. Med., 123, 236 (1966).
27. A. G. Bukrinskaya, V. M. Zhdanov, and G. K. Vorkunova, J. Virol., 4, 141 (1969).
28. A. G. Bukrinskaya, Virology, 52, 344 (1973).
29. A. G. Bukrinskaya, L. V. Agaphono, and T. S. Vovk, Biochim. Biophys. Acta, 312, 173 (1973).
30. A. G. Bukrinskaya, Adv. Virus Res., 18, 195 (1973).
31. B. W. Burge and A. S. Huang, J. Virol., 6, 176 (1970).
32. R. H. Bussell, D. J. Waters, M. K. Seals, and W. S. Robinson, Med. Microbiol. Immunol., 160, 105 (1974).
33. C. M. Calberg-Bacq, M. Reginster, and P. Rigo, Acta Virol., 11, 52 (1967).
34. K. Cantell, Adv. Virus Res., 8, 123 (1961).
35. C. Carter, F. L. Black, and A. Schluederberg, Biochem. Biophys. Res. Comm., 54, 411 (1973).
36. D. L. D. Caspar, Adv. Protein Chem., 18, 37 (1963).
37. R. M. Chanock and R. H. Parrott, in Viral and Rickettsial Infections of Man (F. L. Horsfall, Jr., and I. Tamm, eds.), Lippincott, Philadelphia, 1965, pp. 741-754.
38. C. Chen, R. W. Compans, and P. W. Choppin, J. Gen. Virol., 11, 53 (1971).
39. P. W. Choppin and W. Stoeckenius, Virology, 23, 195 (1964).
40. P. W. Choppin, Virology, 23, 224 (1964).
41. P. W. Choppin, Proc. Soc. Exp. Biol. Med., 120, 699 (1965).
42. P. W. Choppin, A. Scheid, and W. Mountcastle, Neurology, 25, 494 (1975).
43. P. W. Choppin and R. Compans, in Comprehensive Virology, Vol. 4

(H. Fraenkel-Conrat and R. R. Wagner, eds.), Plenum, New York, 1975, pp. 95-178.
44. L. A. Clavell and M. A. Bratt, J. Virol., 8, 500 (1971).
45. L. A. Clavell and M. A. Bratt, Appl. Microbiol., 23, 461 (1972).
46. B. S. Collins and M. A. Bratt, Proc. Natl. Acad. Sci. U.S., 70, 2544 (1973).
47. R. J. Colonno and H. O. Stone, Proc. Natl. Acad. Sci. U.S., 72, 2611 (1975).
48. R. J. Colonno and H. O. Stone, Abstr. Ann. Meet. Am. Soc. Microbiol., New York, 1975, p. 250.
49. R. W. Compans, K. V. Holmes, S. Dales, and P. W. Choppin, Virology, 30, 411 (1966).
50. R. W. Compans and P. W. Choppin, Proc. Natl. Acad. Sci. U.S., 57, 949 (1967).
51. R. W. Compans and P. W. Choppin, Virology, 33, 344 (1967).
52. R. W. Compans and P. W. Choppin, Virology, 35, 289 (1968).
53. R. W. Compans and P. W. Choppin, in Comparative Virology (K. Maramorosch and E. Kurstak, eds.), Academic, New York, 1971, pp. 407-432.
54. R. W. Compans, W. E. Mountcastle, and P. W. Choppin, J. Mol. Biol., 65, 167 (1972).
55. J. Content and P. H. Duesberg, J. Virol., 6, 707 (1970).
56. G. M. W. Cook, Biol. Rev., 43, 363 (1968).
57. R. W. Darlington, A. Portner, and D. W. Kingsbury, J. Gen. Virol., 9, 169 (1970).
58. A. D. Dayan, Proc. Roy. Soc. Med., 67, 1123 (1974).
59. J. Dichfield, A. Zbitnew, and L. W. Macpherson, Can. Vet. J., 4, 175 (1963).
60. Z. Dinter, S. Hermodsson, and L. Hermodsson, Virology, 22, 297 (1964).
61. P. H. Duesberg and W. S. Robinson, Proc. Natl. Acad. Sci. U.S., 54, 794 (1965).
62. P. H. Duesberg and P. K. Vogt, J. Virol., 12, 594 (1973).
63. J. L. East and D. W. Kingsbury, J. Virol., 8, 161 (1971).
64. D. Efron and A. Marcus, Virology, 53, 343 (1973).
65. S. U. Emerson and R. R. Wagner, J. Virol., 12, 1325 (1973).
66. S. U. Emerson and Y.-H. Yu, J. Virol., 15, 1348 (1975).
67. J. R. Etchison and J. J. Holland, Virology, 60, 217 (1974).
68. M. J. Evans and D. W. Kingsbury, Virology, 37, 597 (1969).
69. F. Fenner and D. O. White, Medical Virology, Academic, New York, 1970.
70. J. T. Finch and A. J. Gibbs, J. Gen. Virol., 6, 141 (1970).
71. J. T. Finch, J. Mol. Biol., 66, 291 (1972).
72. H. R. Fischman, Am. J. Epidemiol., 85, 272 (1967).
73. R. M. Franklin, Proc. Natl. Acad. Sci. U.S., 55, 1504 (1966).
74. H. Fukumi, F. Nishikawa, and T. A. Kitayama, Jap. J. Med. Sci. Biol., 7, 345 (1954).

75. P. A. Furman and J. V. Hallum, J. Virol. 12, 548 (1973).
76. W. R. Gallaher, D. B. Levitan, and H. A. Blough, Virology, 55, 193 (1973).
77. W. R. Gallaher and M. A. Bratt, J. Virol., 14, 813 (1974).
78. A. Granoff and W. Henle, J. Immunol. 72, 322 (1954).
79. M. Haire, K. B. Fraser, and J. H. D. Millar, Brit. Med. J., 3, 612 (1973).
80. W. W. Hall and S. J. Martin, J. Gen. Virol., 19, 175 (1973).
81. W. W. Hall and S. J. Martin, J. Gen. Virol., 22, 363 (1974).
82. B. Hammarskjöld and E. Norrby, Med. Microbiol. Immunol., 160, 99 (1974).
83. R. P. Hanson (ed.), Newcastle Disease Virus, University of Wisconsin Press, Madison, 1964.
84. D. H. Harter and P. W. Choppin, J. Exp. Med., 126, 267 (1967).
85. E. A. Haslam, I. M. Cheyne, and D. O. White, Virology, 39, 118 (1969).
86. A. M. Haywood, J. Mol. Biol., 87, 625 (1974).
87. T. T. Hecht and D. F. Summers, J. Virol., 14, 162 (1974).
88. W. Henle and J. F. Enders, in Viral and Rickettsial Infections of Man (F. L. Horsfall, Jr., and I. Tamm, eds.), Lippincott, Philadelphia, 1965, pp. 755-768.
89. L. E. Hightower and M. A. Bratt, J. Virol., 13, 788 (1974).
90. D. S. Hodes, T. J. Schnitzer, A. R. Kalica, E. Camargo, and R. M. Chanock, Virology, 63, 201 (1975).
91. E. J. Hoffman, E. C. Ford, and J. L. Gerin, Abst. Ann. Meet. Am. Soc. Microbiol., New York, 1975, p. 250.
92. M. Homma, J. Virol., 8, 619 (1971).
93. M. Homma and M. Ohuchi, J. Virol., 12, 1457 (1973).
94. M. Homma and S. Tamagawa, J. Gen. Virol., 19, 423 (1973).
95. R. W. Horne, A. P. Waterson, P. Wildy, and A. E. Farnham, Virology, 11, 79 (1960).
96. A. Hornsleth, Acta Pathol. Microbiol. Scand., 76, 637 (1969).
97. F. L. Horsfall, Jr. and R. G. Hahn, Proc. Soc. Exp. Biol. Med., 40, 684 (1939).
98. Y. Hosaka, Virology, 35, 445 (1968).
99. Y. Hosaka and K. Shimizu, J. Mol. Biol., 35, 369 (1968).
100. Y. Hosaka and K. Shimizu, Virology, 49, 627 (1972).
101. Y. Hosaka and Y. K. Shimizu, Virology, 49, 640 (1972).
102. Y. Hosaka, T. Semba, and K. Fukai, J. Gen. Virol., 25, 391 (1974).
103. C. Howe and C. Morgan, J. Virol., 3, 70 (1969).
104. C. Howe, S. A. Milliken, and E. W. Newcomb, Arch. Ges. Virusforsch., 29, 50 (1970).
105. G. D. Hsiung and T. Atoynatan, Am. J. Epidemiol., 83, 38 (1966).
106. A. S. Huang and D. Baltimore, Nature, 226, 325 (1970).
107. A. S. Huang, D. Baltimore, and M. A. Bratt, J. Virol., 7, 389 (1971).
108. A. S. Huang, Ann. Rev. Microbiol., 27, 101 (1973).
109. R. N. Hull, J. R. Minner, and J. W. Smith, Am. J. Hyg., 63, 204 (1956).
110. R. N. Hull, Virol. Monog., 2, 1 (1968).

111. J. E. Hutchinson and B. W. J. Mahy, Arch. Ges. Virusforsch., 37, 203 (1972).
112. M. Iinuma, T. Yoshida, Y. Nagai, K. Maeno, T. Matsumoto, and M. Hoshino, Virology, 46, 663 (1971).
113. D. T. Imagawa, P. Goret, and J. M. Adams, Proc. Natl. Acad. Sci. U.S., 46, 1119 (1960).
114. D. T. Imagawa, Prog. Med. Virol., 10, 160 (1968).
115. R. L. Imblum and R. R. Wagner, J. Virol., 15, 1357 (1975).
116. N. Ishida and M. Homma, Virology, 14, 486 (1961).
117. H. Itoh, Y. Morimoto, S. Okawa, Y. Doi, T. Sanpe, M. Nakajima, T. Katoh, and Y. Sogawa, Jap. J. Med. Sci. Biol., 24, 387 (1971).
118. G. G. Jackson and R. L. Muldoon, J. Infect. Dis., 128, 387 (1973).
119. G. G. Jackson and R. L. Muldoon, J. Infect. Dis., 128, 674 (1973).
120. J. Joncas, L. Berthiaume, and V. Pavilanis, Virology, 38, 493 (1969).
121. S. S. Kalter, J. Ratner, G. V. Kalter, A. R. Rodriguez, and C. S. Kim, Am. J. Epidemiol., 86, 552 (1967).
122. S. S. Kalter and R. L. Heberling, Bacteriol. Rev., 35, 310 (1971).
123. S. L. Katz and J. F. Enders, in Viral and Rickettsial Infections of Man (F. L. Horsfall, Jr., and I. Tamm, eds.), Lippincott, Philadelphia, 1965, pp. 784-801.
124. N. V. Kaverin and N. L. Varich, Arch. Ges. Virusforsch., 35, 378 (1971).
125. N. V. Kaverin, J. Gen. Virol., 17, 337 (1972).
126. N. V. Kaverin and N. L. Varich, J. Virol., 13, 253 (1974).
127. L. Kilham, Proc. Soc. Exp. Biol. Med., 71, 63 (1949).
128. H. W. Kim, J. O. Arrobio, C. D. Brandt, P. Wright, D. Hodes, R. M. Chanock, and R. H. Parrott, Pediatrics, 52, 56 (1973).
129. D. W. Kingsbury, Biochem. Biophys. Res. Comm., 9, 156 (1962).
130. D. W. Kingsbury, J. Mol. Biol., 18, 195 (1966).
131. D. W. Kingsbury, Virology, 33, 227 (1967).
132. D. W. Kingsbury and R. W. Darlington, J. Virol., 2, 248 (1968).
133. D. W. Kingsbury, Curr. Top. Microbiol. Immunol., 59, 1 (1972).
134. D. W. Kingsbury, J. Virol., 12, 1020 (1973).
135. D. W. Kingsbury, Med. Microbiol. Immunol., 160, 73 (1974).
136. H.-D. Klenk and P. W. Choppin, Virology, 38, 255 (1969).
137. H.-D. Klenk and P. W. Choppin, Virology, 40, 939 (1970).
138. H.-D. Klenk, L. A. Caliguiri, and P. W. Choppin, Virology, 42, 473 (1970).
139. H.-D. Klenk, R. W. Compans, and P. W. Choppin, Virology, 42, 1158 (1970).
140. A. Klug and D. L. D. Caspar, Adv. Virus Res., 7, 225 (1960).
141. D. Knipe, J. K. Rose, and H. F. Lodish, J. Virol., 15, 1004 (1975).
142. D. Kolakofsky and A. Bruschi, J. Virol., 11, 615 (1973).
143. D. Kolakofsky, E. Boy de la Tour, and H. Delius, J. Virol., 13, 261 (1974).
144. D. Kolakofsky, E. Boy de la Tour, and A. Bruschi, J. Virol., 14, 33 (1974).

145. D. Kolakofsky and A. Bruschi, Virology, 66, 185 (1975).
146. P. J. Lachmann, Proc. Roy. Soc. Med., 67, 1120 (1974).
147. R. A. Lamb, J. Gen. Virol., 26, 249 (1975).
148. R. A. Lamb and B. W. J. Mahy, in Negative Strand Viruses (B. W. J. Mahy and R. D. Barry, eds.), Academic, New York, 1975, pp. 65-87.
149. E. C. Lazar, L. J. Swango, and L. N. Binn, Proc. Soc. Exp. Biol. Med., 135, 173 (1970).
150. S. Levine and R. Hamilton, Arch. Ges. Virusforsch., 28, 122 (1969).
151. H. F. Lodish, Nature, 251, 385 (1974).
152. J. A. Lucy, Nature, 227, 815 (1970).
153. M. Luczak and M. Korbecki, Acta Virol., 14, 279 (1970).
154. K. Maeno, T. Yoshida, M. Iinuma, Y. Nagai, T. Matsumoto, and J. Asai, J. Virol., 6, 492 (1970).
155. B. W. J. Mahy, J. E. Hutchinson, and R. D. Barry, J. Virol., 5, 663 (1970).
156. P. A. Marx, A. Portner, and D. W. Kingsbury, J. Virol., 13, 107 (1974).
157. P. A. Marx, Jr., C. Pridgen, and D. W. Kingsbury, J. Gen. Virol., 27, 247 (1975).
158. T. Matsumoto and K. Maeno, Virology, 17, 563 (1962).
159. J. L. Melnick, Prog. Med. Virol., 19, 353 (1975).
160. V. ter Meulen, Med. Microbiol. Immunol., 160, 165 (1974).
161. R. H. Miller, A. R. Pursell, and F. E. Mitchell, Am. J. Hyg., 80, 365 (1964).
162. N. F. Moore and D. C. Burke, J. Gen. Virol., 25, 275 (1974).
163. C. Morgan and C. Howe, J. Virol., 2, 1122 (1968).
164. H. R. Morgan, J. F. Enders, and P. F. Wagley, J. Exp. Med., 88, 503 (1948).
165. Y. Morimoto, Y. Doi, and H. Itoh, Jap. J. Med. Sci. Biol., 23, 1 (1970).
166. D. C. Morley, Proc. Roy. Soc. Med., 67, 1112 (1974).
167. T. Morrison, M. Stampfer, D. Baltimore, and H. F. Lodish, J. Virol., 13, 62 (1974).
168. W. E. Mountcastle, R. W. Compans, L. A. Caliguiri, and P. W. Choppin, J. Virol., 6, 677 (1970).
169. W. E. Mountcastle, R. W. Compans, and P. W. Choppin, J. Virol., 7, 47 (1971).
170. W. E. Mountcastle, R. W. Compans, H. Lackland, and P. W. Choppin, J. Virol., 14, 1253 (1974).
171. S. Muthukrishnan, G. W. Both, Y. Furuichi, and A. J. Shatkin, Nature, 255, 33 (1975).
172. T. Nakai, F. L. Shand, and A. F. Howatson, Virology, 38, 50 (1969).
173. H. Nakajima and J. Obara, Arch. Ges. Virusforsch., 20, 287 (1967).
174. A. R. Neurath, Z. Naturforsch., 19b, 810 (1964).
175. A. R. Neurath, Acta Virol., 9, 313 (1965).
176. A. R. Neurath, S. K. Vernon, R. W. Hartzell, and B. A. Rubin, J. Gen. Virol., 16, 245 (1972).
177. A. R. Neurath, S. K. Vernon, R. W. Hartzell, F. P. Wiener, and B. A. Rubin, J. Gen. Virol., 19, 21 (1973).

178. Y. Nonomura and K. Kohama, J. Mol. Biol., 86, 621 (1974).
179. E. Norrby, H. Link, and J. E. Olsson, Arch. Neurol., 30, 285 (1974).
180. E. Norrby and B. Vandvik, Proc. Roy. Soc. Med., 67, 1129 (1974).
181. R. L. Northrop, J. Virol., 4, 133 (1969).
182. Y. Okada, Biken J., 1, 103 (1958).
183. C. Örvell and E. Norrby, J. Immunol., 113, 1850 (1974).
184. P. Palese, K. Tobita, M. Ueda, and R. W. Compans, Virology, 61, 397 (1974).
185. J. C. Parker, Science, 146, 936 (1964).
186. S. M. Perlman and A. S. Huang, J. Virol., 12, 1395 (1973).
187. J. S. Pierce and A. M. Haywood, J. Virol., 11, 168 (1973).
188. P. Platz, B. Dupont, T. Fog, L. Ryder, M. Thomsen, A. Svejgaard, and C. Jersild, Proc. Roy. Soc. Med., 67, 1133 (1974).
189. W. Plowright, Virol. Monog., 3, 25 (1968).
190. L. M. Popa, R. Repanovici, I. Samuel, D. Smelt, and R. Portocala, FEBS Lett., 51, 270 (1975).
191. A. Portner, unpublished data, 1972-1975.
192. A. Portner and D. W. Kingsbury, Nature, 228, 1196 (1970).
193. A. Portner and D. W. Kingsbury, Virology, 47, 711 (1972).
194. A. Portner and R. H. Bussell, J. Virol., 11, 46 (1973).
195. A. Portner, P. A. Marx, and D. W. Kingsbury, J. Virol., 13, 298 (1974).
196. A. Portner, R. A. Scroggs, P. A. Marx, and D. W. Kingsbury, Virology, 67, 179 (1975).
197. C. Pridgen and D. W. Kingsbury, J. Virol., 10, 314 (1972).
198. B. E. Roberts, M. B. Matthews, and C. J. Bruton, J. Mol. Biol., 80, 733 (1973).
199. W. S. Robinson, Nature, 225, 944 (1970).
200. W. S. Robinson, Virology, 43, 90 (1971).
201. W. S. Robinson, Virology, 44, 494 (1971).
202. W. S. Robinson, J. Virol., 8, 81 (1971).
203. J. E. Rodriguez and W. Henle, J. Exp. Med., 119, 895 (1964).
204. R. Rott and C. Scholtissek, Z. Naturforsch., 19b, 316 (1964).
205. L. Roux and D. Kolakofsky, J. Virol., 13, 545 (1974).
206. L. Roux and D. Kolakofsky, J. Virol., 16, 1426 (1975).
207. P. Roy and D. H. L. Bishop, J. Virol., 6, 604 (1970).
208. H. Rubin, R. M. Franklin, and M. Baluda, Virology, 3, 587 (1957).
209. H. Rubin, Virology, 4, 533 (1957).
210. A. A. Salmi, M. Panelius, and E. Norrby, Lancet, 2, 1088 (1972).
211. A. C. R. Samson and C. F. Fox, J. Virol., 12, 579 (1973).
212. A. C. R. Samson and C. F. Fox, J. Virol., 13, 775 (1974).
213. A. Scheid, L. A. Caliguiri, R. W. Compans, and P. W. Choppin, Virology, 50, 640 (1972).
214. A. Scheid and P. W. Choppin, J. Virol., 11, 263 (1973).
215. A. Scheid and P. W. Choppin, Virology, 57, 475 (1974).
216. A. Schluederberg, Biochem. Biophys. Res. Comm., 42, 1012 (1971).

217. A. Schluederberg, C. A. Williams, and F. L. Black, Biochem. Biophys. Res. Comm., 48, 657 (1972).
218. C. Scholtissek and R. Rott, J. Gen. Virol., 4, 565 (1969).
219. J. T. Seto and R. Rott, Virology, 30, 731 (1966).
220. J. T. Seto, H. Becht, and R. Rott, Virology, 61, 354 (1974).
221. K. V. Shah and G. B. Schaller, Bull. Wild. Dis. Ass., 1, 31 (1965).
222. K. V. Shah and C. H. Southwick, Indian J. Med. Res., 53, 488 (1965).
223. H. Shibuta, M. Akami, and M. Matumoto, Jap. J. Microbiol., 15, 175 (1971).
224. K. Shimizu, Y. Hosaka, and Y. K. Shimizu, J. Virol., 9, 842 (1972).
225. K. Shimizu, Y. K. Shimizu, T. Kohama, and N. Ishida, Virology, 62, 90 (1974).
226. L. Shindarov, A. Galabov, V. Vassileva, and N. Runevski, Zentr. Veterinaermed., 16B, 832 (1969).
227. E. H. Simon, Virology, 13, 105 (1961).
228. R. W. Simpson and M. Iinuma, Proc. Natl. Acad. Sci. U.S., 72, 3230 (1975).
229. K. V. Singh and T. I. Baz, Acta Virol., 11, 229 (1967).
230. A. Sishido, K. Yamanouchi, M. Hikita, T. Sato, A. Fukuda, and F. Kobune, Arch. Ges. Virusforsch., 22, 364 (1967).
231. F. Sokol, E. Skačianska, and L. Pivec, Acta Virol., 10, 291 (1966).
232. F. Sokol and H. F. Clark, Virology, 52, 246 (1973).
233. M. Soria, S. P. Little, and A. S. Huang, Virology, 61, 270 (1974).
234. H. O. Stone, P. A. Marx, Jr., and D. W. Kingsbury, unpublished data, 1972-1975.
235. H. O. Stone, A. Portner, and D. W. Kingsbury, J. Virol., 8, 174 (1971).
236. H. O. Stone, D. W. Kingsbury, and R. W. Darlington, J. Virol., 10, 1037 (1972).
237. H. O. Stone and D. W. Kingsbury, J. Virol., 11, 243 (1973).
238. J. H. Strauss, Jr., B. W. Burge, and J. E. Darnell, J. Mol. Biol., 47, 437 (1970).
239. K. Sugamura, H. Tozawa, M. Homma, and N. Ishida, Jap. J. Microbiol., 18, 349 (1974).
240. K. Sugita, M. Maru, and K. Sato, Jap. J. Microbiol., 18, 262 (1974).
241. H. Thacore and J. S. Youngner, J. Virol., 4, 244 (1969).
242. J. M. Tiffany and H. A. Blough, Virology, 37, 492 (1969).
243. J. M. Tiffany and H. A. Blough, Virology, 41, 392 (1970).
244. H. Tozawa, M. Watanabe, and N. Ishida, Virology, 55, 242 (1973).
245. B. Underwood and F. Brown, Med. Microbiol. Immunol., 160, 125 (1974).
246. B. Vandvik and E. Norrby, Proc. Natl. Acad. Sci. U.S., 70, 1060 (1973).
247. G. K. Vorkunova, V. A. Pashova, S. M. Klimenko, B. V. Gushchin, and A. G. Bukrinskaya, Arch. Ges. Virusforsch., 46, 44 (1974).
248. K. Wagner and G. Enders-Ruckle, Zentr. Veterinaermed., 13, 215 (1966).
249. R. R. Wagner, A. H. Levy, R. M. Snyder, G. A. Ratcliff, Jr., and D. F. Hyatt, J. Immunol., 91, 112 (1963).

250. R. R. Wagner, R. M. Snyder, and S. Yamazaki, J. Virol., 5, 548 (1970).
251. R. R. Wagner, in Comprehensive Virology, Vol. 4 (H. Fraenkel-Conrat and R. R. Wagner, eds.), Plenum, New York, 1975, pp. 1-93.
252. M. A. Wainberg and C. Howe, J. Virol., 12, 937 (1973).
253. K. Watanabe and Y. Okada, Biken J., 17, 51 (1974).
254. D. J. Waters and R. H. Bussell, Virology, 55, 554 (1973).
255. D. J. Waters and R. H. Bussell, Virology, 61, 64 (1974).
256. A. P. Waterson, Arch. Ges. Virusforsch., 16, 57 (1965).
257. R. A. Weinberg, Ann. Rev. Biochem., 42, 329 (1973).
258. S. R. Weiss and M. A. Bratt, J. Virol., 13, 1220 (1974).
259. E. F. Wheelock, Proc. Soc. Exp. Biol. Med., 114, 56 (1963).
260. D. O. White, Curr. Top. Microbiol. Immunol. 63, 1 (1974).
261. W. C. Wilcox, Virology, 9, 30 (1959).
262. W. C. Wilcox, Virology, 9, 45 (1959).
263. S. H. Winston, R. Rustigian, and M. A. Bratt, J. Virol., 11, 926 (1973).
264. V. M. Zaides, O. G. Nicolayeva, L. M. Selimova, O. P. Zhirnov, and A. G. Bukrinskaya, Biochem. Biophys. Res. Comm., 59, 1018 (1974).
265. V. M. Zaides, O. G. Nikolaeva, L. M. Selimova, O. P. Zhirnov, and A. G. Bukrinskaya, J. Gen. Virol., 24, 409 (1974).
266. V. M. Zhdanov and M. I. Parfanovich, Arch. Ges. Virusforsch., 45, 225 (1974).
267. Y. Nagai, H.-D. Klenk, and R. Rott, Virology, 72, 494 (1976).
268. A. R. Bräutigam and W. Sauerbier, J. Virol., 13, 1110 (1974).
269. P. B. Hackett and W. Sauerbier, J. Mol. Biol., 91, 235 (1975).
270. K. Glazier, R. Raghow, and D. W. Kingsbury, J. Virol., in press, 1977.
271. A. Portner and D. W. Kingsbury, Virology, 73, 79 (1976).
272. G. Abraham and A. K. Banerjee, Proc. Natl. Acad. Sci. U.S., 73, 1504 (1976).
273. L. A. Ball and C. N. White, Proc. Natl. Acad. Sci. U.S., 73, 442 (1976).
274. L. P. Villarreal, M. Breindl, and J. J. Holland, Biochemistry, 15, 1663 (1976).

Chapter 8

Reoviruses

Robert F. Ramig and Bernard N. Fields

Department of Microbiology
and Molecular Genetics
Harvard Medical School
Boston, Massachusetts

I. Features of the Virion

A. Host Range, Natural Occurrence, and Serology

Reovirus (respiratory enteric orphan virus) was proposed by Sabin [1] as the name for a group of viruses which had originally been classified as ECHO type 10. The cytopathic effects caused by these viruses were different from those caused by the other ECHO viruses and led to the recognition of this new virus type. Subsequently, the reoviruses were confirmed as a separate class of viruses.

Reoviruses have been isolated from a wide range of hosts representing a large cross-section of the higher vertebrates. Among the hosts from which reoviruses have been isolated are humans, horses, kangaroos, cattle, pigs, fowl, mice, sheep, and monkeys [2-10]. Subsequently, other viruses which had previously been described as members of the simian virus series or as hepatoencephalomyelitis virus of mice were found to be reoviruses [9-11]. Reoviruses do not cause overt disease in adult animals but have been isolated in association with disease (see Section 8.I.B.). Reovirus has been isolated from mosquitos, although it probably cannot replicate in insects [12].

The mammalian reoviruses, which share a common complement-fixing antigen [1], were shown to be separable into 3 serotypes by antibody neutralization [2, 3]. The type 2 viruses were further divisible into 4 subgroups by hemagglutination-inhibition [13]. Seventy-seven strains of avian reoviruses were grouped into 5 serotypes, which were not serologically related to any of the other reoviruses [6, 14].

B. The Role of Reoviruses in Human and Animal Disease

Reoviruses are thought to be unimportant as a cause of human disease. Serological surveys reveal that a large number of people have been infected with the virus, but most infections appear to be asymptomatic [15-17]. The various reovirus serotypes have been isolated from patients with a wide range of diseases, among them the "common cold," encephalitis, hepatitis, meningitis, encephalomyelitis, and fatal pneumonia [18-23]. Three deaths in human patients have been attributed to reovirus, but in most cases little evidence indicated that reovirus was the etiologic agent. Reovirus has also been isolated from biopsies of patients with Burkitt's lymphoma [24, 25]. The significance of this finding has not been determined. Recently reovirus-like agents have been isolated from and appear to be an important etiologic agent in infants and young children with acute nonbacterial gastroenteritis [26-29]. The reo-like agent was antigenically related to the epizootic diarrhea of infant mice virus and Nebraska calf diarrhea virus which have been officially classified in genus *Orthoreovirus* [29, 30]. Volunteers given reovirus that was passed in monkey kidney cells developed illness at low frequency, with the majority being asymptomatic. The symptoms in volunteers treated with serotypes 1 and 2 were malaise, nasal discharge, cough, sneezing, pharyngitis, and headache. Serotype 3 recipients developed mild rhinitis. Incubation periods ranged from 1 to 3 days with a duration of 4-7 days [7].

In animal systems, reovirus infections have been extensively studied, with disease in suckling mice being examined in greatest detail, although monkeys, calves, chimpanzees, and puppies have also been infected [31]. No illness has been observed in rabbits, adult mice, guinea pigs, or adult hamsters [32].

Spontaneous disease has been observed in suckling mice and is characterized by hepatitis, encephalitis, steatorrhea, and oily skin [33]. Intranasal infection of weanling mice results in respiratory disease followed by death [34]. Intraperitoneal infection of mice with type 1 resulted in death within 5-7 days. Virus was recovered from every organ system examined [35, 36].

Infection with reovirus type 3 resulted in similar symptoms, but a small percentage of the animals survived and went on to a more chronic illness similar to a runting syndrome [37].

Of considerable interest is a neurotropism exhibited by the reoviruses. Serotypes 1 and 2 produce a clinically silent infection of the central nervous system. Type 3 appears to be far more neurovirulent [38]. Direct intracerebral inoculation of suckling rats with type 3 produces an acute, fatal necrotizing encephalitis [39]. Within CNS lesions the virus showed a predilection to replicate in nerve cells, with glia and ependyma apparently not involved in viral replication [40]. Temperature-sensitive mutants of reovirus type 3 were substantially less neurovirulent in suckling rats and led to a much higher frequency of chronic disease. Infection with the *ts* mutants of class B resulted in a communicating hydrocephalus *ex vacuo* whereas class C mutant-infected animals showed no visible nervous system disease. The *ts* mutants persisted in the central nervous system for 6-8 weeks, after which virus could not be isolated and no antigen could be detected in brain tissue [41]. These findings demonstrated the striking effect of viral mutation in pathogenesis of central nervous system disease. The potential role of viral mutants in disease production has been recently reviewed [44].

Among other animals, reovirus has been associated with central nervous system lesions in monkeys, and rhinitis in chimpanzees [31]. Infection of calves was asymptomatic, but virus was recovered from nasal secretions [8], and macroscopic and microscopic evidence of interstitial pneumonia was found when organs were examined [42]. In puppies, reoviruses have been isolated from animals with respiratory infection [32] and interstitial pneumonia developed in experimentally infected animals [43].

The association of reoviruses with natural and experimental disease in humans and animals has been extensively reviewed [15, 31].

C. Physical Properties and Chemical Composition of the Virion

The buoyant density of reovirus has been determined by a large number of investigators. The wide range of densities observed, 1.32 g/cc [45], 1.36 g/cc [46, 47], 1.37 g/cc [48], 1.38 g/cc [49, 50], 1.39 g/cc [51], and 1.41 g/cc [52], indicates that the nucleic acid content of the virions may vary slightly and may also reflect the use of chymotrypsin in some early purification procedures. Empty virions (top component) have a density of 1.30 g/cc [46] or 1.28 g/cc [47]. Reovirus cores have a density of 1.43 g/cc [46, 47].

The molecular weight of the virion was originally estimated to be 70×10^6 daltons [49, 53] on the basis of a diffusion coefficient of 8.2×10^{-8}

cm^2/sec. Using more sensitive techniques, Farrell et al. [47] have determined that the diffusion coefficient is 4.45×10^{-8} cm^2/sec and estimate the molecular weight of the virion to be 129.5×10^6 daltons. They also determined the diffusion coefficient of the core to be 6.11×10^{-8} cm^2/sec, yielding a molecular weight of 52.3×10^6 daltons. The RNA content of the virion is 15×10^6 daltons [54] or $17.7 - 21.4 \times 10^6$ daltons for whole virions and 15×10^6 for the core [47]. The molecular weight of the protein in virions and cores is 109×10^6 and 37.3×10^6 daltons, respectively [47].

The size of reovirions has been measured by electron microscopy; the diameter of whole virions is 76 nm and that of cores is 52 nm [55]. Light scattering has been used to determine the hydrodynamic radius of virions, top component, and cores. These measurements show that hydrated particles have a much larger radius than dehydrated particles, although the degree of increase depends on ionic strength [56].

The RNA content of virions has been estimated to be 14.6% [49, 53, 57]. The nucleic acid of reovirus has been shown to be double-stranded RNA by a number of criteria: staining of viral nucleic acid [49, 58], G+C : A+U ratios [59], x-ray crystallography [60, 61], and by measurement of RNA width by electron microscopy [62, 63]. The double-stranded RNA of virions is segmented when extracted from the virion [52, 64, 65], the segments falling into 3 size classes. In addition to the double-stranded RNA, virions also contain 15-20% of their RNA as small oligonucleotides varying in molecular weight from 5000 to 20,000 daltons [64, 65]. The oligonucleotides may be located externally to the core [56] and thus be physically separated from the double-stranded genome RNA which is contained in the core. Three classes of oligonucleotides have been identified (see Section 8.II.B).

The 85% of the virion which consists of protein [49] is arranged into a double capsid [58]. The 7 polypeptides which make up the capsid fall into 3 size classes [46].

Reovirions are inactivated by treatment with periodate [66] but contain no carbohydrate [49]. Virions contain no peripheral lipid as shown by their resistance to treatment with ether [1, 67]. Reovirus is highly stable as shown by its resistance to a wide range of chemical treatments—1% H_2O_2, 1% phenol, 3% HCOH, and diethyl ether [1, 58, 68, 69]—a wide range of pH [68, 70, 71], and heat [70-74].

D. Morphology

Reovirions have a double-capsid structure as discerned by electron microscopy (Figs. 8.1 and 8.2) [55, 58, 75-77]. The double-capsid structure consists of a core containing the genome and a closely applied capsid surrounded by a second capsid which is separated from the core by a space of high water content [56].

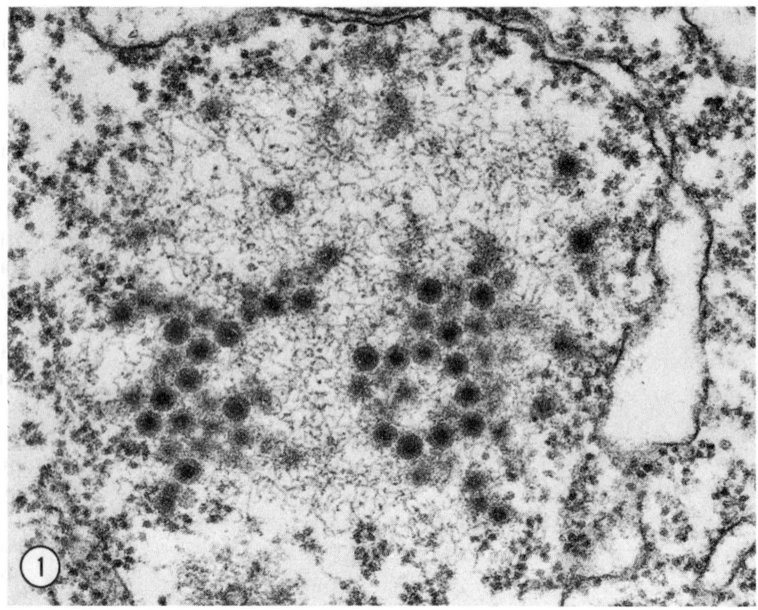

Fig. 8.1 Reovirus type 3 wild type. Thin section of an infected cell showing a viral inclusion. ×42,000. Courtesy of Dr. Cedric Raine.

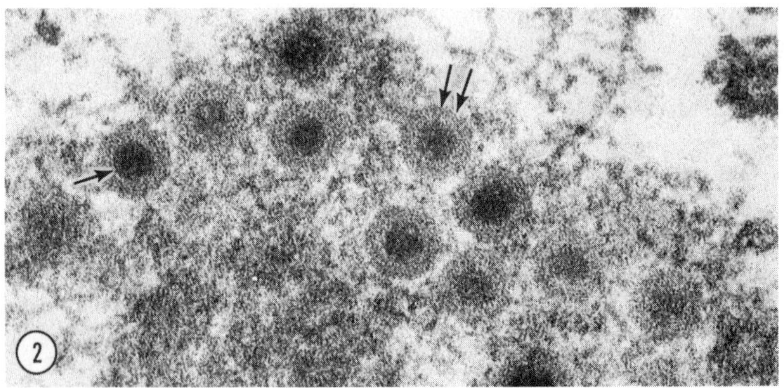

Fig. 8.2 Reovirus type 3 wild type. High magnification of cell-associated virus. The virion is seen to consist of a dense central core of RNA surrounded by an inner capsid (single arrow) and an outer capsid (double arrow). ×165,000. Courtesy of Dr. Cedric Raine.

Fig. 8.3 Longitudinal section of coated microtubules in a reovirus type 3-infected cell. Coated segments are seen between uncoated regions. X90,000. Courtesy of Dr. Cedric Raine.

The diameters of virions and cores have been measured by a variety of means. Diameters ranging from 70 to 75 nm have been obtained by negative staining [76, 78], filtration [1], and thin sections of infected cells [79]. More recently, electron microscopic measurements have yielded diameters of 76 nm for virions and 52 nm for cores [55]. Measurements of hydrated particles have shown them to be much larger; 96.4 nm and 70.0 nm for virions and cores, respectively [56].

Although capsomeres are readily visible on the surface of virions, the precise nature of the capsomeres and their arrangement in the capsid has not yet been determined. Vasquez and Tournier [78] reported that the outer capsid had 80 hexagonal and 12 pentagonal capsomeres, arranged with a 5 : 3 : 2 icosahedral symmetry. They later modified the model [80] so that the surface of the virion had 180 capsomeres, 12 of which were pentamers and the remainder hexamers. Support for a hexameric structure of capsomeres came from examination of virions stored in distilled water [81]. This treatment loosened the structure of the outer capsid and revealed capsomeres with hexagonal symmetry. Mayor et al. [53] proposed an icosahedral structure made up of 92 hollow prismatic capsomeres. Recent studies by Luftig et al. [55] suggest that there must be a total of 127 or fewer capsomeres since the outer capsid has 20 peripheral capsomeres. (The above models would require 18 peripheral capsomeres.)

The core has capsomeres which are 4 nm in diameter. The arrangement of these capsomeres cannot be distinguished but there are 20 on the periphery of the core. Twelve projections are arranged on the surface of the core as if they were at the vertices of an icosahedron. The spikes extend 5.5 nm from the surface of the core, have a diameter of 10 nm, and have central channels with a diameter of 5 nm [55]. Muller et al. [82] proposed that the core had 42 capsomeres.

Both the inner and the outer capsids of reovirus appear to have the same symmetry. Luftig et al. [55] proposed a model in which both capsids have 122 capsomeres arranged as an icosahedral deltahedron for which P = 1 and T = 9.

The outer capsid of reovirus is a relatively labile structure that is easily disrupted by heat or digestion with chymotrypsin. In contrast, the core is highly stable when heated and treated with chymotrypsin or high concentrations of urea, guanidine, dimethyl formamide, DMSO, and SDS [30, 46, 83-85].

E. Biological Features—Hemagglutination, Infectivity

Reoviruses of all three serotypes have been shown to agglutinate human erythrocytes of all ABO blood groups. The order of reactivity at both 4° and 37°C is A > AB > O > B [86]. Reo also agglutinates human RBCs at 23°C, but hemagglutination is rapidly inactivated at 56°C [87]. Virions of type 3 will agglutinate bovine erythrocytes at 4°C, whereas types 1 and 2 will not at either high or low temperature [88]. Virions of types 1 and 3 will elute from group O human erythrocytes upon prolonged incubation leaving the cell receptors intact [66]. Reovirions contain no neuraminidase [66] so the basis for agglutination by the virus is not understood. Hemagglutination by reoviruses is inhibited by periodate oxidation [89], β-glucosidase [90], and N-acetyl-D-glucosamine [91], which suggests that the virus may have a glycopeptide that confers agglutinability. However, Hand and Tamm [92] were unable to detect the presence of glycopeptides in virions by labeling with either glucosamine or fucose. A structural phosphoprotein has recently been identified [93]. The possible relationship of this phosphoprotein to agglutination has not been determined. Hemagglutination is inhibited by p-chloromercuribenzoate; this inhibition is reversed in the presence of reduced glutathione, suggesting that the configuration necessary for agglutination contains free sulfhydryl groups [66]. The hemagglutination receptor on the surface of human erythrocytes is resistant to neuraminidase [10] but the receptor on the surface of bovine erythrocytes is sensitive [94]. A wide variety of chemicals enhances or decreases the hemagglutination titers of virus preparations (see [15] for review).

Hemagglutination can be used as a rapid assay for reovirus. When assaying type 3 with bovine erythrocytes, 1 agglutination unit corresponds to 6.2×10^6 plaque-forming units (PFU) on L-cell monolayers [94]. However, empty virions also agglutinate erythrocytes [95] so the proportion of empty virions in a virus preparation will affect the relationship between PFU and agglutination. Fetal calf serum should be used in such hemagglutination assays as it does not inhibit agglutination as some other sera do [94, 260].

Reovirus stocks can be obtained which have very high infectivity. Particle to PFU ratios as low as 2 : 1 have been reported [71], although ratios of 100 : 1 appear to be much more common. Some preparations of virus treated with the enzyme chymotrypsin have been reported to have particle : PFU ratios of 1 : 1 [96].

The infectivity of virus preparations can be assayed on a number of different cell types and by several methods. Serotypes 1, 2, and 3 produce cytopathic effects on a wide range of cells, including primary *Maccaca* kidney cells, KB cells, HeLa cells, stable human amnion cells, primary *Cercopithecus* kidney cells, BS-C-1 cells, and L-cells [15]. Simple plaque assays have been devised for enumeration and determination of infectivity of all 3 serotypes. Monkey kidney cells (97-99) and mouse L-cells [100] have been used successfully for plaque assays. L-Cells have become the cell line of choice for reovirus plaque assays because of the ease with which they can be maintained. Plaque sizes are determined by the choice of cells as well as being characteristics of the virus strains [99]. Addition of pancreatin to cell monolayers sometimes increases the size of plaques [71]. Assays of infectivity have also been devised which make use of immunofluorescent enumeration of infected cells in monolayers [99].

The infectivity of virus stocks can be estimated spectrophotometrically. The relationship between optical density and viral protein mass measured at 260 nm is 1 OD unit = 2.1×10^{12} virus particles = 185 μg protein [46]. Farrell et al. [47] have developed similar methods for estimating the virus content of various types of solutions. For virus lyzed in 0.5% SDS, 4.7 OD_{250nm} = 1 mg/ml; for intact virus in either $1 \times$ SSC or distilled water, 9.1 OD_{260nm} = 1 mg/ml.

The loss of titer can be prevented in crude stocks of some strains by heating the stock to 50°C in the presence of 1 M $MgCl_2$ [101]. Infectivity of virus stocks can be increased by heating to 50°C in the presence of 2 M $MgCl_2$. However, freezing stocks in the presence of this cation causes extensive inactivation [74]. Heating in the presence of other divalent cations did not enhance infectivity, but freezing in the presence of these other cations caused extensive inactivation. The infectivity of virus preparations can be enhanced by treatment with chymotrypsin [96]. Recently, Borsa et al. [102, 103] have shown that greatest enhancement of infectivity with chymotrypsin occurs when the cation is Na^+ or Li^+. Under these conditions the enzyme only partially degrades the outer capsid of the virion rather than removing it completely.

Infectivity of reovirus is destroyed by the photodynamic interaction of the virus with a number of dyes, including neutral red which is commonly used in plaque assays [104]. Ultraviolet (UV) light also destroys the infectivity of the virus, but multiplicity reactivation has been observed in UV-irradiated virus [105, 106].

II. Virion RNA

A. Genome RNA

The double-stranded nature of reovirus genome RNA (DS RNA) was discovered by Gomatos et al. [58]. They observed that the cytoplasmic inclusions in infected L-cells stained orthochromatically with acridine orange, indicating the presence of double-stranded nucleic acid. The inclusions were susceptible to RNase and were resistant to DNase [49]. A great deal of evidence has accumulated that indicates the genome isolated from virions is also DS RNA. RNA from virions has base ratios that indicate base pairing, e.g., G = C and A = U [59]. Sharp melting profiles are obtained for virion RNA, the T_m depending on ionic strength [49, 107]. x-Ray crystallography showed that the virion RNA was a right-handed double helix with a pitch of 30 Å and either 10 or 11 base pairs per turn. The base pairs are tilted 75-80° to the axis of the helix [60, 61, 108]. The buoyant density of virion RNA in Cs_2SO_4 is 1.61 g/cc which is characteristic of DS RNA [107, 109]. The RNA of the genome resists degradation by DNase, *Escherichia coli* exonucleases I and III, and spleen phosphodiesterase. The genome is degraded to acid-soluble material by venom phosphodiesterase, micrococcal nuclease and RNase III. Denatured genome is rendered susceptible to degradation by ribonuclease and spleen phosphodiesterase [30, 107]. These properties are consistent with the double-stranded nature of the genome RNA.

The double-stranded genome RNA exists in the virion and is isolated as a series of unique pieces, or segments, rather than as a single molecule or chromosome. RNA isolated from virions can be separated into molecules of 3 size classes by centrifugation or chromatography on MAK columns [52, 54, 64, 110]. The 3 size classes (L, M, and S) sediment at 14, 12, and 10.5 S which correspond to molecular weights of 2.7, 1.3-1.4, and 0.6-0.8 $\times 10^6$ daltons, respectively [52, 54, 110]. These molecular weights correspond to 4500, 2300, and 1200 base pairs [110]. The 3 size classes of genome RNA have been further resolved to 10 size classes by polyacrylamide gel electrophoresis (Fig. 8.4) [54, 111, 112]. All segments of dsRNA are present in equimolar amounts in virions [54, 113]. The RNA released from gently lyzed virions has a trimodal size distribution when visualized in the electron microscope, with mean contour lengths of 1.1, 0.6, and 0.35 μm [114, 115]. Occasionally long profiles were observed suggesting that the genome segments might be weakly linked in the virion [115, 237].

A great deal of evidence supports the notion that segmentation of the reovirus genome is real and is not an artifact of isolation procedures. The RNA of the 3 size classes does not cross-hybridize, indicating that the RNA content of each size class represents discrete rather than random segments [64]. The

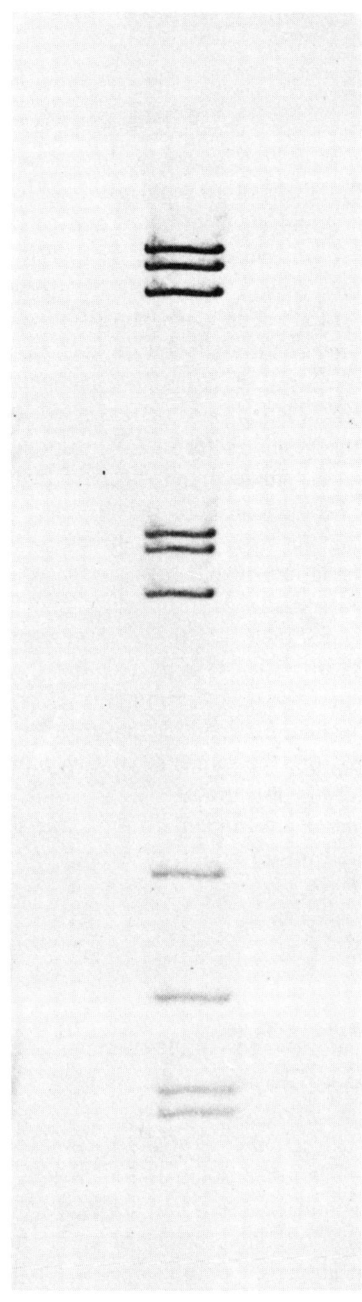

Fig. 8.4 Autoradiograph of double-stranded RNA from virions of reovirus type 3 wild type separated by gel electrophoresis. Note 3 large species, 3 middle species, and 4 small species. The decrease in intensity of the middle and small species is expected with uniformly labeled molecules of differing size which are present in equimolar amounts. Courtesy of Dr. Rise Cross.

segments appear to exist as discrete pieces in the virion because in situ labeling of the 3′ ends uniformly labels all 10 segments [116]. (Parenthetical note: this could also indicate a randomly permuted end to end linkage.) The structure of the 5′ and 3′ termini of the segments also suggests discrete segmentation of the genome. The 5′ termini of all segments contain either ppG . . . or pppG . . . [117]. The 3′ termini of all segments are . . . C_{OH} [118, 119]. The sequences at the termini of the segments suggest that the segments are perfectly base paired, although some evidence has been presented suggesting that the termini of the segments are not perfectly paired [120, 121]. Recently both the 5′ and 3′ termini of genome RNA were shown to be resistant to digestion by the single-strand specific nuclease S_1, under conditions in which the 3′ terminal dinucleotide of tRNA was completely removed [122]. These results indicate that reovirus DS genome RNA does not have single-stranded "tails." The isolation of defective virions which specifically lack the largest genome segment [48, 123] also indicates that the genome is segmented.

It has recently been shown that the 5′ termini of genome RNA exist in a modified form. Miura et al. [124] reported that the 5′ terminus of one strand is GpApUp . . . and the other strand is G*pCp . . . where G* is probably 2′-0-methyl G. Furuichi et al. [125] have shown the 5′ sequence of the (+) strand is blocked and has the sequence $m^7G(5')ppp(5')G^m pCp$. The significance of this structure is not known.

The genome segments are thought to represent single genes which are transcribed and translated into products of exactly corresponding size (see Section 8.III.A.) [46, 126].

B. Oligonucleotides

Reovirions contain, in addition to double-stranded genome RNA, single-stranded RNA (SS RNA) of low molecular weight. This low molecular weight material consists of oligoribonucleotides ranging from 2 to 20 nucleotides long and can comprise as much as 20-25% of the total virion nucleic acid [65, 127, 128, 129]. The oligonucleotides isolated from virions can be separated into 2 classes: oligoadenylates and 5′-G-terminated oligonucleotides [130-132].

The 5′-G-terminated oligonucleotides range in length from 2 to 4 nucleotides and are present at about 2000 copies per virion [131]. The sequences of these molecules have been determined [129, 131].

The oligoadenylates vary in length from 2 to 20 nucleotides and are present at about 850 copies per virion [128, 131]. This species accounts for about 55% of the total oligonucleotides present in the virion. The 5′ termini of these molecules may be tri-, di-, or monophosphorylated, in descending order of abundance.

In addition to the 2 classes of oligonucleotides discussed above, a third class exists. Its sequence has not been determined [131].

The number of phosphates on the 5′ termini of the oligonucleotides has varied from 1 to 3 depending on the investigator [129-131]. This variation may reflect a difference in triphosphate phosphohydrolase activities between the virus strains used [30].

The oligonucleotides have been reported to be contained in the core of the virion [133-135]. Recently it has been suggested that they are external to the core [56].

The function and origin of the oligonucleotides have not been determined. They are not required for infectivity [134, 136], yet cores free of oligonucleotides yield normal oligonucleotide-containing progeny. It has been proposed that the oligonucleotides arise from abortive transcription during the final stages of viral morphogenesis [131, 132].

III. Viral Polypeptides

A. Polypeptides of Virions

When the proteins of reovirions are solubilized and separated by polyacrylamide gel electrophoresis, they are resolved into 3 size classes, large or λ, middle or μ, and small or σ [46, 133]. The 3 polypeptide classes (λ = 1400-1500, μ = 700-800, and σ = 350-400 amino residues) correspond to the coding capacity (L = 4500, M = 2300, and S = 1000-1300 base pairs) of the genome segments [137]. Seven polypeptides species were recognized as components of virions and top component [46]. These polypeptides were designated $\lambda1$, $\lambda2$, $\mu1$, $\mu2$, $\sigma1$, $\sigma2$, and $\sigma3$ in order of increasing mobility on phosphate-urea polyacrylamide gel electropherograms. With the improved resolution offered by the tris-glycine discontinuous gel system, additional virion peptide species have been identified [138a, 139] (Fig. 8.5). However, differences in the properties of these gel systems have made identification of the previously identified proteins difficult and the nomenclature based on mobility has broken down. Both et al. [139] have resolved 10 polypeptide components in virions; 4 λ species, 3 μ species, and 3 σ species. Cross [138a] has identified 8 polypeptide species in virions and top component, adding a third λ species to the 7 species previously resolved by Smith et al. [46].

At least 1 of the virion polypeptides undergoes posttranslational cleavage. Polypeptide $\mu2$ has been convincingly shown to be a cleavage product of the $\mu1$ species by pulse-chase experiments [140]. Evidence has been presented which shows that the amino-terminal portion of $\mu1$ is cleaved off to yield $\mu2$ [141]. Both et al. [139] have suggested that the smallest of the 4 λ polypeptide

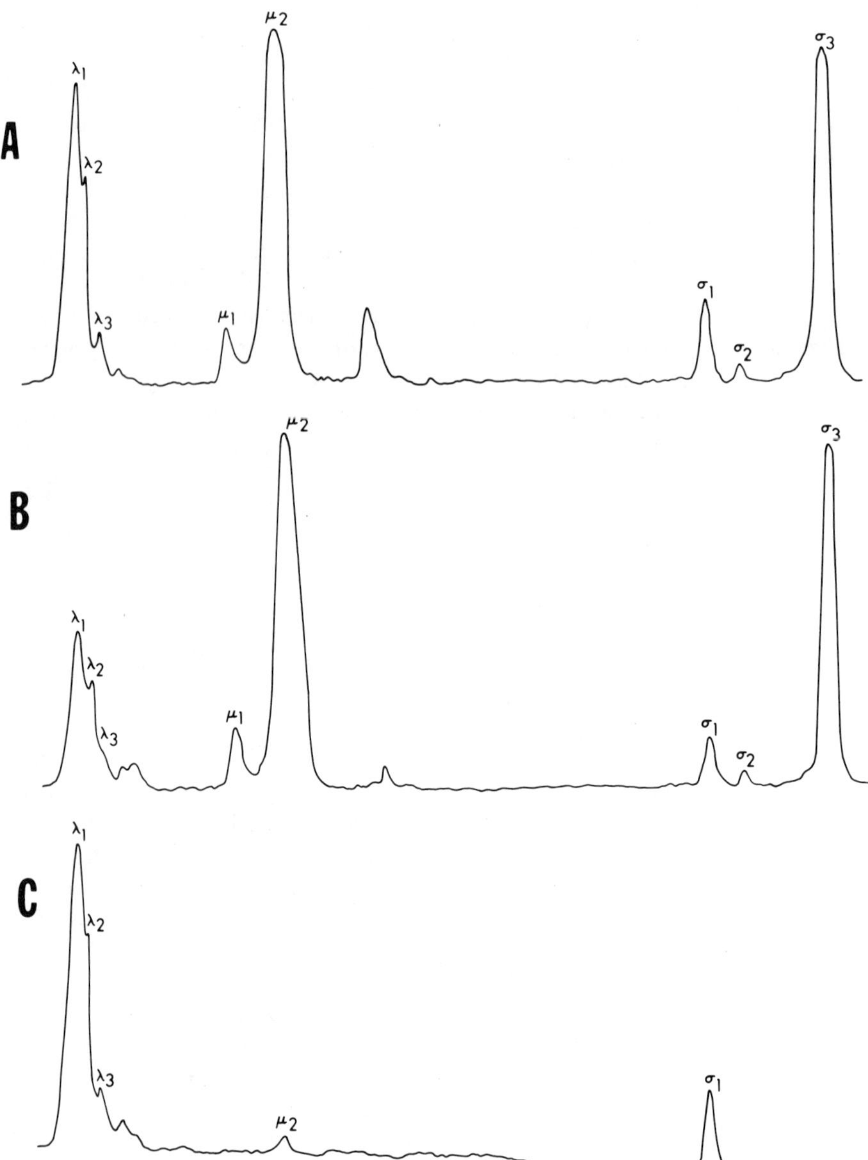

Fig. 8.5 Densitometer tracings of autoradiographs of reovirus type 3 polypeptides from highly purified virions (A), top component (B), and cores prepared with chymotrypsin (C). Electrophoresis was from left to right on tris-glycine gels. From Cross and Fields (138a) with permission.

species they see in virions is not a primary gene product and presumably arises by cleavage of one of the other λ species.

Pett et al. [141] have presented methods by which a number of the polypeptides can be separated by column chromatography. They found no unusual amino acids present but did show that all the polypeptides except μ2 have blocked amino-terminal groups. The carboxyl terminal sequences of some of the polypeptides were determined [σ3 = (val val leu) COOH, μ2 = leu-(arg tyr tyr)-arg-COOH, and λ1 + λ2 = arg-COOH, with the internal sequence different from that of μ2] as well as the amino terminal sequence of μ2 (NH_2-Pro-gly-gly-val-pro . . .). Roy et al. [142] have examined the amino terminal sequences present in the 3 size classes of virion polypeptides and found them all to be identical: NH_2-Pro-gly-gly-val----. These data suggest that reovirus polypeptides may have identical amino termini but dissimilar carboxyl termini. The identity of all the amino termini suggests that the sequence Pro-gly-gly-val may be significant in the posttranslational cleavage of μl and possibly other polypeptide species.

B. Localization of Polypeptides within Virions

The location of the various polypeptides within the virion has been determined by 2 methods: (a) The components of whole virions have been compared to those from cores made in vitro and various subviral particles isolated in vivo and (b) the surface components have been identified by iodination of virions and cores. Cores can be made in vitro by the digestion of virions with chymotrypsin [84]. The removal of the outer capsid by this enzyme leaves a core particle which is morphologically indistinguishable from cores observed in intact virions [55], suggesting that the core polypeptides are completely resistant to degradation by this enzyme. Under carefully controlled conditions the digestion of the outer capsid occurs by a series of well-defined steps [102, 103, 135, 143]. Polypeptide σ3 is removed first via several intermediates. μ2 is then removed by a series of cleavages, and finally σ1 is removed. The cores generated have lost, in addition to the 3 polypeptides, 5 logs of infectivity and the oligonucleotides; in addition the virion transcriptase has been activated. If the conditions used for chymotrypsin digestion are not optimal, not all of the outer capsid polypeptides are removed [102, 103, 135, 143, 144]. The above information suggests that μ2, σ1, and σ3 are the polypeptides of the outer capsid, leaving the balance of the virion polypeptides in the core (λ1, λ2, λ3, μ1, and σ2). Iodination has shown the surface components of the virion to be σ3 and μ2 and that of the core to be λ2 [145, 146].

The ability of reovirus to agglutinate erythrocytes suggests that one or more of the polypeptides of the virion must be modified, probably by the addition of a glucosyl residue [92, 147]. These authors were unable to detect incorporation of either glucosamine or fucose into viral products. The major outer capsid polypeptide $\mu 2$ has recently been reported to be modified by the addition of phosphate groups to serine residues [93]. The relationship of this modified surface polypeptide to hemagglutination has not been investigated.

C. Nonstructural Viral Polypeptides

In addition to examination of structural polypeptides from purified virions, samples can be prepared from infected cells by a number of methods. A method using reo-specific antibodies has been described by Fields et al. [148]. This method is especially good for looking at proteins made at high temperatures where the cellular background is high. Acetone can be used to precipitate total protein from infected cells treated with actinomycin D to turn off host protein synthesis [113]. The large particulate fraction isolated from infected cells [149] or whole infected cell pellets has been used with success for examination of viral polypeptides by polyacrylamide gel electrophoresis [150].

Examination of virus-specific cytoplasmic polypeptides by the above methods has led to the identification of 2 noncapsid polypeptides in infected cells [138a, 140]. These nonstructural polypeptides were designated as $\mu 0$. and $\sigma 2A$ on phosphate urea gels [140] and as $\mu 4$ and $\sigma 2A$ on tris-glycine gels [138a] where the nonstructural polypeptide migrates more rapidly (Fig. 8.6). Both et al. [139] have reported the existence of 6 nonstructural polypeptides in infected cells.

Modification of 1 of the nonstructural polypeptides as well as of $\sigma 1$ and/or $\sigma 2$ is suggested by their behavior in gel electrophoresis. In phosphate-urea gels the nonstructural polypeptide $\mu 0$ is the most slowly migrating of the μ species [140]. In the tris-glycine gel system as described by Laemmli [151] the $\mu 0$ species is the most rapidly migrating of the μ species and has been designated $\mu 4$ [138a]. $\sigma 1$ is the most slowly migrating of the σ species in the phosphate-urea gel system [46], whereas it migrates ahead of $\sigma 2$ in the tris-glycine system [138a]. These major migration differences between the gel systems suggest modifications of these polypeptide species [152]. As stated above, attempts to label glucosyl residues on reovirus polypeptides have not been successful.

The 1 reovirus polypeptide which has been demonstrated to be modified, the phosphoprotein $\mu 2$, does not show migration differences between the 2 gel systems [93, 138a].

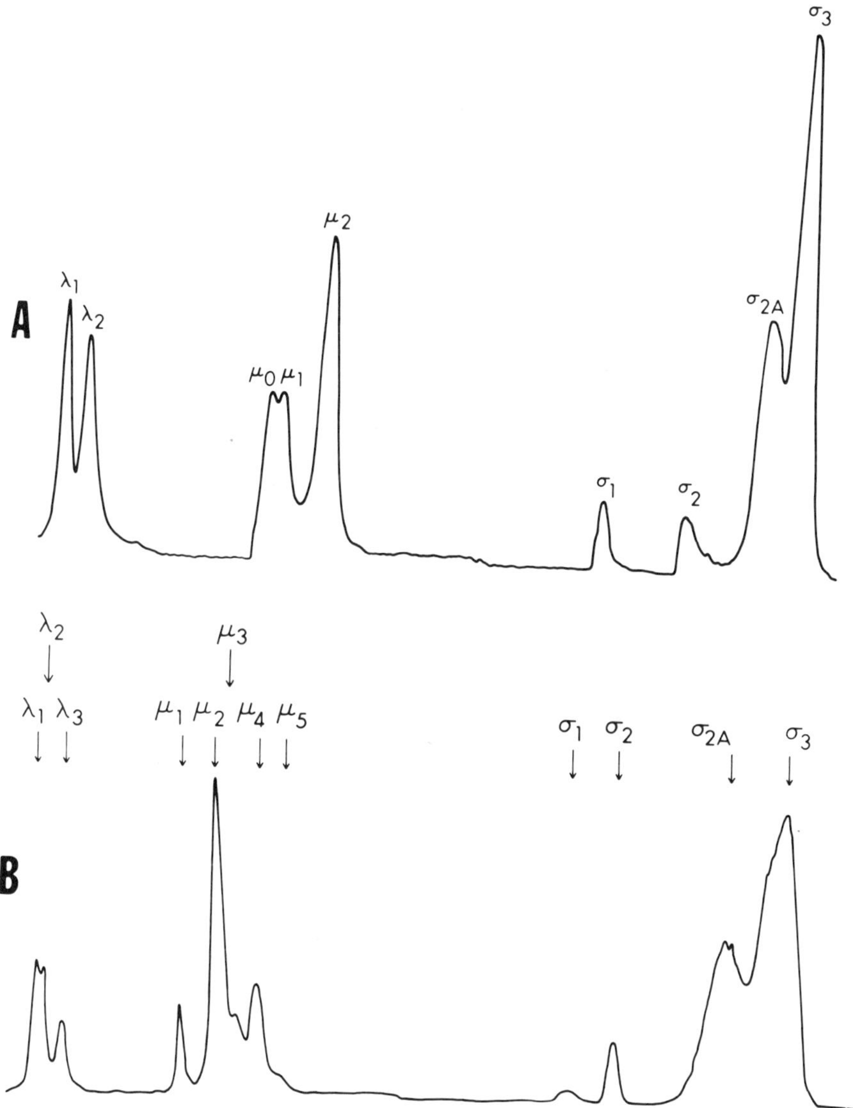

Fig. 8.6 Densitometer tracings of autoradiographs of cell-associated polypeptides of reovirus type 3. Compares patterns seen with the phosphate-urea (A) and tris-glycine (B) gel buffer systems. Modified from Cross and Fields (138a) with permission.

The polypeptides synthesized by reovirus of serotypes 1 and 2 also fall into 3 size classes [153]. However, when the polypeptide species are separated on tris-glycine polyacrylamide gels, reproducible difference in patterns are seen as compared to type 3. The major differences between the serotypes are found in the outer capsid polypeptides $\mu 2$ and $\sigma 3$ [153].

IV. Virion Enzymes

A. Virion Transcriptase

Purified reovirions contain double-stranded RNA-dependent RNA polymerase (transcriptase) in an inactive form. The enzyme can be activated by heating virions [154] or treating them with chymotrypsin [84]. Removal of some of the capsid proteins must be necessary for activation of the enzyme since both heat [83] and chymotrypsin [135] are known to remove protein from virions. The removal or rearrangements of polypeptides by these methods must be specific since the enzyme has been shown to remain inactive unless certain polypeptides are removed or cleaved [135]. Active cores produced by both methods produce identical transcription products [155]. The single-stranded RNA product of the enzyme is copied off of only 1 strand of the genome as shown by its failure to self-hybridize. It will, however, form hybrids with genome RNA [84, 112, 154, 155]. The transcription occurs by a conservative mechanism analogous to that of DNA-dependent RNA polymerase, as shown by the failure of labeled genome strands to appear among the transcription products either in vivo or in vitro [155, 156]. The products made in vitro and in vivo are identical. All 10 genome segments are transcribed in vitro and in vivo as shown by the ability of in vitro SS RNA to compete with in vivo SS RNA for hybridization sites on genome RNA [156, 157]. The transcription products are complete transcripts of the entire length of the genome segments because they form perfect hybrids with genome RNA that coelectrophorese with undenatured genome [112]. Although all 10 species of SS RNA are transcribed by the enzyme, they are not synthesized at the same rate or in the same amounts. The small size class transcripts appear first and are made in the largest amounts; the large class of transcripts are made last and in the smallest amounts [12, 155].

The transcriptase has a temperature optimum between 47° and 52°C but synthesizes the same products at all temperatures in the range 37-52°C [158]. Mn^{2+} is the preferred divalent cation for optimal activity of the enzyme; an absolute dependence for $(NH_4)_2SO_4$ has been demonstrated for the enzyme to synthesize specific products [159]. The enzyme appears to recognize a common initiation sequence on all genome segments since the 5′ terminal base on all transcripts has been identified as G [160-162].

Transcription appears to be intimately associated with the core since exogenously added genome RNA does not serve as a substrate [137]. Actively transcribing cores increase in density from 1.43 to 1.47 g/cc [112, 159]. There appear to be a number of sites of transcription within each core since actively transcribing cores have been shown to have a number of strands extending from them [163]. The transcripts may be extruded from the core through the channels in the center of the core spikes [164].

All attempts to solubilize and isolate the transcriptase have failed, suggesting that it is composed of several polypeptides or that the core structure imposes configurational constraints necessary for activity. Joklik [137] has noted the similarity between the subunit composition of *E. coli* DNA-dependent RNA polymerase and the polypeptide composition of the reovirus core.

B. Other Enzymes

In addition to the transcriptase, reovirions contain several phosphatase activities. Cores have an activity that catalyzes an exchange between the 4 ribonucleoside triphosphates and inorganic pyrophosphate [165]. This may be an alternate activity of the transcriptase since they have the same optimum conditions. Cores also contain a nucleoside triphosphate phosphohydrolase which liberates inorganic phosphate from the 4 ribo- or deoxyribonucleosides. The same catalytic site appears to be responsible for the activity on all substrates but gives the highest rate with ATP and the lowest rate with UTP [166, 167].

An enzymatic activity has been demonstrated in highly purified virions which synthesizes oligo(A) [168]. This activity is present in whole virions and infectious subviral particles but is not present in cores produced by extensive digestion with chymotrypsin.

The fact that SS RNA synthesized in vitro by the viral transcriptase in the presence of S-adenosylmethionine contains blocked 5′ termini with the structure $m^7G(5')ppp(5')G^mpCp$ . . . [169] suggests that reovirions contain additional enzyme activities capable of transfering methyl groups and forming 5′-5′ triphosphate linkages (see Section 8.V.E.). The presence of a viral phosphoprotein in reovirions [93] also suggests that virions may contain a protein kinase activity.

Several of the enzyme activities present in virions may represent alternate activities of the same enzyme. The blocking and modification of the 5′ terminus of SS RNA may be coupled [169], but in the absence of a methyl donor some blocked, unmodified mRNA molecules are obtained [261].

V. Replication

A. Morphology of Infected Cells

Cells infected with reovirus develop characteristic cytoplasmic inclusions called "factories" [58]. These inclusions typically form near the nucleus and then spread throughout the cytoplasm as the infection proceeds. The "factories" have been shown to contain viral double-stranded RNA by staining techniques [58] and binding of fluorescein-coupled poly(I : C) [170]. The presence of viral-specific polypeptides in the "factories" has been shown by fluorescent antibody techniques by a number of investigators [45, 171-174].

Ultrastructural examination of infected cells shows that the factories contain complete and incomplete viral structures, often in crystalline arrays [58, 77, 172, 174-176]. The factories are not bounded by membranous structures but do appear to be intimately associated with arrays of microtubules and spindle tubules [177]. Dales [177] noted that during infection the tubules are coated with virus-specific material in regions where they pass near or through the factories (Fig. 8.3). However, intact tubules are not necessary for normal infection and yields since the mitotic inhibitor colchicine, which causes disassembly of tubules, does not affect yields but does cause the factories to take on a more diffuse appearance [172, 177, 178].

B. Adsorption, Penetration, and Uncoating

Reovirus adsorbs efficiently to L-cells at 37°C in the presence of 0.02 M Mg^{2+} and 1% fetal calf serum. Under these conditions, 38, 52, and 65% of the input virus is adsorbed after 7.5, 15, and 30 min, respectively, giving an adsorption rate constant of 2.5×10^{-9} cm^3/min/cell [135]. The rate of adsorption at 4°C is similar to that at 37°C, so 4°C is often used to initiate synchronous infections [77]. L-Cells can adsorb enough virus to approach close packing over the surface of the cell [135].

Penetration into the cell appears to be by a process of invagination and engulfment similar to phagocytosis [77, 176, 179]. Shortly after adsorption virions are observed in lysosomes which move toward the center of the cell. Uncoating appears to take place in the lysosomes.

The uncoating process involves the removal of some of the outer coat protein. Approximately 1/2 of the outer coat is ultimately removed, presumably by lysosomal enzymes [51]. The subviral particle generated by the uncoating process has lost all of polypeptide $\sigma 3$ and an 8000 dalton fragment of the $\mu 2$ polypeptides [180]. The removal of these polypeptides activates the virion transcriptase [51, 149, 156, 180]. The genome is never completely uncoated but remains within the subviral particle throughout the infectious cycle [149]. Late in the infectious cycle capsid polypeptides are added back to the subviral particle, and these recoated virions are liberated with the progeny [30, 181]. It is not known if the recoated virions are infectious. Uncoating appears to be a passive process because it will occur in the presence of cycloheximide, an inhibitor of cellular and viral protein synthesis [149]. However, subviral particles are never recoated in the presence of this drug indicating that de novo synthesis of viral polypeptides is required for recoating.

C. The One-Step Growth Cycle

The growth cycle of reovirus in L-cells has been demonstrated to consist of 2 main components: an eclipse period of approximately 6 hr followed by a rise in virus-specified synthesis beginning about 9 hr postinfection [179, 182]. The eclipse period has been subdivided into an early period, during which uncoating occurs and no virus-specific products are formed, and a subsequent period during which synthesis of virus-specific products begins at a low rate [64, 126, 183, 184]. The length of the eclipse depends on 2 variables: (a) temperature and (b) multiplicity of infection. The low rates of synthesis observed in the last portion of the eclipse are thought to primarily represent transcription catalyzed by the transcriptase contained in the parental virions.

At the end of the eclipse a rapid increase is noted in the rates of synthesis of single-stranded RNA and protein, as well as the initiation of synthesis of the double-stranded RNA of progeny genomes [54, 126, 183, 184]. The rapid rise in synthesis after the eclipse requires the de novo synthesis of enzymes (transcriptase and replicase) since it is blocked by cycloheximide [183-185]. In the presence of cycloheximide, the synthesis of single-stranded RNA continues at the rate observed in late eclipse and the synthesis of double-stranded RNA never begins. It thus appears that the end of the eclipse is marked by the formation of an amplification mechanism for the synthesis of single-stranded RNA. This mechanism is envisioned to be the formation of new subviral particles which contain active transcriptase and catalyze the synthesis of single-stranded RNA [185].

Progeny virus are first formed during the period of rapid synthesis. The progeny, however, remain cell-associated until very late in the replicative cycle. Burst sizes of approximately 2000 PFU/cell have been observed [179].

Cytocholasin-enucleated cells will support limited amounts of reovirus replication [186]. Both viral RNA and proteins are made, but RNA synthesis seems to be suppressed to only about 10% of normal. Virus yields can be increased by lowering the temperature, which indicates that the limited life of the enucleated cells is the factor which limits reovirus yields.

D. Transcription In Vivo

The single-stranded reovirus-specific RNA made in infected cells represents a complete transcript of a single strand of each genome segment since the in vivo SS RNA hybridizes with denatured genome RNA to give 10 segments but does not self-anneal [184]. This single-stranded transcription product serves as the viral messenger because it can be isolated from polysomes in infected cells [64, 184, 187] and directs the synthesis of authentic viral polypeptides in in vitro protein-synthesizing systems [139, 188-190]. Early SS RNA synthesis is thought to be catalyzed by the virion-associated transcriptase; this enzyme has been shown to be activated as early as 1 hr after infection [51, 180]. However, although the transcriptase is active, no single-stranded RNA can be detected in infected cells until at least 2 hr after infection [191]. This may simply indicate that the transcribing subviral particles have not yet been liberated from lysosomes so that the transcripts are degraded.

As noted previously, late transcription can be blocked with inhibitors of protein synthesis [126, 185], indicating that new transcriptase must be synthesized in infected cells. The enzyme that catalyzes both early and late transcription appears to be the same. The enzyme has not been purified in a soluble form from infected cells but is always associated with a particulate fraction [159, 192].

Stoltzfus et al. [193] have been unable to demonstrate the presence of polyadenylic acid on reovirus-specific single-stranded RNA from infected cells.

The question of regulation of transcription in vivo involves conflicting observations. Zweerink and Joklik [113] found that the relative rates of synthesis of all the transcripts was constant throughout the infectious cycle, although the rates of transcription from the different segments are variable [30]. Control of transcription, i.e., a qualitative difference in the transcripts made early and late in the infectious cycle, has been repeatedly reported by the group of investigators working with Graham [120, 121, 194-197]. They reported that reovirus-infected cells could be "locked" into an early mode of transcription by treatment with cycloheximide. Under these conditions only ℓ1, m3, s3, and s4 mRNA species are made. At late times after removal of the drug all 10 transcripts were made, and an apparently controlled transition from the early to the late pattern was seen. From these observations they proposed that the host cell contained a repressor that prevented transcription of the late mRNAs. The function of 1 of the early viral products would be to inactivate the repressor, thus derepressing the transcription of late mRNA species [121]. The controversy over regulation of transcription remains unresolved. See reviews by Joklik [30] and Millward and Graham [121] for discussions of both views.

E. Transcription In Vitro

Reovirus transcription can easily be studied in vitro by using cores made with chymotrypsin. The products of the enzymatic reaction are identical to those made in vivo, although control elements that act in vivo do not act in vitro. Cores made from virions with chymotrypsin transcribe the genome into single-stranded RNA [84, 112, 135]. Such cores have been demonstrated to transcribe the same strand of the genome in vitro as is transcribed in vivo [157]. Alternatively, the synthesis catalyzed by subviral particles isolated from infected cells can be used to study transcription in vitro [149, 156, 180]. When measured by incorporation of substrate into acid precipitable material, transcription by cores is linear for several hours [84, 112]. A sensitive assay, depending on the binding of ethidium bromide to single-stranded reaction products, has been developed which allows continuous monitoring of the reaction [198]. In vitro transcription products can also be assayed by hybridization to denatured genome RNA followed by polyacrylamide gel electrophoresis [112, 155]. The in vitro transcription products assayed in this way form complete hybrids with genome RNA indicating that the genome segments are transcribed in their entirety. All 10 genome segments are transcribed in vitro [112, 155, 157, 199]. Recently a technique has been developed which allows the resolution of all 10 species of single-stranded transcripts on polyacrylamide gels without first hybridizing them to genome RNA [200].

The in vitro transcription reaction has a very high temperature optimum (47-52°C) but is rapidly inactivated at 55°C [112, 158]. Besides the triphosphates of the 4 usual bases, A, G, C, and U, the transcriptase will incorporate 5Br-UTP and 5Br-CTP [158]. The rates of polynucleotide chain elongation have been measured and reported to be 7-8 nucleotides/sec [112] and 60 nucleotides/sec [155].

The in vitro transcription reaction appears to be uncontrolled. All 10 species of SS RNA are made by cores, although they are not made in equal numbers [112]. The order of appearance of transcripts and the relative numbers of each that are made appear to reflect a constant rate of synthesis of all segments, with the smallest being transcribed more frequently and being completed first simply because of their smaller size [112].

All cores appear to be active in in vitro assays since the density of the active cores increases uniformly by 0.02 g/cc [112]. Electron microscopic examination of actively transcribing cores indicates that all 10 segments must be transcribed simultaneously because many RNA molecules are seen extending from cores [163, 164].

The terminal sequences of the in vitro transcription products have been examined by a number of investigators. The 5′ terminal nucleotide of all the transcripts is G (160-162], the same as has been observed for transcripts made in vivo [160]. The 5′ terminus of the transcripts contains a modified G, possibly 2′-0-methylguanosine [201]. The 5′ terminus is blocked when the methyl donor S-adenosylmethionine is provided in the reaction mixture; its sequence has been determined to be $m^7G(5')ppp(5')G^m pCp$ . . . [169]. In the absence of a methyl donor, the 5′ terminus is blocked with much lower efficiency and has predominantly the sequence ppGpc [169]. Modification of the penultimate G appears to follow methylation at the 7 position of the terminal G [261]. The 3′ terminal base of all in vitro transcripts is cytosine [118, 119].

When certain suboptimal conditions are used for in vitro transcription only 1 of the small RNA species (s4) is made [162]. The 5′ terminal sequence of this message has been determined to be:

(p)ppGCCAUUUUUGCU(C,U)UCCAGACGUUG

This sequence resembles the untranslated sequence of the RNA of R17, Qβ, and MS2 and may represent a regulatory region.

The binding of ethidium bromide to transcription products indicates that they have a high degree of secondary structure [198, 202].

F. Translation In Vivo

The rate of protein synthesis in infected cells falls slowly over the course of infection [182, 203]. The synthesis of reovirus-specific polypeptides begins at a low rate and increases concomitantly with the synthesis of late RNA [113]. Early in infection the synthesis of viral-specific proteins is obscured by the background of cellular synthesis, making it necessary to use actinomycin D to lower the cellular background [113] or antibodies directed against reo polypeptides as a probe for reo-specific protein synthesis [149, 190]. Antibody precipitation is also useful for examination of polypeptides made at high temperatures where actinomycin D treatment leaves a very high cellular background.

All 10 pieces of reovirus-specific mRNA are found on polysomes isolated from infected cells [184, 187]. All 10 species of SS RNA are found in both large (~30) and small (5-8) polysomes [204, 205]. However, the large polysomes contain the RNA species in a ribonucleoprotein complex with reovirus capsid proteins, whereas the RNA of the small polysomes is not associated with virus-specific protein. Christman et al. [206] have shown that deprivation of infected cells for essential amino acids causes the polysomal RNA (reo specific) to condense into a ribonucleoprotein particle which contains no virus-specific polypeptides. These ribonucleoprotein particles are rapidly converted to functional polysomes upon restoration of the amino acids.

Infected cells make, in addition to the 7 structural polypeptides, essential and nonessential noncapsid virus-specific polypeptides [113, 140]. The essential noncapsid polypeptides are always made in large amounts in infected cells. The nonessential noncapsid polypeptides are seen only in cells infected at high temperatures and may represent artifactual cleavage of some of the viral peptides [140]. Cross and Fields [147] have detected 11 virus-specific polypeptides in infected cells (Fig. 8.6). Both et al. [139] reported the presence of 16 virus-specific polypeptides in infected cells, which they have divided into 10 structural species and 6 nonstructural species. One of the structural polypeptides, $\mu 1$, is cleaved in vivo, yielding a secondary product $\mu 2$ [113, 140]. This cleavage has been shown to involve the removal of a fragment from the amino terminus of $\mu 1$ [141].

The virus-specific polypeptides are synthesized in widely varying amounts in infected cells [113, 140]. However, the patterns and amounts of polypeptides synthesized early and late did not differ, leading them to conclude that protein synthesis was not regulated into an early and late pattern in vivo. Graham's group has been able to lock infected cells into a pattern of protein synthesis where only 4 species of single-stranded RNA and proteins are made by the use of cycloheximide and cordycepin [121, 197]. They interpret this pattern to represent an early phase of protein synthesis since it cannot be obtained if the drugs are used late in infection. Thus, the question of regulation of protein synthesis in reovirus-infected cells remains unsolved.

G. Translation In Vitro

A number of cell-free systems has been examined for their ability to synthesize reovirus polypeptides. A system made from infected L-cells was able to support the completion of polypeptides that had been initiated but was unable to initiate new rounds of synthesis. Eight polypeptides were made by this system [188]. A system made from uninfected L-cells and programmed with single-stranded reovirus RNA made in vitro was able to initiate the synthesis of polypeptides but was unable to make more than small fragments of 10,000-15,000 daltons. The incorporation of this system could be increased by pretreatment of the message with formaldehyde which presumably lowered the secondary structure of the mRNA [207]. An L-cell system programmed with m or s size mRNA was able to synthesize only μ and σ sized proteins [190]. McDowell et al. [189] examined the ability of several cell-free systems to synthesize reovirus-specific polypeptides. They found that systems made from L-cells, HeLa cells, and CHO cells could direct the synthesis of only 5-7 virus-specific polypeptides when programmed with mRNA made in vitro. Systems made from rabbit reticulocytes or Krebs II ascites cells could direct the synthesis of 8 polypeptides. Recently, Both et al. [139] have reported that a cell-free system made from wheat germ made 10 polypeptides which they call the 10 primary gene products of reovirus.

In vitro studies of initiation have shown that reovirus mRNA forms initiation complexes with ribosomal subunits, f-met tRNA, and GTP. This indicates that reovirus mRNAs must have an AUG initiation codon [208]. The 25 base sequence that has been determined at the 5′ terminus of 1 of the messages does not contain an AUG codon [162]. This suggests the presence of a regulatory sequence on that message.

Recent studies have shown the 5′ terminus of reovirus message to have the structure $m^7G(5')ppp(5')G^m pCp$ . . . [169]. Reovirus mRNA with this 5′ structure is efficiently translated in a cell-free system derived from wheat germ. However, removal of m^7G by β elimination abolishes the ability of reovirus message to direct the synthesis of polypeptides [209]. When wheat germ systems are programmed with unmodified and unblocked reovirus mRNA (5′ structure ppGpCp . . .) synthesis of polypeptides occurs only if the methyl group donor S-adenosylmethionine is added [210]. In such a system the 5′ terminus of the mRNA is blocked and methylated before translation proceeds [261]. Further studies have shown that only mRNA molecules with the 5′ terminal m^7G are capable of participating in the initiation of protein synthesis [236]. A short sequence at the 5′ terminus of mRNA is protected by 80 S ribosomes from digestion by T_1 or pancreatic ribonuclease. About 35% of RNase resistant sequences include the 5′ terminal m^7G, suggesting that 5′ terminal m^7G may function as a recognition signal for ribosome binding, and also indicating that ribosomes bind very close to the 5′ termini of some of the reovirus mRNAs [236].

H. Replication of Double-Stranded RNA

The fact that the reovirus genome is never totally uncoated in infected cells [51] presents a problem of how the genetic information is transferred from parent to progeny. The (+) strand transcripts must, besides functioning as message, serve as templates for the synthesis of the double-stranded progeny genome RNA. Schonberg et al. [211] provided further evidence for this sort of mechanism when they showed that the synthesis of (+) and (–) strands is asynchronous in the infected cell. The (+) strands were shown to be synthesized primarily early in the infectious cycle, whereas the (–) strands were synthesized only late in the cycle, with (+) strands serving as template for the synthesis of the complementary (–) strands. The (+) strands are presumably synthesized by the transcriptase, while the (–) strands are synthesized by a second viral enzyme, the replicase (SS → DS RNA polymerase).

The synthesis of double-stranded RNA has been studied mostly using in vitro systems derived from infected cells. Acs et al. [212] showed that the synthesis of (–) strands could be catalyzed by a large particulate fraction isolated from infected cells. The replicase activity could not be separated from the particulate fraction, and the (+) strands which served as template for the replicase could be removed only with ribonuclease. This suggested that the (–) strand was synthesized on a particulate structure into which the single-stranded (+) strand template RNA had previously been gathered. All 10 segments were present in the DS RNA product of the in vitro reaction, indicating that all 10 segments of single-stranded RNA were present in the particles prior to the reaction. Sakuma and Watanabe [213-215] also found that replicase activity was bound to the template in the particulate fraction of infected cells. These particles they described as resembling viral cores. They found that the (–) strand was synthesized in a $5' \rightarrow 3'$ direction with a single initiation point for synthesis on each segment. The replicase would support only a single round of (–) strand synthesis and the (–) strand remained associated with the (+) template strand, presumably as a hydrogen-bonded double helix [213-216]. Support for the notion that the replicase catalyzed only a single round of complementary (–) strands came from Zweerink et al. [217], who showed that the amount of (–) strand synthesized never exceeded the amount of (+) strand template associated with the particles.

Zweerink et al. [217] found that particles having replicase activity sedimented heterogenously (300-600 S), although they had a uniform density of 1.34 g/cc. After the reaction was complete, the RNAse-sensitive template had been converted to RNAse-resistant double-stranded RNA, and the density of the particles had shifted to that of cores (1.43 g/cc). Further studies showed that if the heterogeneously sedimenting replicase active particles were fractionated, S, M, and L segments of DS RNA were formed sequentially in particles of 300, 450, and 550 S, respectively [218]. These particles were corelike with the exception that the 300 S particle lacked the λ2 polypeptide and they all contained the nonstructural polypeptide μ0.

I. Synthesis of Oligonucleotides

The synthesis of oligonucleotides in reovirus-infected cells presents a dilemma since the oligonucleotides appear to be a nonessential viral product. Indeed, it has been proposed that the oligonucleotides represent nothing more than an artifact caused by the presence of a virion transcriptase.

The oligonucleotides of reovirus have been shown to be synthesized late in the infectious cycle, at the same time or after the synthesis of double-stranded RNA [65, 127]. The 5′-G-terminated oligonucleotides have been suggested to be the product of abortive transcription during the final stages of viral morphogenesis [131, 132, 162]. This notion is supported by the fact that the 5′-G-terminated oligonucleotides resemble the 5′ terminal sequence of reovirus transcripts made in vitro. These abortive transcripts are presumably sealed in the virion by the completion of the outer coat. Thus the synthesis of the 5′-G-terminated oligonucleotides would represent an alternate activity of the virion transcriptase.

Stoltzfus and Banerjee [130] suggested that the oligoadenylates [poly(A)] might arise from a method of template selection whereby transcriptase products destined to become template were marked by the addition of a poly(A) sequence which was cleaved during replication and trapped within the virion. This mechanism seems unlikely since Stoltzfus et al. [193] have reported that reovirus single-stranded RNA does not have poly(A). Recently, Stoltzfus et al. [168] found a poly(A) polymerase activity associated with purified virions, and Silverstein et al. [219] found that a poly(A) polymerase activity present in infected cells was associated with particles having the properties of completed or nearly complete virions. Johnson et al. [220] have shown that none of the temperature-sensitive mutants of reovirus make poly(A) at the restrictive temperature. This suggests that the poly(A) synthetic event is so late as to be preceeded by the blocking defect in all the mutants. The above data suggest that the synthesis of oligo(A) is a very late event which occurs in complete or nearly complete virions. The synthesis of oligo(A) has been suggested to be an alternative activity of the virion transcriptase and to occur by abortive transcription [168, 219]. Nichols et al. [162] have suggested that oligo(A) synthesis could be caused by the slippage of the transcriptase on a poly(U) sequence near the 5′ terminus of a genome segment. They have demonstrated the presence of a short run of U in 1 segment.

J. Morphogenesis

Very few studies have yet been directed toward determining the methods and pathways by which reovirus is assembled. The fact that reovirus has only 10 genes yet incorporates the products of at least 8 of these gene products into the virion (Fig. 8.5) suggests that morphogenic information must be programmed into at least a number of the polypeptides. However, the fact that reovirus also codes for at least 2 nonstructural proteins (μ4 and σ2A; see Fig. 8.6) leaves open the possibility that these noncapsid proteins may have morphogenetic function.

A number of subviral particles which may represent intermediates on the morphogenetic pathway have been identified in infected cells. Some of these particles seen in cells infected with temperature-sensitive mutants [174, 221, 222] undoubtedly represent the accumulation of immature virions the further maturation of which is blocked by the temperature-sensitive lesion. Particles isolated from infected cells which have replicase activity or poly(A) polymerase activity [212, 213, 218, 219] also probably represent intermediates on the morphogenetic pathway. At this point, however, these data have not been assembled into a meaningful pathway.

Some evidence has been presented suggesting that the addition of polypeptide $\sigma 3$ to the outer capsid may be the final step of assembly [181].

In studying morphogenesis, one must remain aware of the possibility that structures accumulated in the presence of specific blocks may represent artifactual side reactions rather than true morphogenetic intermediates. An example which has been noted is the increase in the yield of empty virions (top component) when lysine is omitted from the growth medium [223] or when ethidium bromide is added to the growth medium [224].

K. Effect on the Host Cell

The synthesis of protein by reovirus-infected cells falls slowly over the course of the infective cycle. The synthesis of total proteins in infected cells falls at approximately 8 hr after infection in monolayer cells and slightly earlier in suspension cells [182, 183, 203, 225]. Examination of host-specific protein synthesis has shown that it falls 3-6 hr before total protein synthesis falls, but this inhibition of the host is masked by rapid virus-specific protein synthesis [113]. The precise time after infection that inhibition of host protein was first observed depended on the multiplicity and the temperature of the infection. It was noted that the decline in host protein synthesis corresponded closely with the phase of rapid viral mRNA synthesis and suggested that inhibition of the host-specific synthesis represented a competition for some component of the protein-synthesizing apparatus.

The synthesis of DNA by infected cells is inhibited several hours before total protein synthesis declines [203]. The inhibition of DNA synthesis appears to result from a block of the initiation of new rounds of synthesis rather than a block at some other stage [226-228]. It is thought that no new initiations occur because some host cell factor necessary for initiation is not made (this could be because host cell messages appear to be completed away from ribosomes at this time). It had been suggested that viral proteins might specifically inhibit host cell DNA synthesis, but the failure of high multiplicities of top component to inhibit host synthesis showed that capsid polypeptides did not cause the inhibition [224, 229]. The inhibition of host DNA synthesis does occur in cells infected with virions which have received large doses of ultraviolet irradiation [224, 230]. The fact that virions the infectivity of which has been abolished by irradiation but which still have transcriptase activity inhibit host DNA synthesis suggests that active synthesis of viral products is necessary for inhibition of host DNA synthesis.

L. Inhibitors

The synthesis of cellular DNA can be inhibited by a number of drugs without lowering the yield of reovirus. 5-Fluorodeoxyuridine (FUdR), 0.25 μg/ml; 5-bromodeoxyuridine, 100 μg/ml; mitomycin C, 1 μg/ml [58]; and cytosine arabinoside, 10 μg/ml [231, 232] had no effect at low concentrations. Higher levels of mitomycin C or cytosine arabinoside did lower the yields from infected cells.

Host RNA synthesis in infected cells is sensitive to low concentrations of actinomycin D, whereas virus yields are unaffected at low concentrations [183, 233, 234]. Higher levels of actinomycin D do prevent viral RNA synthesis. Yields from infected cells are also markedly reduced by azauridine [235]. Cordycepin can also be used to block viral RNA synthesis [121].

Viral protein synthesis is blocked by cycloheximide or puromycin. The block with cycloheximide is reversible [185].

Ethidium bromide appears to have no effect on viral transcription or translation but blocks the synthesis of double-stranded RNA, giving a high yield of empty particles [224].

VI. Genetics of Reovirus

A. Genetic Studies

The genetics of reovirus has several unique features not commonly found in other viral systems. Of particular interest are the segmented nature of the viral DS RNA genome and the conservative manner in which genetic information is passed from parental to progeny virus. These unique features lead to a number of possibilities for the function of the genetic apparatus of reovirus. Specifically, one recognizes the possibility for genetic exchange to occur either via traditional intramolecular recombination or via a reassortment of genome segments. The identification and isolation of viral mutants allows one to study the process of genetic exchange as well as to determine the biochemical defects, and possibly morphogenetic defects, in the specific mutants.

Conditional lethal, temperature-sensitive (*ts*) mutants of reovirus type 3 have been isolated in 2 laboratories. Ikegami and Gomatos [238, 239] isolated 14 *ts* mutants which fell into 2 classes on the basis of phenotype.

Virions of 1 class possessed lower hemagglutinating activity per infectious unit and had increased thermolability relative to the other class. Fields and Joklik [240] isolated approximately 60 *ts* mutants which fell into 5 groups by "reassortment" testing (see below). Fifteen additional *ts* mutants were subsequently isolated, which upon "reassortment" testing increased the number of mutant groups to 7 [85]. Although the temperature-sensitive mutants isolated have the "leakiness" commonly found in *ts* mutants, the difference in plating efficiencies at permissive (31°C) and restrictive (39°C) temperatures is sufficiently great to make the mutants useful for genetic and biochemical studies. Prototype strains and other useful mutants are listed in Table 8.1. The relationship of the mutants isolated by Ikegami and Gomatos to the 7 groups identified by Fields and Joklik has not been determined.

Early genetic experiments with reovirus mutants were directed at determining whether the generation of ts^+ (wild type) progeny from cells infected with 2 different *ts* mutants occurred by intramolecular recombination events or resulted from reassortment of genome segments. One would expect that intramolecular recombination events would occur at a very low frequency because of the small size of the genome segments. In contrast, reassortment would be expected to give rise to high frequencies of ts^+ "recombinants." Thus low recombination frequencies were expected among mutations lying on the same genome segment (or no recombination at all if DS RNA molecules did not undergo intramolecular recombination) and high recombination frequencies were expected if the mutations lay on different, reassorting segments. Cells were infected and the virus was allowed to grow at permissive temperature with single mutants or with pairwise combinations of mutants. The progeny were plated at 31° and 39°C and the percentage of recombinants was determined by the following formula [240]:

$$\% \text{ recombinants} = \frac{\text{yield }(A \times B)^{39^\circ C} - \text{yield }(A)^{39^\circ C} - \text{yield }(B)^{39^\circ C}}{\text{yield }(A \times B)^{31^\circ C}} \times \frac{100}{1}$$

Analysis of such 2-factor crosses showed that all of the crosses gave 1 of 2 results [240]: (a) No recombinants could be detected above background levels or (b) high frequencies of recombinants resulted (Table 8.2). The differences in recombination frequency between those crosses yielding recombinants were not statistically significant [241, 242]. The "all or none" nature of recombination was interpreted to mean that (a) intramolecular recombination did not occur at detectable levels and (b) ts^+ recombinants were generated by a mechanism of reassortment of genome segments. Those mutant pairs which failed to generate ts^+ recombinants were assumed to lie on the same genome segment, i.e., within the same gene. This type of analysis has been used to group the mutants into the groups shown in Table 8.1, i.e., mutants within a group

Table 8.1 Selected *ts* Mutants of Reovirus Type 3

Group	*ts* Strain	Mutagen[a]	EOP 39/31°C[b]
A	201[c]	PRO	0.10
	234	NTG	0.06
	270	PRO	0.02
	279	PRO	0.25
	290	PRO	
	340	NA	0.001
	438	NTG	0.025
	470	NTG	0.3
	474	NTG	0.008
B	271	PRO	0.005
	352[c]	NA	0.0025
	405	NA	0.001
C	447[c]	NTG	0.0004
D	357[c]	NA	0.02
	585	NTG	0.006
E	320[c]	NA	0.3
F	556[c]	NTG	0.02
G	453[c]	NTG	0.0002

[a]NA: nitrous acid; PRO: proflavine; NTG: M-methyl-N′-nitro-N-nitrosoguanidine.
[b](Number of plaques at 39°C/number of plaques at 31°C) × 100.
[c]Standard prototype strain.

yield no ts^+ progeny, whereas those in different groups yield ts^+ progeny at high frequencies.

Additional experiments demonstrated that maximal recombination frequencies were obtained early in the infectious cycle and were maintained throughout the entire cycle [241, 242]. This suggested that recombination was occurring during a single cycle of mating. The frequency of recombinants among the progeny was shown to depend on multiplicity of infection (MOI), with maximal frequencies being obtained at MOI of 10 PFU/cell or greater. Since all stocks contain a high proportion of noninfectious particles (50-150 particles/PFU), these results suggested that noninfective particles do not take part in genetic interactions [241, 242].

The 2-factor crosses suggested strongly that no linkage existed between markers lying on different genome segments. Three-factor crosses, using the technique of the unselected third outside marker, would be much more sensitive in the detection of linkage between genome segments. A stable, independent, nontemperature-sensitive mutation has only recently been

Table 8.2 Presence and Absence of Genetic Recombination Between Several *ts* Mutants at the Permissive Temperature (31°C)[a]

Virus strain used in infection	Yield at 24 hr (PFU/ml) 39°C	31°C	Efficiency of plating 39°C/31°C	Presumed wild type recombinants (% of 31°C titer)
Wild type	3×10^7	3×10^7		
*ts*201	3×10^4	1×10^8	3×10^{-4}	–
*ts*352	$<1 \times 10^3$	1×10^8	1×10^{-5}	–
*ts*447	2×10^3	5×10^7	4×10^{-5}	–
*ts*234	$<1 \times 10^3$	2×10^8	5×10^{-6}	–
*ts*201 X 352	7×10^6	1×10^8	7×10^{-2}	7
*ts*201 X 447	3×10^6	3×10^7	1×10^{-1}	10
*ts*201 X 234	2×10^4	3×10^8	7×10^{-5}	0
*ts*352 X 447	2×10^7	1×10^8	2×10^{-1}	20
*ts*352 X 234	3×10^6	1×10^8	3×10^{-2}	3
*ts*447 X 234	3×10^6	4×10^7	8×10^{-2}	8

[a]From Fields and Joklik [240] with permission.

identified, making 3-factor crosses possible [138b, 138c]. This mutation results in the more rapid migration of the $\mu 1$, and the $\mu 2$ polypeptide in a high-resolution polyacrylamide gel electrophoresis system containing tris-glycine buffers, than is seen for wild type $\mu 1$ and $\mu 2$. Aberrantly migrating $\mu 1/\mu 2$ complexes have been identified in 4 of the mutant classes: A (*ts*201), D (*ts*357), F (*ts*556), and G (*ts*453) (Fig. 8.7). The presence of these mutations, designated $\mu 1^-/\mu 2^-$, has allowed crosses of the following type to be analyzed: *ts*X$^+$ *ts*Y $\mu 1^+/\mu 2^+$ X *ts*X *ts*Y$^+$ $\mu 1^-/\mu 2^-$. *ts*$^+$ recombinants are selected from among the progeny of such a cross and are subsequently analyzed for the presence of $\mu 1^+/\mu 2^+$ or $\mu 1^-/\mu 2^-$. Only a limited number of clones from 4 3-factor crosses have been analyzed. Two crosses (*ts*D X *ts*B and *ts*D X *ts*C) yielded approximately 50 : 50 segregation of the $\mu 1/\mu 2$ phenotype among the *ts*$^+$ progeny [138b, 138c]. These preliminary results are currently interpreted to indicate that no linkage can yet be demonstrated among mutants of the different groups [138b, 138c].

Complementation between reovirus mutant pairs from different groups appears to be very inefficient [240]. Ito and Joklik [243] have assayed complementation by measuring the synthesis of RNA in cells mixedly infected at 39°C, as compared to the synthesis directed by infections with single mutants. However, since reassortment occurs at 39°C [244], appreciable numbers of *ts*$^+$ are clearly generated. The contribution of these *ts*$^+$ progeny

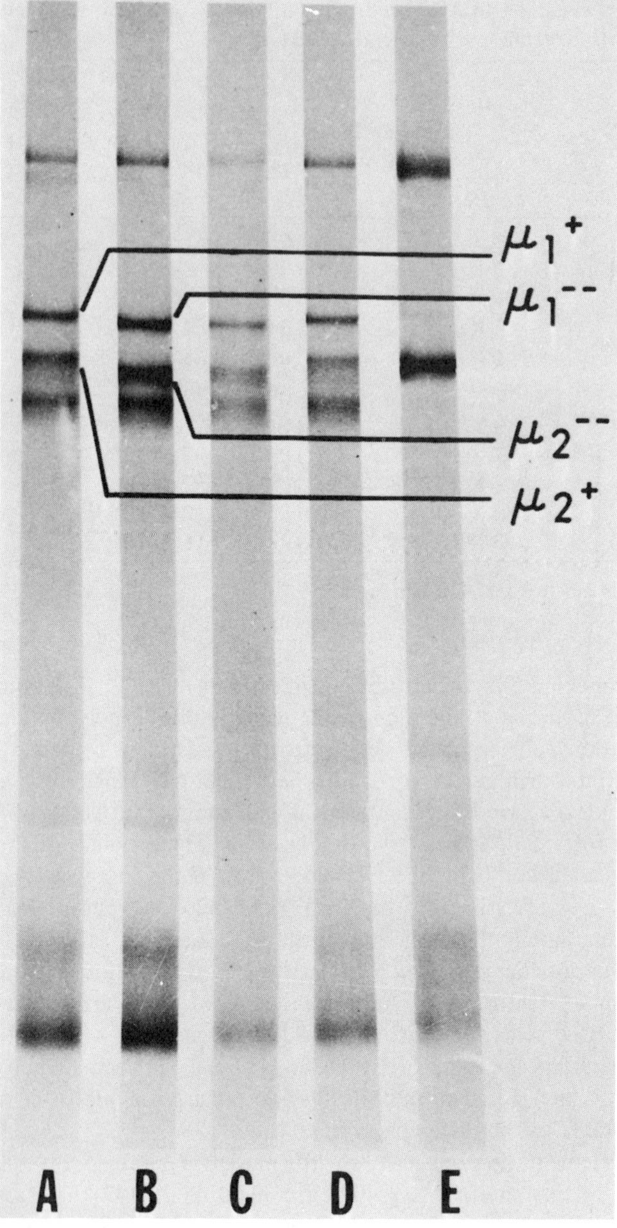

Fig. 8.7 Autoradiograph of viral polypeptides induced by wild type and a group D (*ts*357) mutant. Electrophoresis was from top to bottom on a tris-glycine polyacrylamide gel. Note the difference in migration between wild type $\mu 1^{+}/\mu 2^{+}$ (A and D) and *ts*357 $\mu 1^{--}/\mu 2^{--}$ (B and C). (D) is purified reovirions. From Cross and Fields [138c] with permission.

to the increase in synthesis measured in such a test may be so substantial as to obscure any increases caused by true complementation. Complementation has recently been demonstrated between deletion and temperature-sensitive mutants of reovirus [245, 246] (see Section 8.VI.C.).

Multiplicity reactivation has been shown to occur following the exposure of reovirions to large doses of ultraviolet irradiation [106]. Multiplicity reactivation has also been shown to occur between the 3 reovirus serotypes, suggesting that reassortment can occur in heterologous infections with different serotypes [106].

Deletion mutations have also been demonstrated in reovirus. Graham et al. [48, 123, 195, 196] showed that if the outer protein coat was removed from high-passage wild type virus, followed by buoyant density centrifugation in CsCl, 2 density populations of cores resulted. The lighter population specifically lacked the largest (L1) genome segment. The percentage of the population with the L1 deletion increased with passage up to the ninth passage, when it stabilized at a defective : nondefective ratio of 4 : 1. Subsequently, Schuerch et al. [247] have shown that newly cloned stocks of 2 mutant groups, C(*ts*447) and F(*ts*556), when passaged at the permissive temperature, give rise to particles with deletions at a lower passage level than wild type. *ts*447 immediately gives rise to L3 deletions, and L1 deletions at higher passages. *ts*556 gives rise to particles lacking either L3 or both L1 and L3.

Several nongenetic variables could possibly affect the results of genetic studies with reovirus. Among these variables are (a) aggregation of complementing *ts* mutants; (b) stable heterozygosity; and (c) temperature at which virus stocks are propagated. Extensive sonic vibration and filtration, known to disperse virus aggregates, have no significant effect on the percentage of recombinant progeny from a mixed infection [242]. This strongly suggests that ts^+ progeny plaques do not result from the infection of a single cell with a clump of complementing mutants. The possibility of stable heterozygosity has been eliminated as an important mechanism in nonpermissive plaque formation by showing that virus of all densities has the same specific infectivity [242]. Additionally, a few $\mu1^+/\mu2^+/\mu1^-/\mu2^-$ heterozygous clones found among the progeny of the 3-factor crosses were shown to segregate at the first passage into 1 or the other of the phenotypes [138c, 147]. The temperature at which the virus stocks were grown can affect the percentage recombination observed in genetic crosses. Wild type virus grown at 39°C has an efficiency of plating (EOP) (39°C/31°C) of 0.6-1.0 whereas virus grown at 31°C has an EOP of 0.3-0.6 [242]. Since the temperature-sensitive mutants are grown at 31°C, the EOP at 39°C is reduced, thus decreasing the frequency of recombinants actually observed. This may account for some of the deviation of observed recombination frequencies from those expected if reassortment of genome segments were totally random.

B. Physiology of Temperature-Sensitive Mutants

The 7 mutant groups defined by "reassortment" tests fall into 3 general types on the basis of their physiological behavior. Group A and F mutants rapidly uncoat; synthesize normal amounts of SS RNA, DS RNA, and viral polypeptides; and finally assemble morphologically normal virions. The synthesis of SS and DS RNA is delayed and significantly reduced in B and G group mutants. Viral polypeptides are also synthesized in reduced amounts and subviral particles lacking the outer coat are assembled. Mutants of groups C, D, and E synthesize only very small amounts of SS RNA and virtually no DS RNA. Synthesis of viral polypeptides is correspondingly reduced and empty virions accumulate (Table 8.3).

Transcription by the various mutants has been studied both in vitro and in vivo [85, 199, 243, 248]. The in vitro synthesis of SS RNA by cores does not differ significantly from that of wild type for all mutant groups except G. With the G mutant in vitro, SS RNA synthesis is reduced at all temperatures. In vivo A and F mutants produce normal amounts of RNA but the maximal rate of synthesis is obtained somewhat earlier than with wild type. B and G mutants synthesize only 20-30% of the SS RNA produced by ts^+. C, D, and E mutants synthesize only minute amounts of SS RNA, ranging from 3 to 8% of wild type yields. In every group all 10 SS RNA species can be detected by hybridization to unlabeled genome RNA. No mutant group reproduces the selective transcription pattern (only 4 SS RNAs made) observed when an "early" mode of transcription is "locked in" by the presence of cycloheximide [194].

Three groups of mutants, C, D, and E, are blocked in the synthesis of DS RNA [85, 243]. The other groups synthesize amounts comparable to wild type (A and F) or somewhat reduced (B and G). The DS RNA made by those mutants capable of DS RNA synthesis represents the full complement of 10 segments. Temperature shift experiments reveal that among those groups blocked in DS RNA synthesis, once synthesis has been initiated, shiftup causes an abrupt cessation of synthesis with C and D mutants, but once synthesis has initiated in E mutants, shiftup has no effect [243].

The study of viral polypeptide synthesis at nonpermissive temperatures is obscured by high backgrounds of cellular synthesis. Fields et al. [148] developed a method for immune precipitation of viral polypeptides which circumvents this problem. In general, the amounts of virus-specific polypeptides synthesized by each mutant class is proportional to the amount of SS RNA made by that mutant group [148]. Analysis by high-resolution gel electrophoresis suggests that all of the viral polypeptides are made by all of the mutant groups. Four of the mutant classes (A, D, F, and G) synthesize $\mu1$ and $\mu2$ polypeptides which migrate slightly faster than the corresponding products of wild type (Fig. 8.7). These aberrantly migrating polypeptides are

Table 8.3 Properties of *ts* Mutant Groups

Group (prototype)	Percent synthesized			Morphogenesis	Cell killing	Aberrant DS RNA hybrid	Aberrant proteins
	DS RNA	SS RNA	Protein				
A (*ts*201)	100	100	100	Full virion	Like wild	L2	$\mu 2^{-}$
B (*ts*352)	25-50	25-50	25	Corelike	Minor	–	–
C (*ts*447)	<0.1	~5	5-10	Outer capsid shell	Minor	–	–
D (*ts*357)	<0.1	~5	10-20	Mixed capsid shell	Intermediate	M2	$\mu 2^{--}$
E (*ts*320)	<1	~5	5-10	Uncertain	Intermediate	S3	–
F (*ts*556)	50-100	50-100	50-100	Full virion	Like wild	–	$\mu 2^{-}$
G (*ts*453)	15-25	~20	25	Corelike	Like wild	–	$\mu 2^{--}$

synthesized at both restrictive and nonrestrictive temperatures [138b, 147]. Ito and Joklik [249] describe an altered $\mu 2$ polypeptide in the D class mutants detected by limited chymotrypsin digestion. The relationship of the altered $\mu 2$ described by Ito and Joklik and the aberrant migration of the $\mu 1/\mu 2$ complex in classes A, D, F, and G, described above, has not yet been investigated.

Analysis of the synthesis of viral oligonucleotides reveals that whereas all mutants synthesize them at permissive temperature, none makes them at restrictive temperature [220]. These data support the notion that synthesis of oligonucleotides is a very late event in the viral replication cycle.

Studies of the accumulation of viral antigen in infected cells shows mutants of groups A and F are identical to wild type in that large coalescing masses of antigen fill the cytoplasm. The D mutant gave intermediate size masses of antigen, whereas groups B, C, and G induced a pattern of small, discrete masses of uniform size. The leaky E mutant gave variable results [174]. Ultrastructural examination of infected cells showed that A and F mutants assembled particles morphologically identical to wild type virions. B and G mutants accumulated cores with a diffuse outer coat, whereas C and D mutants accumulated empty virions, with the C particles appearing as empty outer shells and D particles a heterogeneous mixture of empty single and double shells. The E class mutant gave morphologically complete particles at high MOI and virtually no structures at low MOI [174].

Biochemical examination of morphogenesis in the mutants has only just begun. Morgan and Zweerink [221] found that B and G mutants accumulate particles sedimenting at 400 S and having a density of 1.40 g/cc. These particles contain all 10 segments of DS RNA as well as the 4 core polypeptides $\lambda 1$, $\lambda 2$, $\mu 1$, and $\sigma 2$. They differ from enzymatically prepared cores in that they contain variable amounts of the outer capsid polypeptides $\mu 2$ and $\sigma 3$, as well as the nonstructural polypeptide $\mu 0$. These particles appear to be normal morphogenetic intermediates. The C group mutant *ts*447 when infected at 33°C accumulates empty particles which have a bimodal distribution in density gradients [222]. The denser component resembled top component from wild type having all 7 structural polypeptides. The less dense component specifically lacked 2 of the core polypeptides, $\lambda 1$ and $\sigma 2$.

Infection of cells with wild type virus causes the coating of microtubules in the vicinity of the viral inclusions. Similar coated microtubules are found in cells infected with mutants of groups A, C, D and F. Cells infected with B or G mutants have markedly reduced numbers of coated microtubules [174].

C. Specific Gene Lesions

The assignment of specific mutations or groups of mutations to specific genome segments has not yet been done with any certainty. Attempts have been made by both genetic and biochemical studies.

Ito and Joklik [199] noted that certain hybrid RNA molecules composed of (+) RNA strands from *ts* mutants and (–) strands from wild type migrate more slowly on polyacrylamide gels than the reciprocal hybrids or parental DS RNA molecules. Twelve group A mutants have anomalous L2 RNAs and 1 group A mutant (*ts*256) has anomalous L2 and S3 RNAs; 1 group D mutant (*ts*357) has anomalous L1 and M2 RNAs; the group E mutant (*ts*320) has anomalous L2 and S3 RNAs. No mutants in classes B, C, F, and G showed alteration in electrophoretic migration rate. From these data and the electrophoretic behavior of revertants, Scheurch and Joklik [248] have proposed that the class A mutation was on the L2 segment, the class E on the S3 segment, and the class D on the M2 segment.

Spandidos and Graham [246] showed that virus with the L1 segment deleted was complemented by *ts* mutants from all groups except C. They therefore assigned the group C mutation to the L1 genome segment.

Mixed infections between *ts* mutants of reovirus type 3 and ts^+ of serotypes 1 or 2 promise to be helpful in gene assignments. Since the size of numerous polypeptides as resolved by high-resolution polyacrylamide gel electrophoresis differs between the serotypes [153], selection of ts^+ recombinants should reveal if the ts^+ phenotypes of serotype 3 progeny is linked to replacement of a specific polypeptide with a corresponding ts^+ peptide of differing size.

Although the following approach does not lead to the assignment of specific mutant groups to specific genome segments, it has been helpful in correlating gene products with the genome segments in which they are encoded. Comparison of the SS RNAs made in the presence of a cycloheximide block (ℓ1, m3, s3 and s4) [194] to the polypeptides synthesized immediately after release from the drug (λ2, μ0, σ2A and σ3) [121] strongly suggest the assignment of specific polypeptides to the segments which code for them is as follows. Polypeptide λ2 is encoded in ℓ1 mRNA; μ0 in m3 mRNA, σ2A in S3 mRNA, and σ3 in S4 mRNA [138a, 147]. This example illustrates the potential error in assigning gene products to genome segments on the basis of size as determined by polyacrylamide gel electrophoresis. The polypeptide μ0 is the most slowly migrating polypeptide in the phosphate-urea gel system, suggesting that it is encoded in the largest M genome segment M1. However, in the tris-glycine gel system, μ0 migrates significantly more rapidly, suggesting that its molecular weight is less than it appeared to be in the phosphate-urea system. Thus, on the basis of behavior in the tris-glycine system, one can suggest that this polypeptide is encoded in the smallest of the M genome segments, M3 (see Fig. 8.6).

VII. Other Double-Stranded RNA Viruses

Reovirus is a member of a large family of viruses the genome structure and morphology of which are similar. This family, the Reoviridae, consists of 4 groups. (a) The genus *Orthoreovirus* contains mammalian and avian reoviruses. (b) The genus *Orbivirus* contains many viruses originally classified as arboviruses because of their association with insect vectors. Included in the group are bluetongue virus, African horsesickness virus, and Colorado Tick fever virus, among others. (c) The cytoplasmic polyhedrosis virus group contains numerous serotypes of cytoplasmic polyhedrosis virus. (d) The plant virus group includes wound tumor virus, rice dwarf virus, Fiji disease virus, and rice black-streaked dwarf virus [30]. The criteria which place these viruses in the family Reoviridae are (a) a quasi-spherical morphology exhibiting some elements of icosahedral symmetry and a particle diameter of 60-80 nm. (b) The genome consists of double-stranded RNA having a molecular weight of about 15×10^6 daltons distributed among 10-12 molecules or genome segments. The Reoviridae have, in addition, a double capsid, although variations in capsid structure occur throughout the family.

Two other virus groups have been identified as having double-stranded RNA genome. One of these groups includes a number of viruses isolated from fungi, the mycoviruses. This group has a single capsid which resembles the core of reovirus. The number of genome segments is smaller than in the Reoviridae. The segments appear to be encapsidated individually, giving rise to multicomponent profiles on gradients and apparently requiring a full complement of particles for infection. This unique encapsidation gives rise to the possibility of extracellular genetic mixing [250].

The second nonreoviridae group has a single member, the *Pseudomonas phaseolicola* phage $\phi 6$. The virion is polyhedral, has only a single capsid, and has an envelope which is required for infectivity. The genome has been fractionated into 3 molecular weight species [250].

The majority of the work done on DS RNA viruses, except reovirus, has been descriptive. A number of recent reviews have discussed the double-stranded viruses [30, 250, 251]. Numerous other reviews have been written on specific viruses of this group: reovirus [15, 30, 70, 137, 232, 252-254], bluetongue virus [255], cytoplasmic polyhedrosis virus [256], wound tumor virus [257, 258], and maize rough dwarf virus [259].

Wood [250] has divided the DS RNA viruses into 3 groups on the basis of virion morphology. The first group, including those with double capsids, can be subdivided into three categories: (a) viruses having 2 distinct capsids of similar symmetry—reovirus, wound tumor virus, and possibly rice dwarf virus; (b) viruses having double capsids, the outer capsid being diffuse and having little distinct structure—Bluetongue virus, African horsesickness virus, and probably Colorado tick fever virus; (c) viruses having a double capsid with projections or spikes extending from the outer capsid—cytoplasmic polyhedrosis virus and maize rough dwarf virus. The second morphological group consists of those viruses with a single capsid, the mycoviruses. The third group, those viruses with membranes that are required for infection, has a single member, $\phi 6$.

The double-stranded RNA viruses share several common biological properties. They all replicate in the cytoplasm of the host cell, in ill-defined aggregates of virus-specified material called "factories." The factories generally are associated with arrays of microtubules or spindle fibers, although in the case of reovirus the association with tubules is not necessary for replication [177]. Many of these viruses have been shown to have virion-associated RNA-dependent RNA polymerases. This enzyme may be a necessary component of all DS RNA viruses since known host enzymes do not transcribe double-stranded RNA.

Recent Developments

Recently, the reovirus-like agents associated with acute nonbacterial gastroenteritis have been intensively studied. Since these viruses (examples: SA-11, "O" virus, epidemic diarrhea virus of infant mice, human rotavirus) differ morphologically and in genome structure from both *Orthoreoviruses* and *Orbiviruses,* a new virus group *Rotavirus* has been suggested [262-264].

A new model for the structure of the reovirus capsid has been presented [265] in which capsid subunits are shared. Activation of the virion transcriptase in cores has been shown to have no gross effect on the physical properties of the core [266]. Selective removal of protein from cores at high pH has been used to show that polypeptide $\lambda 2$ is the spike protein [267]. Examination of particles from infected cells has shown that transcriptase particles contain polypeptides $\lambda 1$, $\lambda 2$, $\mu 0$, and $\sigma 2$ and replicase particles contain $\lambda 1$, $\lambda 2$, $\mu 2$, $\sigma 2$, and $\sigma 3$ [268]. Particles that sediment at 400 S and 600 S have been shown to be intermediates in reovirus morphogenesis [269].

The structural polypeptide $\mu 2$ has been shown to be glycosylated in addition to being phosphorylated [270]. Polypeptide $\sigma 3$ has been demonstrated to strongly bind to DS RNA and $\sigma 2A$ to strongly bind to SS RNA [271].

Samuel and Joklik [272] have shown that reopolypeptides made in an in vitro system are initiated with formyl-methionine. Sen et al. [273] have shown that protein-synthesizing extracts made from interferon-treated cells do not methylate reo mRNA.

Persistent infections of CHO cells with reovirus have been reported [274].

Spandidos and Graham [275] have developed an infectious center assay to measure recombination and complementation. These authors have found additional evidence for a cellular repressor of reovirus transcription. Avian reovirus does not permissively infect L cells, but does make the 4 early mRNAs. Coinfection of L cells with avian and mammalian reo allows the avian virus to synthesize all 10 mRNAs. However, no avian reovirus appears in the yield of such an infection, suggesting that avian and mammalian reoviruses are genetically isolated [276-278]. In addition, Spandidos and Graham have been unable to demonstrate recombination between an L1 deletion mutant and *ts* mutants, although complementation does occur [275].

The polypeptides and genome RNAs synthesized by the 3 reovirus serotypes can be distinguished following electrophoresis [153]. Efficient reassortment of genome segments between *ts* mutants of type 3 and serotypes 1 and 2 have been demonstrated [279]. Analysis of recombinants between *ts* mutants and other serotypes allows construction of a map correlating genome segments between serotypes and assignment of *ts* lesions to genome segments. The following assignments have been made: *ts*A resides on genome segment L2, *ts*C on S2, *ts*D on L1, *ts*E on S3, and *ts*G on S4 [279, 280]. Some of these gene assignments conflict with those made previously (see Section 8.VI.C).

A revertant of the *ts*A mutant *ts*201 has been shown to be a pseudorevertant in which the *ts* phenotype of the group A lesion is suppressed by an extragenic suppressor mutation [281].

References

1. A. B. Sabin, Science, 130, 1387 (1959).
2. L. Rosen, Am. J. Hygiene, 71, 242 (1960).
3. L. Rosen, Ann. N.Y. Acad. Sci., 101, 461 (1962).
4. N. F. Stanley and P. J. Leak, Am. J. Hygiene, 78, 82 (1963).
5. N. F. Stanley, P. J. Leak, G. M. Grieve, and D. Perret, Aust. J. Exp. Biol. Med. Sci., 42, 373 (1964).
6. H. Kawamura, F. Shimiza, M. Mueda, and H. Tsubahara, Natl. Inst. Anim. Hlth. Quart., 5, 115 (1965).
7. L. Rosen, H. E. Evans, and A. Spickard, Am. J. Hyg., 77, 29 (1963).
8. L. Rosen and F. R. Abinanti, Am. J. Hyg., 71, 250 (1960).
9. R. N. Hull, J. R. Minner, and J. W. Smith, Am. J. Hyg., 63, 204 (1956).
10. N. F. Stanley, Nature, 189, 687 (1961).
11. R. N. Hull, J. R. Minner, and C. C. Marcoli, Am. J. Hyg., 68, 31 (1958).
12. D. H. Simpson, A. J. Haddow, J. P. Woodall, M. C. Williams, and T. M. Bell, E. Afr. Med. J., 42, 708 (1965).

13. J. W. Hartley, W. P. Rowe, and J. B. Austin, Virology, 16, 94 (1962).
14. H. Kawamura and H. Tsubahara, Natl. Inst. Anim. Hlth. Quart., 6, 187 (1966).
15. L. Rosen, in Virology Monographs, Vol. I (S. Gard, C. Hallauer, and K. F. Meyers, eds.) Springer-Verlag, Vienna, 1968, pp. 73-107.
16. W. D. Leers and K. R. Royce, Can. Med. Ass. J., 94, 1040 (1966).
17. I. Spigland, J. P. Fox, L. R. Elveback, F. E. Wasserman, A. Kelter, C. D. Brandt, and A. Kogan, Am. J. Epidemiol., 83, 413 (1968).
18. G. G. Jackson, R. L. Muldoon, G. C. Johnson, and H. F. Dowling, Am. Rev. Resp. Dis., 88, 120 (1963).
19. L. Krainer and B. E. Aronsen, J. Neuropathol., 18, 339 (1959).
20. R. A. Joske, D. D. Keall, P. J. Leak, N. F. Stanley, and M. H.-I. Walters, Arch. Int. Med., 113, 811 (1964).
21. E. DeLavergne, D. Olive, and M. T. LeMoyne, La Presse Med., 73, 951 (1965).
22. J. R. Tillotson and A. M. Lerner, N. Eng. J. Med., 276, 1060 (1967).
23. J. H. Edmonson, S. J. Millian, M. Goodenow, and S. L. Lee, J. Infect. Dis., 121, 438 (1970).
24. T. M. Bell, A. Massie, and M. G. R. Ross, Br. Med. J., 1, 1212 (1964).
25. T. M. Bell, A. Massie, and M. G. R. Ross, Br. Med. J., 1, 1514 (1966).
26. R. F. Bishop, G. P. Davidson, I. H. Holmes, and B. J. Ruck, N. Eng. J. Med., 289, 1096 (1973).
27. T. H. Flewett, A. S. Bryden, and H. Davies, Lancet, 1973-II, 1497 (1974).
28. R. Bortolussi, M. Szymanski, R. Hamilton, and P. Middleton, Pediatr. Res., 8, 379 (1974).
29. A. Z. Kapikian, H. W. Kim, R. G. Wyatt, W. J. Rodriguez, S. Ross, W. L. Cline, R. H. Parrott, and R. N. Chanock, Science, 185, 1049 (1974).
30. W. K. Joklik, in Comprehensive Virology (H. Frankel-Conrat and R. R. Wagner, eds.), Plenum, New York, 1974, pp. 231-334.
31. G. G. Jackson and R. L. Muldoon, J. Infect. Dis., 128, 811 (1973).
32. T. Y. Lou and H. A. Wenner, Am. J. Hyg., 77, 293 (1963).
33. M. N.-I. Walters, R. A. Joske, and P. J. Leak, Br. J. Exp. Pathol., 44, 427 (1963).
34. T. R. Hobbs and C. C. Mascoli, Proc. Soc. Exp. Biol. Med., 118, 847 (1965).
35. S. A. Hassan, E. R. Rabin, and J. L. Melnick, Exp. Mol. Pathol., 4, 66 (1965).
36. W. D. Kundin, C. Liu, and J. Gigstad, J. Immunol., 97, 393 (1966).
37. N. F. Stanley, P. J. Leak, M. N.-I. Walters, and R. A. Joske, Br. J. Exp. Pathol., 45, 142 (1964).
38. B. N. Fields, N. Eng. J. Med., 287, 1026 (1972).
39. G. Margolis, L. Kilham, and L. Gonatas, Lab. Invest. 24, 91 (1971).
40. C. S. Raine and B. N. Fields, J. Neuropathol. Exp. Neurol., 32, 19 (1973).
41. C. S. Raine and B. N. Fields, Am. J. Pathol., 75, 119 (1974).

42. P. H. Lamont, J. H. Darbushire, P. S. Dawson, A. R. Omar, and A. R. Jennings, J. Comp. Pathol., 78, 23 (1968).
43. E. L. Massie and E. D. Shaw, Am. J. Vet. Res., 27, 783 (1966).
44. O. T. Preble and J. S. Younger, J. Infect. Dis., 131, 467 (1975).
45. J. S. Rhim, L. E. Jordan, and H. D. Mayor, Virology, 17, 342 (1962).
46. R. E. Smith, H. J. Zweerink, and W. K. Joklik, Virology, 39, 791 (1969).
47. J. A. Farrell, J. D. Harvey, and A. R. Bellamy, Virology, 62, 145 (1974).
48. M. Nonoyama, Y. Watanabe, and A. F. Graham, J. Virol., 6, 226 (1970).
49. P. J. Gomatos and I. Tamm, Proc. Natl. Acad. Sci. U.S., 49, 707 (1963).
50. M. T. A. Fouad and R. Engler, Z. Naturforsch., 216, 706 (1966).
51. S. C. Silverstein, M. Schonberg, D. H. Levin, and G. Acs, Proc. Natl. Acad. Sci. U.S., 67, 275 (1970).
52. A. R. Bellamy, L. Shapiro, J. T. August, and W. K. Joklik, J. Mol. Biol., 29, 1 (1967).
53. H. D. Mayor, R. M. Jamison, L. E. Jordan, and M. van Mitchell, J. Bacteriol., 89, 1548 (1965).
54. A. J. Shatkin, J. D. Sipe, and P. Loh, J. Virol., 2, 986 (1968).
55. R. B. Luftig, S. Kilhaus, A. J. Hay, H. J. Zweerink, and W. K. Joklik, Virology, 48, 170 (1972).
56. J. D. Harvey, J. A. Farrell, and A. R. Bellamy, Virology, 62, 154 (1974).
57. R. Engler and M. T. A. Fouad, Arch. Ges. Virusforsch., 20, 29 (1967).
58. P. J. Gomatos, I. Tamm, S. Dales, and R. M. Franklin, Virology, 17, 441 (1962).
59. P. J. Gomatos and I. Tamm, Science, 140, 997 (1963).
60. R. Langridge and P. J. Gomatos, Science, 141, 694 (1963).
61. S. Arnott, F. Hutchinson, M. Spencer, M. H. F. Wilkins, W. Fuller, and R. Langridge, Nature, 211, 227 (1966).
62. P. J. Gomatos and W. Stoeckenius, Proc. Natl. Acad. Sci. U.S., 52, 1449 (1964).
63. A. K. Kleinschmidt, T. H. Dunnebacke, R. S. Spendlove, F. L. Schaffer, and R. F. Whitcomb, J. Mol. Biol., 10, 282 (1964).
64. A. R. Bellamy and W. K. Joklik, J. Mol. Biol., 29, 19 (1967).
65. A. J. Shatkin and J. D. Sipe, Proc. Natl. Acad. Sci. U.S., 59, 246 (1968).
66. A. M. Lerner, J. D. Cherry, and M. Finlay, Virology, 19, 58 (1963).
67. L. Rosen, J. F. Hovis, F. M. Mastrota, J. A. Bell, and R. J. Heubner, Am. J. Hyg., 71, 258 (1960).
68. N. F. Stanley, Br. Med. Bull., 23, 150 (1967).
69. A. D. McCrae, Ann. N.Y. Acad. Sci., 101, 455 (1962).
70. C. J. Gauntt and A. F. Graham, in The Biochemistry of Viruses (H. B. Levy, ed.), Marcel Dekker, New York, 1969, pp. 259-291.
71. C. Wallis, J. L. Melnick, and F. Rapp, J. Bacteriol., 92, 155 (1966).
72. J. S. Rhim, K. O. Smith, and J. L. Melnick, Virology, 15, 428 (1961).
73. R. S. Spendlove and F. L. Schaffer, J. Bacteriol., 89, 597 (1965).
74. C. Wallis, K. O. Smith, and J. L. Melnick, Virology, 22, 608 (1964).
75. L. E. Jordan and H. D. Mayor, Virology, 17, 597 (1962).
76. P. C. Loh, H. R. Hohl, and M. Soergel, J. Bacteriol., 89, 1140 (1965).

77. S. Dales, P. Gomatos, and K. C. Hsu, Virology, 25, 193 (1965).
78. C. Vasquez and P. Tournier, Virology, 17, 503 (1962).
79. J. M. Papadimitriou, Am. J. Pathol., 50, 59 (1967).
80. C. Vasquez and P. Tournier, Virology, 24, 128 (1964).
81. Y. Amano, S. Katagiri, N. Ishida, and Y. Watanabe, J. Virol., 8, 805 (1971).
82. G. Muller, C. C. Schneider, and D. Peters, Arch. Ges. Virusforsch., 19, 110 (1966).
83. H. D. Mayor and L. E. Jordan, J. Gen. Virol., 3, 233 (1968).
84. A. J. Shatkin and J. D. Sipe, Proc. Natl. Acad. Sci. U.S., 61, 1462 (1968).
85. R. C. Cross and B. N. Fields, Virology, 50, 799 (1972).
86. M. M. Brubaker, B. West, and R. J. Ellis, Proc. Soc. Exp. Biol. Med., 115, 1118 (1964).
87. P. Halonen, Ann. Med. Exp. Finn., 39, 132 (1961).
88. H. J. Eggers, P. J. Gomatos, and I. Tamm, Proc. Soc. Exp. Biol. Med., 110, 879 (1962).
89. J. R. Tillotson and A. M. Lerner, Proc. Natl. Acad. Sci. U.S., 56, 1143 (1966).
90. A. M. Lerner, E. J. Bailey, and J. R. Tillotson, J. Immunol., 95, 111 (1966).
91. L. D. Gelb and A. M. Lerner, Science, 147, 404 (1965).
92. R. Hand and I. Tamm, J. Gen. Virol., 12, 121 (1971).
93. G. Krystal, P. Winn, S. Millward, and S. Sakuma, Virology, 64, 505 (1975).
94. P. J. Gomatos and I. Tamm, Virology, 17, 455 (1962).
95. M. T. A. Fouad and R. Engler, Z. Natursforsch., 216, 706 (1966).
96. R. S. Spendlove, M. E. McClain, and E. H. Lennette, J. Gen. Virol., 8, 83 (1970).
97. J. S. Rhim and J. L. Melnick, Virology, 15, 80 (1961).
98. J. S. Rhim and J. L. Melnick, Tex. Rep. Bio. Med., 19, 851 (1961).
99. M. E. McClain, R. S. Spendlove, and E. H. Lennette, J. Immunol., 98, 1301 (1967).
100. R. M. Franklin, Proc. Soc. Exp. Biol. N.Y., 107, 651 (1961).
101. C. Wallis and J. L. Melnick, Virology, 16, 504 (1962).
102. J. Borsa, M. D. Sargent, D. G. Long, and J. D. Chapman, J. Virol., 11, 207 (1973).
103. J. Borsa, M. D. Sargent, T. P. Copps, D. G. Long, and J. D. Chapman, J. Virol., 11, 1017 (1973).
104. C. W. Hiatt, Trans. N.Y. Acad. Sci., 23, 66 (1960).
105. A. M. Rauth, Biophys. J., 5, 257 (1965).
106. M. E. McClain and R. S. Spendlove, J. Bacteriol., 92, 1422 (1966).
107. A. J. Shatkin, Proc. Natl. Acad. Sci. U.S., 54, 1721 (1965).
108. S. F. Arnott, M. H. F. Wilkins, W. Fuller, and R. Langridge, J. Mol. Biol., 27, 525 (1967).
109. W. J. Iglewski and R. M. Franklin, J. Virol., 1, 302 (1967).

110. Y. Watanabe and A. F. Graham, J. Virol., 1, 665 (1967).
111. L. Prevec, Y. Watanabe, C. J. Gauntt, and A. F. Graham, J. Virol., 2, 289 (1968).
112. J. J. Skehel and W. K. Joklik, Virology, 39, 822 (1969).
113. H. J. Zweerink and W. K. Joklik, Virology, 41, 501 (1970).
114. C. Vasquez and A. K. Kleinschmidt, J. Mol. Biol., 34, 137 (1968).
115. N. Granboulon and A. Niveleau, J. Microsc., 6, 23 (1967).
116. S. Millward and A. F. Graham, PNAS, 65, 422 (1970).
117. A. K. Banerjee and A. J. Shatkin, J. Mol. Biol., 61, 643 (1971).
118. A. K. Banerjee, R. L. Ward, and A. J. Shatkin, Nature, New Biol., 232, 114 (1971).
119. A. K. Banerjee and M. A. Grace, Biochem. Biophys. Res. Comm., 45, 1518 (1971).
120. S. Millward and M. Nonoyama, Cold Spring Harbor Symp. Quant. Biol., 35, 1518 (1971).
121. S. Millward and A. F. Graham, in Viruses, Evolution, and Cancer (A. Kustak and K. Maramorsch, eds.), Academic, New York, 1974, p. 651.
122. S. Muthukrishnan and A. J. Shatkin, Virology, 64, 96 (1975).
123. M. Nonoyama and A. F. Graham, J. Virol., 6, 693 (1970).
124. K.-I. Miura, K. Watanabe, M. Sugiura, and A. J. Shatkin, Proc. Natl. Acad. Sci. U.S., 71, 3979 (1974).
125. Y. Furuichi, S. Muthukrishnan, and A. J. Shatkin, Proc. Natl. Acad. Sci. U.S., 72, 742 (1975).
126. Y. Watanabe, L. Prevec, and A. F. Graham, Proc. Natl. Acad. Sci. U.S., 58, 1040 (1967).
127. A. R. Bellamy and W. K. Joklik, Proc. Natl. Acad. Sci. U.S., 58, 1389 (1967).
128. A. R. Bellamy and L. V. Hole, Virology, 40, 808 (1970).
129. A. R. Bellamy, L. V. Hole, and B. C. Baguley, Virology, 42, 415 (1970).
130. C. M. Stoltzfus and A. K. Banerjee, Arch. Biochem. Biophys. 152, 733 (1972).
131. J. L. Nichols, A. R. Bellamy, and W. K. Joklik, Virology, 49, 562 (1972).
132. A. R. Bellamy, J. L. Nichols, and W. K. Joklik, Nature, New Biol., 238, 49 (1972).
133. P. C. Loh and A. J. Shatkin, J. Virol., 2, 1353 (1968).
134. C. Carter, C. M. Stoltzfus, A. K. Banerjee, and A. J. Shatkin, J. Virol., 13, 1331 (1974).
135. W. K. Joklik, Virology, 49, 700 (1972).
136. R. M. Krug and P. J. Gomatos, J. Virol., 4, 642 (1969).
137. W. K. Joklik, J. Cell. Physiol., 76, 289 (1970).
138a. R. K. Cross and B. N. Fields, J. Virol., 19, 162 (1976).
138b. R. K. Cross and B. N. Fields, J. Virol., 19, 174 (1976).
138c. R. K. Cross and B. N. Fields, Virology, 74, 345 (1976).
139. G. W. Both, S. Lavi, and A. J. Shatkin, Cell, 4, 173 (1975).
140. H. J. Zweerink, M. J. McDowell, and W. K. Joklik, Virology, 45, 716 (1971).

141. D. M. Pett, T. C. Vanaman, and W. K. Joklik, Virology, 52, 174 (1973).
142. D. Roy, W. D. Graziadei, P. Lengyel, and W. Konigsberg, Biochem. Biophys. Res. Comm., 46, 1066 (1972).
143. J. Borsa, T. P. Copps, M. D. Sargent, D. G. Long, and J. D. Chapman, J. Virol., 11, 552 (1973).
144. A. J. Shatkin and A. J. LaFiandra, J. Virol., 10, 698 (1972).
145. L. J. Lewandowski and B. L. Traynor, J. Virol., 10, 1053 (1972).
146. S. A. Martin, D. M. Pett, and H. J. Zweerink, J. Virol., 12, 194 (1973).
147. R. Cross and B. N. Fields, unpublished data, 1976.
148. B. N. Fields, R. Laskov, and M. D. Scharff, Virology, 50, 209 (1972).
149. S. C. Silverstein, C. Astell, D. H. Levin, M. Schonberg, and G. Acs, Virology, 47, 797 (1972).
150. R. F. Ramig and B. N. Fields, unpublished data, 1976.
151. U. K. Laemmli, Nature, 227, 680 (1970).
152. U. K. Laemmli, personal communcation, 1974.
153. R. F. Ramig, R. K. Cross, and B. N. Fields, J. Virol., submitted for publication, 1977.
154. J. Borsa and A. F. Graham, Biochem. Biophys. Res. Comm., 33, 895 (1968).
155. A. K. Banerjee and A. J. Shatkin, J. Virol., 6, 1 (1970).
156. D. H. Levin, N. Mendelsohn, M. Schonberg, H. Klett, S. Silverstein, A. M. Kapular, and G. Acs, Proc. Natl. Acad. Sci. U.S., 66, 890 (1970).
157. A. J. Hay and W. K. Joklik, Virology, 44, 450 (1971).
158. A. M. Kapuler, Biochemistry, 9, 4453 (1970).
159. P. J. Gomatos, J. Virol., 6, 610 (1970).
160. A. K. Banerjee, R. Ward, and A. J. Shatkin, Nature, New Biol., 230, 169 (1971).
161. D. H. Levin, G. Acs, and S. C. Silverstein, Nature, 227, 603 (1970).
162. J. L. Nichols, A. J. Hay, and W. K. Joklik, Nature, New Biol. 235, 105 (1972).
163. S. Gilles, S. Bullivant, and A. R. Bellamy, Science, 174, 694 (1971).
164. N. M. Bartlett, S. C. Gilles, S. Bullivant, and A. R. Bellamy, J. Virol., 14, 315 (1974).
165. J. T. Wachsman, D. H. Levin, and G. Acs, J. Virol., 6, 563 (1970).
166. A. M. Kapuler, N. Mendelsohn, H. Klett, and G. Acs, Nature, 225, 1209 (1970).
167. J. Borsa, J. Grover, and J. D. Chapman, J. Virol., 6, 295 (1970).
168. C. M. Stoltzfus, M. Morgan, A. K. Banerjee, and A. J. Shatkin, J. Virol., 13, 1338 (1974).
169. Y. Furuichi, M. Morgan, S. Muthukrishnan, and A. J. Shatkin, Proc. Natl. Acad. Sci. U.S., 72, 362 (1974).
170. S. C. Silverstein and P. H. Shur, Virology, 41, 564 (1970).
171. V. Drouhet, Ann. Inst. Pasteur, 98, 618 (1960).
172. R. S. Spendlove, E. H. Lennette, C. O. Knight, and J. H. Chin, J. Immunol., 90, 548 (1963).

173. H. K. Oie, P. C. Loh, and M. Soergel, Arch. Ges. Virusforsch., 18, 16 (1966).
174. B. N. Fields, C. S. Raine, and S. G. Baum, Virology, 43, 569 (1971).
175. A. B. Jensen, E. R. Rabin, C. A. Phillips, and J. L. Melnick, Am. J. Pathol., 47, 223 (1965).
176. M. Anderson and F. W. Doane, J. Pathol. Bacteriol., 92, 433 (1966).
177. S. Dales, Proc. Natl. Acad. Sci. U.S., 50, 268 (1963).
178. R. S. Spendlove, E. H. Lennette, J. N. Chin, and C. O. Knight, Cancer Res., 24, 1826 (1964).
179. S. C. Silverstein and S. Dales, J. Cell Biol., 36, 197 (1968).
180. C.-T. Chang and H. J. Zweerink, Virology, 46, 554 (1971).
181. C. Astell, S. C. Silverstein, D. H. Levin, and G. Acs, Virology, 48, 648 (1972).
182. P. J. Gomatos and I. Tamm, Biophys. Biochem. Acta, 72, 651 (1963).
183. H. Kudo and A. F. Graham, J. Bacteriol., 90, 936 (1965).
184. A. J. Shatkin and B. Rada, J. Virol., 1, 24 (1967).
185. Y. Watanabe, H. Kudo, and A. F. Graham, J. Virol., 1, 36 (1967).
186. E. A. C. Follet, C. R. Pringle, and T. H. Pennington, J. Gen. Virol., 26, 183 (1975).
187. L. Prevec and A. F. Graham, Science, 154, 522 (1966).
188. M. J. McDowell and W. K. Joklik, Virology, 45, 724 (1971).
189. M. J. McDowell, W. K. Joklik, L. Villa-Komaroff, and H. F. Lodish, Proc. Natl. Acad. Sci. U.S., 69, 2649 (1972).
190. W. D. Graziadei and P. Lengyel, Biochem. Biophys. Res. Comm., 46, 1816 (1972).
191. L. Wiebe and W. K. Joklik, unpublished; cited in Ref. [30].
192. P. J. Gomatos, J. Mol. Biol., 37, 423 (1968).
193. C. M. Stoltzfus, A. J. Shatkin, and A. K. Banerjee, J. Biol. Chem., 248, 7993 (1973).
194. Y. Watanabe, S. Millward, and A. F. Graham, J. Mol. Biol., 36, 107 (1968).
195. A. F. Graham and S. Millward, in Nucleic Acid-Protein Interactions and Nucleic Acid Synthesis in Viral Infections (D. W. Ribbons, J. F. Woesner, and J. Schultz, eds.), North-Holland, Amsterdam, 1971, pp. 333-352.
196. M. Nonoyama, S. Millward, and A. F. Graham, Nucl. Acids Res., 1, 373 (1974).
197. R. Y. Lau, D. VanAlstyne, R. Berckmans, and A. F. Graham, J. Virol., 16, 470 (1975).
198. A. M. Kapuler, Biochem. Biophys. Acta, 238, 363 (1971).
199. Y. Ito and W. K. Joklik, Virology, 50, 202 (1972).
200. R. W. Floyd, R. H. Stone, and W. K. Joklik, Anal. Biochem., 59, 599 (1974).
201. A. J. Shatkin, Proc. Natl. Acad. Sci. U.S., 71, 3204 (1974).
202. R. C. Warrington, C. Hayward, and A. M. Kapuler, Biochem. Biophys. Acta, 331, 231 (1973).

203. W. D. Ensminger and I. Tamm, Virology, 39, 357 (1969).
204. R. L. Ward, A. K. Banerjee, A. LaFiandra, and A. J. Shatkin, J. Virol., 9, 61 (1972).
205. R. L. Ward and A. J. Shatkin, Arch. Biochem. Biophys., 152, 378 (1972).
206. J. K. Christman, B. Reiss, D. Kyner, D. H. Levin, H. Klett, and G. Acs, Biochem. Biophys. Acta, 294, 153 (1973).
207. D. H. Levin, D. Kyner, G. Acs, and S. C. Silverstein, Biochem. Biophys. Res. Comm., 42, 454 (1971).
208. D. H. Levin, D. Kyner, and G. Acs, Proc. Natl. Acad. Sci. U.S., 69, 1234 (1972).
209. S. Muthukrishnan, G. W. Both, Y. Furuichi, and A. J. Shatkin, Nature, 255, 33 (1975).
210. G. W. Both, A. K. Banerjee, and A. J. Shatkin, Proc. Natl. Acad. Sci. U.S., 72, 1189 (1975).
211. M. Schonberg, S. C. Silverstein, D. H. Levin, and G. Acs, Proc. Natl. Acad. Sci. U.S., 68, 505 (1971).
212. G. Acs, H. Klett, M. Schonberg, J. K. Christman, D. H. Levin, and S. C. Silverstein, J. Virol., 8, 684 (1971).
213. S. Sakuma and Y. Watanabe, J. Virol., 8, 190 (1971).
214. S. Sakuma and Y. Watanabe, J. Virol., 10, 628 (1972).
215. S. Sakuma and Y. Watanabe, J. Virol., 10, 943 (1972).
216. Y. Watanabe, C. J. Guantt, and A. F. Graham, J. Virol. 2, 869 (1968).
217. H. J. Zweerink, Y. Ito, and T. Matsuhisa, Virology, 50, 349 (1972).
218. H. J. Zweerink, Nature, 247, 313 (1974).
219. S. C. Silverstein, C. Astell, J. Christman, H. Klett, and G. Acs, J. Virol., 13, 740 (1974).
220. R. B. Johnson, R. Soeiro, and B. N. Fields, Virology, 73, 173 (1976).
221. E. M. Morgan and H. J. Zweerink, Virology, 59, 556 (1974).
222. T. Matsuhisa and W. K. Joklik, Virology, 60, 380 (1974).
223. P. C. Loh and H. K. Oie, J. Virol., 4, 890 (1969).
224. M.-H. Lai, J. J. Werenne, and W. K. Joklik, Virology, 54, 237 (1973).
225. H. Huismans, Virology, 46, 500 (1971).
226. W. D. Ensminger and I. Tamm, Virology, 39, 935 (1969).
227. W. D. Ensminger and I. Tamm, J. Virol., 5, 672 (1970).
228. R. Hand, W. D. Ensminger, and I. Tamm, Virology, 44, 527 (1971).
229. R. Hand and I. Tamm, J. Virol., 11, 223 (1973).
230. J. E. Shaw and D. C. Cox, J. Virol., 12, 704 (1973).
231. S. Silagi, Cancer Res., 25, 1446 (1965).
232. A. J. Shatkin, Adv. Virus Res., 14, 63 (1969).
233. A. J. Shatkin, Biochem. Biophys. Res. Comm., 19, 506 (1965).
234. P. C. Loh and M. Soergel, Proc. Natl. Acad. Sci. U.S., 54, 857 (1965).
235. B. Rada and A. J. Shatkin, Acta Virol., 11, 551 (1967).
236. G. W. Both, Y. Furuichi, S. Muthukrishnan, and A. J. Shatkin, Cell, 6, 185 (1975).
237. T. H. Dunnebacke and A. K. Kleinschmidt, Z. Naturforsch., 226, 159 (1967).

238. N. Ikegami and P. J. Gomatos, Virology, 36, 447 (1968).
239. N. Ikegami and P. J. Gomatos, Virology, 47, 306 (1972).
240. B. N. Fields and W. K. Joklik, Virology, 37, 335 (1969).
241. B. N. Fields, Virology, 46, 142 (1971).
242. B. N. Fields, in Virus Research (F. Fox, ed.), Academic, New York, 1973, p. 461.
243. Y. Ito and W. K. Joklik, Virology, 50, 189 (1972).
244. B. N. Fields, unpublsihed data, 1976.
245. D. A. Spandidos and A. F. Graham, J. Virol., 15, 954 (1975).
246. D. A. Spandidos and A. F. Graham, J. Virol., 16, 1444 (1975).
247. A. R. Scheurch, T. Matushia, and W. K. Joklik, Intervirology, 3, 36 (1974).
248. A. R. Scheurch and W. K. Joklik, Virology, 56, 218 (1973).
249. Y. Ito and W. K. Joklik, Virology, 50, 282 (1972).
250. H. A. Wood, J. Gen. Virol., 20, 61 (1973).
251. D. W. Verwoerd, Prog. Med. Virol., 12, 192 (1970).
252. W. K. Joklik, in Viral Replication and Cancer (J. L. Melnick, S. Ochoa, and J. Oro, eds.), Editorial Labor, S. A., Barcelona, 1973, pp. 123-152.
253. W. K. Joklik, in Virus Research (C. F. Fox and W. S. Robinson, eds.), Academic, New York, 1973, p. 105.
254. A. J. Shatkin, Bacteriol. Rev., 35, 250 (1971).
255. P. G. Howell and D. W. Verwoerd, in Virology Monographs (S. Gard, C. Hallaeur, and K. F. Meyer, eds.), Springer-Verlag, New York, 1971, pp. 37-74.
256. H. Aruga and Y. Tanada (eds.), The Cytoplasmic Polyhedrosis Virus of the Silkworm, Univ. Tokyo Press, Tokyo, 1971.
257. L. M. Black, in Encyclopedia of Plant Physiology (W. Ruhland and A. Long, eds.), Springer-Verlag, New York, 1965, pp. 236-266.
258. L. M. Black, Prog. Tumor Res., 15, 110 (1972).
259. I. Harpaz, Maise Rough Dwarf, a Plant Hopper Disease Affecting Maize, Rice, Small Grain and Grasses. Israel University Press, Jerusalem, 1972.
260. N. J. Schmidt, J. Dennis, M. N. Hoffman, and E. H. Lennette, J. Immunol., 93, 377 (1964).
261. A. J. Shatkin and G. W. Both, Cell, 7, 305 (1976).
262. G. N. Woode, J. C. Bridger, J. M. Jones, T. H. Flewett, A. S. Bryden, H. A. Davies, and G. B. B. White, Infection and Immunity, 14, 804 (1976).
263. T. H. Flewett, A. S. Bryden, H. Davies, G. N. Woode, J. C. Bridger, and J. M. Derrick, Lancet, Saturday 13 July, 61 (1974).
264. R. D. Schnagl and I. H. Holmes, J. Virol., 19, 267 (1976).
265. E. L. Palmer and M. L. Martin, Virology, 76, 109 (1977).
266. A. R. Bellamy and J. D. Harvey, Virology, 70, 28 (1976).
267. C. K. White and H. J. Zweerink, Virology, 70, 171 (1976).
268. E. M. Morgan and H. J. Zweerink, Virology, 68, 455 (1975).
269. H. J. Zweerink, E. M. Morgan, and J. S. Skyler, Virology, 73, 442 (1976).
270. G. Krystal, J. Perrault, and A. F. Graham, Virology, 72, 308 (1976).
271. H. Huismans and W. K. Joklik, Virology, 70, 411 (1976).

272. C. E. Samuel and W. K. Joklik, Virology, 74, 403 (1976).
273. G. C. Sen, S. Shaila, B. Leblen, G. E. Brown, R. C. Desrosiers, and P. Lengyel, J. Virol., 21, 69 (1977).
274. R. Taber, V. Alexander, and W. Whitford, J. Virol., 17, 513 (1976).
275. D. A. Spandidos and A. F. Graham, J. Virol., 18, 1151 (1976).
276. D. A. Spandidos and A. F. Graham, J. Virol., 19, 968 (1976).
277. D. A. Spandidos and A. F. Graham, J. Virol., 19, 977 (1976).
278. D. A. Spandidos, G. Krystal, and A. F. Graham, J. Virol., 18, 7 (1976).
279. A. S. Sharpe, R. F. Ramig, T. A. Mustoe, and B. N. Fields, submitted for publication, 1977.
280. R. F. Ramig, A. S. Sharpe, T. A. Mustoe, and B. N. Fields, submitted for publication, 1977.
281. R. F. Ramig, R. M. White, and B. N. Fields, Science, 195, 406 (1977).

Chapter 9

Biology of RNA Tumor Viruses

Raymond V. Gilden

NCI Viral Oncology Program
Frederick Cancer Research Center
Frederick, Maryland

I. Introduction

The history of the RNA tumor virus field is rather like the life cycle of a lytic virus: infection at low multiplicity, followed by eclipse with few signs of activity, and finally by explosion into a tremendous burst of activity. The explosion phase dates from the discovery of the RNA-dependent DNA polymerase [1, 2] (reverse transcriptase); this provided vindication of the heretical concepts of Temin [3], who had suggested along with Bader [4] that the RNA tumor viruses multiplied through a DNA intermediate. Of equal importance and coming in the same time frame (the last decade) has been the orderly development of routine tissue culture methods for assay of the relevant viruses, thus permitting many of the purification and characterization studies currently in vogue.

Other conceptual advances include the formulation of the oncogene theory by Huebner and Todaro [5], which states in its simplest form that information for virus synthesis resides in cellular DNA and may be totally or partially expressed depending on repressors, activators, etc., essentially according to the classical operon concepts of Jacob and Monod. A similar idea had been proposed by Bentvelzen [6] in the Netherlands to account for experimental results with mammary tumor viruses in mice. Genes controlling synthesis of viral structural components and those controlling synthesis of hypothetical oncogenic porteins are presumably linked but may be controlled in noncoordinate fashion [7]. In contrast to the concept of preexisting viral and oncogenic sequences, Temin [8] has recently proposed that these, especially the latter, arise by a series of information transfers mediated by intracellular reverse transcriptase; thus, the final oncogenic sequence may be formed in any given cell on a chance basis. This concept, termed the *protovirus theory,* thus differs from the oncogene theory in not requiring preexisting discrete oncogenic sequence but does require plasticity in generating these sequences. This concept is readily distinguished from the *provirus theory,* originally proposed by Temin [3], which suggests that viral information is integrated into cell DNA upon infection by RNA tumor viruses. Once integrated, expression of these sequences is controlled as for other cellular genes. As will be seen the oncogene and provirus theories have considerable experimental support and are not mutually exclusive. The inherited genome concept has stimulated considerable discussion and experimentation and thus, regardless of one's opinion, has beyond question served the most useful functions of any theory. In fact, the experimental data derived from nucleic acid hybridization and virus induction experiments have lent considerable support to the general theory which itself has been derived from observations on the widespread occurrence of viral proteins in tissues of certain species, primarily chickens and mice, in the absence of overt infectious virus [9]. The conceptualization and popularization of this concept is largely attributed to Huebner, whose insight has stimulated tumor virus research for many years. As with Temin, many of Huebner's concepts, including a probable role for type C viruses in normal development, were thought to be "far out." Now these ideas are widely discussed and it is noteworthy to see the frequency of "rediscovery" of these concepts in current publications.

Among the recent events of critical importance to understanding of the field is the ability of Rowe and associates to map a genetic locus for 1 leukemia virus to a specific linkage group in the mouse [10], and in addition the demonstration that this gene represents the physical viral genome by molecular hybridization assays [11]. Regulatory genes have also been mapped and thus the beginnings of the unraveling of a complex of genes controlling synthesis and ultimately malignant potential are in hand.

The discovery of xenotropic (xeno = foreign; tropic = turning) viruses in mice [12] and other species [13] is also a crucial new finding important to the understanding of natural disease. These viruses do not infect cells of their natural host but frequently have a wide host range in heterologous cells. Details of the above key areas along with descriptions of virus structure, mode of replication, interviral relationships, and biological studies are discussed in this chapter. I will attempt to give a broad view of the field and point out important ideas of current research including what appear at present to be areas of controversy. Some of these will no doubt be resolved by the time that this book reaches print; however, at worst this will give some historical perspective.

One should not forget that the key discoveries which initiated the field were made over 60 years ago in the chicken by Rous, Ellerman, and Bang demonstrating the existence of sarcoma- and leucosis-inducing viruses respectively. In the eclipse phase where such viruses were viewed as curiosities came the work of Bittner in 1942, showing the existence of mouse mammary tumor virus, and in 1951 the work of Gross, showing the existence of a mouse leukemia virus. The tremendous efforts of these individuals operating with difficult assay systems are often unappreciated yet they form the cornerstone of today's research. After the virus is clearly established the rest is relatively easy, although of course not without hard work and innovative technology. For a thorough description of the early history of the field the reader is referred to Ref. [14].

The outstanding hope is that all of the information which is accumulating can lead to positive intervention in the area of human malignancy; either prevention, interruption, or cure. The solid efforts of many researchers is gradually increasing the probability of this outcome. To me, who entered the field at a time when Temin was considered a heretic by many because of the concept of replication of RNA tumor viruses through a DNA intermediate and when the only mammalian species known to have an RNA tumor virus was the mouse, the events of the past 10 years amply justify the financial support the field has been given. We now have viruses in rats, hamsters, baboons, gibbons, pigs, cattle, and guinea pigs which are tangible and can stand the trip from laboratory to laboratory. We have the reports of virus antigens and nucleic acid sequences in man (1 area of promise and controversy). Thus, in all of these advances we have the hope for the eventual application to the human condition. There is obviously much more to be found and with the expected continued exponential increase in knowledge; the next 10 years should prove to be exciting indeed.

II. Classification

A. Principles and Current Grouping of Type C and Related Viruses

We will first consider some principles of virus classification and indicate current groupings of the reverse transcriptase-containing viruses. Although subject to varying terminologies I favor the term "retraviruses," as proposed by Parks [15], to be a broad family name for those RNA viruses which multiply via a DNA intermediate. "Retra" (reverse transcriptase-containing) indicates the key feature, e.g., mode of reproduction, as the dominant unifying factor. Temin has proposed "ribodeoxy (rnadna)" as a family name [8]. Either of these terms may be equally prevalent in current literature. This grouping includes a number of viruses which are not related to any malignant disease: visna and maedi (responsible for neurologic diseases of sheep), the syncytia-forming viruses (foamy viruses) found in many species, and certain as yet unclassified agents. The term "oncornaviruses" should perhaps be restricted at present to those viruses that either have clearly demonstrated oncogenic potential or are genetically related to known oncogenic agents by virtue of antigenic cross-reactivity or nucleic acid sequence homology. This designation is utilized primarily for classification purposes only and does not necessarily mean that all viruses designated "oncornaviruses" are oncogenic. This is also true for other viruses named for pathogenic potential, e.g., polioviruses, where all members are clearly not possessed of the same pathogenic potential. Other categories of classification include candidate oncornaviruses, which have no clear relationship to known oncornaviruses other than reverse transcriptase (RT). The prime example in this category at present is the Mason-Pfizer virus which was isolated from a rhesus monkey mammary carcinoma. Table 9.1 lists the current scheme based on a meeting chaired by Dalton and Melnick [16] in Washington and revised according to more recent information and my preferences.

The major features of oncornaviruses are (a) the presence of a high molecular weight RNA sedimenting at 60-70 S which converts to a 30-35 S species on denaturation, thus indicating subunit structure or substantial conformational change; (b) presence of an external envelope comprised of glycoprotein(s); and (c) presence of an internal "core" which is bounded by a major internal viral protein commonly referred to as the group-specific or gs antigen. This protein is cross-reactive among type C mammalian viruses and thus constitutes 1 major criterion for subgrouping of these viruses. The various viruses mature at the cell surface and bud with the central nucleoid complete (type B) or incomplete (type C). Dimensions and representative micrographs of type C viruses are shown in Figs. 9.1 and 9.2.

As regards the classification scheme itself, one should be aware that any such effort depends on the observer's bias. Accurate classification depends upon the concepts of common ancestry, which, of course, is difficult to ascertain in the absence of fossil records. We are limited in the visual sense of seeing things that look alike and can perhaps be fooled as would be an extraterrestrial being upon first sighting of a table and a cow without any other reference points. Does the presence of 4 legs constitute a reasonable basis for classification?

As mentioned above a common mode of replication seems to indicate common ancestry or derivation by a similar process from cellular components. For the type C viruses common gross morphology is built on common subunit morphology (discussed in Section 9.III.A), and for the mammalian type Cs antigenic relatedness provides the best grouping criteria available. For all of the other viruses there are no clear data of interrelatedness and very little critical data indicating absence of same. The resolution of these uncertainties will depend upon detailed primary structural analyses of both protein and nucleic acid components of the viruses.

With this brief description of classification principles we will turn to a more detailed structural description. Since the only members of the retravirus family which qualify as oncornaviruses are those of the type C and B categories (the latter restricted to mice), the principle focus of this chapter is on these viruses.

Table 9.1 Classification of Reverse Transcriptase-containing (Retra) Viruses

I. Type C Viruses[a]: Centrally located nucleoid formed during budding process, Mn^{2+} preferring reverse transcriptase in presence of synthetic templates (e.g., $rAdT_{12}$) for mammalian viruses; however, avian group enzyme shows Mg^{2+} preference. Cause leukemia or sarcoma in their hosts.
   A. Mammalian: Common polypeptide pattern, cross-reactive antigenic determinants on major internal polypeptide (p30). Isolates from mice (MuLV), rats (RaLV), Syrian (HaLV) and Chinese hamster, pig, cat (2 distinct families) represented by RD114 and feline leukemia virus (FeLV), baboon (BaLV), woolly monkey (SSV-1) and gibbon ape (GaLV).
   B. Avian: Isolates from chickens, pheasants, turkeys, and ducks: 2 distinct families with Rous sarcoma virus (chicken) as 1 prototype—at least 6 subgroups—and reticuloendotheliosis (turkey) as the second.
   C. Reptilian: One isolate from a cell line of a Russell's viper.

II. Type B Viruses: Nucleoid formed in cytoplasm, Mg^{2+} preferring reverse transcriptase in presence of synthetic templates.
   A. Mouse mammary tumor virus (MMTV)[a]: Causes mammary carcinoma in mice.
   B. Mason Pfizer Monkey virus (MP-MV).[b]
   C. Bovine leukemia virus (BLV).[b]
   D. Slow viruses of sheep, e.g., Visna, Maedi, Progressive pneumonia virus.
   E. Foamy (syncytia forming viruses): many species, primate, bovine, hamster, cat.
   F. Guinea pig endogenous virus.

[a] Generally considered oncornaviruses

[b] Candidate oncornaviruses. This classification is not intended to be definitive. The grouping of the type B category is based to a large extent on cation preference of reverse transcriptase which may or may not indicate genetic relationship. In this category MMTV is the only virus with proved oncogenic potential. Bovine and guinea pig viruses have previously been referred to as type C by some authors but are morphologically distinct, do not share interspecies determinants with mammalian type C viruses, and have Mg^{2+} preferring polymerases. For this reason they are placed in the "B" category.

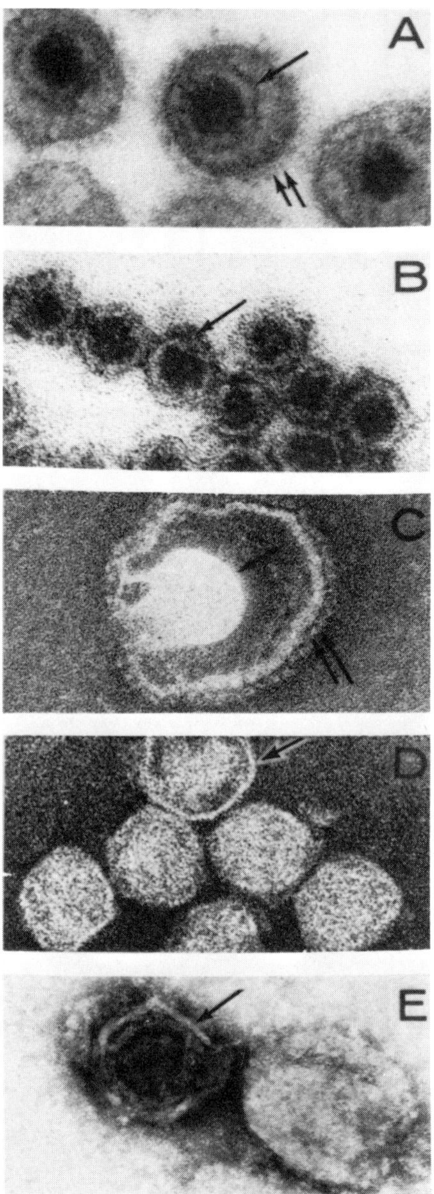

Fig. 9.1. Electron micrographs of avian myeloblastosis virus (AMV) by the thin-section technique (A and B), and the negative-stain technique using either 2% potassium phosphotungstate (C and D) or 2% uranyl acetate (E). Intact AMV virions (A, C, and right side of E) are compared with the AMV core component (B, D, and left side of E) to illustrate the peripheral projections of the envelope (double arrows) and the shell of the core (single arrow) which surrounds the coiled ribonucleoprotein subcomponent (E, left side). Electron micrographs were provided by Dr. Kurt Stromberg. Magnification of each is × 180,000.

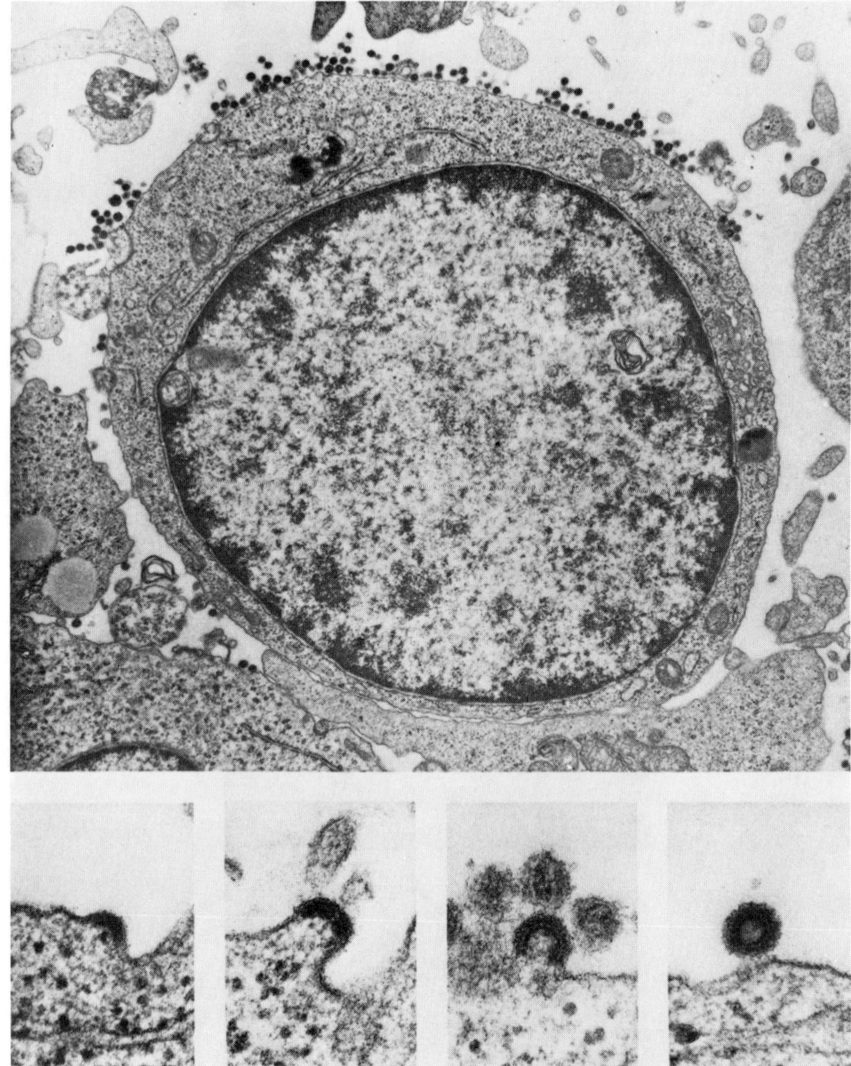

Fig. 9.2 Mouse type C virus budding from infected human lymphoblastoid cell. Above: note lack of intracytoplasmic structures. Below: stages in the budding process. Photograph courtesy of Dr. Berge Hampar.

Type C viruses have been described from species within 3 vertebrate classes: mammals, birds, and reptiles. The number of species revealing such isolates are evidently limited mainly by the intensity of experimental investigation. In most species studied type C viruses share a common group-specific antigenic reactivity which is associated with a major internal viral protein (the 1 significant exception to this rule being the cat, as will be detailed in Section 9.IV.A.). Assays for this protein thus provide an initial basis for typing of new isolates and are obviously useful for detecting possible contamination in routine tissue culture systems. Not surprisingly, the number of assays employed in contemporary studies is as varied as immunological procedures allow. For example, the routine test for multiplication of avian type C viruses in culture, the complement fixation test for avian leukosis (COFAL) [17], used complement-fixation methodology. In this test dilutions of test sample are added to chick embryo fibroblasts. At 3 weeks, after several subcultures, cells are harvested and tested for viral antigen (usually the gs antigen). Presence of antigen indicates virus replication. While this is an economical test in terms of reagent usage and ease of assay of multiple samples, gel diffusion, fluorescent antibody, radioimmunoassay, and passive hemagglutination procedures have also been used with success. Each of the latter procedures has value for special experimental situations.

B. Sarcoma Viruses

Type C viruses are divided into 2 major pathogenic categories which generally correlate both in vivo and in vitro. The sarcoma viruses (as Rous's original isolate) produce sarcomas in vivo and convert normal tissue culture fibroblasts into malignant cells in vitro. With virus dilution, transformation occurs in focal areas (Fig. 9.3) and thus provides a biological in vitro method for quantifying sarcoma viruses. The transformed foci may be isolated and shown to produce sarcomas upon inoculation into suitable hosts. In chickens there are examples of sarcoma viruses which are competent for replication functions and there are those which are unable to replicate in the absence of helper virus. In mice the sarcoma viruses studied in detail all appear to be replication defective, which provides 1 interesting class difference. The sarcoma viruses of cats (FeSV) and the woolly monkey sarcoma virus (SSV-1), which constitute the totality of known type C sarcoma viruses, can transform cells under conditions which do not allow progeny formation and thus are probably also replication defective. These virus stocks all contain excess helper virus.

C. Leukemia Viruses

The leukemia-inducing viruses isolated in vivo do not transform fibroblasts in vitro and at present there are no in vitro systems to give pathogenic correlates. The leukemia viruses are detected by their ability to induce virus-specific structural proteins in cultured cells (generally detected by immunological methods),

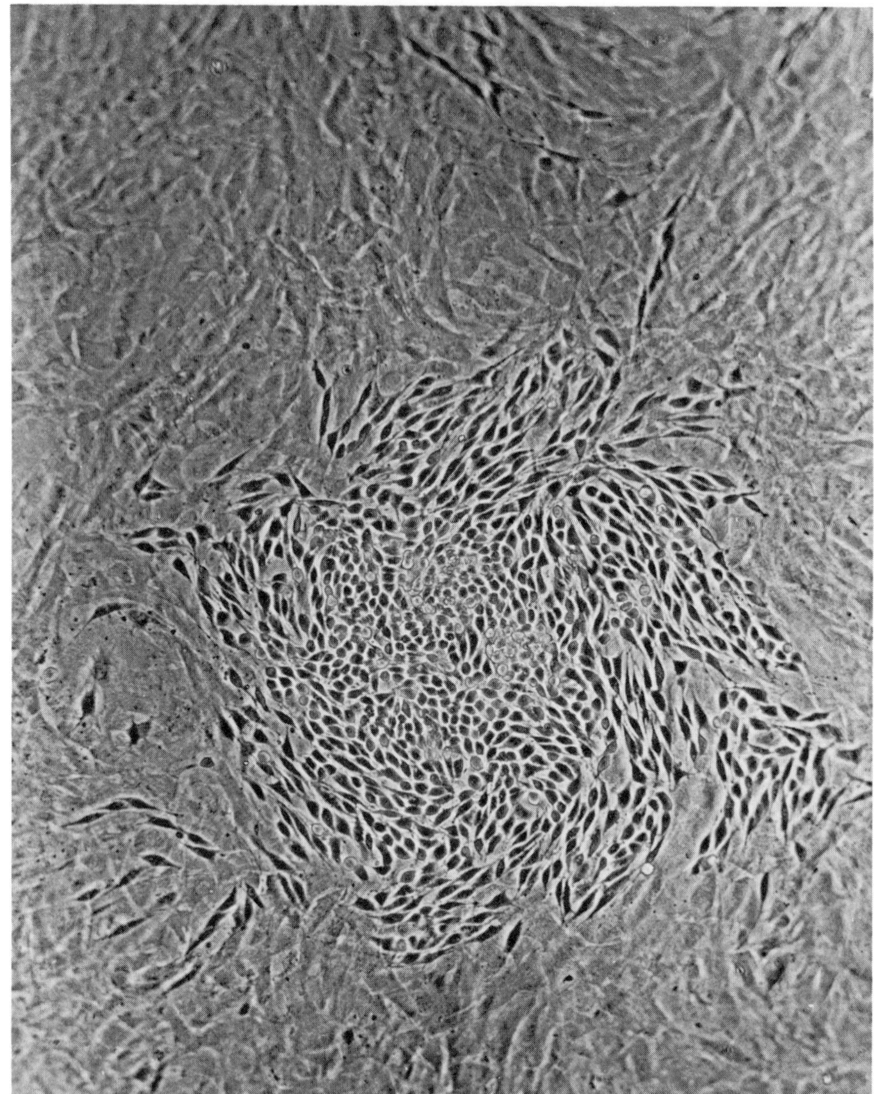

Fig. 9.3 Transformed cell focus induced by the Kirsten sarcoma virus in rat cells. Note the ease in discriminating the central area from the surrounding flat confluent fibroblastic cell sheet. Photograph courtesy of Dr. Stuart Aaronson.

and also by ability to provide helper activity for defective sarcoma viruses. At present there are numerous examples of type C viruses which can perform these 2 functions without exhibiting transforming activity. Tests for leukemogenic activity in vivo for many of these viruses have not been made, although it is routine jargon to refer to these as leukemia viruses. Thus, in reading current literature one frequently finds the terms "helper virus" and "leukemia virus" used interchangeably without prejudice as to in vivo activity.

It appears from a variety of analyses that defective chicken sarcoma viruses are unable to provide major structural proteins, particularly the envelope glycoprotein, and thus dual infection with helper virus is necessary for sarcoma virus replication. Cells transformed by defective sarcoma virus in the absence of helper virus contain the sarcoma genome which can be "rescued" by the addition of helper viruses. The resulting viruses are termed pseudotype sarcoma viruses carrying the structural proteins of the helper virus. Since cell transformation assays are rapid (e.g., about 5-7 days) and are easily quantified, the pseudotype sarcoma viruses are very useful for studies of virus host range and neutralizing properties. These viruses are written by convention with the helper in parentheses, e.g., MSV(RLV) or MSV(FeLV), which in turn designates the Rauscher and feline leukemia virus pseudotypes of the Moloney strain of murine sarcoma virus. One obvious question is the origin of sarcoma viruses in general and of the competent sarcoma viruses specifically. One possibility seems to be origin by recombination of host genes with helper viruses, as has almost certainly occurred in the case of origin of Kirsten (KiSV) and Harvey (HSV) sarcoma viruses (discussed in Section 9.VI.).

Another useful assay method which utilizes pseudotype sarcoma viruses is the development of type-specific interference to transformation in cells infected with helper viruses of the same envelope type. With the chicken viruses subgrouping based on this property parallels host range and neutralization results and has led to the distinction of at least 6 major subgroups [18], while in mice interference seems to be more broadly cross-reactive than the other procedures. Interference with the type C viruses is independent of interferon and is conceptually conceived of as a blockage of receptor sites by budding viruses of the same type. Culture-wide interference generally takes several weeks after infection with low multiplicities, e.g., 100 infectious units per plate, which is the same time as required for maximal antigen inducing activity by the helper virus. In interference assays among the mammalian viruses there is cross-reactivity between woolly monkey and gibbon ape (GaLV) viruses and also between endogenous cat (RD114 type) and baboon viruses. Relationships among these viruses are discussed in more detail in later sections.

Relationships among the chicken viruses have been worked out in great detail by several laboratories, especially in those of Hanafusa, Vogt, and Weiss [18, 19]. These will be described in some detail to illustrate some of the points made above and to acquaint the reader with the term "endogenous virus," which is often used to convey different meanings by workers in the field. In general, a virus which is genetically inherited in the host genome of all members of a species should be considered endogenous to that species. Because of manipulations and uncertain histories many of the laboratory viruses are mixtures and possibly recombinants of several types. For this reason many restrict the term "endogenous" to viruses which can be induced from virus-free cultures by various chemical treatments: halogenated pyrimidines, e.g., iododeoxyuridine (IUdR), bromodeoxyuridine (BrdU), have been used with great effectiveness for this purpose in several species [20, 21]. The viruses thus induced in chickens tend to be of one envelope type (designated subgroup E) and are represented in the genome of every chicken tested regardless of what other subgroup may also be found [22]. Similarly, in mice 1 specific envelope type is found in all strains and has the special property of growth only in heterologous cell species [12]. This virus is thus termed "xenotropic." However, there is no reason to exclude other viruses of mice or chickens from the endogenous category as, for example, the AKR mouse virus (a mouse-tropic virus) is certainly represented in the genome of AKR mice and can be induced from virus-free cells in that species.

The grouping of chicken viruses into subgroups and host range characteristics are shown in Table 9.2 and 9.3. Chicken cell types are designated as C (chicken)/letter (indication of resistance to infecting subgroup); thus, C/A are chicken cells resistant to subgroup A and C/E are chicken cells resistant to subgroup E. Resistance and susceptibility are under complex genetic control; 4 different loci have been described and certainly involve surface receptors. Subgroups A and B include the usual field isolates, while subgroup E includes the widespread chicken endogenous virus which is distinct from the well-known laboratory strain of chicken type C viruses.

The groupings are upheld by neutralization tests using chicken antisera. In a later section (Section 9.VI.) describing relationships of virus to host cells we will consider the complex story of subgroup E endogenous viruses. Of importance is the ability of this virus to recombine with viruses of other subgroups even in cells not overtly expressing virus. This can lead and has led to complexity in the extreme and finally seems to be well understood. The ability to complement defective sarcoma virus in virus-negative cells depends on a factor designated "chick helper factor" (*chf* or *f*) which correlates with a state of viral protein expression and presence of viral RNA in such cells. Phenotypic and genotypic mixing occur with high frequently in *chf*+ cells infected with appropriate viruses and one can thus see the possibility for extreme difficulty in attempting to unravel laboratory viruses. For example, at least 6 separate helper viruses have been isolated from the Bryan strain of RSV which itself (in nonproductively transformed cells) is replication defective.

The mouse leukemia viruses (MuLV) also show subgroup specificity based on neutralization tests and host-range assays. Interference properties seem broader and may well reflect the same problems of recombination and ubiquitous presence of an endogenous virus common to all mouse strains. As 1 illustration

Table 9.2 Subgroups of Avian RNA Tumor Viruses[a]

	Virus subgroup[b]					
	A	B	C	D	E	F
Sarcoma virus	SR-RSV-A	SR-RSV-B		SR-RSV-D		
	PR-RSV-A	PR-RSV-B	PR-RSV-C			
	MH-RSV		B-77	CZ-RSV		
			(BH-RSV)[c]	(HA-RSV)		
			(BS-RSV)	(FU-SV)		
Leukosis virus	RAV-1	RAV-2	RAV-7	RAV-50	RAV-0	RAV-61
	RAV-3		RAV-49		RAV-60	
	RAV-4				(*chf*)[d]	
	RAV-5					
	FAV-1					
		AMV-2				
	RIF-1	RIF-2				
	RPL-12	AEV				
	MAV-1	MAV-2				
	MC29-A	MC29-B				

[a]Courtesy of H. Hanafusa, Ref. [18].

[b]RSV, Rous sarcoma virus; SR, Schmidt-Ruppin strain; PR, Prague strain; CZ, Carr-Zilber strain; MH, Mill Hill strain; B-77, avian sarcoma virus strain B-77; BH, Bryan high-titer strain; BS, Bryan standard strain; HA, Harris strain; FU-SV, Fujinami sarcoma virus; RAV, Rous-associated virus, FAV, Fujinami-associated virus; AMV, avian myeloblastosis virus; RIF, resistance-including factor (lymphoid leukosis virus); RPL-12, avian leukosis virus strain RPL-12; MAV, myeloblastosis-associated virus; AEV, avian erythroblastosis virus; MC29, avian myelocytoma virus strain MC29.

[c]BH-RSV, BS-RSV, HA-RSV, and FU-SV are defective in the glycoprotein determining subgroup specificity. These viruses can be produced as pseudotypes of any one of the subgroups depending on the associated leukosis virus, for example, BH-RSV(RAV-1) and HA-RSV(RAV-2).

[d]*chf* (chicken helper factor) is not a virus but a product of endogenous viral genes in chicken cells. The *chf* can provide the envelope glycoprotein for defective RSV.

Table 9.3 Avian RNA Tumor Virus Subgroup Specificity: Host Range and Interference Properties[a]

	Virus Subgroup					
	A	B	C	D	E	F
Host Range						
Chicken C/O	S[b]	S	S	S	S	S
C/A	R	S	S	S	S	S
C/AE	R	S	S	S	R	S
C/BE	S	R	S	S	R	NT
C/ABE	R	R	S	S	R	NT
C/E	S	S	S	S	R	S
Japanese quail	S	R	R, S[c]	S	S	S
Ringneck pheasant	S	R	R, S	S	S	S
Duck	R	R	R, S	S	R	S
Interference						
Chicken ○ cells preinfected with:						
Virus A	R	S	S	S	S	S
Virus B	S	R	S	Partially R	Partially R	S
Virus C	S	S	R	S	S	S
Virus D	S	Partially R	S	R	NT	S
Virus E	S	Partially R	S	Partially R	R	S
Virus F	S	S	S	S	R	R

[a]Courtesy of H. Hanafusa.
[b]S, susceptible or not interfered; R, resistant or interfered; NT, not tested. Thus C/O cells are susceptible to all viruses, C/A resistant to A viruses but not B-F.
[c]Members of subgroup C have different host range for quail, pheasant, and duck cells.

of the complexity, antisera prepared against purified envelope glycoprotein of the Rauscher leukemia virus (RLV) neutralize all strains of MuLV including the xenotropic virus to approximately equivalent titer. When the xenotropic virus is isolated in heterologous cells (e.g., human or rabbit), antisera to it will not neutralize RLV but are highly specific for xenotropic virus. This would indicate that RLV from tissue culture is either a mixture or a recombinant with envelope components from the xenotropic virus. Several workers have noticed a tendency to attenuation of pathogenicity of leukemogenic viruses in tissue culture for prolonged periods. This could clearly result from selective growth of nonpathogenic variants or recombination with endogenous viruses. With conventional MuLVs an assay of great utility involves the use of the XC cell line, a Rous sarcoma virus transformed rat cell line which does not produce Rous virus but contains the RSV genome and which forms syncytia when cocultivated with MuLV-producing cells [23]. This permits a quantal assay for MuLVs and allows plaque purification procedures.

In addition to the multiplicity of chicken and mouse viruses, at least 3 subgroups of feline viruses are known as described by Sarma [24] and by Jarrett. These are distinguished based on interference, neutralization, and host range. At present there is no clear distinguishing feature for viruses from other species, including hamster, rat, and baboon. These viruses form single groups within their respective species.

III. Structure

A. Gross and Subunit Morphology

All of the oncornaviruses have a similar chemical composition containing about 1-2% RNA, 60-70% protein, 30-40% lipid, and less than 1% carbohydrate. In sucrose gradients run to equilibrium the viruses have a buoyant density of 1.16-1.18 g/cc. In thin sections both types C and B viruses have a diameter of 100-120 nm with an outer membrane and an inner core. The outer unit membrane is formed during maturation, which occurs by budding through the cell membrane. The inner core is centrally located in the case of type C viruses and has an eccentric localization in the case of type B viruses. Both cores are surrounded by an additional membrane now referred to as the core shell (Fig. 9.1). The outer membrane contains projections which are called spikes or knobs. For the B particle the length of spikes is about 100 Å and some report these to be localized in hexagonal arrays with a distance between spikes of 74 Å. These are clearly seen by negative staining. There is a clear difference in maturation of B and C particles. For C particles the first indications in viral assembly are the appearance of a crescent-shaped structure with concave orientation toward the cell center

(Fig. 9.2). Two shells are discerned of diameter ~ 50 nm (inner) and 70 nm (outer). The crescent closes at the time of budding to form a hollow sphere. The B particle core in contrast is complete before budding and particles designated intracytoplasmic A particles are seen in abundance in cells replicating MMTV.

A representation of the complete virus structure is given in Fig. 9.4 [25]. Additional features of this model are the presence of subunit structures on the intermediate membrane. This was visualized in specially prepared viral cores using the technique of freeze drying and shadowing. The repeating globular

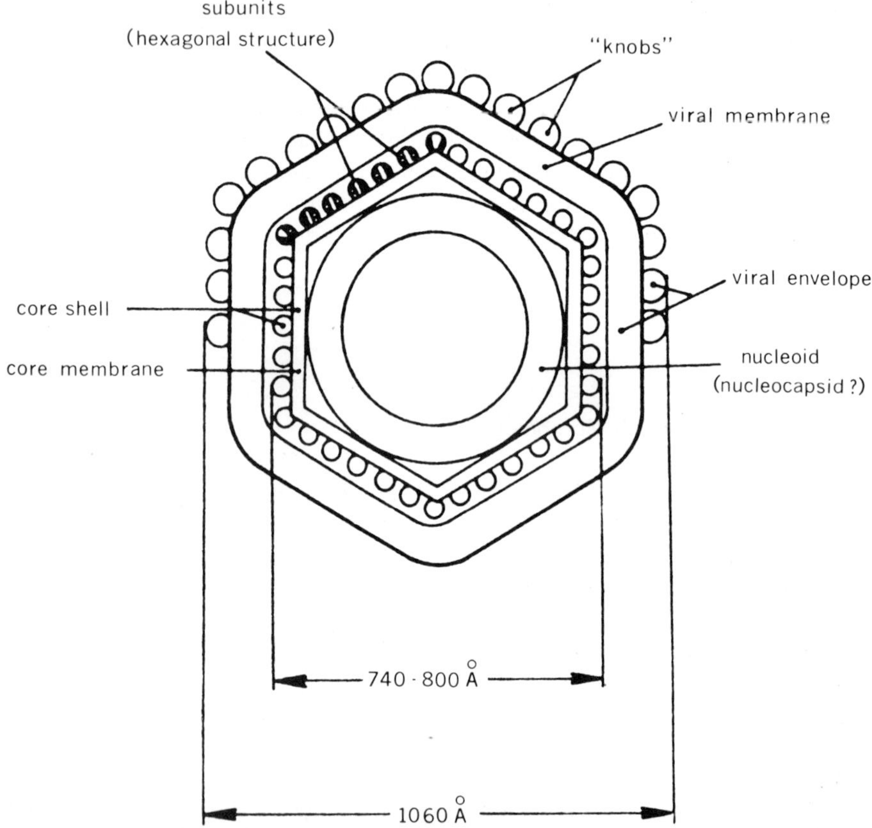

Fig. 9.4 Schematic representation of a MuLV virion in transverse section. From Ref. [25].

subunits show cubic symmetry, are 60 Å in diameter, and have a center to center distance of 75 Å. This structure is considered to be the viral capsid. This entire core structure has proved difficult to visualize and depends on delicate technique. So far the subunit structures of the core shell have been obtained by only 1 laboratory and therefore at this time are viewed as tentative. In addition, in biophysical purification attempts many workers have had difficulty in isolating core structures containing the intermediate or capsid membrane. Internal to the capsid is located the viral RNA, the RNA-dependent DNA polymerase, and 1 of the low molecular weight polypeptides. The viral nucleoprotein is found in a helical nucleocapsid that is coiled in a hollow spherical structure.

To date a large number of attempts have been made to define and characterize the subunit components that make up the composite structure of types C and B particles. These methods involve disruption of virus by detergents, reduction of disulfide bonds, and denaturation usually in urea or guanidine-HCl. Individual proteins are separated by gel filtration or other chromatographic procedures and are generally analyzed for purity by polyacrylamide gel electrophoresis (PAGE) in the presence of sodium dodecyl sulfate (SDS). The SDS-PAGE method is routine in virtually every contemporary laboratory and can be used with radioactive virus or by staining with protein, carbohydrate, or lipid stains. Isolated proteins are analyzed by immunological and biophysical methods thus forming the basis for intra- and interspecies comparisons. We will defer discussion of immunological estimates of relatedness and limit ourselves here to structure and known location of individual components. We will use the nomenclature agreed upon by a group of workers meeting at Sloan-Kettering Cancer Center in June of 1973 [26]. Viral proteins are designated by a lower case p and glycoproteins by gp according to their molecular weight in thousands. Thus p30 is a protein of 30,000 molecular weight. The antigenic properties of the isolated proteins are referred to as species (cross-reactive among multiple viruses from a single species), type (specific for a single isolate or group of isolates), and interspecies (broadly reactive among viruses from different species).

A summary of components obtained by the various separation procedures is shown in Table 9.4 and actual separations are shown in Figs. 9.5 and 9.6. The avian viruses yield 7 clear protein species; 2 glycoproteins, gp 85 and gp 35; and 5 low molecular weight polypeptides, p27, p19, p15, p12, and p10. The mammalian viruses are very similar and although a p19 is not resolved, recent data indicate the probable presence of a p19 homolog. Data on amino acid composition of the various components are now available in the literature; however, since they are not critical to the overall subject they are not given here. The key features are that individual components are readily distinguished based upon both compositional data and the number and composition of tryptic peptides obtained from the various proteins. By gel filtration in guanidine-HCl mammalian viruses give superimposible patterns; however, some workers report missing or extra

Table 9.4 Designation of Avian and Mammalian Type C Viral Proteins

	Avian	Mammalian
Glycoproteins	gp85, gp35	gp70 (or 69/71), gp45
Nucleoprotein	p12	p10
Internal core shell	p27[a]	p30[a]
Surface	p10	p12 or p15[b](p15E)
Other components with uncertain location	p15, p19	p12 or p15

[a]Approximately 3000 molecules per virion, 25-30% of total protein.
[b]There may be strain and species differences in the size of this surface component.

components. I would emphasize that single "peaks" in any separation do not necessarily indicate homogeneity and that several proteins may actually make up any 1 peak. This is evidently the case in the p15 region of the Friend strain of MuLV where at least 3 proteins have been found. This is obviously of importance in comparison of results from lab to lab since initially p15 was thought to be a core protein (1 of them probably is); however, there is now evidence of a p15 (or p17) envelope component. There are certainly other proteins in viral preparations than those listed in Table 9.4, for example see Fig. 9.7. Which if any of these additional components are viral is not certain. The problem of homogeneity may well apply to other components; however, the p30 component at least shows a single amino terminal end group and no evidence of structural heterogeneity in sequence studies.

Substructures of virus particles are obtained by selective digestion of the surface glycoproteins by proteases or detergent disruption followed by centrifugation. This is a well-known procedure for enveloped RNA viruses and is known to remove surface projections. Alternate approaches giving complementary results involve staining the virion using specific antisera labeled with the electron-dense substance ferritin. By these procedures it seems clear that the glycoproteins are located at the virus surface and form the projections referred to as knobs or spikes. The major internal proteins, p27 (avian) and p30 (mammalian), are located at the core shell. Some evidence has been accumulated to show that the p15s are also core proteins while the avian p19 cannot be clearly localized. Note also (above) that a component in the p15 range has been located at the surface of MuLVs. This component is designated p15 (E) [27] and is identical to a component originally designated p17 by Ihle and colleagues [28] using sera from normal mice in precipitation assays. p12 (avian) and p10 (mammalian) components are associated with the viral nucleic acid, evidently as a nucleoprotein

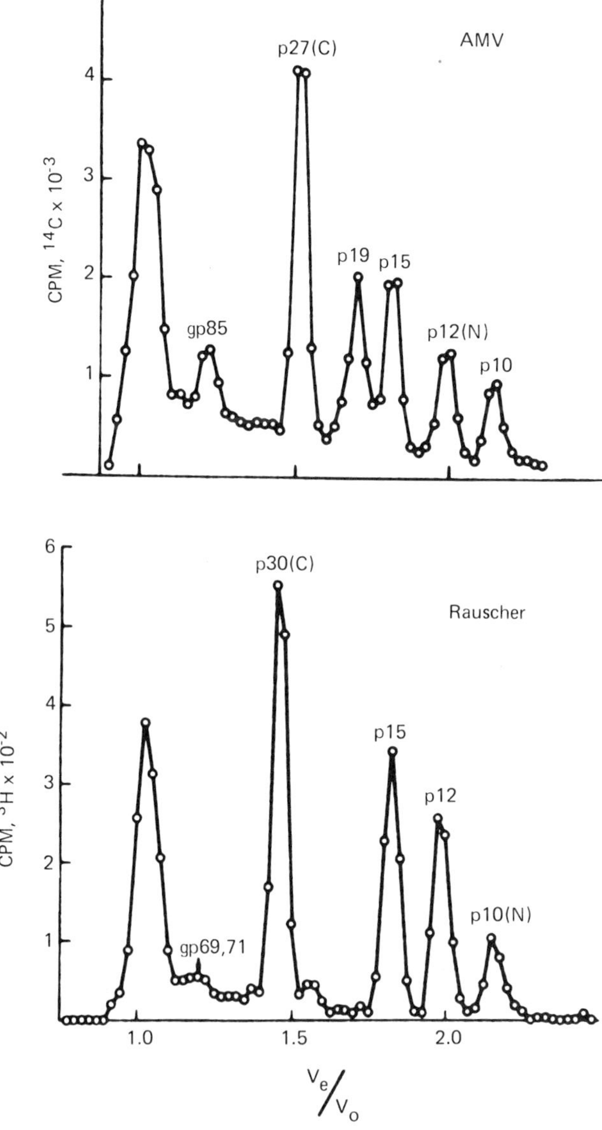

Fig. 9.5. Polypeptide profiles obtained by gel filtration in 6 M guanidine hydrochloride of avian myeloblastosis virus (AMV) and Rauscher MuLV labeled in vivo with [^{14}C]-labeled amino acids and [^{3}H]-labeled amino acids, respectively. In this procedure proteins are eluted from the gel based on relative size. Vo = column void volume. Ve = elution volume. The material at Ve/Vo = 1 (column void volume) contains multiple components which are not resolved by this procedure. From Ref. [26].

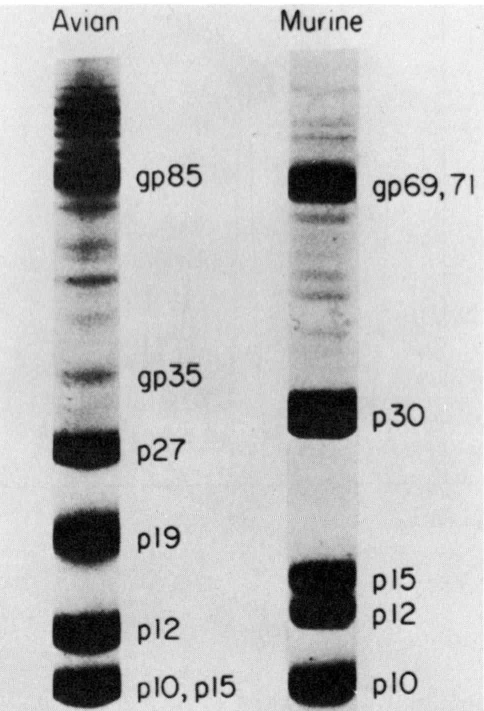

Fig. 9.6. Polypeptide composition of avian myeloblastosis virus and Friend murine leukemia virus obtained by polyacrylamide gel electrophoresis in sodium dodecyl sulfate. The location of the various proteins is shown by staining gels with a specific protein stain. Molecular weights are based on position of standard marker proteins in parallel gels (log molecular weight plotted vs. relative mobility). From Ref. [26].

complex. p10 (avian) is of uncertain origin and may be a cell surface-derived component.

In addition to these major structural components, virions contain an RNA-dependent DNA polymerase (reverse transcriptase, RT) which is obviously essential for virus replication (for detailed references see [29] and [30]). In mammalian viruses this enzyme has a molecular weight of ~ 70,000, while in avian viruses the enzyme is comprised of 2 subunits of molecular weights 110,000 and 70,000. The purified RT also possesses an enzymatic activity (ribonuclease H) which degrades the RNA portion of RNA-DNA hybrids. This activity is considered to be that of a processive exonuclease. There is some controversy at present as to the ability to separate RT and RNase activity: the bulk of the evidence favors location on a single molecule with possible difference in stability of the 2 activities. In contrast

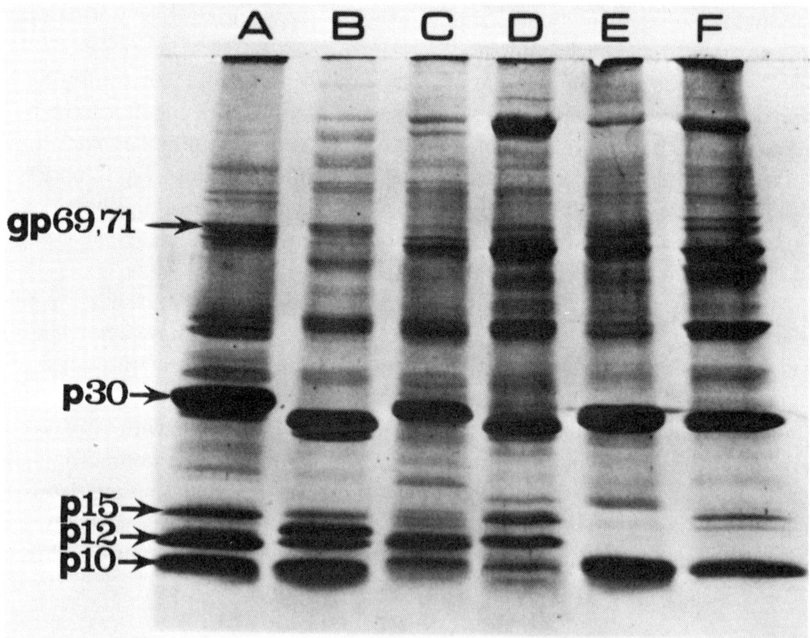

Fig. 9.7. Polyacrylamide gel electrophoresis of C-type viruses from different mammalian species in a high resolution, 7.5-20% gradient polyacrylamide slab gel in the presence of 0.1% sodium dodecyl sulfate. (A) Kirsten murine sarcoma virus, 66 μg, propagated in normal rat kidney cells. (B) Rat leukemia virus, 68 μg, propagated in Lewis rat embryo fibroblasts. (C) Hamster leukemia virus, 72 μg, propagated in hamster embryo fibroblasts. (D) Theilen feline leukemia virus, 70 μg, propagated in the FL74 cat lymphoblastoid cells. (E) RD114 virus, 68 μg, propagated in human rhabdomyosarcoma (RD) cells. (F) SSV-1 (woolly monkey virus), 70 μg, propagated in NC-37 human lymphoblastoid cells. This analysis was kindly provided by Mette Strand, Dept. of Molecular Biology, Albert Einstein College of Medicine. From Ref. [26].

to p30, which may comprise 20-30% of the total virion protein and thus is represented several thousand times per virion, RT probably occurs at about 10 molecules per virion. In addition to RT, purified virions evidently contain a number of other enzyme activities, including protein kinase, DNA ligase, and even some enzymes associated with carbohydrate metabolism. There is no evidence that any of these are viral-coded products and thus we cannot consider them in any detail. Certainly there is justification for having a ligase activity, and even a role for protein kinase is easily imagined, but for some of the other activities contamination by cellular enzymes appears most likely. Extensive characterization of viral RT has been made; however, discussion in this chapter is concerned with functionality rather than detailed studies of template specificities and other properties of interest to the enzymologist.

The polypeptide composition of MMTV is given in Table 9.5. Multiple components are resolved and again the same considerations as for type C virus components apply. gp52 and gp36 appear to be constituents of the spikes since they are (a) preferentially labeled by lactoperoxidase in intact particles (gp52 only) and (b) removed preferentially from the virus surface by protease digestion, leaving bald particles. p27 is a major internal component, as is p14. The remaining proteins remain to be evaluated for virus specificity.

The polypeptide patterns of the spleen necrosis virus (SNV) isolated from ducks and the reticuloendotheliosis virus isolated from turkeys have been determined. These viruses form a second antigenic group of avian retraviruses clearly distinguishable antigenically and by molecular hybridization from the Rous group of chicken type C viruses. Table 9.6 shows a summary of results with SNV as determined by Ruechert and colleagues, utilizing the SDS-PAGE method of analysis with labeled virus. Localization of the polypeptides in the virus was determined also as described above: lactoperoxidase iodination of intact particles, proteolytic digestion (bromelain) to remove the envelope glycoproteins, and then

Table 9.5 MMTV Major Structural Proteins[a]

Glycoproteins	gp52, gp36	Located at virion surface
Proteins	p27	Major internal protein
	p14, p10	Uncertain localization

[a]Six or 7 other minor components have been noted by SDS-PAGE. Modified from Refs. [31] and [32].

Table 9.6 Major Polypeptides of Spleen Necrosis Virus[a]

Designation	Number of chains/virion	Localization
gp71	800	Surface
p30	5180	Internal
gp22	2180	Surface
p14	6860	Internal

[a]The components listed account for ~ 90% of recovered radioactivity of amino acid-labeled virus after SDS-PAGE. The remaining activity was distributed over 6 components. Note the general similarity to other retraviruses, i.e., external glycoproteins and a major internal protein of ~ 30,000 molecular weight. From Mosser et al. [33].

analysis of the remaining core structure after purification. While the pattern obtained is distinct from chicken type C viruses, there are obvious overall points of similarity to the retravirus group as a whole. Of 10 resolved components, 2 are glycoproteins associated with virus surface (gp71 and gp22); the remaining are interior and several are possibly cellular contaminants. Major low molecular components, p30 and p14, are strikingly similar in size to the mammalian virus p30 and p15 and occur in a molar ratio very similar to 1:1 which is precisely the case for mammalian viruses and the p27 and p15 of chicken viruses. It seems to me that these similarities must reflect common ancestry and that inability to detect relationships by antigenic analysis is a reflection of the limitations in sensitivity of such methods. While we have seen that the envelope glycoproteins of type C viruses are in general larger than the remaining polypeptides, the existence of a surface component of lower molecular weight than the major internal protein (p27-p30) has been noted for these viruses as well. One expects that certain estimates of size and composition will continue to be refined as methodology improves. Apparent molecular weight in gels is subject to confirmation by detailed analytical procedures, such as amino acid analyses. This has profound effects on molecular weight estimates of glycoproteins. For example, the gp70 of MuLVs contains approximately 30% carbohydrate and has a molecular weight of approximately 60,000 based on analytical procedures as opposed to the PAGE values. Variations in gel composition and even in use of different standards can lead to differences which are more apparent than real.

In general, similar polypeptide patterns are obtained for other retraviruses with some individual variations. Thus, both MP-MV and BLV contain a major internal protein, which is p27 (MP-MV) [15] and p24 (BLV) [34], and glycoproteins of 50,000-60,000 molecular weight. The guinea pig endogenous virus was resolved into 5 components, gp70, gp36, and p24, p18, and p16 [35].

As a general summary, the oncornaviruses all possess a similar gross morphology with 2 high molecular weight glycoproteins and probably 1 small surface polypeptide or glycopeptide comprising surface projections, several internal proteins comprising a core shell, and an internal nucleoprotein complex containing RT and probably 1 low molecular protein. The major protein components discussed all appear viral coded based on similar biophysical and immunological properties when grown in cells of different species.

B. Nucleic Acid

The genomes of RNA tumor viruses are generally thought to be aggregates of 3-4 high molecular weight subunits (~$3.0\text{-}3.5 \times 10^6$ daltons) and several classes of low molecular weight RNAs. The total complex sediments at 60-70 S in sucrose gradients and is converted to the smaller forms upon heat denaturation or dimethylsulfoxide treatment. While electron microscopic examination of the avian virus 70 S RNA has generally yielded long single strands, consistent with a MW of ~10^7, recent micrographs of RD114 and baboon virus RNA [36] have shown 2 associated strands with loops which apparently lie in close proximity at the 5′ end of the RNA (Fig. 9.8). Since the number of strands in the 70 S complex bears directly on the question of ploidy of the genome it is necessary to accurately know the complexity of the genome in terms of unique sequence nucleotides and to determine definitively if the 70→35 S conversion is really a dissociation of multiple units or a conformational change resulting from loss of the small associated RNAs. The current literature contains evidence that the viral genome has a complexity of ~10^7 based on hybridization kinetics and the precise alternative, ~3×10^6, based on both chemical and hybridization analysis. The hybridization assays utilized the rate of reassociation of cDNA to viral RNA in the presence of excess RNA so that the complexity of RNA would determine reaction rate. There is evidence to show that the rate of cDNA-RNA hybridization can vary depending on base composition of the RNA, thus introducing an uncertainty factor into the calculation, which itself is based on reassociation rates of standards. A more compelling argument that the genome is actually 3.0-3.5 $\times 10^6$ is based on analyses of oligonucleotides purified from uniformly labeled RNA[^{32}P] after T-1 ribonuclease digestion. This enzyme cleaves RNA at the 5′ end of guanosine residues. The actual experimental scheme is to digest with T-1 and separate the resulting oligonucleotides by 2-dimensional electrophoretic or chromatographic methods. Fractions

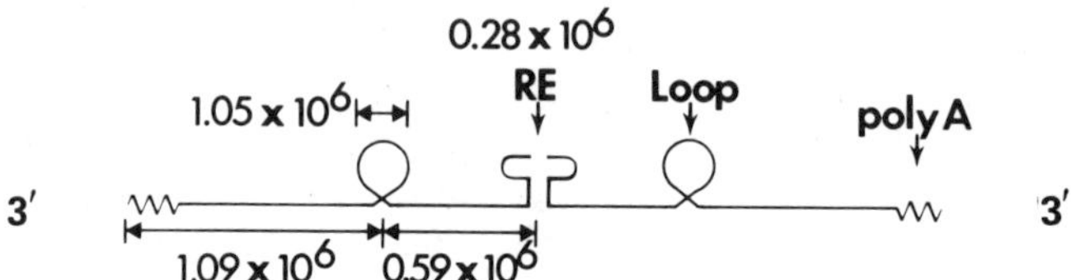

Fig. 9.8. Structure of RD114 RNA. The molecule has a molecular weight of 5.75 $\times 10^6$ and a sedimentation coefficient of 52 S. It contains 2 reproducible secondary structure features: (A) a central Y- or T-shaped structure with a molecular weight of 0.3 $\times 10^6$ (the "rabbit ears"); (B) 2 loops, symmetrically disposed around the center. By EM poly(A) mapping, there is a poly(A) stretch on the 2 outside ends of the molecule. The loops melt out under moderately denaturing conditions. The rabbit ears duplex melts under strongly denaturing conditions (equivalent to 110°C in aqueous 0.2 M Na^+, or about 63% G+C), resulting in dissociation of the 5.74 $\times 10^6$ dalton molecule into 2.28 $\times 10^6$ dalton subunits. Thus, RD 114 RNA consists of 2 2.8 $\times 10^6$ dalton subunits, joined together at their 5′ ends to give the rabbit ears feature. From Ref. [36].

containing clearly separated components are analyzed for radioactivity and then further for precise base composition by digestion with ribonuclease A (a pyrimidine-specific ribonuclease). Based on the amount of radioactivity compared to the input and the total number of bases in isolated oligonucleotides, one can calculate the precise number of nucleotides in the original RNA preparation. The calculations are improved by assay of multiple oligonucleotides, even though they represent collectively only a portion of the total genome. There are clear controls for the assay: (a) only a single guanosine residue per oligonucleotide is permitted if the oligonucleotide is pure; and (b) other bases should occur in precise molar ratio multiples of guanosine. Reports from 3 separate laboratories, which included individual and average values from at least 10 unique oligonucleotides in each study, gave the same result; namely, a molecular weight of 3.3-3.5 $\times$ 10^6. An example of data from analysis of the Prague strain of RSV is given here for 2 particular oligonucleotides (Table 9.7; [37]). This type of analysis appears highly convincing to most workers and thus we will conclude that the genomic molecular weight of the type C viruses is in the 3.0-3.5 $\times$ 10^6 MW range. The remaining decision is therefore whether there is 1 of these units per 70 S RNA or whether the 70→35 S conversion is based on conformational changes. There is no persuasive evidence on this point at present.* Careful electrophoretic analysis of cloned avian viruses has shown only a single subunit size per genome. Of interest in the chicken system is the finding that competent transforming viruses have a greater genome complexity than nontransforming viruses [38]. Recent studies on the apparent "sarcoma-specific" portion of the genome are discussed in Section 9.VI.) This finding is exactly in the opposite direction of results with mammalian sarcoma viruses in which the sarcoma-specific subunit is smaller [39] (~30 S vs. 35 S for helper viruses). This presumably relates to the replication defective nature of the mammalian sarcoma viruses.

Genetic analysis of the RNA tumor viruses, particularly the chicken viruses, give evidence of crossing over without substantial changes in subunit size [37]. This is interpreted as evidence for polyploidy although the result is not decisive. Presumably this issue will be decided by the time that this text is published.

In addition to the genomic RNA (35 S subunit) there are also several low molecular weight species of RNA which appear complexed in the 70 S unit. These include a ribosomal 5 S RNA and 10-15 species of 4 S RNA which are functional tRNAs. One of these tRNA species is of special interest since it apparently serves as the primer for RNA-directed DNA synthesis. This tRNA, which for the avian viruses appears to be a tryptophan tRNA and which can also accept methionine, is only dissociated by heating at high temperature (80°C), whereas the remaining tRNA species dissociate at much lower temperature (63°C). The laboratories of Bishop and Dahlberg [40, 41] have contributed most significantly to this area of

*The structure shown in Fig. 9.8 has now been found for other type C viruses, thus a diploid genome (2 identical subunits) seems to be the current conclusion.

Table 9.7 Genome Complexity of RSV Based on Oligonucleotide Fingerprinting[a]

Spot number	^{32}P counts spot/input (%)	RNase A products	Number of bases in genome	Complexity of genome ($\times 10^{-6}$)[b]
5	$2840/1.16 \times 10^6$ (0.24)	$U_4C_4G\ (AC)_3(AU)_3(A_3C)$	$25 \times 100/0.24 = 10{,}400$	3.3
14	$1790/1.15 \times 10^6$ (0.15)	$U_2C_4G\ (AC)_2(AU)(AAC)$	$16 \times 100/0.15 = 10{,}600$	3.4

[a]The data shown are representative of a large number of individual oligonucleotides assayed independently in 3 laboratories. Taken from the work of Beeman et al. [37].

[b]Calculated from average molecular weight of 323/nucleotide.

research. The primer appears to be identical for various strains of chicken viruses and appears to be a product of normal chicken cells.

Thirty-five S RNA subunits have little or no template activity for purified viral reverse transcriptase. Initially Duesberg and colleagues showed that 4-5 S RNA was able to reconstitute the template activity by mixture with genome RNA. At an appropriate temperature (74°C in 0.6 M NaCl) primer RNA will specifically reassociate with 35 S RNA to give an active template-primer complex as measured by ability to synthesize DNA in the presence of reverse transcriptase. The primer physically binds to the 35 S RNA with kinetics, suggesting 2-4 binding sites per dissociated 70 S genome (1 site per 35 S subunit) [43]. This at first seems to contradict the finding that endogenous reverse transcriptase products are of very small size, ~150 nucleotides, suggesting multiple initiation sites. However, very recent information indicates the possibility of synthesis of much longer sequences in vitro.* The native 70 S RNA appears to be completely saturated in terms of primer sites. Primer RNA or cellular tryptophan tRNA will bind specifically to reverse transcriptase while other tRNAs will not. In similar experiments, other tRNAs did not show binding activity or reconstitute template activity for reverse transcriptase. This lends considerable weight to the suggestion that these tRNA species are nonspecifically absorbed onto the 70 S complex during extraction. This is an obviously important technical consideration in studies where labeled 70 S RNA is used as a probe to detect viral sequences in cellular DNA. For this reason the most specific probes are those utilizing only 35 S RNA or its complementary DNA (cDNA). Residual primer RNA would not seem a problem since at most it would constitute a small portion of the total RNA.

In addition to the primer and 35 S RNAs, which are intrinsic components of the RNA of mature virions, total nucleic acid extracts of purified virus reveal ribosomal RNA, cellular mRNA in the case of certain viruses (e.g., hemoglobin mRNA in Friend virus), and even small amounts of DNA. One major problem here is that methods of viral purification are not as satisfactory as for many other viruses, and frequently "purified" virus contains cell membranes, cytoplasmic vesicles, etc. This is not to rule out the possibility that incorporation into virions is a natural phenomenon based on maturation at the cell membrane. *Nevertheless, the only species for which essential functions are known are the genomic RNA and the primer RNA.*

The nature of primer RNAs for mammalian type C viruses has not yet been reported and one awaits comparative data here with great interest.

One interesting structural feature of the genomic RNA is the presence of a relatively long (~200) stretch of adenylic acid sequences at the 3′ terminus [43]. Of interest is the report that the primer RNA attachment site is at the 5′ terminus [42], which is surprising in view of the findings that the direction

*Very recently full length infectious DNA has been synthesized in vitro.

of template transcription is 3′ to 5′. This could indicate a coiled structure in which the 5′ and 3′ ends are brought together. The significance of this preliminary finding will doubtless become clearer in the near future.* Another point of interest is the finding that sarcoma-specific sequences in avian viruses are found near the 3′ end (Section 9.VI.B.).

IV. Interspecies and Intraspecies Antigenic Relationships

A. Mammalian Type C Virus p30

As noted in Section 9.III, the mammalian type C viruses contain a major protein, provisionally assigned to the core shell, which has a molecular weight of ~30,000, thus p30. This protein has proved most useful for determining interviral relationships since it is a genetically stable marker which shows no alterations on passage of virus in heterologous hosts [44]. These proteins are assayed by a variety of methods which give supplementary results as detailed in this section.

1. p30 Antigenic Determinants: Interspecies, Type

Immunization of rats, guinea pigs, rabbits, or goats with p30 purified by a variety of techniques (isoelectric focusing has proved most efficient in my laboratory) leads to antisera useful for a variety of immunoassays. The guinea pig antisera has proved most useful for complement-fixation (CF) tests (also gel diffusion) and tend to be highly "species-specific," i.e., they react preferentially and exclusively at appropriate dilutions with viruses from the same species with several notable exceptions.

With such sera prepared against 1 strain of MuLV, all other strains may be detected without differences in sensitivity when ordinary CF microtitration or gel diffusion tests are employed. These sera are restricted in reaction to MuLV strains without exception, i.e., no reaction with p30s of any other type C viruses. This pattern is true of hamster, rat, and the conventional cat (FeLV type) p30 antisera and seemed at one point to permit the expectation that this would be a generality for type C viruses from all species. Determinants detected by these antisera using routine gel or CF tests are called group specific, or more recently species specific [26], to indicate sharing by members of a species. The finding in cats of a second type C virus group represented by RD114 virus broke this pattern completely. This virus group is distinct from

*A short identical sequence (~20 bases) of nucleotides has now been found at both 3′ and 5′ ends of viral RNA. This provides one means for generating a circular molecule of DNA known to be a critical intermediate in the replication cycle.

FeLV in a variety of properties accounting for the original thought that it could have been an activated human virus [13]. Most striking and a problem for all orderly inheritance theories was the unexpected finding that by our criteria the p30 of baboon endogenous viruses falls into a category with RD114. Thus, identity reactions are formed in gel diffusion with reciprocal reagents and either antiserum can be used to titrate antigen induction in tissue culture by homologous and heterologous viruses of the 2 types [45]. While these 2 p30s can be distinguished in radioimmunoassays (RIA), estimates of relatedness by quantitative CF were in accord with those expected of 2 members of the same virus strain (i.e., no greater than differences between various MuLVs). Another key deviation from the previous concept of 1 p30 group per species was the finding that p30s of gibbon ape (GaLV) and woolly monkey viruses (SSV-1) also carried species-specific determinants; i.e., sera prepared in guinea pigs against 1 virus detected not only itself but the other primate virus as well in CF, gel diffusion tests, and radioimmunoassays [44, 46]. Of great recent interest is the finding that endogenous viruses of the pig and a southeast Asian rodent (*Mus caroli*) contain p30s which are more closely related to the SSV-1/GaLV group than to any other group [47]. The *Mus caroli* result is in keeping with the molecular hybridization analyses, indicating a possible horizontal transmission from a rodent species to the 2 primates [48]. Thus, we find that 1 species can have 2 p30 groups and 2 distinct species can share 1 group. When other tests for interviral relationships, such as interference, neutralization, or molecular hybridization, are carried out they give results consistent with the p30 analyses. At present, therefore, with the exception of Chinese hamster type C p30 for which reagents have not yet been prepared but which do not react with known species-specific antisera, we can recognize 6 groups of p30 species determinants (Fig. 9.9). The results shown in the gel figure are readily duplicated in CF tests, thus allowing routine infectivity titrations using appropriate sera. The various p30 molecules contain interspecies determinants (Fig. 9.10) as well which are of obvious critical importance for identification and classification purposes. These were originally designated gs-3 and thought to be physically separable from species (originally gs-1) determinants, but subsequent studies have all agreed with the conclusions from our laboratory based on extensive analyses, that both categories of determinants were present on the same structure. p30 interspecies determinants are not found in avian or reptilian type C viruses nor by sensitive RIAs in non-type C reverse transcriptase-containing viruses. Thus, at present the presence of interspecies determinants appears to be restricted to mammalian type C viruses.

In the past 2 years, several laboratories have found that p30 molecules, at least for MuLVs, possess type-specific determinants as well. Both RIA and quantitative CF techniques show these differences for MuLV, while for Syrian hamster p30s no such differences were seen [46]. The quantitative CF technique is especially useful for such studies based on extensive use in comparative

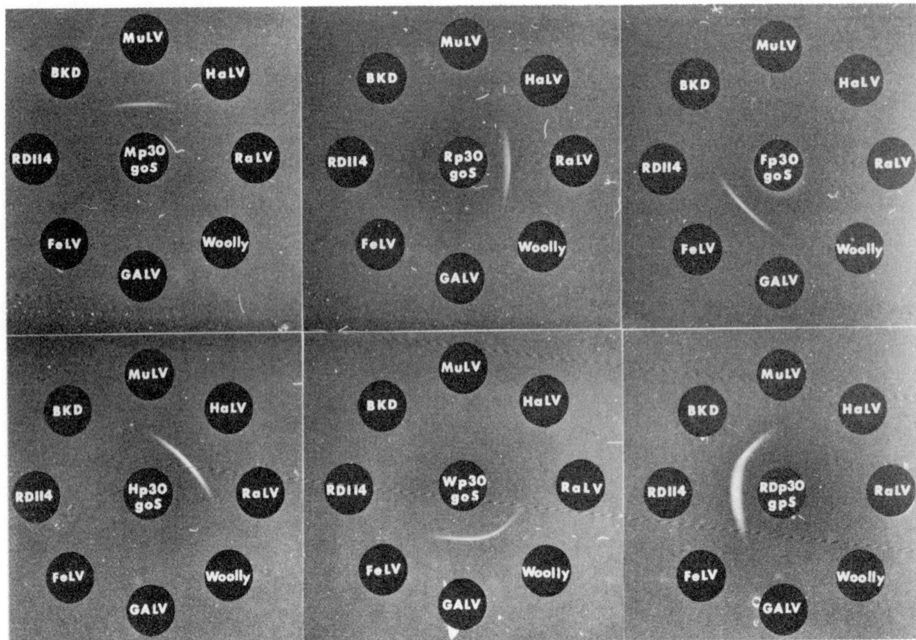

Fig. 9.9 Species-specific determinants of mammalian type C virus p30s. Purified viruses (~0.2 mg/ml) disrupted by Tween 80-ether treatment were placed in circumferential wells: Mu, mouse; Ha, hamster; Ra, rat; Ga, gibbon ape; Fe, feline; RD114 (endogenous feline); BKD, baboon. Antisera prepared in goats (go) or guinea pigs (gp) against purified p30 were placed in center wells and diffusion at room temperature in 0.8% agarose gels on microslides allowed for 48 hr. The 6 groups described in the text are clearly shown by these analyses and by complement-fixation tests with 4 units of homologous antisera. Pig endogenous viruses and the virus from *Mus caroli* also fall in the woolly, GaLV p30 group.

studies of homologous proteins. By comparison of immunological distances (related directly to peak CF with graded serum dilutions) with known primary structure differences of model proteins, we estimate that p30s of MuLV strains are 95-98% related while a similar degree of relationship exists in the SSV-1/GaLV comparison and the baboon-RD 114 comparison. The specificity of the antisera used for these studies of type-specific differences is such that they did not show cross-reactions with other p30s at dilutions which would indicate relationship of approximately 65%. The guinea pig antisera underestimate

Fig. 9.10. Interspecies determinants of mammalian type C virus p30s. Purified disrupted viruses were diffused against a goat antiserum prepared by sequential immunization with FeLV, RD114, and gibbon p30s. The left hand panel shows that this serum (R2198) reacts with all p30s including those not included in the immunization schedule. A single absorption with MuLV (center panel) leaves a reduced reaction with FeLV and RD114 and baboon p30s. Absorption with HaLV leaves only a greatly diminished reaction with FeLV. Thus, the major antibody population in this serum has interspecies reactivity. This serum does not react with avian or reptilian type C viruses nor with non-type C viruses, such as MP-MV, MMTV, bovine leukemia virus, guinea pig endogenous virus, or Visna virus.

relatedness between p30s and thus when cross-reactions occur they are highly significant. These type-specific differences among strains from within a species might be especially useful in genetic studies. Using the RIA technique this laboratory has studied 3 hamster viruses attempting to detect type-specific differences among p30s. Two of the viruses were derived from apparent in vivo rescue of the MSV genome by endogenous hamster viruses; thus it seemed possible that some relationship to MuLV p30 might be found. This was not the case, however, and the p30s of these viruses exclusively have hamster specificity (Fig. 9.11) [49].

The key points of this section are that all mammalian type C viruses contain a major structural protein of approximately 30,000 daltons with antigenic determinants which are broadly cross-reactive (interspecies reactivities), cross-reactive among different isolates from within a species (species reactivities): however, note that SSV-1-GaLV and baboon-RD114 share "species" determinants. p30s from within a specific group can exhibit type specificity as well. Quantitative estimates of type-specific differences indicate that these correspond to less than 5% difference in primary structure. For species identification it should be noted that the detection of interspecies determinants is assurance of quantitative ability to detect species determinants corresponding to known groups, if present in an "unknown" virus.

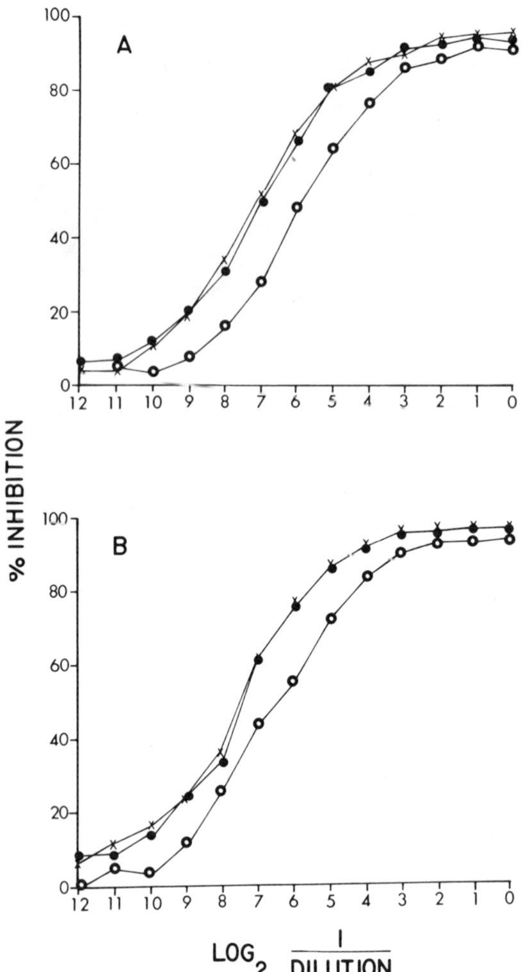

Fig. 9.11. Inability to detect type-specific determinants on hamster type C virus p30s. Competition radioimmunoassays using 20% V/V cell suspensions from cell lines producing three strains of hamster type C virus designated B-34, GLOH⁻, and G-HaLV. Two-fold dilutions were mixed with 1 ng of [^{125}I]-p30 and p30 antibody (diluted to give ca. 50% precipitation of the labeled antigen) for 3 hr at 37°C and overnight at 4°C. Anti-species gamma globulin was then added to precipitate any p30-antibody complexes. Phenylmethanesulfonyl fluoride was included to minimize proteolytic digestion of the trace labeled antigen. (A) Homologous assay using [^{125}I]-B-34 p30 and goat anti-B-34 p30. (B) Heterologous assay using [^{125}I]-B-34 p30 and guinea pig anti-GLOH⁻ p30. Symbols: B-34 (●–●); GLOH⁻ (X–X); G-HaLV (O–O). In both assays the 3 virus strains were indistinguishable based on slope of the inhibition curves and final extent of inhibition. At similar concentrations no inhibition is seen with heterologous type C viruses. Assays with analogous specificity can be established for any of the p30 species groups shown in Fig. 9.9. From Charman et al. [49].

2. *p30 Determinants: Heterogeneity of Categories*

The previous section dealt with the various reaction categories as if they were single homogeneous determinants. While the categories may conveniently be treated in this fashion, each may consist of multiple determinants in a single p30 and may be heterogeneous in the case of the interspecies category. When individual p30s are partially degraded by trypsin or modified by chemical treatment, e.g., citraconylation, multiple determinants can be detected by the use of a variety of antisera. At present we find at least 3 distinct specificities of species and interspecies nature on MuLV p30s [50]. In comparisons of various p30s there are distinct differences in interspecies determinants. For example, in RIA inhibition tests using [^{125}I]-MuLV p30 and anti-FeLV serum, MuLV, HaLV, and RaLV give complete displacement with similar inhibition slopes. In contrast, RD114, SSV-1, and GaLV give only partial inhibition of that test system [51]. Similar results are seen with [^{125}I]-HaLV p30 substituted for [^{125}I]-MuLV p30; in this case however, MuLV p30 appears to compete less efficiently than RaLV, HaLV, or FeLV but more efficiently than RD114, SSV-1, and GaLV (Fig. 9.12) [49]. For very broad RIA interspecies tests, one can use RD114 or SSV-1-[^{125}I] p30 and anti-FeLV, or as we have done prepare a broadly reacting interspecies serum by consecutive injection of different p30s. Such a serum can be used with any labeled p30 to give broad interspecies tests ($>$ 70% maximum displacement with all inhibiting antigens), and thus is of practical value. I emphasize that inhibition tests using the very broad serum are completely negative with all of the reverse transcriptase-containing viruses with the exception of the mammalian type C category.

In gel diffusion, patterns of interspecies cross-reactivity tend to agree with RIA results with some exceptions. A potent anti-FeLV goat serum gave clear reactions with all viruses with the exception of SSV-1 and GaLV, yet other goat antisera to FeLV and MuLV showed clear cross-reactivity with these viral p30s at equivalent concentration. This further indicates the heterogenity of interspecies determinants. In screening large numbers of individual bleedings from animals immunized with the various p30s, we have occasionally noted variations in cross-reactivity. These do not necessarily reflect any special genetic relationships but are probably based on restricted similarity in immunodominant antigenic sites. We emphasize the importance of absorptions to indicate the nature of a given reaction (species or interspecies) and the number of distinct antigenic reactivities associated with a given molecule. There is no doubt of heterogeneity of interspecies determinants and the practical need of the broadest test system to verify the presence or absence of such determinants in new suspected type C isolates. Serum prepared by successive injections of related antigens has the advantage of increase in titer toward those determinants which are related. With appropriate immunization schedules, this should yield

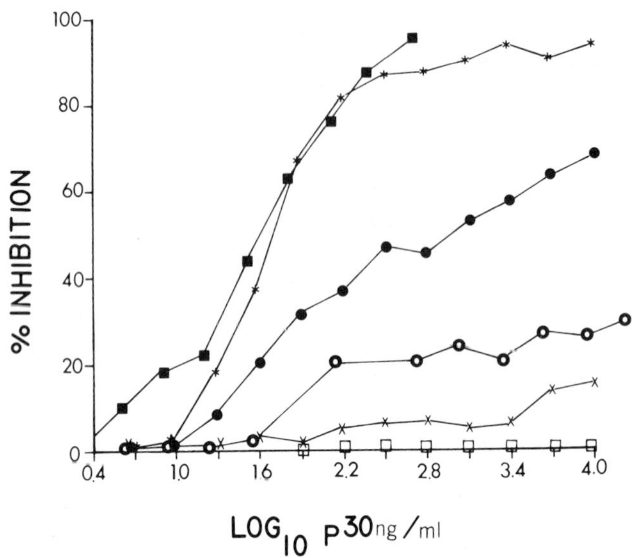

Fig. 9.12. Heterogeneity of p30 interspecies determinants. Heterologous interspecies inhibition assays with [^{125}I]-B-34 p30 (hamster) and goat anti-FeLV p30 were made with banded type C viruses treated with 0.1% Triton X-100 adjusted to estimated p30 concentrations of 10-20 μg/ml (total viral proteins 40-80 μg/ml). Symbols: hamster (■—■); FeLV (*—*); mouse (●—●); RD114 (○—○); gibbon (X—X); MP-MV (□—□). The relatively low reaction extent and shallow slopes seen with RD114 and gibbon viruses in this assay system is similar to that seen with [^{125}I]-MuLV and anti-FeLV as originally shown by Parks and Scolnick. From Ref. [49].

a serum with higher interspecies titer than that specific for the last antigen in the sequence. This prediction seems to have been met by an interspecies goat reagent prepared in this manner.

A final consideration regarding multiplicity of determinants involves the choice of terms based on different test systems. Thus, the presence of species and interspecies determinants as qualitatively different reaction categories on the same p30 molecule was established by rigorous absorption procedures. Cross-reactions seen in RIA tests, while referred to as interspecies, are not necessarily based on the same determinants seen in gel diffusion. The greater sensitivity of RIA tests allows for detection of more distant relationships; thus, for example, some "interspecies" cross-reactions could be based on detection of reactions with homologous molecular regions with 1 or more amino acid substitutions. These same regions might appear to be species determinants with less sensitive test procedures.

3. *p30: Primary Structure*

The key assumption in the use of immunological criteria to indicate species relationships is the direct correspondence to primary sequence and thus gene base sequence (Fig. 9.13). This assumption was implicit, even if precise knowledge of colinearity of gene and protein was not available, in the earliest studies of this type. More recently the relationship between primary structure and immunological cross-reactivity has been studied with precision, and it is most impressive to see the high degree of correlation which is obtained. Since both of these techniques correlate well with phylogenetic history and, in fact, are now widely used correlates of paleontological data, they are directly applicable to studies of virus relationships.

To date primary structure data is only available for comparative purposes for a limited region at the NH_2 terminal region of p30 molecules [46, 52, 53]. Even these limited data, 10-15% of the p30 molecule, allow some definitive conclusions which are generally supportive of the immunological data. Thus, of 24 residues compared, several strains of MuLV have shown single-residue differences, interestingly at the same position, number 4. If this pattern were repeated throughout the molecule, the extent of homology would be approximately 96% which correlates well with the immunological test results. This is likely to be a relatively accurate estimate since peptide maps of the various MuLV p30s show only several differences out of 35-40 resolved peptides. There is obvious homology among all the mammalian p30s, i.e., identical initial tripeptide, long stretch of homologous sequences, positions 11-24 in all. The region 4-10 (4-17 in GaLV) appears hypervariable and is an obvious candidate for species and/or type-specific immunological reactivity. The long stretch of homologous sequences provides the possibility for development of broad interspecies reagents suitable for RIA tests (2 tyrosines in 11 residues). While this may not be detected by current immunological reagents, e.g., it may be a silent region, coupling to carrier molecules should yield an immunogenic product extremely useful for probing human tissues.

The RIA result that RD114 and primate p30s are distinct in p30 reactivity seems borne out of the sequence data for these p30s. We see in RD114 and baboon virus evidence of a gap at position 5 and for GaLV insertion of a large sequence of 7 residues in this region. Although not directly related to immunological reactivity at present, these features do provide a good general fit with those results. We are thus reasonably confident that the inferred p30 relationships will be supported by a complete structure analysis when available.

We emphasize again that the only viruses to show p30 interspecies determinants are mammalian type C viruses. None of the other viruses listed in Table 9.1 shows this cross-reactivity and thus it seems a most potent criteria for subgrouping this group of viruses. In cases where negative results have suggested that the viruses in question are not typical type C, e.g., bovine and

	Position																													
	1				5					10					15					20					25					30
Cat																														
FeLV	P	L	R	E	G	P	N	N	R	P	Q	Y	W	P	F	S	A	S	D	L	Y	N	W	K	S	H	B	P	P	F
RD 114	P	L	R	T	*	V	N	R	T	V	Q	Y	W	P	F	S	A	S	D	L	Y	N	W	K	T	H	N	P	X	F
Baboon																														
BAB8-K*	P	L	R	T	*	V	N	R	T	V	Q	Y	W	X	F															
Mouse																														
R-MuLV	P	L	R	L	G	G	N	G	Q	L	Q	Y	W	P	F	S	S	S	D	L	Y	N	W	K						
AKR	P	L	R	L	G	G	N	G	Q	L	Q	Y	W	P	F	S	S	S	D	L	Y									
G-MuLV	P	L	R	L	G	G	N	G	Q	L	Q	Y	W	P	F	S	S	S	D	L	Y									
WMLV	P	L	R	S	G	G	N	G	Q	L	Q	Y	W	P	F	S	S	S	D	L	Y	N	W	K						
SLV	P	L	R	A	G	G	N	G	Q	L	Q	Y	W	P	F	S	S	S	D	L	Y	N	W	K						
M-MSV	P	L	R	A	G	G	N	G	Q	L	Q	Y	W	P	F	S	S	S	D											
Rat																														
MSV(RaLV)	P	L	R	Q	G	A	X	G	X	M	Q	Y	W	X	F															
Gibbon Ape																														
GaLV	P	L	R	A	I	G	P	P	A	E	Q	Y	W	P	F	X	X	A	X	L	Y									
										↑ (X N G L V P L)																				

Fig. 9.13. NH_2-terminal amino acid sequences of p30s from several type C RNA viruses. The 1-letter code used for amino acids is: A, alanine; B, aspartic acid or amide; D, aspartic acid; E, glutamic acid; F, phenylalanine; G, glycine; H, histidine; I, isoleucine; K, lysine; L, leucine; M, methionine; N, asparagine; P, proline; Q, glutamine; R, arginine; S, serine; T, threonine; V, valine; W, tryptophan; X, unidentified; Y, tyrosine. *Gap inserted to maintain homology at position 11 and beyond. From Ref. [52].

guinea pig viruses, other properties correlate with this conclusion. These include gross morphology and cation preference of virion reverse transcriptase [34]. Whether or not these viruses and in fact all retraviruses are related can only be established by detailed primary structure analyses of suspected protein homologs. The immunological tests are clearly too insensitive to permit detection of distant relationships, with a lower limit which requires ~60% primary structure relatedness. Current structure analyses do not allow detection of relationship between the avian type C virus p27 (an obvious candidate), but this could be based on insufficient data for comparative evaluation. This suggests, however, that relationships between type C viruses and other retraviruses will not be found and thus the grouping of retraviruses will ultimately depend only on the similar mode of replication of its members. p30 appears to be the most broadly cross-reactive of the viral proteins and even though interspecies determinants have been found for other proteins these do not extend to all type C viruses. This of course is entirely dependent on available antisera and by following an immunization protocol which emphasizes cross-reactive determinants, much broader reactive sera will be produced.

B. Mammalian Type C Virus Envelope Glycoproteins

As discussed in Section 9.III., the surface protrusions of oncornaviruses, "knobs" in the case of mouse leukemia viruses–knobbed spikes in the case of avian type C viruses, consist of 2 glycoprotein species gp85 and gp35 (avian), and gp70 and gp45 (murine). There is further evidence that an additional low molecular protein may also have surface localization (Section 9.IV.C.). There is clear evidence that the major components, gp85 and gp70, are involved in the various manifestations of surface interactions with cells and with neutralizing antibody. One may expect these components to show a high degree of type specificity since the initial basis of distinction among viruses is neutralization specificity. In considering mouse type C viruses there have historically been 2 major subgroups, G (Gross) and FMR (Friend, Moloney, Rauscher). While these can clearly be distinguished by selected sera in neutralization tests, there are clearly cross-reactions and in tissue culture interference tests, also indicators of type specificity, the 2 groups show cross-interference. This is a clear distinction from the 6 avian virus subgroups where interference is a potent tool for subclassification. More recently new groups of MuLV have been isolated, including the xenotropic viruses from all strains and an amphotropic virus (both mouse and heterospecies host range) from wild mice. The xenotropic viruses have envelope properties which are clearly distinct from the G and FMR subgroups; e.g., they do not give cross-interference and are not affected by FV-1 alleles (Section 9.VII.) introduced into foreign cells by somatic cell hybridization. The recognition of the widespread distribution of the xenotropic class of viruses in mice and observations of high-frequency recombination among avian viruses suggest the possibility of envelope recombinants among laboratory viruses. One method of activating x-tropic virus is infection with a mouse tropic virus and thus possibilities of contamination are raised. It is not surprising that anti-RLV gp70 neutralizes not only RLV but viruses of the G subgroup and xenotropic group as well. When such an antiserum is used near its end point (50% reaction) in competition radioimmunoassays with RLV gp70-[^{125}I], type-specific assays may be established. In such tests we have found that Moloney virus gives significant but not complete competition and other MuLVs have been ineffective competitors. Other mammalian viruses were clearly negative in these assays. By using a heterologous MuLV antiserum near its end point (e.g., anti-Gross) with the RLV-gp70-[^{125}I], one could detect all strains of MuLV, including the xenotropic viruses. Such an assay is termed a "heterologous intraspecies assay." Consistent with lack of cross-neutralization of nonmouse type C viruses, no inhibition of the latter assay was obtained with such viruses with 1 significant exception, this being a partial inhibition with FeLV. Strand and August [54] had originally reported the existence of interspecies determinants associated with gp70 (used here interchangeably with August's designation of gp69/71) based on similar observations. Moreover, selected anti-FeLV antisera can precipitate MuLV gp70, thus allowing development of an interspecies assay which also detects all MuLVs. Even in this assay, viruses of the SSV-1, GaLV, and baboon RD114 groups are not competitive, which indicates a high degree of relationship between cat and mouse viruses. One notes here that on other grounds, reverse transcriptase inhibition, especially MuLV, HaLV, RaLV, and FeLV, seem to be more related to each other than to the other viruses. We have not seen cross-neutralization between viruses of this group; thus the precipitation and inhibition assays appear more sensitive or equally plausibly capable of measuring relatedness over areas of the molecule not important for neutralization. Antisera to the Friend virus gp70 (actually designated gp71) appeared to cross broadly in CF tests among the mouse, rat, hamster, and feline viruses and did not react with RD114 [55]. If indeed all gp70s are products of homologous genes, one should expect to find evidence of a high degree of structural similarity as seen for p30. Selection and preparation of appropriate antisera are crucial in such studies. Recent reports by Aaronson and colleagues indicate that anti-FeLV can also precipitate SSV-1 gp70 and thus provide another interspecies assay. Among the other virus groups one notes that SSV-1 and GaLV cross-interfere and cross-neutralize but are distinguishable based on other criteria. Baboon and RD114 viruses also cross-interfere; however, while some cross-neutralization has been seen, sera may be selected which are type-specific. To the extent that information is available it is clear that other members of the retravirus family do not compete in interspecies gp70 assays.

Thus, by way of summary, mammalian gp70s are antigenic mosaics, as p30, and contain type, species, and interspecies determinants. The interspecies reactions are not generally detected in neutralization and interference tests based on either or both sensitivity and test property considerations. There are analogous situations wherein a heterologous antiserum may bind to but not neutralize a virus. In such cases addition of an antibody to the globulin fraction of the serum donor can effect neutralization.

C. Mammalian Type C Virus: Low Molecular Weight Polypeptides

As described in Section 9.III., there are 3 size classes of polypeptides with a molecular weight less than 30,000. These are designated p15, p12, and p10. Unfortunately for the student and me, there are serious literature discrepancies regarding antigenic characteristics of some of these components and a balanced view will be attempted.

One possible source of discrepancy probably is the presence of multiple unique components in the p15 region of guanidine•HCl separations. Schafer and colleagues purified 1 of 3 components in this region and found a high degree of interspecies reactivity in gel diffusion and CF tests [56]. In fact, in categorizing this protein's reactivity it was described as mainly interspecies. In contrast, the independent studies of other workers (August and Parks [57]) categorized the reactivity of p15 as mainly type-specific; in fact, p15 was not detected by Parks in 1 strain of MuLV (Kirsten) [58] and in other studies no p15 was detected in the woolly monkey virus. My colleague, S. Oroszlan, also detected multiple NH_2 terminal amino acids in similarly "purified" p15 and noted that 1 component in this size range could be preferentially eluted from the virion surface under conditions which allowed the retention of gp70. Recently others have also described a p15 molecule at the virion surface which, however, does not appear to be resolved by gel filtration in guanidine•HCl [27].

From a different approach completely, Ihle and Hanna have found that many mice produce natural antibody to envelope components of MuLV. These include gp70, gp45, and most important for this discussion a component with a molecular weight of ~17,000. This protein (possibly also a glycoprotein) was found to be identical to the p15 of Schafer and was found at the surface of intact MuLV [28]. Fleissner has termed this component p15(E) [27] as noted in Section 9.III. It thus seems possible that reasons for the differences in antigenic reactivity are based on isolation of different molecular species.

Both August and Parks did find evidence of interspecies reactivity (anti-FeLV vs. MuLV p15), but this was considered minor. The p15 reactivity in all cases was not related to determinants of gp70, p30, p12, or p10.

The low molecular polypeptides have proved most interesting and have been studied in some detail by the laboratories of Aaronson and Parks [58, 59]. Aaronson and colleagues noted that p12 was an excellent protein to utilize in type-specific assays. Various strains of MuLV could easily be distinguished and 2 classes of xenotropic viruses were also delineated. The ability to distinguish xenotropic viruses of BALB/c and NIH Swiss was also found possible by molecular hybridization. Other significant contributions of Aaronson and colleagues using the p12 assays were (a) ability to discriminate among the woolly monkey and gibbon ape viruses, in fact the two gibbon isolates studied could also be distinguished; and (b) also to distinguish RD114 from baboon viruses. We have carried out collaborative studies by molecular hybridization on viruses which by p30 analyses could have been either SSV-1 or GaLV in several cases and RD114 or baboon in others. The unknown viruses were either known or potential laboratory contaminants. In each case the viruses clearly typeable by molecular hybridization were identified precisely by the p12 assay. Because of this extreme type specificity the p12 assays are valuable for a variety of genetic studies and serve as useful monitors of new or established isolates.

Parks et al. performed similar experiments which have provided confirmatory and supplementary information. Most surprisingly it has been noted that for Kirsten and Moloney MuLVs, p12 and p10, show a high degree of intrastrain cross-reactivity in radioimmunoassay [60]. The p12 and p10 polypeptides were highly purified and based on specific activity one could safely rule out cross-contamination. Interspecies determinants on p12 or p10 were not noted in these studies; however, low-level cross-reactions are not ruled out. The levels of p12/p10 reactivity for 10 viruses of several subgroup specificities based on neutralization and cytotoxic typing assays are recorded in Table 9.8 (from Ref. [58]). The levels of reaction are expressed relative to p30 content (expected to be constant) using Kirsten and Moloney p12 or p10 assays and the Moloney p15 assay. The Kirsten virus is an N-tropic member of the Gross-AKR subgroup and as shown the 5 members tested react equally. They do not show any cross-reactivity in the Moloney p12/10 or p15 assays. In contrast, all viruses give equivalent reactions in p30 assays. Friend, Rauscher, and Moloney, which constitute the FMR subgroup, all showed reactions in both Moloney p12/10 and p15 assays; however, only Moloney virus gave complete displacement. This suggests, as has been done on other grounds, that the Moloney virus is only "distantly" related to Friend and Rauscher within the FMR group. Note that FMR viruses do not inhibit the Kirsten p12/p10 assay, nor do the xenotropic viruses of BALB/c or NIH Swiss (bottom 2 entries) give positive results in any of these test systems. Thus, taking the results of Aaronson et al.,

Table 9.8 Type-Specific Immunoassay Classification of Representative Isolates of MuLV[a]

Subgroup	Virus strain	Mouse strain of isolation	Ratio of p30 to low molecular weight immunoreactivity			
			p30[b]	K p12/10[c]	M p12/10[c]	M p15
Gross	Gross	AKR	1.0	0.12	NR	NR
	Kirsten	C3H	1.0	0.12	NR	NR
	S2C13	BALB/c	1.0	0.2	NR	NR
	AKV-1	AKR	1.0	0.06		
	AKV-2	AKR	1.0	0.06		
FMR	Friend	DBA	1.0	NR	Partial	Partial
	Rauscher	Swiss	1.0	NR	Partial	Partial
	Moloney	BALB/c	1.0	NR	0.1	0.1
Xenotropic	S16cl 10	BALB/c	1.0	NR	NR	NR
	ATS 124	NIH Swiss	1.0	NR	NR	NR

[a]From Ref. [58].

[b]Determined as in [^{125}I]-labeled Rauscher p30 assay with anti-Rauscher p30 which measured all murine viruses tested with comparable sensitivity (0.4 ng/assay) and slopes. The variation in ratios was within experimental variation (2-fold). Based on the nanograms of p30 measured, calculations of expected low molecular weight reactivity were made based on known standards.

[c]K, Kirsten; M, Moloney; NR = nonreactive

that the 2 xenotropic viruses can be distinguished based on p12 reactions, one sees the possibility of 5 mouse virus groupings: AKR-Gross, FR, M, xeno-BALB, and xeno-NIH. In terms of interviral hybridization our own data indicate a high degree of similarity among Gross-type viruses (AKR V-1, AKR V-2, Gross, Kaplan radiation leukemia virus); clear differences between these and all others; ability to distinguish Rauscher and Moloney; and also, as stated, Todaro and colleagues distinguish 2 classes of xenotropic virus [61]. Thus, the low molecular weight polypeptide assays have shown the clearest correlation with hybridization assays in terms of a high degree of type specificity. No cross-reactions have been seen between p10/p12 and any other viral proteins. While Parks and colleagues found p10 and p12 to be cross-reactive there were differences allowing the molecules to be distinguished.

Another of my colleagues, C. Long, has recently purified a p10 molecular species by ability to bind to DNA cellulose columns. Previous studies have indicated that p10 is associated with viral nucleic acid [61]. p12 does not bind to DNA-cellulose under the same conditions; thus, these proteins are functionally distinct. At this time it must be admitted that the p12-p10 relationship was unexpected; however, in FeLV evidence was provided that p15 and p12 were cross-reactive. It remains to be tested whether the p10 isolated by functional properties shows the same serological specificity as seen in guanidine·HCl purified proteins. The possibility of multiple p10s and production of one class from p12 by proteolysis must be considered as 1 possible explanation of the data of Parks et al. It will be most interesting to see primary structure data on p12 and p10.*

D. Avian Virus Polypeptides

Among the avian viruses there is evidence that the envelope glycoprotein, gp85, can be distinguished by radioprecipitin methods as expected from neutralization tests [63]. By gel diffusion analyses the low molecular proteins (p27, p19, p15, and p12) were found to be antigenically unique and to exhibit group specificity, i.e., complete cross-reactivity among various envelope subgroups [64]. By means of RIA, p19 was found to exhibit subgroup specificity while the other proteins were not distinguished. p19 is an interesting protein since it appears to be located at the amino terminus of a polyprotein precursor found in infected cells. p19 has a blocked amino terminus which is similar to the situation in reovirus precursor proteins. In studies of tryptic peptides between 2 avian viruses, p19 was found to show a single major difference, while the other proteins (p27, p15, and p12) were very similar [65]. In the avian viruses, p12 appears to be a homolog of mammalian virus p10, both being associated with viral nucleic acid. In summary, each of the

*Recent studies indicate considerable difference between p10 and p12 in both biophysical and immunologic properties; thus the evidence of relationship is suspect.

avian proteins is unique, both structurally and antigenically, and none shows cross-reactions with mammalian type C viruses.

E. Virion Reverse Transcriptase (RT)

In addition to reactivities of major structural proteins, studies of several other viral proteins have provided information useful in grouping viruses. The virion RT has been utilized to prepare specific antisera for use in enzyme inhibition studies. The results obtained, as a generality, fit well with groupings suggested by the p30 data [46]. Thus, antisera prepared against MuLV will completely inhibit all strains of MuLV and will also inhibit FeLV, RaLV, and HaLV polymerase activity. The quantitative aspects of this inhibition suggest preferential inhibition of the homologous species enzyme (species determinants), although the use of species and interspecies reactivities in this case seems less warranted. Significantly, SSV-1 and GaLV form a second cross-reactive group showing no inhibition of other viral polymerase activities at matched concentrations. More recently the pig type C virus RT has also been found to be related to the SSV-1, GaLV group as is its p30. RD114/baboon viruses from a third crossreactive group among the mammalian type C viruses.

A clearly reproducible result has been the negative reactions in comparisons of the 3 mammalian type C virus subgroups with any other of the RT-containing viruses. Thus, while the chicken type C virus polymerases show cross-reactions among themselves, they show no cross-reaction with any of the mammalian viral polymerases. At present no cross-reactions have been reported among the viruses with type B virus RTs, e.g., MMTV, MP-MV, BLV, and guinea pig virus; thus there is no evidence suggesting any genetic relationship among these viruses.

When the viral anti-RT sera were tested against normal cell polymerases the general result appeared to be complete lack of cross-reaction [8]. However, recent studies indicate the possibility of some relationship to certain specific cellular polymerases raising the obvious question of viral origin. Specifically, anti-SSV-1 antisera inhibited an enzyme activity purified from human leukemia cells by Gallo and colleagues which biochemically also appears to resemble a viral enzyme. This enzyme activity has not been found in normal blood lymphocytes stimulated with phytohemagglutinin and thus to date is highly suggestive of a viral origin. This most interesting result is subject to confirmation from other laboratories.

In a surprising qualification of findings obtained in direct assays, using serum blocking methodology (blocking by test enzyme of the ability of antiserum to inhibit a standard enzyme preparation), Temin and colleagues have shown a degree of relationship between chicken viral RT, RT of the reticuloendotheliosis virus group, and a cellular RT-like activity from chicken cells. These enzyme activities are readily distinguishable in ordinary inhibition tests so that the blocking test appears capable of detecting more distant relationships [8].

These data may be considered at several levels. For ranking purposes it seems clear that as for the p30 proteins, viruses from within a species appear indistinguishable except for the RD114-FeLV situation. Where p30 assays indicated unexpected close relationships, e.g., SSV-1-GaLV and RD114-BaLV, polymerase inhibition assays gave similar indications of close relationships. Distant relationships are detected with less certainty because of technical problems inherent in enzyme inhibition assays, e.g., serum stimulation and nonspecific inhibitors. Considering thc relationships of viral to cell enzymes which are generally not clear (negative) except for the situations described above, possibilities of distant relationships emphasized by conservation of active site regions or evolutionary convergence may be raised. Thus, the finding of RT activity in human cells inhibited by anti-WoLV serum could be interpreted as (a) detection of a virus-specific enzyme; (b) detection of a cell enzyme related in the distant past to viral enzyme, i.e., these derive from a common ancestral enzyme activity; or (c) inadvertent inclusion of cell RT in preparations used to prepare the SSV-1 anti-RT serum. Possibility (a) would indicate virus activation or horizontal transmission of an infectious agent as postulated by Gross for the ultimate origin of all tumor viruses. Possibility (b) is related to the consideration of the origin of viruses which I feel to be an extremely likely possibility, as in the protovirus theory. Possibility (c) seems to be ruled out by the lack of appropriate enzyme in normal cells.

F. Relationship Among Mouse Mammary Tumor Virus (MMTV) Strains

There are at least 5 strains of MMTV recognized as listed in Table 9.9 which are classified based mainly on their biological properties. The question has frequently been raised as to whether these strains can be distinguished on immunological grounds or not. Claims to do so have been reported; however, studies which identify the type-specific components have not been reported as yet. In fact, the major viral proteins studied to date, gp52, p27, and p14, seem to be group specific [67]. Surprisingly an antigenic reactivity originally found to be group specific and thought to reside in the nucleoid has been localized on the envelope component, gp52. There is clearly a need for comparative studies in this area; however, limitations in virus availability make this a challenging task.

Table 9.9 Properties of Strains of Mouse Mammary Tumor Viruses[a]

	Common name	Natural host	Characteristic	Transmission
MMTV-S	Bittner virus	C3H	Virulent	Milk
MMTV-P	Muhbock virus	GR	Virulent; induces hormone-dependent plaques	Milk, egg, sperm
MMTV-L	Nandi virus	C3HF	Low oncogenicity	Egg, sperm
MMTV-O	Van Leeuwenhoek virus	BALB/c	Very low oncogenicity	Egg, sperm
MMTV-X	Timmermans virus	020	Induced by x-irradiation	Egg, sperm
MMTV-Y		C57BL	Only antigens detected	

[a]Taken from Tooze [19]. So far these strains are not differentiated by type-specific immunoassays with purified polypeptides.

MMTV proteins have shown no immunological relationship with any of the other oncornaviruses, including the Mason-Pfizer virus and other viruses with magnesium-preferring reverse transcriptase.

G. Mason Pfizer Monkey Virus

The virus designated MP-MV was isolated from a rhesus monkey mammary carcinoma. This virus is of current interest because of reports of ability to transform selected cell types and presence of viral nucleic acids in human malignant breast tissue [68]. One could thus consider this a candidate oncornavirus. Recently a virus with the morphology of MP-MV was detected in certain sublines of HeLa cells (a long-term tumor cell line containing a characteristic enzyme marker) and a series of tumor cell cultures derived by workers in Russia. As part of a joint U.S.-U.S.S.R. collaborative program these viruses have been analyzed by molecular hybridization and immunoassays. By means of gel diffusion it was shown that these isolates share a common reactivity associated with a molecule of 27,000 molecular weight, p27 [15]. In our experience, in collaboration with Parks and Scolnick, it was repeatedly found that the "Russian" viruses gave complete inhibition of the MP-MV p27 radioimmunoassay with no obvious kinetic differences from MP-MV. These results were repeated in 1 other laboratory; however 1 of the isolates appeared to give an incomplete reaction. This result appears anomalous based on the finding that the various cell lines all contain HeLa cell markers.

At present the origin and significance of MP-MV is uncertain.

H. Bovine Leukemia Virus (BLV)

BLV is included as an oncornavirus or candidate oncornavirus based on (a) its association with lymphosarcoma in cattle; and (b) the reports of cell-free transmission of leukemia to sheep. Naturally and experimentally infected cattle and sheep produce precipitating and CF antibodies to a major internal protein of BLV designated p24. These immune sera do not cross-react with any other member of the retravirus family. Similarly a guinea pig antiserum to BLV p24 did not react with type C viral p30 in very sensitive radioimmunoassays. BLV p24 was nonreactive in broad interspecies radioimmunoassays for mammalian virus p30s.

The lack of interspecies determinants and the presence of a Mg^{2+} preferring polymerase in BLV virions distinguish BLV from mammalian type C viruses [34]. This obviously does not reduce the significance of the virus which could be an important factor in leukemia in cattle.

I. Summary

Among the oncornaviruses the major structural protein components may exhibit type, species, and interspecies reactivity. The virion reverse transcriptase shows similar properties. The chicken type C viruses vary in specificity of the envelope glycoprotein, gp85, in keeping with subgroup specificity based on neutralization and interference. In tests to date only p19 among low molecular polypeptides has shown type specificity; other components, p27, p15, and p12, seem exclusively group-specific. This conclusion is substantiated by amino acid analyses and peptide mapping. The components are all structurally and antigenically unique. No cross-reactions with any other virus group are seen, with the possibility of a distant relationship with the polymerase of the reticuloendotheliosis group and normal chicken cell enzyme.

The mammalian type C virus proteins have a similar complex of determinants associated with their structural proteins and reverse transcriptase. The internal protein, p30, has proved most useful for subclassification purposes and at present 6 "species" groupings are possible based on gel diffusion. Structural correlates of the immunological results have been obtained which are consistent with the existence of 3 major subgroups. The envelope glycoprotein, gp70, gives broader reactivity in immunoassay than in neutralization or interference tests but can be used to develop type-specific assays. The low molecular weight polypeptides seem especially useful for development of type-specific assays and show a high degree of correlation with the type specificity seen in interviral molecular hybridization assays. Multiple components in the p15 size range perhaps account for differences in extent of interspecies reactivity seen by different workers; 1 of these has surface localization. This component may be missing in certain viruses or not resolved by purification techniques or lost in purification. One laboratory has consistently shown cross-reactions between p12 and p10; however, these components can be distinguished. With this major exception the mammalian type C polypeptides are antigenically unique and also are not related to other members of the retravirus family.

The multiple strains of MMTV seem to contain identical major proteins, designated gp52 and p27. Polypeptides bearing type-specific determinants have not yet been clearly identified, although their existence has been inferred. MMTV reagents have shown no cross-reactions with other well-established members of the retravirus family.

A virus morphologically similar to MP-MV has been found in certain HeLa cell sublines and in cultures derived by workers in Russia. Based on radioimmunoassay with the major internal protein p27 component of MP-MV, these viruses appear indistinguishable. Other evidence indicates that all positive cell lines have HeLa cell markers. The origin of MP-MV is at present uncertain.

MP-MV reagents do not crossreact with any other members of the retravirus family.

The bovine leukemia virus, BLV, which is not a typical type C particle, has a major internal protein, p24, which does not cross-react with reagents for type C viruses, MMTV, or MP-MV. Antisera to BLV p24 also are completely BLV specific.

In many cases the lack of relationship may be a reflection of the inherent limits in immunological tests. Thus below 60% homology in primary structure of well-studied proteins does not allow detection of cross-reactivity. The need for evaluation by determination of primary structure is thus clearly indicated.

V. Interviral Relationships Based on Molecular Hybridization

In addition to the viral protein studies described in Section 9.IV., extensive studies of viral relationships have been made using the technique of molecular hybridization with viral nucleic acids (Fig. 9.14). This requires the availability

	Method 1 Excess RNA	Method 2 Excess cDNA	Method 3 HA Chromatography
1. Add:	Homologous or heterologous cDNA*	Homologous or heterologous RNA*	Either methods 1 or 2
2. Incubate 18 hr, 67°C to form heteroduplexes with homologous or related nucleic acids			
3. Treat with single-strand specific nuclease or ribonuclease			Add mixture to column at 60°C and wash with buffer—single-strands elute
4. Precipitate with trichloroacetic acid—duplex structures insoluble; degraded nucleic acids soluble			Raise temperature in increments with heated buffer—collect fractions
5. Collect precipitate on filter: determine radioactivity			Determine radioactivity. Construct thermal elution profile

Fig. 9.14. Mechanics of nucleic acid hybridization. Temperature, incubation time and reagent concentration determine sensitivity of assays. Conditions shown are typical. (*) Denotes radioactivity. Thermal stability of duplexes can also be determined by tube assays as well.

of a copy of the genomic RNA which was made technically feasible by the discovery of reverse transcriptase [1, 2]. To produce the DNA complement (cDNA) of viral RNA, purified virions are disrupted by detergent and incubated together with the 4 deoxyribonucleotide triphosphates, generally 1 being radioactive. The optimal product would be a complete genome copy which could then be purified and utilized as desired. With the exception of very recent reports it appears that most cDNA products of the endogenous (no added template) reverse transcriptase reaction are of low molecular weight, 100-150 nucleotides, and do not represent the viral genome in 1:1 fashion. The specificity of such cDNAs is generally shown by ability to hybridize to the homologous (and related) viral RNA, but not to unrelated viral or cellular RNAs. One approach frequently employed to insure specificity is to prepare cDNA from virus grown in cells from a different species than the normal host. This cDNA should not as a rule show any reactivity with nucleic acids of the foreign cell. There are several approaches to measurement of viral relationships using the molecular hybridization technique and many variations also imaginable. These are (a) addition of saturating amounts of unlabeled RNA to a standard preparation of radioactive (tritiated thymidine is generally used) cDNA of high specific activity; (b) addition of excess unlabeled cDNA to radioactive (^3H or ^{32}P) viral RNA; or (c) competition assays in which a heterologous RNA or cDNA is utilized in a first reaction to effectively attempt to eliminate the control positive reaction. The hybridization reaction itself may be monitored by protection of radioactive reactants from nuclease digestion, density shifts in cesium sulfate gradients or specific elution characteristics on hydroxylapatite columns. The use of single-strand specific nuclease, such as the S-1 enzyme from *Aspergillus oryzae*, has gained wide usage since many assays can be done with ease and without requirement for centrifuges, columns, etc. Since the products of S-1 nuclease digestion are soluble in trichloroacetic acid and the intact reactants are not, this provides a most convenient procedure for evaluating the extent of reaction. The hybridization reaction has 2 main parameters with respect to the evaluation of relationship; these are reaction extent and degree of base sequence matching. These are discussed below.

The first method (excess RNA) has the intrinsic uncertainty of percentage of genome copied and degree of redundancy. Thus, for example, some cDNAs have been described in which 80-90% of the DNA synthesized represents only 10-15% of the genome. Improved representation is achieved by inclusion of actinomycin D in endogenous reactions. This prevents the synthesis of (+) strand DNA (see Section 9.VI.A.), which seems to then allow a more complete transcript to be formed.

The second method involves the use of cDNA in excess using labeled genomic RNA. Here the extent of reaction can be monitored by ribonuclease or S-1 digestion. One can then select for comparison concentrations of cDNA known to completely protect viral RNA regardless of redundancy problems.

These systems are also then amenable to competition hybridization in which an unlabeled RNA can compete with labeled RNA for protection by cDNA. This in principle is similar to competition or inhibition immunoassays.

Hybrid formation may be measured by the various techniques described above and under a variety of salt concentrations, all of which have an effect on the final result and thus on the interpretation of relatedness. These methods have all been utilized in my laboratory using cDNA prepared by ability to cosediment with purified 70 S RNA. This insures virus specificity and freedom from extraneous nucleic acid sequences copied during the endogenous reaction.

The S-1 nuclease procedure is relatively stringent and under usual salt concentrations (e.g., 0.1-0.5 M NaCl) requires a high degree of base pairing fidelity to allow detection of hybrid structures. When thermal stability of hybrid molecules is evaluated (usually by HA chromatography) it has been shown with model reactants that a 1°C lowering of the midpoint (T_m) of the melting curve results from a 1.5% difference in base pairing. When extent of reaction with S-1 nuclease is compared to T_m with sequences showing an evolutionary pattern of relatedness, there is an operative reaction range of 12-14°C. Thus, sequences related by 80% may fail to register as hybrids by the S-1 assay. It is important to recognize that 50% cross-hybridization (reaction extent) does not necessarily mean 50% relatedness. When sequences of unknown origin are being compared this could result from 50% complete identity in sequences or from sequences which have diverged through evolution only by ~ 10%. The HA technique is especially suitable for determinations of thermal stability since the total hybrid formed can be absorbed to a column at low temperatures (50-60°C), and the amount of reactant (usually radiolabeled cDNA) eluted determined at each increment of temperature rise. Well-matched RNA-DNA hybrids will "melt" sharply with a T_m of 80-88°C.

In evaluating results from many laboratories, it does seem that there is a tendency to overlook the critical relationship between reaction extent and T_m. Current data indicate that where interviral relationships have been described within a species the differences are entirely consistent with evolutionary divergence. However, there are clear cases where reaction extent is influenced by sequence loss, e.g., the replication and transformation defective mutants of avian sarcoma viruses (see Section 9.VI.B.). There are also clear cases involving interspecies pseudotype sarcoma viruses where the fidelity of hybridization allows the specification of the origin of the discrete sequences in the recombinant virus. Section 9.VI.C. describes relationships between specific viruses and their natural hosts by similar hybridization methodology. In those cases key quantitative parameters for measurement of sequence concentration are CoT (concentration of DNA nucleotides × time of incubation) and CrT (concentration of ribonucleotides × time of incubation). The student is presumably aware that rates of reassociation of hybrid molecules are directly proportional to the above parameters and that DNA present at single copy level per cell will reassociate with a $CoT_{1/2}$ of ~3000. To determine concentration of viral RNA sequences per cell one uses $CrT_{1/2}$ of purified viral RNA as standard (compared to $CrT_{1/2}$ of sample assuming the concentration of cellular RNA to be 10^{-5} μg/cell).

A. Intraspecies Variability

The large body of data suggesting close relationship among viruses from within a species (e.g., Section 9.IV.) did not predict the ease of distinction among the subtypes by molecular hybridization. For example, in the case of MuLVs the Gross subgroup (AKR, RadLV, Gross) show similar reaction extents and thermal dissociation profiles, i.e., they are essentially indistinguishable (Table 9.10, Fig. 9.15). In contrast, viruses of the FMR group and the xenotropic viruses are easily distinguishable by reaction extent and T_m. This is clearly indicative of evolutionary divergence since absence of specific sequences would give the same T_m but lower reaction extent. In other comparisons Rauscher and Moloney viruses are also readily distinguishable, leading to a clear distinction of at least 4 subgroups. Note that these data correlate perfectly with the

Table 9.10 Relationship of Gross Virus to Other Strains of MuLV[a,b]

70 S RNA	% Hybridized	T_m (°C)	ΔT_m
Gross[c]	92.4	83.3	
Akv-1	93.2	83.5	+0.2
Rauscher	66.8	75.0	-8.3
Moloney	83.1	74.8	-8.5
G-HaLV	14.0	68.7	-14.6

[a]Note the correlation in ability to distinguish Gross and FMR viruses by hybridization and low molecular weight polypeptide assays (Table 9.8). From Ref. [69].

[b]70 S RNAs (6 μg/ml) were incubated with a [^{3}H]-DNA transcript (8000 CPM) of Gross virus and assayed by HA chromatography. The extent of hybridization based on the percentage of input radioactivity eluted after first washing the column at 60°C and then raising the temperature in 5°C increments is higher by about 20% than estimates based on use of single-strand specific nucleases. Using conditions of 0.3 M NaCl and S-1 nuclease, Gross [^{3}H]-DNA did not show significant hybridization with G-HaLV (an endogenous hamster virus).

[c]The Kaplan radiation leukemia virus was indistinguishable from Gross and Akv-1 in similar experiments (data not shown).

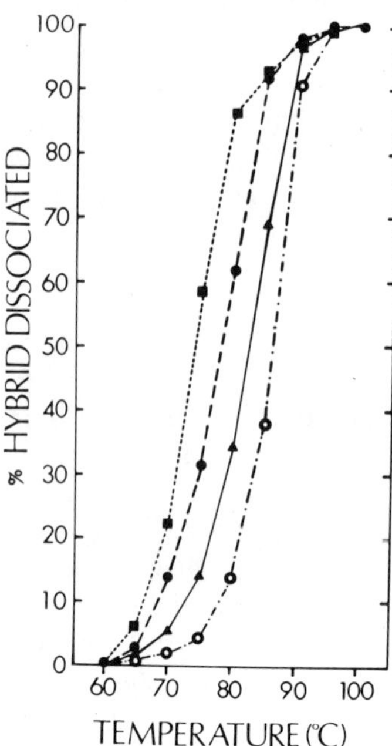

Fig. 9.15. Integral thermal dissociation profile of hybrid molecules formed with the DNA transcript of Gross virus and RNA from Gross, Akv-1, RLV, and H-MSV. Symbols: GLV (○—○); Akv-1 (▲---▲); RLV (●-·-·-●); H-MSV (■---■). Reaction extents and T_ms are given in Table 9.10. In this typical form of analysis the cumulative amount of hybrid eluted as a percent of the total eluted in the 60-100°C range is plotted vs. the elution temperature. H-MSV is derived from Moloney-MuLV. From Ref. [69].

immunological data for the low molecular weight virion polypeptides (Table 9.8). Stephenson and Aaronson have recently shown that there are 2 distinct p12 subgroups among the xenotropic viruses. In this context the recent studies by Callahan et al. [61] are of interest. These workers categorized xenotropic viruses from 9 mouse strains and an asian feral mouse subspecies.

In agreement with the p12 data, hybridization analyses showed 2 subclasses. One subclass, designated MuLV-X^a, includes isolates from BALB/c, C_{57}BL/6, C58, AKR, CBA, DBA/2, and the asian feral mouse *M. musculus molossinus.* The other subgroup, MuLV-X^b, consists of viruses from NZB and NIH Swiss mouse strains. Of further interest was the correlation between relative growth properties in heterologous cells: X^b viruses replicate preferentially in primate cells, while X^a replicate better in rabbit cells. The X^a class occurs in mouse strains which also contain the viruses of the AKR class which are also endogenous in mice (see Section 9.VII.A.). The X^b group occurs in strains from which only this group of viruses has been isolated. In terms of degree of relationship the mouse tropic and X^a viruses are more closely related to each other than either is to the X^b class. The lack of sequences closely related to AKR in the NIH Swiss strain was critical for the genetic experiments of Rowe (Section 9.VII.A.). A possible model for the origin of the 3 sets of endogenous viruses is shown in Fig. 9.16. This model shows ancestral sequences (Vo) diverging to lines leading to X^a and X^b families. The M (AKR subgroup) sequences derive from X^a after this divergence. Viruses of the FMR group are all highly passaged laboratory strains with complex histories and high potential for genetic alterations. Viruses of this type have not been induced from virus-free cells in tissue culture. While these differences are sufficient to allow intraspecies subgrouping, the MuLVs as a whole are clearly closely related; for example, hamster and rat type C viruses show no significant hybridization to MuLVs under stringent conditions. By relaxing conditions hamster-mouse interactions are seen (Table 9.10). The ΔT_m here is ~14°C compared to a maximum of ~8°C among MuLVs.

With reference to the p30 "species" groups described in Section 9.IV., we note a general correlation with results of hybridization analyses. Thus, FeLV and RD114/BaLV, HaLV, RaLV, and SSV-1/GaLV show no reactions with viruses outside their species group, while SSV-1/GaLV show "interspecies" hybridization as do RD114 and BaLV. RD114 and BaLV show about 85% sequence matching based on a ΔT_m of ~8°C, again in keeping with discrimination

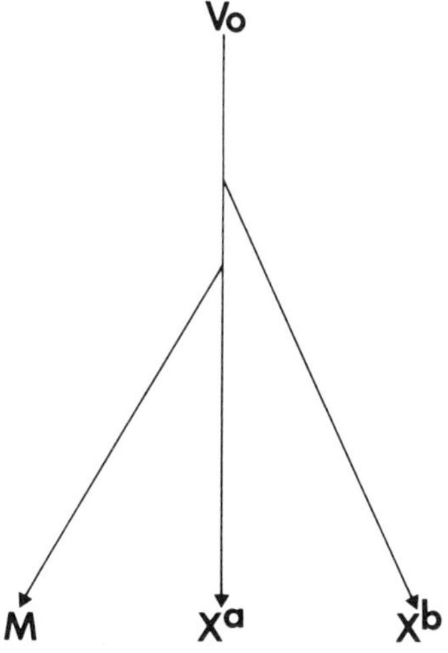

Fig. 9.16. Evolution of mouse type C viral sequences. From Ref. [61].

by p15 assays. Recent studies [70] indicate that SSV-1/GaLV group can be subdivided into at least 4 subclasses (Table 9.11). Virus prototypes of these 4 subclasses are the original GaLV isolate from a gibbon at the San Francisco zoo, a second isolate from Thailand (SEATO colony) which has been reported by Kawakami to induce myelogeneous leukemia in newborn gibbons, 3 isolates from gibbon brains frozen 8 years prior to virus isolation, and the single woolly monkey isolate. Of interest, the single woolly monkey isolate appears more closely related to the first 2 gibbon viruses than to the gibbon brain isolates. Where p12 comparisons have been done, all prototype viruses can be distinguished; however, even in p30 RIAs these viruses appear identical. There is 1 striking exception to the species rule, namely the repeated observation that viruses of the SSV-1/GaLV group hybridize to MuLVs with ΔT_m suggestive of evolutionary relationship (Table 9.12). Even more striking and essentially required was the demonstration that these viruses hybridize to mouse DNA but not gibbon apes, woolly monkey, or any primate DNA [48]. This strongly suggests origin of the group of viruses "recently" from mice or close relatives. The recent detection of a virus in a southeast asian feral mouse, *M. caroli,* with proteins similar to SSV-1/GaLV adds strongly to this hypothesis.

Similar to the MuLVs, various subclasses of avian viruses while related can also be distinguished by molecular hybridization analyses. Non-type-C viruses have shown no interrelationships with each other or with any type C virus.

Table 9.11 Designation of 4 Subgroups Among Members of the WoLV/GaLV Group by Molecular Hybridization[a]

	% Hybrid with cDNA from:			
Viral RNA (in excess)	GaLV-1	GaLV-S	GaLV-B	WoLV
GaLV-1 (San Francisco Zoo)	100	88	61	68
GaLV-SEATO (S)	83	100	57	70
GaLV-Brain (B)	55	62	100	66
SSV-1 (SSAV)	70	78	73	100
Baboon	2	3	3	2

[a]Data of Benveniste and Todaro [70]. Three separate brain isolates could not be distinguished from one another. These assays were with S-1 nuclease and data are normalized to 100% for the homologous virus. Note the excellent agreement in reciprocal assays and the unusual finding that the gibbon brain viruses seem more distantly related to the GaLV-1 and GaLV-S than is the virus isolated from a woolly monkey tumor. This virus originally designated SSV-1 (simian sarcoma virus type 1) has a large excess of helper virus which we designate SSAV (simian sarcoma-associated virus). None of the Wo/Ga viruses show any relationship to the endogenous virus of the baboon by molecular hybridization analyses.

Table 9.12 Cross-Hybridization Between Gibbon Virus Nucleic Acids and Those of Hamster and Mouse Type C Viruses[a]

	[³H]-cDNA		
70 S RNA	GaLV	G-MuLV	G-HaLV
GaLV-1	100	19	7
R-MuLV	20	38	<2
G-MuLV	27	100	<2
W-MuLV	18	40	
H-MSV	22	35	
BaLV	1.1		
RaLV	1.5	0.7	<2
FeLV	2.4		
RD 114	3.0	0	
G-HaLV	13.0	0	100

[a]70 S RNAs used at saturating concentrations (0.1-0.5 μg/assay) based on homologous reactions. Homologous reactions were taken as 100%, actual values were 80-95% of input activity. R = Rauscher, G = Gross, W = wild mouse, H = Harvey (all strains of mouse type C virus), Ba = baboon, Ra = rat, GaLV = gibbon virus, G-HaLV = endogenous hamster virus. These values, obtained with S-1 nuclease give a minimal estimate of relatedness as discussed in the text. Note that mouse and hamster viruses do not show any relationship when assayed under stringent conditions. This serves to further emphasize the reaction of gibbon with mouse and hamster viruses.

B. Interspecies Hybrid Viruses

One of the interesting developments, based on ability to readily discriminate between viruses from different species, has been the detection of interspecies hybrid viruses. For example, the Moloney strain of mouse sarcoma virus with a history of restriction to mouse passage does not hybridize with any of several endogenous rat viruses; however, both the Kirsten (Ki) and Harvey (H) strains of MSV which arose on passage of murine leukemia virus in rats contain rat-specific nucleic acids. These sequences are retained on repassage of the viruses in mouse cells; thus, KiSV and HSV are stable recombinants containing mouse and rat type C nucleic acids. This, according to Scolnick and colleagues [71], provides a model for origin of sarcomagenic viruses; a hypothesis which I support and will discuss further. Other similar findings were the presence of rat and mouse sequences in the virus designated MSV(O) by Ting and more recently MSV(RaLV) in our laboratory based on immunological data. This virus originated in a rat tumor induced originally by M-MSV. The virus recovered from this tumor was rat specific in host range, had rat type C viral proteins, and showed no evidence of mouse proteins. We supposed that this virus arose as a result of in vivo rescue of the defective MSV genome by endogenous rat virus. This seems verified by the hybridization data (Table 9.13).

Viruses derived from hamsters in the same manner as MSV(RaLV) was obtained from the rat have now been found to contain both mouse and hamster sequences. Most interesting was the virus designated B-34 or HaSV originating in a hamster tumor induced by H-MSV. H-MSV, as we point out above, should be considered best as RaSV(M-MuLV) in accord with the convention of placing the viral protein contribution in parentheses. B-34 is a hamster-specific virus with hamster (HaLV) proteins and nucleic acid sequences of hamster, mouse, and rat origin (Table 9.14). Thus, the viral population has 3 distinct sets of nucleic acid sequences encapsulated in virions which appear to have only hamster type C p30s. The exact manner in which these distinct sequences are incorporated into genomic RNA in the same or distinct virions in the population is not clear.

The ability to specify sequences of the original mouse virus in these interspecies virus pseudotypes is of obvious significance. A clear example is the virus designated GLOH$^-$, a nontransforming virus which was isolated by end point dilution from a sarcomagenic virus isolated from a hamster tumor induced by the Gross pseudotype of MSV. GLOH$^-$ has hamster proteins and

Table 9.13 Ability to Specify the Mouse Sequences in the Interspecies Pseudotype MSV(RaLV) as Being Derived from M-MSV[a]

	$[^3H]$-cDNA		
70 S RNA	M-MSV	Gross	RaLV
M-MSV	94.4 (86.4)	77.0 (72.6)	Negative
M-MSV(RaLV)	93.9 (84.3)	61.7 (72.0)	92.0 (84.5)
Gross	64.9 (76.4)	92.4 (83.3)	Negative
RaLV	Negative	Negative	93.0 (85.0)

[a]The figures are actual % hybridization at saturating concentrations of RNA determined by HA chromatography. The T_m (°C) of the DNA:RNA hybrid molecules is shown in parentheses. One sees that based on reaction extent and T_m there are sequences in M-MSV(RaLV) which are still highly related to M-MSV and easily distinguished from Gross virus. The lack of any reaction between mouse and rat viruses makes the detection of sequences related to each species relatively easy. Note that while M-MSV(RaLV) has both mouse and rat nucleic acids it only contains rat virus structural proteins. From Ref. [69].

Table 9.14 Three Sets of Nucleic Acid Sequences in B-34, a Virus Derived from Hamster Tumors Induced by H-MSV

	[^{3}H]-cDNA		
70 S RNA	M-MSV	HaLV	RaLV
M-MSV	+ (35 and 30 S)[a]	–	–
HaLV	–	+ (35 S)	–
RaLV	–	–	+ (35 and 30 S)
H-MSV	+ (35 and 30 S)	–	+ (30 S)
B-34	+ (30 S)	+ (35 S)	+ (30 S)

[a]+ = Positive hybridization using S-1 nuclease. In parentheses are sedimentation values of subunit RNAs found in virions and in virus-producing cells. Note that H-MSV contains a 30 S RNA of rat origin which is distinguished from helper virus 35 S RNA. This virus originated from passage of Moloney MuLV in rats. In all cases the presence of a 35 S RNA subunit seems to specify the species of structural proteins possessed by the virus. From Ref. [72].

mouse and hamster nucleic acid sequences. One could specify sequences related to MSV, not GLV, in this nontransforming virus which except for hybridization analyses would be considered simply HaLV. In terms of additional complexity the B-34 virus seems to preferentially synthesize MuLV nucleic acid sequences in endogenous reactions. Here the immunologist clearly designates B-34 a hamster virus, while the molecular biologist would call this a mouse virus which shows close relationship to hamster and rat viruses. We make this point to emphasize the critical need for virus histories before extensive characterization of any virus.

C. Summary

In summary, viruses of a p30 species group will cross-hybridize and can be subgrouped in accord with immunoassays of low molecular weight polypeptides. The relationships obtained between naturally occurring viruses from different species are based on evolutionary divergence from hypothetical common ancestors. An unusual finding is the close relationship of the woolly-gibbon subgroup with MuLVs. RD114 (endogenous cat) and baboon viruses also cross-hybridize. The latter relationships are discussed further in Section 9.VI. Interspecies "hybrid" viruses contain sequences of both parental viruses which retain fidelity after several years in culture.

VI. Relationship of Viruses to Host Cells

A. Replication of Oncornaviruses

1. Cycle of Nondefective Type C Virus

The replication cycle of nondefective type C viruses in host cells can be summarized as shown in Table 9.15. As with any virus, initial steps involve the attachment of the virus via specific receptors to the cell surface. Attachment is mediated via the envelope glycoproteins which have been shown in certain cases to possess hemagglutinating activity and binding activity to the cell surface. Methods of purification can often inactivate viral attachment proteins; however, there is clear evidence that soluble envelope subunits can prevent attachment of intact virus. For example, a soluble factor of ~70,000 MW was isolated from RD114 virus which, when added to a cell line which formed syncytia upon direct infection with RD114, prevented syncytia formation. Activity of the inhibitor for this cell line (KC) was inactivated by neutralizing antibody. Treatment with disulfide reducing agents quickly inactivated

Table 9.15 Events in Replication of Type C Viruses[a]

1-3	Attachment, penetration, uncoating
4	Synthesis of (–) strand DNA
5-6	Synthesis of (+) strand DNA and degradation of RNA template
The above events occur in the cytoplasm with the possibility of completion of events 5-6 in the nucleus	
7	Transport of double-stranded DNA to nucleus and conversion to circular form
8	Integration of provirus DNA into cellular DNA
9	Transcription of DNA into mRNA (viral + strand)
10	Attachment to polyribosomes
11	Synthesis of precursors to virion structural proteins
12	Cleavage of precursors, assembly (maturation)
13	Budding from cell surface

[a]For transformation by sarcoma viruses specific proteins are required. These have not been identified directly as yet.

this syncytial inhibition activity, thus emphasizing the need for gentle isolation conditions to insure retention of biological activity [73].

In studies of murine virus attachment and entry, clear evidence for fusion with the cell membrane followed by dissolution of the membrane and direct penetration into the cytoplasm or cytoplasmic vesicles was obtained. In contrast, entrance of avian virus into cells appeared to be based completely on a phagocytic process. There are obvious difficulties in interpretation of divergent results: perhaps both modes of entry are correct. There is also the possibility of virus to virus variation. The real question is what is the biologically relevant process. To answer this, one ideally needs to use preparations in which the physical:infectious particle ratio is 1–systems in which cells infected with a single particle can be studied. This is an impossibility for this group of viruses and for electron microscopic technique. In comparison to the rest of the virus life cycle, the early steps up to uncoating, while essential, are not novel. What is clearly novel are the steps involved from this point on. For ~10 years it was known that inhibitors of DNA synthesis and transcription would prevent both replication and transformation by RNA tumor viruses. This formed the basis of Temin's provirus hypothesis, stating that replication of this group of viruses occurred via a DNA intermediate. The requirement for DNA synthesis is a relatively early event in the virus life cycle as is illustrated by the following experiment carried out by G. Lovinger in the laboratory of my colleague, M. Hatanaka. Mouse embryo cells were infected with murine sarcoma virus and treated at various intervals with cytosine arabinoside (araC) which rapidly and reversibly inhibits DNA synthesis. Short 15 min pulses of araC were given using concentrations of drug, 10μg/ml, which when removed left no residual toxicity. At 4-5 days postinfection cultures were assayed for replicating virus by measuring reverse transcriptase activity in the culture medium using synthetic templates specific (in these circumstances) for viral RT and for cell transformation. Sample data, shown in Table 9.16, indicate that in this system maximal inhibition occurred 45-60 min postinfection even though effects could be seen at later time periods. This result and similar findings with avian viruses led to an extensive search for the hypothetical DNA intermediates with a long period of negative experience. The discovery of reverse transcriptase (independently by Baltimore and Mizutani and Temin) provided new tools and a fresh determination which has eventually led to success in such efforts. My colleague, M. Hatanaka, with K. Kakefuda, provided evidence by autoradiography that DNA synthesis took place very early after infection in the *cytoplasm* [74]. This was initially disputed but in the last 1-2 years numerous reports of isolation of virus-specific DNA from the cytoplasm of infected cells have appeared. One obvious reason for early controversy is the extreme rapidity of early events, e.g., within several hours the final product of replication, which appears to be a circular DNA, is found in the nucleus where integration into

Table 9.16 Timing of Proviral DNA Synthesis[a]

Time Postinfection	Number of transformed foci/plate	Reverse transcriptase (counts)
15-30	66	5000
30-45	42	3000
45-60	28	1000
60-75	33	1000
75-90	47	3000
90-105	61	4500
Control	143	7000

[a]Mouse embryo cells infected with M-MSV were incubated with cytosine arabinoside (ara C) an inhibitor of DNA synthesis for 15 min intervals followed by medium change. Ara C effects are readily reversible. At 4 days culture media were assayed for reverse transcriptase as an indicator of virus replication and at 5 days foci were enumerated. The data show that 45-75 min postinfection is a critical period in replication and transformation cycle requiring DNA synthesis. This indirect evidence was verified by ability to isolate the expected DNA intermediates during this time period.

cellular DNA can occur. Within 45-70 min after infection of mouse embryo cells by RLV labeled in its RNA by tritiated uridine Takano and Hatanaka found a transcient, synchronized shift of acid-insoluble radioactivity from a region of RNA density to 1 intermediate between that of DNA and RNA [75]. Such experiments are readily done using gradients of cesium sulfate in which DNA will band at ~1.42 g/cc and viral RNA at ~1.7 g/cc. Simple mixing of labeled virus with uninfected cells before extraction of nucleic acids resulted in recovery of the label at 1.70 as expected for viral RNA. At 45, 60, and 70 min after infection a shift in the density from 1.70 to lower values intermediate between viral and cell DNA was observed. The final density reached appeared to be 1.56 g/cc, which is equivalent to about 50% RNA in the complex. The density of the hybrid molecules was not affected by heat denaturation, indicating covalent attachment of RNA to DNA. This is analogous to the finding with in vitro synthesized DNA in the endogenous reaction; namely, that a portion is found covalently attached to viral RNA. By 90 min this hybrid structure was no longer detectable. The localization of these hybrid molecules was then determined by fractionation of nucleus and cytoplasm before nucleic acid extraction. At 45 and 60 min, these molecules were found predominantly in the cytoplasm and were similar in density to those recovered from unfractionated cells. By 70 min a large peak appeared in the nuclear fraction at a density of 1.48 g/cc; this component was not seen in the cytoplasm at any time point. The simplest explanation of these results is that initial transcription of the viral genome occurs in the 45-60 min time frame and that hybrid molecules may then be transported to the nucleus.

Considering the viral 70 S RNA to be (+) strand, then both (-) and (+) strand DNAs have now been isolated from infected cells [76]. The (-) strand has been detected by hybridization with viral RNA and the (+) strand by hybridization with cDNA prepared in the endogenous reaction and purified by cosedimenting with 70 S RNA [(-) strand cDNA]. Peak synthesis of (-) strand DNA occurred 40-60 min after infection of mouse cells with RLV while (+) strand synthesis occurred between 50 and 70 min postinfection. The (-) strand was found in 2 major molecular forms; 1 with an estimated molecular weight of $\sim 1 \times 10^5$ and the other at 3×10^6 (the size of a viral genome subunit. It was possible to accumulate (-) strand DNA in the cytoplasm in the presence of hydroxyurea, thus providing support for kinetic data indicating initial synthesis of (-) strand DNA in the cytoplasm. Regarding timing, it is noteworthy that similar experiments in the avian system also indicated that the critical time period for provirus synthesis is ~1 hr postinfection. The mechanism of synthesis of the (-) strand in vitro requires a primer and is known to proceed in the 5′ to 3′ direction (DNA), and transcribes the RNA in the 3′ to 5′ direction. Recent evidence indicates the natural primer for avian viruses, tryptophan tRNA to be located near the 5′ end of the RNA [42] which seems to be the exact opposite of the expected location. This leads to the suggestion that the 5′ end of the RNA is brought into contact with the 3′ end under biological conditions of replication.* In vitro synthesis has so far yielded only small nucleotide stretches, 100-150 bases, which however can represent the entire genome but not necessarily in 1:1 fashion.† Since there is only 1 primer per molecule of 35 S RNA, one expects only several initiation sites per 70 S molecule. One possible explanation of the in vitro results is that the 5′ end can be brought to the rest of the RNA at various positions, thus giving the effect of multiple initiation sites. At each position new synthesis begins from the 3′ end of the template RNA. Most of the discussion on the mechanism of synthesis is conjecture at present, the important clear point is that (-) strand DNA of genomic size is synthesized in the cytoplasm in the time frame predicted from drug inhibition experiments. While again there is some current experimental discrepancy, the next step in synthesis is a combined degradation of genomic RNA by the RNase H activity associated with reverse transcriptase and synthesis of the plus (+) strand, thus leading to a double stranded DNA.

Synthesis of (+) strand DNA can occur in enucleated cells and can also be demonstrated in the cytoplasm of infected mouse cells in carefully timed experiments. The double-stranded DNA is very quickly transported into the

*See page 457.
†See page 458.

nucleus where it may be found in closed circular form which is then integrated into host cell DNA. Intermediates in the formation of double-stranded DNA which contain a complete (–) strand and short segments of (+) strand have also been described.

A brief discussion of recent experiments by Guntaka et al. [77] will illustrate the key points that (a) integration into cell DNA requires a supercoiled viral DNA intermediate; and (b) failure to integrate also results in failure of virus replication. The latter point has been widely discussed with claims for lack of requirement for integration and vice versa being brought forward. The experiments discussed here seem to answer the question in critical fashion. Duck cells were infected with the B77 strain of avian sarcoma virus and viral DNA synthesis and integration and virus production measured in the presence and absence of ethidium bromide (EB). This dye interferes with the formation of closed circular DNA molecules. The duck system is ideal in that no sequences homologous to B77 are found in uninfected cells. Virus-specific DNA could be quantitated by acceleration of the rate of reannealing of labeled double-stranded viral DNA. Integration was measured by ability to localize viral DNA by its presence in "networks" of reiterated cell DNA. Initial experiments established that infected cells with or without EB contain 1.2 copies of viral DNA per cell; thus, the drug at 1 μg/ml did not affect total viral DNA synthesis. One and 1/10 copies per cell were found in networks (integrated) in untreated cells while only 0.15 copies were integrated in EB-treated cells, the remainder being found in unintegrated form in the supernatant. When virus-specific DNA was analyzed at 9 hr postinfection, 1 hr prior to the beginning of integration, by methods used to demonstrate closed circles in other systems, ca. 20% of the DNA was found in this form; EB treatment reduced this fraction by a factor of 4-5. Other forms of viral DNA, including a fraction of genomic size, were unaffected by EB treatment. In cesium chloride-ethidium bromide density gradients supercoiled DNA bands at a higher density than open circular or linear DNA; this difference is based on relative dye binding. By this technique a fraction corresponding to that expected for supercoiled DNA (based on SV40 form I as marker) was readily isolated from 9 hr infected duck cells. A shift to lower density of this fraction was obtained by introducing single strand breaks by limited treatment with pancreatic DNase. This serves to validate the conclusion that the high-density fraction contains closed circular DNA. Cells infected in the presence of EB showed >80% reduction in recovery of high-density B77-specific DNA. These results strongly suggest that formation of closed circular DNA is a critical event in the replication cycle and that prevention of this process eliminates the possibility of integration. The closed circular intermediate is a classical form required for integration of bacteriophage DNA and papovavirus DNA in animal cells.

These observations allow a direct consideration of the requirement for integration on replication. Both by measure of total virion synthesis and biological assay of viral infectivity it was shown that EB effectively inhibits replication in direct proportion to its effectiveness in prevention of integration.

Similar results to the above have been obtained with mammalian type C viruses and suggestions of closed circles have been obtained by electron microscopy. This is a far cry from the early days in which attempts to detect the provirus directly generally met with frustrating failure.

Once the DNA is integrated, transcription occurs via cellular RNA polymerase and the viral proteins are synthesized by usual translational mechanisms [30]. Thus, viral RNA has been localized on polyribosomes of infected cells, generally in a size class similar to genome subunit size, but with smaller size classes as well. The significance of these smaller RNAs (sedimentation rate ~20 S) is not clear. Synthesis of virus-specific proteins occurs on polyribosomes and after assembly of the structural products the virion departs the cell by a budding process (Fig. 9.2). Clear evidence has been obtained that the virion proteins are first found as polycistronic products [78]. A precursor for the low molecular weight polypeptides has been found for both avian and mammalian viruses (MW 60,000-75,000) which by immunological and biophysical analysis contains at least several of the proteins in the p12-p27 range (avian) and p10-p30 range (murine). For the avian precursor polyprotein p19 appears to be the NH_2 terminal component. A separate precursor for the envelope glycoprotein has been described (MW ~90,000) and components much larger than p60-p75 have tentatively been identified. Considering briefly coding capacity if the genome is polyploid with 3.5×10^6 MW identical subunits (10,500 nucleoides), a total amino acid potential of 3500 is available or roughly 350,000 daltons of protein. Referring to the known polypeptide composition and adding the polymerase brings one within the limits allowable by this calculation. This would seem to indicate little room for nonvirion viral-coded polypeptides.

2. *Replication of Sarcoma Viruses*

Replication of avian and mammalian sarcoma viruses presents some points of contrast. Chicken viruses contain members which replicate and transform independently of helper virus and also examples of replication defective viruses which when isolated free of all helper lack the major envelope glycoprotein, gp85, and thus cannot infect cells directly [18, 19]. In addition, these viruses cannot recombine genetically to form infectious virus. These defective viruses do produce progeny lacking in gp85 when introduced into susceptible cells by a fusion factor from Sendai virus. The mammalian sarcoma viruses with 1 possible exception are all replication defective and will not produce infectious progeny unless a helper virus is present. The murine sarcoma viruses (MSVs), however, can infect cells directly without helper and produce foci of transformed cells by cell multiplication. This is readily demonstrated with contact-inhibited cells typified by the 3T-3 lines of Aaronson and Todaro, where the transformed cells have a growth advantage. In this case cell transformation by MSV is a 1-hit process, i.e., 1 particle produces 1 transformed focus. See Fig. 9.3 for an example of a typical MSV focus. In contrast, when normal mouse embryo cultures are infected with MSV, the kinetics of focus formation follows a 2-hit curve. This suggested initially a requirement for coinfection with helper virus; however, careful analysis revealed 2 patterns of focus formation. Initial rapidly appearing foci (several days postinfection) result from coinfection with helper which permits recruitment of adjacent cells. Later appearing foci derive from single infection with MSV and foci are a result of cell multiplication. The lack of growth advantage for the transformed cells in mouse embryo cultures evidently was the determining factor in generating the 2-hit pattern. In addition to the contact-inhibited cell lines, plating of infected cells in semisolid agar which specifically selects for transformed cells will also give a 1-hit curve for transformation by MSV. It should be clear that defectiveness of MSV is not based on lack of envelope glycoproteins since entry into the cell occurs and MSV infectivity under 1-hit assay conditions is inhibited by antiserum to the envelope type of the helper virus present in all MSV stocks, usually in considerable excess to the MSV.

Replication defectiveness for MSV (and other mammalian sarcoma viruses) may be absolute in that transformed cells may produce neither virions nor virion antigens. They do express viral RNA and contain integrated proviral DNA. The genome of MSV can be incorporated into any mammalian type C virus to produce a pseudotype bearing structural proteins of the rescuing virus. This can be accomplished by direct infection of the nonproducer (NP) cells, or by cocultivation with helper virus-infected cells, or by cell fusion (somatic cell hybridization) with helper virus-infected cells. MSV pseudotypes are extremely useful for tissue culture titrations since a relatively rapid quantitative focus forming assay can be developed. In contrast, assays for helper virus by antigen induction may require 21 days, although plaque assays for some of the viruses are now available (XC test for MuLV, KC for RD114, BaLV). The production of true NP cells differs from the defective avian sarcoma virus system where virions, even if noninfectious, are always produced.

The formation of NP cells by solitary infection with MSV is clear evidence that virus stocks carry a replication defective, transforming agent. Yet there are transformants which synthesize noninfectious particles exactly as will be described for the defective Bryan strain of Rous sarcoma virus, B-RSV(-), in Section 9.VI.B. Cells carrying such particles were initially referred to as

S+ (sarcoma^{positive}), L– (leukemia^{negative}) since they carry sarcoma genome in the absence of infectious helper virus. This is a confusing nomenclature since originally these were thought to be true NP cells (when first named), and now it is realized that these cultures actually produce noninfectious virions. The deficiency in this case does not seem to be synthesis of structural proteins but rather a deficiency in reverse transcriptase. S+L– virions do not transform mouse cells even when fused into mouse cells with inactivated Sendai virus and thus differ in this property from B-RSV(–). Certain of these cells are also unusual in that they form foci when infected with MuLVs. The lines before infection contain a mixed population of normal and transformed appearing cells. These cells are thus useful for MuLV titrations. S+L– cells contain a rescuable sarcoma genome and may be obtained by infection of cells from various species. At present the basis for the difference between true NP and S+L– transformed cells is not clear. The important point to be emphasized is that a 1-hit curve for transformation can be obtained independently of ability to produce progeny virus. Like B-RSV(–) it would appear that MSVs do not recombine genetically to produce competent sarcoma viruses.

In addition to several authentic murine sarcoma viruses (Moloney, FBJ, Gazdar), the other mammalian sarcoma viruses are transspecies pseudotypes (Harvey-SV, Kirsten-SV) or natural isolates (SSV-1, FeSV–several strains). All have been reported to give rise to NP cells and are thus replication defective.

In contrast to the above, Ball and colleagues have described what appears to be a competent strain of M-MSV which itself is negative for MuLV in the XC test but does segregate MuLV on growth in mouse cells. This is thought to be the counterpart of competent avian sarcoma virus. This virus, free of helper, would be exceedingly valuable for structural studies and attempts to detect sarcoma-specific nucleic acid sequences as will be evident from Section 9.VI.B.

In summary, mammalian sarcoma viruses are generally replication defective but can penetrate and transform cells without coinfection with helper. Their expression can vary from lack of virion production, to production of defective (transforming and replication) virions, to possibly complete competence. Their genome is easily transferred to mammalian nontransforming type C viruses which contribute structural proteins to the newly formed pseudotype sarcoma viruses.

B. Recombination of Oncornaviruses (Endogenous Viruses, Sarcoma-Specific Sequence)

By far the most detailed knowledge concerning interactions among oncornaviruses has been derived from studies of the chicken viruses. These studies involving several key laboratories (often in collaboration) and viewed from hindsight have been technically and intellectually among the most impressive undertakings in contemporary virology. They have led to discovery of the endogenous chicken virus, clarified the nature of defectiveness (at least for some sarcoma viruses), and most recently have led to the probable detection of sarcoma-specific nucleic acid sequences in the class Aves. Much of this involves a logical progression of findings and thus will be considered together in this section. Because of the complexity and importance of this section a detailed summary is presented at its conclusion.

Most of the strains of RSV studied today owe their origins to chicken sarcoma 1, the original virus-yielding tumor described by Peyton Rous in 1908 [14]. Through the years the virus has undergone travel through many laboratories, many chickens, and many tissue cultures; thus the relation of contemporary strains to the original virus is beyond current means of evaluation. Studies of 1 of these latter day descendants, namely the Bryan (B) strain or RSV (B-RSV), has yielded a body of information almost beyond comprehension. A large number of helper viruses designated RAVs (rous associated virus) were isolated from B-RSV and some years ago the key observation was made that the transforming agent itself did not produce progeny infectious for chicken cells. This observation depended upon careful selection of virus dose and isolation of transformed cells free of helper virus. The transformed cells thus selected were called nonproducer (NP) cells and when infected with RAV again yielded infectious virus bearing the envelope properties of the particular RAV used for rescue of the RSV genome. Rescued viruses are termed pseudotypes and are abbreviated RSV(RAV-1), for example, if the rescuing helper is RAV-1. This methodology and terminology is widely used in oncornavirology; mammalian type C viruses will not rescue RSV; however, all mammalian viruses can rescue defective murine sarcoma virus.

Subsequent studies by electron microscopy and assay of culture supernatant fluids for physical particles revealed that the RSV-NP cells were in fact producing virus and further studies showed that this virus was infectious for Japanese quail and certain chicken cells. Thus one could consider RSV obtained in this way to be simply a host-range varient. Careful observations by Weiss and Hanafusa [18] showed that RSV could in fact only replicate in chick embryo cells which were positive for the group-specific antigen (presumably mainly p27) in CF tests. Negative cells could be transformed but would not produce progeny. The gs+ phenotype, however, was not always correlated with replication but cells could be classified as + or - for ability to replicate RSV based on a hypothetical factor designated chick helper (*chf* or *f*). The key findings after these original observations (see [19] for refs.) were that:

1. RSV introduced into *chf*-negative cells in the absence of helper using a fusion entry technique mediated by Sendai virus can transform cells and produce physical particles which lack gp85; thus *chf*

donates the envelope glycoprotein to RSV. Replication defective B-RSV grown in *chf*- cells is designated RSV(-).

2. RSV grown in *chf*+ cells is designated RSV(*chf*) and the original virus found in NP (actually virus producing) cells is designated RSV(O). Both pseudotypes RSV(O) and RSV(*chf*) have the same host range and envelope specificities and being distinct from all other known chicken viruses were placed in a new subgroup designated "E."
3. Certain inbred chicken lines spontaneously produce a virus of subgroup E specificity, which is generally termed RAV-O based on the envelope similarity to RSV(O), the original subgroup E virus.
4. *chf*+ cells synthesize gp85 of subgroup E specificity but with the exception of embryo cultures from only a few lines physical virions are not produced in most *chf*+ cultures.

With the development of molecular hybridization technology it was possible to study the occurrence and transcription of subgroup E nucleic acid sequences in both *chf*+ and *chf*- cells. All chickens were found to have comparable numbers (~2) of viral genome equivalents integrated into cellular DNA but only *chf*+ cells transcribed these sequences into virus-specific RNA. Most recent evidence indicates that the majority of the viral sequences are transcribed yet physical particles are not made in most *chf*+ cells. This appears to be a fruitful system for further study of control of virus maturation. *chf*+ cells do not contain detectable levels of reverse transcriptase which thus may be the key controlling deficiency. With these findings and new tissue culture efforts, Weiss et al. [22] were able to demonstrate activation of infectious RAV-O in *chf*- cells by treatment with carcinogens and mutagenic agents; this provided one of the contemporary definitions of endogenous viruses, namely ability to induce from virus-free cells. This finding and similar findings in mammalian species (mouse, rat, cat, hamster, pig, baboon) give proof to the concept of viral genomes being a part of the inheritance of most (probably all?) higher vertebrates [5].

Further points of interest are that:

1. RSV(*chf*) is not a stable genetic recombinant of RSV(O) and RSV(-) but rather a phenotypically mixed population. Thus, cloning of RSV(*chf*) on quail cells will yield RSV(-) again. B-RSV is thus replication defective and defective in ability to acquire envelope genes from *chf*.
2. Any virus infecting *chf*+ cells will activate and yield recombinants with RAV-O. A virus designated RAV-60 was isolated by Hanafusa in this manner after infection of *chf*+ cells by a subgroup B virus

RAV-2 and thought to be a potential recombinant based on ability to replicate to high titer on quail cells compared to RAV-O. Very recent evidence indicates [79] that RAV-60 contains RNA sequences corresponding to both RAV-O and RAV-2 indicating that RAV-60 was not simply an activated endogenous virus. Recombination in *chf*+ cells occurs at a frequency of 10^{-3} and even in *chf*- cells can be detected at very low frequency, 10^{-6}-10^{7}. Thus, the infection of any chicken cell with an exogenous chicken type C virus can result in phenotypic and genotypic recombinants. Weiss also demonstrated that infection of *chf*+ cells with competent (nonreplication defective) sarcoma virus of subgroup B specificity would yield recombinants of host-range and reverse transcriptase markers.

3. A mutant of RSV lacking both gp85 and reverse transcriptase has also been described. This virus, RSV(-), does seem able to acquire polymerase genes from a second virus and from *chf*+ cells [17].

There are competent chicken sarcoma viruses which have yielded information crucial to the question of recombination and the nature of sarcoma virus-specific sequences. Such viruses yield a high percentage of transformation defective (*Td*) "mutants" upon passage so that frequent cloning is essential to maintain the population in adequate configuration for biophysical studies. Such viruses and their *Td* variants have been analyzed in detail by laboratories of Vogt and Duesberg, often in collaboration [37, 38]. In general, the distinction between replication defective and competent sarcoma viruses can be made based on analyses of titration patterns in tissue culture. A key requirement is for adequate dispersion of a viral stock to avoid clumping of particles. Competent sarcoma viruses will show a 1-hit titration pattern; that is the number of transformed cells per standard number of cells will be directly proportional to the concentration of input virus. Replication defective viruses require dual infection with helper virus to produce transformation. This is especially noticeable under conditions in which foci result from recruitment of cells adjacent to the initial infected cells and transformed cells show no growth advantage over normal. The titration pattern here will be 2-hit since in the absence of replicating helper recruitment will not occur. Actual data for a defective virus will show that transformation decreases with the square of the dilution. Addition of excess helper will return the pattern to 1-hit.

Analyses of the subunit RNA size of avian sarcoma viruses initially revealed 2 subunits (termed a and b) based on size. After realization that *Td* variants occurred with high frequency on passage of sarcoma viruses in culture, analyses were made with newly cloned populations. Here a striking result was obtained. All cloned sarcoma viruses contain the a size class subunit which is ~10% larger in size than the b subunit (molecular weight difference ~3 × 10^5), whilc *Td* variants and naturally occurring leukosis viruses all contain subunits of the b size class. Uncloned sarcoma viruses would show a and b subunits in ratios as high as 1:1. In one study [38] 3 sarcoma viruses and their *Td* derivatives were analyzed by an oligonucleotide fingerprinting procedure using T-1 ribonuclease as described in Section 9.III.B. It was found that all of 20-25 large T-1RNase-resistant oligonucleotides present in class a subunits were present in the corresponding class b subunits of the homologous *Td* variant. Class a subunits contained in addition 1 or 2 additional oligonucleotides not present in class b subunits. Comparison between the 3 strains studied, B77, SR-RSV, and Pr-RSV, showed considerable differences. The conclusion from these studies is quite clear: the a subunit may be represented as b + x, with x being the transforming or sarcoma-specific nucleotide sequence.

The oligonucleotide fingerprinting method clearly provides 1 potent approach to analysis of replication defective viruses and recombinants. The results described above also pointed the way to a possibility for isolation of sarcoma-specific sequences of avian sarcoma viruses. We will consider these areas of investigation in succession.

The RNAs of a replication defective (*Rd*) mutant of the nondefective (*Nd*) SR-RSV strain (subgroup A) and *Nd* SR-RSV, and *Td* mutant of SR-RSV were compared by oligonucleotide fingerprinting [80]. B-RSV(-) was also included in this study. The *Rd* mutant of SR-RSV, designated SR-N8, is very similar to B-RSV(-) in lack of envelope glycoprotein (gp85), ability to transform cells by use of Sendai fusion or as a pseudotype with helper virus, and inability to genetically recombine with helper viruses to produce a competent (*Nd*) sarcoma virus. The subunit genomic size of the *Rd* virus was found to be 21% smaller (8 × 10^5 daltons) than the *Nd* parental virus. All 14 large clearly separated T-1 oligonucleotides of the *Rd* virus were shared with the *Nd* parent, while the parent and the *Td* virus had 6 oligonucleotides not found in *Rd* virus. The *Rd* and *Nd* viruses also had 2 oligonucleotides not found in the *Td* virus; thus it may be concluded that the defects in replication and transformation reside in different portions of the genome. B-RSV(-) had a different pattern than SR-N8(*Rd*) which rules out origin of the latter virus by inadvertent contamination with RSV(-). Since *Rd* viruses are deficient in gp85, it may be useful to calculate the coding capability lost by deletion of 8 × 10^5 daltons of RNA. This represents 2500 nucleotides (average MW 320 per nucleotide) and thus a coding potential for 833 amino acids. This is roughly equivalent to 80,000-100,000 daltons of protein and thus consistent with coding for gp85 or a slightly larger precursor. Mapping of delection mutants on a large scale is now possible by a technique which utilizes distance from the poly(A) terminus as the reference point. The transforming sequences (discussed below) were found to reside 5-15% from the poly(A) end. These methods will obviously be applied to viruses other than the avian viruses and results are awaited with great interest.

Given a genome which consists of several identical subunits, recombinants could occur by either reassortment of subunits or crossing-over between homologous segments of nucleic acid (presumably at the provirus DNA level). In 1 experiment [37] 5 recombinants each of *Nd* sarcoma viruses, PrA (subgroup A), were selected based on the transforming markers from the *Nd* viruses and the host-range marker from the RAVs. All viruses were cloned before study. The subunit size of each recombinant was clearly in the range of the a subclass with only slight differences from the parental *Nd* subunit. This presents clear evidence against the reassortment possibility since the host-range marker must derive from a b size class subunit. The small differences in mobility (equivalent to about 70,000 daltons) were about 1/5 the size difference of a and b subunits and were stable on passage of virus in different cells. These minimal size differences could occur by crossing over at different positions in the 2 genomes among the various recombinants. Evaluation of the viruses by oligonucleotide mapping showed differences in 2-3 of 20 resolved spots, which is smaller than the difference between parential *Nd* and RAV viruses. This favors the opportunity for crossing-over to occur at multiple genomic sites. Consistent with a cross-over model and the electrophoretic analyses, the genomic complexity (Section 9.III.8.) of 1 recombinant was found to be 3.3×10^6. Reassortment would obviously require increase in complexity of genomic RNA. Since analysis of the recombinants yield unique fingerprints (5/5) in each cross, there is no logical way to accomplish this by reassortment with 4 or less subunits. Recombination is assumed to occur at the DNA level after the initial formation of heterozygous virions with different genomes in a 70 S complex. This would argue for a polyploid genome which is the model currently most popular.

Studies of Weiss et al. [81] clearly demonstrate recombination among exogenous and endogenous tumor viruses and the existence of a heterozygous stage in the formation of stable genetic recombinants. The techniques used were analysis of progeny virus from cells (initiator cells) infected with single particles using an infectious center technique. Yields from single initiator cells were obtained by preventing infected cells from dividing by use of mitomycin C and overlay with selected indicator cells characterized by susceptibility or resistance to viruses of the parental subgroups. Foci which appeared in the indicator cells were reassayed by direct focus and repeat of the infectious center assay. Spread of virus was prevented by use of a semisolid agar overlay of the mixed cultures. The complexity of the various cell types and virus populations will not be completely detailed here but the reader should appreciate the extensive characterization required before carrying out such experiments. In 1 experiment SR-RSV (B subgroup), previously cloned in *chf*- cells and unable to plate on quail cells (Q/B) which are resistant to B subgroup viruses, was used to infect *chf*+ C/E cells. Three days later sonicated culture medium was used to infect Q/B cells which were used as both initiator and indicator cells in the infectious center assay. Note that Q/B cells will not support focus formation by SR-RSV(B) but RSV/*chf*) can form foci on Q/B cells. Thus any foci in the indicator cells would result from recombination or dual infection. Several arguments against dual infection were provided: (a) failure to isolate a helper virus of RAV-E specificity even on passage in highly susceptible pheasant cells resistant to B subgroup and (b) single-hit kinetics in the direct focus and infectious center assays on Q/B cells under conditions in which B-RSV(*chf*) gives a 2-hit titration pattern. Indicator cell foci taken from plates with small numbers of foci were analyzed further. Six out of 8 foci picked for plating on Q/B cells yielded virus which plated on both Q/B and C/E cells. On repeated cycling through Q/B cells the ability to grow on C/E was lost and a stable Pr–RSV(*chf*), based also on neutralization as well as host range, was thus obtained. Passage through a system using C/E as indicator cell resulted in a loss of ability to grow on Q/B cells. Thus it appears that the initial progeny carry host range genes of both parental viruses but by appropriate selective pressure 1 or the other can be eliminated. By use of a temperature-sensitive RSV mutant it could be shown that the remote possibility of activation of an endogenous sarcoma virus in *chf*+ cells did not occur. This experiment and a number of similar ones clearly indicate a heterozygous intermediate state which precedes the formation of a stable genetic recombination with host-range genes from *chf*+ cells or other RAVs and transforming genes from *Nd* sarcoma virus.

These studies also emphasized the ability to detect with high sensitivity viruses of varying host range in noncloned populations. These may occur at low frequency but are readily amplified on passage through selectively resistant or susceptible cells. This is of importance for understanding the appearance of apparently new virus markers on passage in heterologous cells.

Formation of heterozygotes should be the result of an initial reassortment, i.e., at least 1 subunit from each parental virus in the 70 S RNA complex. This would indicate a polyploid genome (multiple subunits) since the final stable recombinant shows only a single size class of RNA. Recombination would then occur after viral DNA synthesis but whether before or after integration is not known.

One of the areas of great interest is the existence and distribution of sarcoma specific nucleic acid sequences. The evidence that *Td* variants of *Nd* viruses lacked small amounts of RNA allows for preparation of cDNA specific for this region. Such a cDNA ($cDNA_{sarc}$) can be prepared by absorption of *Nd* cDNA with *Td* RNA in excess and isolation of $cDNA_{sarc}$ by its failure to hybridize to *Td* RNA. When such a cDNA was prepared results which were exciting in the extreme were obtained [87].

Thus sarc sequences were found in all avian sarcoma viruses, including Pr, SR, and B77 strains, while leukosis viruses and *Td* variants of chicken sarcoma viruses were negative. These data were obtained by annealing $cDNA_{sarc}$ to various RNAs and analysis of hybrid formation by S-1 nuclease digestion. This specific sequence was found in the DNA of other avian species, including chicken, quail, turkey, and emu, with evidence of evolutionary divergence (lowered T_m of heterologous hybrids) and also in the DNA of mammalian cells transformed by RSV. The latter cells were true NP cells; i.e., no virus production occurred but the genome could be demonstrated by rescue procedures.

Transcription of the sarcoma-specific sequence was not found in avian embryos or cultured cells but was found in explants of quail tumors induced by methylcholanthrene (a chemical carcinogen).*

Thus we see here a dramatic demonstration of the probable reality of the oncogene theory, namely the existence of specific sequences which are inherited and which can be activated to contribute in a yet unspecified way to malignant conversion. From other studies with temperature-sensitive mutants of RSV it is known that retention of the transformed cell phenotype requires continued synthesis of a viral product (Section 9.VII.D.). At this writing the number of experimental models to be studied with $cDNA_{sarc}$ probes is unlimited and this will be a vigorous area of research in the immediate future.

As might be expected the avian sarcoma-specific sequences were not found in normal mammalian cells or mammalian sarcoma viruses. This could be based on extent of evolutionary divergence or existence of sequences with diverse ancestry. This work should obviously be repeated in several laboratories before complete acceptance. There is in the literature 1 apparent contradiction from another laboratory [83], namely that sequences in PR-RSV not shared with RAV-O were not found in chicken cell DNA. This would have been interpreted exactly opposite to the above, that sarcoma sequences did not exist in cell DNA. This difference will doubtless be resolved in the near future. My bias, based on the developments preceeding detection of sarcoma sequences by hybridization, and the general excellence of the workers referred to above, and the completeness of their study is for the reality of existence and distribution of these sequences.

I presented above evidence that the subunit RNA of *Nd* sarcoma viruses could be represented as a = b + x, where b represented replication sequences and x the sarcoma specific sequences. B-RSV(–) is deficient in some portion of b sequences, but does contain x, thus it has a subunit size in the b range. Current evidence for MSVs, including the Harvey and Kirsten interspecies pseudotypes, suggests a situation which in principle is similar to B-RSV(–),

*More recent evidence indicates transcription of sarc in chicken cells independent of malignancy.

but with a proportionately greater loss of helper sequences because loss of genes necessary for replication. Thus, stocks of M-MSV with high sarcoma: leukemia activity contain subunit RNA which sediments at 30 S compared to helper virus subunits which sediment at 35 S. If the helper virus subunit is represented as b, which included y replicating function, then the sarcoma sequence-containing subunit is most likely b - y + x, where x = sarcoma specific sequences. For HSV and KiSV the situation is more complex since both rat-derived and mouse-derived sequences appear to be present on a small subunit of ~30 S. The formation of this molecule must be from a recombinational event.

3. Summary

1. Viruses with long, frequently unknown passage histories may be and are complex mixtures of distinctive and recombinant viruses.
2. Avian sarcoma viruses include those which are replication defective (*Rd*) such as B-RSV(-), and those which are competent nondefective (*Nd*) such as SR or Pr-RSV.
3. *Rd* avian sarcoma viruses lack the major envelope glycoprotein but may still transform when introduced into susceptible cells by coinfection with helper virus or by fusion with cell membranes using Sendai virus.
4. *Nd* sarcoma viruses give rise with high frequency to transformation defective (*Td*) viruses upon tissue culture passage.
5. *Nd* viruses possess a single subunit RNA which is 10-20% greater in size than the single subunit RNA of *Td* and *Rd* viruses. By oligonucleotide fingerprinting all *Td* and *Rd* sequences are present in *Nd* RNA, but *Td* and *Rd* RNAs have qualitatively different deletions.
6. Recombination between *Nd* sarcoma viruses, but not *Td* and helper viruses, has been demonstrated. *Nd* viruses also recombine with chick helper factor (*chf*), while *Td* viruses do not.
7. All chicken cells possess equivalent genome copies of an endogenous viral genome of subgroup E specificity. Expression via viral RNA, *chf*, and gs antigen synthesis is under genetic control (discussed in Section 9.VI.); however, a subgroup E virus can be induced from all chicken cells by treatment with chemical or physical carcinogens.
8. Recombination occurs via crossing over, presumably at the proviral DNA level. Initial events involve production of heterozygous virions which indicate a polyploid genome for the RNA tumor viruses.
9. Sarcoma specific sequences are localized 5-15% from the poly(A) terminus and by use of specific DNA transcripts, these sequences seem to be present in all avian sarcoma viruses, in the DNA of the class Aves, and in mammalian cells transformed by RSV. Leukosis viruses do not contain these sequences.

10. The mammalian sarcoma viruses show interesting similarities and differences from their avian counterparts. Being replication defective (in general), their genomic RNA is smaller than that of helper viruses. If b represents helper sequences, y those sequences required for replication, and x the sarcoma sequences, the mammalian sarcoma virus subunit is represented by $b - y + x$.

11. Evidence exists suggesting that interspecies pseudotype sarcoma viruses isolated from rats after MuLV infection are created by recombination among the smaller subunit which carries sequences related to viruses from the 2 species [72].

C. Virus-Host Relationships

1. *Viral Proteins in Normal Cells and Tissues*

Prior to the development and application of sensitive molecular hybridization techniques to studies of type C viruses, considerable evidence was available indicating an intimate relationship between type C viruses and their natural hosts. The classical experiments of Kaplan demonstrating virus "activation" by x-irradiation in the low incidence C_{57} black mouse strain provided clear evidence for latent viral information [84]. More recently, viral group-specific antigens were found in tissues of infectious virus-negative chickens and mice using complement-fixation [9] or immunofluorescence tests. Analysis of antigen expression by Payne and Chubb [85] in chickens provided clear evidence for a genetic component inherited in a simple Mendelian pattern. Expression of antigen in 2 highly inbred lines was found to be controlled by a single autosomal dominant gene (Section 9.VII.). In crosses of AKR (high tumor incidence) and $C_{57}L$ (low tumor incidence mouse strains), Meier and colleagues [86] found that expression of gs antigen was controlled by 2 autosomal dominant genes. In contrast [87] the $C_{57}BL$-10 Snell strain appears to carry a dominant gene which prevents gs antigen expression in F-1 hybrids with high incidence strains. Even in low tumor incidence mouse strains evidence of p30 expression in embryos was obtained by serological tests even though infectious virus could not be isolated in many cases. The evident expression of viral proteins in the absence of full infectious particles was a key element in the development of the virogene-oncogene hypothesis. These findings also led Huebner [9] to suggest that these proteins might play a role in embryogenesis. While the genetic studies mentioned above do not indicate whether structural or regulatory genes were being measured, both gene classes are clearly involved in virus expression (Section 9.VII.). Thus, by appropriate chemical treatments, virus-negative mouse cells can be induced to produce virus, indicating both presence and control of expression. Further, individual mouse strains may harbor multiple separable type C viruses with distinct host ranges, i.e., N, B, and X-tropic viruses have all been isolated from the BALB/c mouse strain. The chromosomal localization of MuLV structural genes is described in Section 9.VII.

In the NIH Swiss mouse strain, where high levels of p30 were found in embryos in the absence of infectious virus [9], more recent studies show that this strain carries a virus of xenotropic host range; i.e., it is only able to infect cells of species other than mouse [12]. Expression of p30 or envelope antigens as inferred from findings of high levels of neutralizing antibody to the xenotropic virus in many mouse strains shows that this virus is evidently highly prevalent in mice.* We should note that in genetic crosses, expression of p30 in spleens (based on CF tests) at an early age is highly predictive of eventual tumor development. In addition to the general correlation between virus expression and cancer in mouse strains, these data are strongly supportive of the frequent statements that type C viruses are determinants of natural cancer in their host species.

When sensitive RIA techniques were applied to studies of p30 expression in mice or chickens, the results were in general agreement with those described above [46]. As might be expected, low levels of p30 can be found in many supposedly negative strains probably reflecting synthesis of components of the xenotropic virus found in all mouse strains and the subgroup E (RAV-O) virus found in all chicken strains. While occasional "absolutely negative" animals may still be found, this could be based on quantitative considerations as opposed to an indication of complete lack of antigen expression. Insofar as we are aware, no mouse or chicken has been found which lacks type C virus information either by demonstration of infectious virus, viral proteins, or virus nucleic acids.

In other species RIA procedures have also been shown widespread prevalence of p30 proteins. Thus, in hamsters where ubiquitous type C virus was suspected because of the relative ease in pickup or activation by in vivo passage of tumors, p30 was readily found in embryo and adult tissues with the exception of muscle tissue [49]. Recently, the baboon type C virus p30 has also been found in tissues of baboon and other old world monkeys [88]. At present the precise correlation between viral protein expression and synthesis of virions is difficult to establish. This is primarily because of insensitivity of electron microscopic techniques and the presence of multiple virion populations with different host ranges. Nevertheless, one cannot fail to be impressed with the widespread prevalence of type C viral proteins in those species where adequate sensitive tests have been carried out. These test procedures are by no means the ultimate, and future technology may indicate a still broader distribution.

In summary, the CF test appears most useful (at least in mice) in detecting levels of antigen predictive of tumor development. RIA techniques indicate

*This conclusion is supported by type-specific hybridization and immunoassays; however, the neutralizing factor in mouse sera is not an immunoglobulin.

more general expression of viral proteins which most probably represent low-level expression of 1 or more endogenous xenotropic viruses. Multiple genes may affect viral protein expression by limiting synthesis or intrahost spread of virus. These factors all influence ability to detect viral proteins even with highly sensitive tests.

2. *Viral Nucleic Acids in Normal Cells and Other Tissues: Detection by Molecular Hybridization*

One of the most compelling arguments in favor of the oncogene-virogene hypothesis is the general finding of nucleic acid sequences in cell DNA and RNA which are virus specific. Detection of viral sequences in cellular DNA has been accomplished both by use of labeled viral 70 S RNA and use of radiolabeled DNA prepared in the endogenous reverse transcriptase reaction. While technical and biological considerations often preclude conclusions that 100% of the genome is represented in cellular DNA, the data are sufficient and of such specificity to permit such a deduction. In many instances ability to induce virus by halogenated pyrimidines or other chemical treatment from otherwise negative cells provides the clinching evidence for full genome equivalents in cells. With the observation of intrastrain variability when viruses are compared, hybridizations of less than 100% (compared to virus as control) may reflect strain differences compared to origin of the test virus. Viruses with a long history of passage either in vivo or in vitro may lose, gain, or have modified sequences, all of which may influence the final extent of hybridization. Nonetheless, it is possible to specify the strain of origin of most type C viruses without ambiguity based on hybridization to cell DNA. If cDNA transcripts or 70 S RNAs are prepared with care, reactions are highly species specific and rarely extend to closely related species. Thus, while all mouse virus cDNAs will hybridize to mouse (*Mus musculus*) cellular DNA, these probes fail to hybridize with DNA from other rodents or any other nonrodent species tested to date. This high specificity is seen for B-type virus of mice, MMTV, as well which shows no reactivity with rodents other than *Mus musculus.* Strain variations can also occur, as, for example, in comparisons of AKR mouse cell DNA with NIH Swiss mouse cell DNA when hybridized to AKR or Gross viral transcripts (see Section 9.VII.). Current evidence indicates only a portion of AKR viral DNA can be found in NIH Swiss mouse embryo cells (see also Section 9.VII. for a discussion of xenotropic mouse viruses). This result is most striking in view of essential similarity of bulk DNA from these mouse strains. The implications of this result may in fact be more than strain variation in homologous genes and may indicate the absence of certain loci. At present, only xenotropic viruses have been isolated from NIH Swiss mice, thus the possibility of unequal representation of mouse viruses in all strains of mice is raised. This has profound implications for evolutionary considerations of the type C virus family as a whole. Rat and hamster type C viruses which are each represented by unique p30 families are also represented in cellular DNA of their host species in highly specific fashion with no cross-hybridization under stringent conditions with DNA of related species [46].

The domestic cat provides a most interesting animal for study since it contains representatives of 2 "species" of type C virus as indicated by immunological and hybridization criteria [46]. Conventional FeLV sequences have been found in cellular DNA by several laboratories; however, because of the widespread occurrence of this virus one could not be sure that these did not result from natural infection (see Section 9.VIII.). Crucial tests of FeLV-negative cats maintained under pathogen-free conditions also show the presence of virus-specific sequences in all DNA. So far, FeLV has not been induced from virus-negative cultures (Section 9.VI.C.3.), nor is viral RNA detected in such cells. RD114, the prototype of the second cat type C virus species, presents a highly interesting natural history. This virus is readily induced from negative cells and such cells produce large amounts of viral RNA. The distribution of homologous DNA sequences among Felidae and other mammals is shown in Table 9.17, taken from the work of Benveniste and Todaro [89]. Note that although unique sequence DNA of all "cats" are highly related, only a select few show cross-hybridization with a RD114 cDNA transcript. This dramatic finding is completely inconsistent with a normal evolutionary relationship and suggested the possibility of transmission to the domestic cats immediate ancestors in 1 restricted geographical location (North Africa, Mediterranean basin) from an outside source. Table 9.17 also shows the striking result of detection of RD114-related sequences in cellular DNA of old world monkeys and to a lower extent in chimpanzees and gorilla.

The picture was given considerable clarity by the finding of viruses immunologically related to RD114 in 2 species of baboon with different geographical locations. Molecular hybridization analyses with the *Papio cynocephalus* isolate are shown in Table 9.18 (from Ref. [19]). These provided the first evidence for orderly evolution of a set of viral sequences among a mammalian order. Here one sees that the viral cDNA transcript is actually a sensitive indicator of evolutionary relationship among the primates. The rate of evolution of these viral sequences is about 4 times that of bulk unique sequence cellular DNA. This explains the ready ability to discriminate among *P. cynocephalus* and *P. hamadryas* isolates (data not shown) even though these species are indistinguishable by cellular DNA hybridizations. Since the rate of evolution of cellular DNA sequences appears to be proportional to generation time and not geological time, one can easily rationalize inability to detect related sequences among the mouse and rat type C viruses even if a common

Table 9.17 Nucleic Acid Homology and Thermal Stability Between Domestic Cat Unique Sequence Cellular DNA and RD114 Viral DNA and the DNA from Various Species[a]

Species[b]	Geographical range	Unique Sequence domestic cat cellular [^{3}H]-DNA[c]		RD114/CCC viral [^{3}H]-DNA[d]	
		% Hybrid	ΔT_m	% Hybrid	ΔT_m
Carnivores					
Felidae					
Felis					
Domestic cat (*F. catus*)	North Africa[e]	100	<0.5	100	<0.5
European wildcat (*F. sylvestris*)	Europe, Asia Minor	>95	<0.5	95	<0.5
Sand (*F.* margarita)	North Africa, Arabia	>95	<0.5	81	<1.0
Jungle cat (*F. chaus*)	Nile valley, Asia Minor	>90	<1.0	70	1.5
Leopard cat (*F. bengalensis*)	Southeast Asia	>90	<1.0	<2	–
Golden cat (*F. temmincki*)	Southeast Asia	>90	<1.0	<2	–
Geoffroy's cat (*F. geoffroyi*)	South America	>90	<1.0	<2	–
Lynx					
Caracal cat (*L. caracal*)	Africa, Asia Minor	>90	<1.0	<2	–
Bobcat (*L. rufus*)	North America	>90	<1.0	<2	–
Panthera					
Lion (*P. leo*)	Africa, Asia	>90	<1.0	<2	–
Tiger (*P. tigris*)	Asia			<2	–

Leo					
Leopard (*L. pardus*)	Africa, Asia	>90	<1.0	<2	–
Uncia					
Snow leopard (*U. uncia*)	Central Asia	>90	<2.0	<2	–
Acinonyx					
Cheetah (*A. jubatus*)	Africa, Asia Minor			<2	–
Other carnivores					
Mink		23	10	<2	–
Dog		19	11	<2	–
Primates					
New world monkeys (Cebidae)					
Capuchin		2	–	<2	–
Old world monkeys (Cercopithecoidea)					
Baboon		2	–	20	10.5
Mangabey		3	–	15	11.0
Patas		2	–	11	11.0
African green		3	–	11	11.0
Macaque:					
Rhesus			–	18	11.0
Stumptail		1		17	10.5

Table 9.17 (Continued)

Species[b]	Geographical range	Unique Sequence domestic cat cellular [^{3}H]-DNA[c]		RD 114/CCC viral [^{3}H]-DNA[d]	
		% Hybrid	ΔT_m	% HybridΔ	ΔT_m
Apes and man (Hominoidea)					
Chimpanzee		2	–	7.5	NT[f]
Gorilla		2	–	8.5	NT[f]
Gibbon		1	–	<2	–
Human		1	–	<2	–
Other mammals					
Mouse		2	–	<2	–
Cow		2	–	<2	–

[a]Data of Benveniste and Todaro [89]. Note the surprising result that RD114 sequences are found in only 4 cat species even though all Felidae are highly related. In addition, the presence of RD114 sequences in old world monkeys is clearly shown.

[b]Cellular DNA was extracted from various tissues of the species listed.

[c][^{3}H]-Thymidine-labeled unique sequence domestic cat cellular DNA was hybridized to the cellular DNA of the various species. The percent hybrid is the saturating normalized value obtained after digestion of the hybrids with S-1 nuclease. The actual final extent of domestic cat cell DNA hybridization to the cellular [^{3}H]-DNA probe was 73%. The temperature at which 50% of the hybrids are dissociated (T_m) was 85°C for the homologous domestic cat:domestic cat hybrid; the ΔT_m is the difference in T_m between the other DNA:DNA hybrids and the T_m of the homologous hybrid. The T_m of hybrids that hybridize less than 3% cannot be determined because of insufficient homology.

[d]A [^{3}H]-DNA probe prepared from RD114 virus grown in a human cell line was used for the hybridizations to feline DNA, and a [^{3}H]-DNA probe prepared from the highly related CCC virus grown in a canine thymus cell line (Cf2Th) for the hybridizations to primate DNA. The T_m of the homologous RD114 :domestic cat DNA hybrid was 88.5°C.

[e]Since its domestication this cat has traveled with man all over the world. All breeds of the domestic cat (e.g., Burmese, Siamese, Abyssinian) belong to the same species.

[f]NT, not tested.

Table 9.18 Thermal Stability of Hybrids Formed Between Baboon Unique Sequence Cellular DNA and Type C Viral DNA and the DNA from Various Primate Species[a]

	Unique sequence baboon cellular [^{3}H]-DNA[c]		Baboon type C viral [^{3}H]-DNA[d]	
Species[b]	% Hybrid	ΔT_m	% Hybrid	ΔT_m
Old world monkeys				
Baboon				
P. cynocephalus	100	<0.5	100	>0.5
P. papio			81	1.5
P. hamadryas	100	<0.5	74	2.0
Mangabey			48	5.5
Patas			38	6.0
African green			30	7.0
Macaque				
Stumptailed	91	1.5	25	8.5
Pigtailed			27	9.0
Crab-eating			22	9.0
Rhesus			23	9.0
Apes and man				
Chimpanzee	61	4.5	9	NT[e]
Gibbon	66	4.5	7	NT
Man	62	4.0	7	NT
New world monkeys				
Capuchin	31	7.5	2	>15
Howler	29	8.0	2	>15
Nonprimates				
Mouse	<2	>15	<2	>15
Dog	<2	>15	<2	>15
Domestic cat	<2	>15	19	11.0

[a]Data from Benveniste and Todaro. Note the evolutionary pattern of baboon viral sequences in the various old world monkeys and the highly important positive results in the higher apes.

[b]Cellular DNA was extracted from various tissues of the species listed. The scientific names of the primate species not given in the table are: pigtailed macaque, *Macaca nemestrina*; crab-eating macaque, *M. fascicularis*; rhesus monkey, *M. mulatta*; gibbon ape, *Hylobates lar*; and howler, *Alouatta spp.*

[c][^{3}H]-Thymidine-labeled unique sequence baboon cellular DNA was hybridized to the various species. The percent hybrid is the normalized value obtained after digestion of the hybrids with S-1 nuclease. The temperature at which 50% of the hybrids are dissociated (T_m) was 83°C for the homologous baboon:baboon hybrid; the ΔT_m is the difference in T_m between the other DNA:DNA hybrids and the T_m of the homologous hybrid.

[d][^{3}H]-DNA probe was prepared from the baboon virus grown in the canine thymus cell line (Cf2Th). The T_m of the homologous baboon:baboon hybrid was 88°C.

[e]NT, not tested.

ancestral sequence for these 2 viruses is assumed. Table 9.18 also shows that higher apes also possess a low but definitive degree of hybridization with baboon virus cDNA. This is discussed in more detail in Section 9.IX. This table also shows the reciprocal of the RD114 old world monkey relationship, namely the presence in cats of DNA sequences related to the baboon virus on an evolutionary basis (ΔT_m 11°C). These data can be interpreted in several ways; however, the bulk of the evidence seems to favor horizontal transmission occurring 5-10 × 10^6 years ago from progenitors of old world monkeys to the immediate ancestors of the domestic cat. Not only is horizontal transmission required but one must also add a rapid widespread infection followed by integration into the germ line. Based on current relative lack of infectivity of RD114 for cat cells, one assumes that the original virus has been modified in envelope properties. In Section 9.IV. I have described the sequence relatedness of BaLV and RD114 p30s, including an apparent gap event at position 5 from the amino terminal. This being a relatively rare event in protein evolution as compared to simple replacements argues for existence of the p30 gene prior to the hypothetical horizontal transmission to cats.

A major surprise is the lack of homology between the other primate viruses, SSV-1 and GaLV and their suspected host species [91]. Both labeled RNA and cDNA techniques have been used and both have detected viral sequences in infected and transformed nonproducer cells so that single copies of viral DNA could have been detected. I noted previously that primate and mouse viruses showed some degree of cross-hybridization. This is paralleled by ability of primate virus probes to hybridize to mouse DNA [48], leading to suggestions of mouse or related rodent to woolly and gibbon ape transmission. This could be an analogy for the early stage of the supposed old world monkey progenitor to domestic cat ancestor transmission of RD114-like viruses at a stage before integration into the new host cellular genome. Data from our laboratory suggest that this may be much more complex since a GaLV transcript is found to give significant levels of hybridization with an endogenous hamster virus RNA (10-15%). This could be explained based on a model such as depicted in Fig. 9.17. Here sequences related to mouse, hamster, and primate viruses are shown to be derived from a common ancestor and, since divergence, shows different rates of evolution. This model predicted detection of the mouse-hamster relationship which has actually been found under relaxed conditions of hybridization [69]. The SSV-1-GaLV problem must occupy a major part in any comprehensive theory of type C virus evolution. The pig virus, which we will not discuss in detail, is also represented in cellular DNA of its species of origin and therefore is considered an endogenous virus. Proteins of this virus are highly related to those of SSV-1 and GaLV which is possibly suggestive of a propensity for interspecies transmission of viruses of this group.

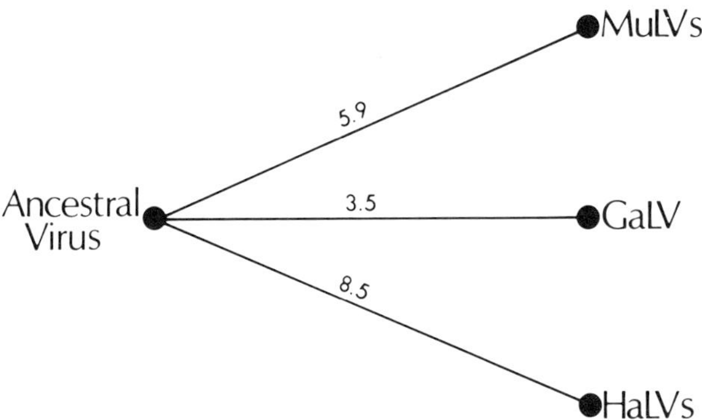

Fig. 9.17. Hypothetical evolution of GaLV, mouse, and hamster type C viruses from a common ancestor. This figure is a prediction of mouse-hamster viral genome relationships obtained using the T_m data for their relationship to GaLV, assuming some evolution of GaLV from a common ancestral genome for the 3 viruses and setting the limit of GaLV evolution just beyond the ability to detect hybridization between mouse and hamster viruses by S-1 nuclease study. Thus, the limits of the 2 predictions indicate a ΔT_m range of 14.4-21.4°C for the mouse-hamster relationship. In fact, using less stringent conditions a mouse-hamster virus relationship was detected showing a ΔT_m of 14.6°C (Table 9.10).

Finding copies of viral information in cellular DNA naturally raises the question of how many. These answers are generally arrived at by kinetic analysis comparing the rate of reassociation of labeled single-stranded viral DNA to that of unique sequence cellular DNA or, alternatively, the increase in rate of hybridization of double-stranded viral DNA in the presence of cellular DNA. A general, simple summary of all the results obtained indicates that a portion of the viral genome is represented only several times per cell genome. This is true of MMTV as well as avian and mammalian type C viruses. In addition, some investigators report that a portion of the viral genome is highly reiterated, perhaps 100-200 copies per genome in the case of RD114 in cat cells and subgroup E avian viruses. The biological evidence for certain mouse strains is certainly consistent with multiple complete genomes per cell.

A topic considered in the previous section on recombination was the detection and distribution of sarcoma virus-specific sequences in chicken viruses. The most recent data suggest that these sequences are at present in cell DNA of many members of the class Aves and are related in those species in an evolutionary manner [82]. This contrasts with the leukemia viruses which show a high degree of species specificity. This would seem to suggest a distinct mode of origin of these 2 sets of sequences. One obvious thought, consistent with the relative low frequency of isolation of sarcomagenic virus from spontaneous sarcomas in both chickens and mice is that sarc sequences are transduced into the genome of the leukemia virus on a chance basis or because of location near a proviral integration site. A possible model for such an event is found in the origin of the in vivo produced pseudotypes of mouse-rat or mouse-hamster origin (see Section 9.V.B.). While these pseudotypes were produced experimentally, the possibility of such a situation occurring naturally cannot be excluded.

In the case of mouse viruses where nonproducer cells carrying only the sarcoma genome can be prepared, nucleic acid sequences common to the helper virus are readily detected. However, the sedimentation coefficient of the intracellular viral RNA in such NP cells was found to be 30 S in contrast to 35 S RNA subunits found in helper viruses (after DMSO or heat treatment of 70 S RNA), or helper virus-infected cells. A striking finding which parallels the ability to detect viral proteins in "virus-free" tissue culture lines of various mouse strains is the ability to detect viral-specific RNA in the same cells. This now appears to be a generality for proved endogenous viruses of rats, hamsters, mice, chickens, baboons, and cats; namely, that many but not necessarily all tissue culture cells and organs contain easily measurable levels of viral RNA. This appears independent of whether the cells are normal or altered by transforming agents.

This background of viral RNA synthesis is strong support for the theory of natural inheritance of virus-specific information and the deduction that this may be of import in differentiation. To complete the picture on endogenous virus presence in cell DNA and transcription in normal cells are (a) failure to find viral RNA in normal primate cells homologous to GaLV or SSV-1 cDNA (as expected); and (b) lack of expression of FeLV RNA in cultured cat cells even though this virus is present in cell DNA. In fact, the cat expresses large amounts of RD114 RNA even in cells actively producing FeLV. Thus, the evidence that FeLV can spread horizontally (Section 9.IV.) as opposed to RD114 which is essentially xenotropic seems to correlate with lack of expression in normal cells.

Another point of great interest concerning expression of RNA emerged from studies of cultured rat cells. As noted in Section 9.V., KiSV and HSV are now known to be pseudotypes including rat and mouse nucleic acid sequences. In 1 study using virus activated by BrdU treatment of KiSV transformed NP mouse cells to prepare cDNA probes, Tsuchida, Hatanaka, and I demonstrated the presence of mouse- and rat-specific sequences in activated virus [92]. The mouse sequences were removed by absorption of cDNA with mouse virus RNA leaving a rat virus-specific probe. This KiSV cDNA probe detected a 30 S RNA species in normal rat cells and failed to react with 35 S RaLV-specific RNA in infected cells. Infected cells also produce a 30 S RNA species. Thus, it appears that normal rat cells produce RNAs with considerable relatedness to those sequences found in a sarcoma virus. To date, however, in vitro attempts to produce a sarcoma virus by superinfection of rat cells in culture have not proved successful even though pickup of rat sequences has been shown; thus some activation step or chance event may be necessary to produce the recombinant or reassorted sarcoma virus. Very recent evidence [72] suggests that the 30 S RNA of HSV and KiSV is a recombinant of rat and mouse sequences. This would seem to be a crucial step in formation of these sarcoma viruses. Other hybridization experiments established that rat and mouse 30 S RNA sarcoma sequences are not closely related so that these sequences like the helper sequences are readily distinguished.

The natural history of the mouse mammary tumor virus, MMTV, has been studied by both immunological and molecular hybridization techniques. MMTV is present in cellular DNA of mice with both high and low incidence of mammary cancer (several copies/per cell) [91, 93]. Expression of MMTV has long been known to result from a complex interplay of genetic, hormonal, and viral factors. Virus-specific RNA was readily found in large amounts in both virus-producing tumors and lactating mammary glands from high-incidence strains (up to 10^3 genome equivalents per cell) and in low amounts (1-2 copies per cell) in virus-free tumors from low-incidence strains [93]. Cultured cells producing MMTV show enhanced synthesis of MMTV RNA when incubated with the synthetic steroid dexamethasone. Studies of hormone receptors and transcriptional controls with this single known mammary cancer virus are current areas of intense investigation. This system seems to provide an excellent correlation to natural disease expressed in molecular terms in an in vitro system.

Among the other retraviruses, the only other member clearly represented in cellular DNA is the endogenous guinea pig virus [94]. This virus is clearly not a type C virus, based on Mg^{2+} preferring reverse transcriptase and lack of p30 interspecies determinants. MP-MV does not appear to be an endogenous virus of the rhesus but may be of significance in man. Discussion of this virus is found in Section 9.IX.

Thus at least certain members of the retravirus family are found in cellular DNA and their information is transcribed and translated (Table 9.19). Precisely what this means to cellular function and cancer requires a full understanding of viral gene functions and ability to measure their expression.

3. *Biological Detection of Viruses in Normal Cells: Induction and Transfection*

While we have presented convincing evidence by molecular hybridization that infection by oncornaviruses results in integration of a copy of the viral genome and that normal cells contain full genomes, the ultimate evidence for this

Table 9.19 Relationship of Type C Viruses to Their Species of Origin

Virus designation[a]	Species of isolation	Induction[b]	V-RNA[c]	V-DNA[d]
MuLV	Mouse	+	+	+
RaLV	Rat	+	+	+
FeLV	Cat	–	–	+
RD114	Cat	+	+	+
HaLV	Hamster	N.D.[e]	N.D.	+
GaLV	Gibbon ape	–	–	–
SSV-1	Woolly monkey	–	–	–
BaLV	Baboon	+	+	+
Pig LV	Pig	+	+	+

[a]The designation "LV" does not necessarily indicate pathogenicity.
[b]From cells of the homologous species using halogenated pyrimidines.
[c]Virus-specific RNA in normal cells using DNA product of reverse transcriptase.
[d]Virus-specific DNA in cellular DNA of normal cells using DNA product of reverse transcriptase.
[e]ND, not determined.

assertion has come from biological experiments. The "easiest" technically are those involved in treating virus-free cells with chemicals, especially halogenated pyrimidines. Initial experiments were performed by the laboratories of Rowe [20], Klement [23], and Weiss [22] using mouse, rat, and chicken cells, respectively. The experimental techniques are quite straightforward; drug is added, and after the appropriate time (only a few days) cultures are monitored for virus expression by the desired technique. Most useful has been simply to assay culture media for reverse transcriptase activity using synthetic templates, for example, polyriboadenosine (rA):oligodeoxythymidine (dT) and [^{3}H]-TTP as substrate. The results of such experiments have proved positive in virtually every case where the presence of endogenous virus is suspected. In addition to bromodeoxyuridine and iododeoxyuridine as inducers, cyclohexamide has been found to efficiently induce the xenotropic virus in BALB/c mouse embryo cells. In addition to these artificial inducers, we pointed out that in MMTV-containing cultures the synthetic steroid dexamethasone increased virus synthesis. This depended on preexistence of some level of virus synthesis which supports the conclusion of Rowe that inducers act to increase the probability of a normally occurring event. One particular case where this type of experiment was particularly useful was with the endogenous cat virus, RD114. At a time when convincing evidence as to the origin of this virus was not yet in hand, it was shown that virus-negative cat cells treated with BRdU or IUdR would regularly yield a virus indistinguishable from RD114. This was augmented by other types of evidence but for many, induction provided the crucial evidence. As might be expected, the technique is widely used at present.

Another dramatic piece of evidence for the integrated virus concept was the demonstration by Hill and Hillova [95] that DNA extracted from RSV transformed rat cells could transform chicken cells. This technique, termed "transfection," has been confirmed with many variations, including use of such special markers as temperature sensitivity. While this is clearly a repeatable technique for cells infected with virus, so far no reports of transfection with normal chicken cell DNA have appeared. In the mammalian system Nicolson and McAllister have transfected indicator cells with DNA from RD114 and gibbon ape virus-producing cells [96] and Karpas and Milstein [97] have transfected with the MSV genome from NP cells. This latter required passage through MuLV-producing cells to produce sarcoma virus. Scolnick and Bumgarner [98] have recently reported that endogenous xenotropic viruses of mouse and cat could be isolated by the transfection technique. This procedure provides a means for isolation of endogenous virus without induction.

VII. Genetic Control of Virus Expression

A. Chromosomal Localization of Viral Genes in Mice

The fact that expression of viral genes is under genetic control was established by cross-breeding experiments from many laboratories. The experiments of Rowe and colleagues [10, 11] are especially noteworthy and provide the strongest association between a genetic locus for expression and physical presence of the viral genome. This structural locus is distinguished from regulatory loci, such as Fv-1, which controls the spread of virus presumably based on cell surface receptors. Of special importance to the question of oncogenicity is the demonstration by Lilly et al. [99] that levels of virus expression of the genetically inherited virus in early life correlate with leukemogenesis in later life in the same mice. Because of their critical significance we will describe the experiments of Rowe and colleagues in some detail. These experiments utilized crosses of the AKR inbred mouse strain to various other mouse strains which differ in overt virus expression and normal incidence of leukemia. The AKR strain was specifically bred to produce a high incidence of leukemia, mainly involving the thymus, which reaches ~100% in 8-11 months. It is from this strain that Gross made the initial isolation of a mouse leukemia virus and thus it is perhaps fitting that it is this virus and strain which occupy a critical place in contemporary viral oncology. The AKR strain is homozygous for a regulatory gene designated Fv-1^{N} which allows cell to cell spread of viruses with N tropism. Mouse strains are of 2 types with regard to this property and are either Fv-1^{N} or Fv-1^{B} The N cells are resistant to B (not to be confused with MMTV B particles) tropic type C viruses, and B-type cells are resistant to N-tropic viruses. Table 9.20 (from Ref. [100]) lists N and B characteristics of a number of mouse strains. The resistance factor is dominant since N × B f-1 hybrids are resistant to both N and B viruses. Among mice the only exceptions seem to be feral mice which can yield cell lines capable of replicating both N and B tropic viruses. The resistance is not absolute but is a factor of 10^2 or 10^3 difference. While natural isolates of MuLV seem to be either N or B tropic, many laboratory strains, e.g., RLV, are NB tropic, thus indicating mixtures or recombinants resulting from laboratory manipulation. The xenotropic mouse viruses are discussed elsewhere in this

Table 9.20 Mouse Strain Distribution of the Alleles at the Fv-1 Locus[a]

Fv-1^{n}	Fv-1^{b}
Swiss (NIH, Ha/ICR, SIM)	BALB/c
DBA/1 and DBA/2	A
DDD	I
C3H/He and C3H/Bi	PRI
CBA and CBA-T6	C57BL/6
129	C57BL/10
NZB	B10.BR/Sn
AKR	
RF	
ST	
CE	
C57BR	
C57L	
C58	

[a]N-Tropic viruses grow in Fv-1^{n} cells and not Fv-1^{b} cells. B-Tropic viruses grow in Fv-1^{b} and not in Fv-1^{n} cells. From Ref. [100].

chapter; this class of MuLV does not infect either N or B cells but only cells of different species. Clearly, 1 reason for the high incidence of virus expression and leukemia in the AKR strain is the presence of an inherited N-tropic leukemogenic virus and N-type cells; thus, any initial virus synthesized will rapidly spread to other cells. Important technical innovations in Rowe's studies were first the development of quantitative assay for MuLV (the XC plaque test), and the observation that the tail bone provided a reliable tissue for assay of the virus status of individual mice.

The initial experiments involved crosses of AKR with low-incidence mice also homozygous for Fv-1^{N}. Among the strains tested were C_{57}brown (Br), C_{57}leaden (L), DBA/2, and a special noninbred strain, NIH Swiss, which so far has not yielded either a N- or B-tropic virus but only 1 of the xenotropic class. This particular strain will be seen to be of critical importance in establishing the physical basis of 1 of the viral expression genes. Tests of tails for virus at 6 weeks of age revealed 100% positive in the AKR strain (26/26) while 91 samples from the combined 4 strains including tests on C_{57}L at 4-6 months were completely negative. In the F-1 generation produced by mating AKR mice with each of the low-incidence strains, 100% of the progeny were positive for virus at 6 weeks of age. Thus in these crosses inheritance of the AKR virus expression phenotype was dominant. In addition no evidence of sex relatedness was noted, i.e., the same results were obtained whether the AKR parent was male or female.

The F-1 progeny were then back-crossed to the low-virus parental strain and in some cases to a nonparental low-incidence strain mouse with similar results. Of a total of 475 mice tested at 6 weeks of age, 345 (73%) were positive for virus. These results are clearly consistent with a 2 independent gene inheritance model. Expected results for a single gene would be a 1:1 virus positive to virus negative ratio, 2 genes should give a 3:1 ratio, and a 3-gene model would predict 7:1 virus positives to negatives.

The F-1 progeny of the C_{57}BR × AKR cross were allowed to mate and thus produce an F-2 generation. According to a 2-gene model one should expect a ratio of 15:1, positive to negative. The actual results were 92% positive (49/53), in close agreement with the predicted ratio. The genes thus identified were termed v1 and v2 with the stipulation that from these experiments alone one could not be sure of whether they were regulatory or structural loci.

The next step in these experiments was the isolation of the 2 loci in separate families of mice. This was accomplished by a second backcross to a low-incidence parent using unselected first backcross mice. Families of the second backcross would be expected to carry 2, 1, or 0 AKR-V genes in the ratio 1:2:1. Among 19 families tested, 3 showed a 3:1 ratio in their progeny (2 gene), 12 had a 1:1 ratio (1 gene), 1 family was not clearly in either category, and 3 yielded progeny with no virus expression. This result obviously further fulfills the 2-gene model and allows the possibility of separation of the 2 loci.

In the initial F-2 generation produced by C_{57}brown × AKR (an albino strain) it was noted that white mice were consistently positive for virus by 2 weeks of age. The possibility of linkage was evaluated for a hemoglobin marker (Hbb locus), closely linked to the albino gene (C) located on linkage group 1 which is part of chromosome 7 of the mouse.

The observation was made that backcross-1 mice with the C-Hbb region from AKR were consistently more virus positive than those without these linkage markers. Obviously the problem of 2 independent segregating genes would make it difficult to see linkage clearly and thus the second backcross mice are critical. In these experiments, 4 families with a 1-gene segregation pattern showed clear-cut linkage to group I and 4 did not. The locus on linkage group 1 is formally designated Akv-1 and has been localized 12 map units from another group I linkage marker, glucose phosphate isomerase (Gpi), which is expressed in tissue culture cells and thus is a valuable marker for many in vitro studies. The gene order appears to be C (Albino)-Gpi-Akv-1. The v-2 gene has not yet been mapped.

Interestingly enough, expression of the GIX antigen, which appears now to be the envelope glycoprotein, is controlled by 2 genes, Gv-1 on linkage group IX and Gv-2 on linkage group I. Gv-2 is found on the opposite side of Gpi-1 than Akv-1 with the gene order: Akv-1, Gpi-1, Hbb, Gv-2. Fv-1 has been localized on linkage group 8 (chromosome 4) and is not linked to either v-1 or v-2.

The key question remaining was whether Akv-1 was a structural or regulatory locus. For several reasons, it did appear likely that this was a structural locus, although convincing evidence was not in hand. However, this was clearly provided by molecular hybridization experiments. By way of background, it was shown that [^{3}H]-cDNA prepared from AKR viral genomic RNA by reverse transcriptase could be used to qualitatively distinguish cellular DNAs from high-, medium-, and low-incidence mouse strains. This is in terms of calculated percent of genome present, number of copies per genome, and thermal dissociation kinetics of DNA-DNA hybrids. While perhaps subject to alternate explanations in terms of evolutionary divergence, nevertheless operationally one could unambigously distinguish the insertion of AKR viral genes onto a NIH Swiss mouse background. The experiment involved crossing of Akv-1-positive mice from the families described above with NIH female mice for 6 generations. At each generation the male to be bred was heterozygous for Akv-1, Gpi, and C genes. Nine embryos from the fifth generation of mating were tested. After generation 6, brother-sister matings were made to produce partially congenic mice; these mice were selected to be homozygous for Akv-1 by utilizing the Gpi-1 marker. Thus, 3 markers, Akv-1+ (from AKR) and Gpi-1^a and C+ (both from C_{57}brown), were introduced into a NIH Swiss genetic background. NIH markers are Akv-1-, Gpi-1^b - C. Presence of Akv-1 is scored by virus induction using IUdR in tissue culture (NIH cells are noninducible), Gpi-1 on embryo cells, and C by pigment presence or absence in the embryo.

In the 9 embryos tested from the fifth mating generation, 5 were positive for Akv-1 and 4 were negative. DNA from the 5 positive embryos hybridized with more of the [^{3}H]-AKR cDNA preparation than the negatives and these also showed thermal stability characteristics expected of insertion of AKR viral DNA onto NIH, e.g., comparable to DNA of an AKR × NIH Swiss F-1 mouse. Interestingly, among the 9 embryos there were 5 recombinants for the linkage markers but the hybridization characteristics correlated only with Akv-1.

The partially congenic lines provided more material for assay and allowed quantitation of an operationally distinct set of sequences between AKR and NIH. There are 3-4 sets of these sequences in AKR DNA and 1-2 in the F-1 hybrid of AKR × NIH. The Akv-1 congenic DNA give results equivalent to that of the F-1 hybrid, indicating presence of about 1/2 the sequences present in the AKR mouse DNA.

The obvious conclusion from these studies is that the Akv-1 locus is a structural locus for virus genetic information. It would be of some interest to insert the V-2 gene into the Akv-1 cogenic mice and then to compare hybridization characteristics with AKR itself. This perhaps could determine if Akv-1 and Akv-2 constitute the total number of AKR viral loci in the AKR mouse. We have mentioned previously that in other cross-breeding experiments the level of overt virus expression in early life was predictable of leukemia development. Here, therefore, we see a clear demonstration that an inherited genome which can be expressed as an infectious agent is responsible for malignancy. This is the central theme of the viral oncogene theory.

B. Regulatory Genes Controlling Expression and Synthesis of Mouse Leukemia Viruses

In Section 9.VII.A. I detailed the experimental evidence indicating that nucleic acid sequences corresponding to the AKR virus were located at a genetic locus controlling expression of this virus. This locus, Akv-1, is thus clearly a structural locus. There are in addition multiple loci which have a regulatory function, the effects of which are manifest in either spontaneous or experimentally induced disease [100]. For naturally occurring viruses of the Gross (AKR) type the most important of these appear to be a gene closely linked to H-2, the major histocompatibility locus of mice, and Fv-1, the locus described in Section 9.VII.A. which controls spread of virus. In addition there are a number of special situations which control expression by such factors as limitations in availability of target cells and possibly immune response capability.

H-2 is a complex locus on the IXth linkage group of the mouse which carries a number of functions of immunobiological interest, e.g., immune response genes, Ir-1, in addition to the histocompatibility factors. It was noted early on that mice of the $H\text{-}2^k$ haplotype were highly susceptible to leukemogenesis by Gross virus (strains AKR, C58, C3H, C_{57}brown), while other strains were resistant, e.g., C_{57}black ($H\text{-}2^b$) and I ($H\text{-}2^1$). Cross-breeding experiments indicated that susceptibility behaved as a recessive trait; e.g., the F-1 generation of K × b crosses were resistant and backcrosses to the susceptible parent showed a high correlation of H-2 type with leukemia. Similar results were obtained in spontaneous leukemia in these crosses: viz., $H\text{-}2^k$ homozygotes from the backcross generation were similar to the AKR parent in age of onset and incidence, whereas heterozygotes showed a lower incidence with longer latent period. The responsible gene in the H-2 region was designated Rgv-1. Among the alleles at H-2, a and d are also correlated with susceptibility to leukemia viruses, although less so than $H\text{-}2^k$. Congenic mouse strains which are extremely valuable for measuring single-gene effects were also prepared by Lilly. C3H ($H\text{-}2^k$) and C3H-SW ($H\text{-}2^b$) strains which differ only at the H-2 locus were found to differ radically in response to Gross virus as predicted from the H-2 type. One should emphasize here that H-2 type does not correlate with presence of the Akv-1 gene which is located on a different chromosome.
With regard to other isolates which induce mainly thymic lymphomas, such as B/T-L and RadLV strains, these also follow the resistance-susceptibility pattern shown for Gross virus and H-2 type. We noted above that a gene which governs immune response capability is located in close association with $H\text{-}2^k$ in mice. This gene has been identified by virtue of ability of mice to produce antibody in response to small synthetic polypeptides. Aoki and colleagues provided evidence that antibodies to a Gross virus-associated antigen could be detected among progeny of an AKR × C_{57}B1 cross in direct correlation to presence of the resistant H-2 type ($H\text{-}2^b$). At present this correlation seems to provide the strongest lead to the mechanism of the H-2 effect. Further analysis seems warranted in view of possible associations between the human HL-A system and disease.

As noted in Section 9.VII.A., the major genetic system involved in regulation of virus replication in tissue culture and in vivo is designated Fv-1. At present all inbred strains are designated as $FV\text{-}1^N$ (resistant to B, susceptible to N-tropic virus) or $Fv\text{-}1^B$ (susceptible to B, resistant to N-tropic virus), while a wild mouse cell line has been identified as susceptible to both types. The identification of this locus and the Fv designation derives originally from studies of the Friend virus. This virus is clearly a distinct pathogenic entity from the Gross and related viruses which mainly cause thymic lymphoma.

Friend virus has an erythroid precursor as target cell and induces neoplasms involving spleen, bone marrow, and liver. There is some variability in pathogenicity of current strains of Friend virus, some yielding cells arrested at an earlier stage of maturation. The latter are being studied intensively since by special treatment they can be induced to synthesize hemoglobin. A convenient assay for Friend virus in vivo described originally by Axelrod and Steeves is its ability to produce macroscopic focal lesions in the spleen of susceptible mice. In a large variety of cross-breeding experiments it was determined that 2 genes influence Friend virus leukemogenesis. One, now designated Fv-1, controls the replication of a virus which occurs in association with Friend spleen focus-forming virus (SFFV) and is clearly a helper virus of the lymphatic leukemia type (LLV). The latter virus confers host-range properties on SFFV exactly as in the case of defective sarcoma viruses and thus multiple strains have been created by animal passage. The LLV component is readily isolated free of SFFV, but the converse has not been found. Fv-2, the second genetic locus, is independent of Fv-1 and controls the ability of the host to respond to SFFV. Before considering Fv-1 in detail, we should emphasize that effects of SFFV are restricted to in vivo assays since a suitable in vitro assay has not been developed. With a recent exception, the Abelson leukemia virus (to be discussed in Section 9.VIII.), there is no ready in vitro correlate for leukemia virus activity. Being aware of helper viruses, need for appropriate target cells, etc., provides 1 ready explanation for failure to specify a virus for each type of malignancy. I will not discuss here other genes influencing Friend virus susceptibility but simply note that alleles at W (dominant spotting), S1 (steel), and F (flexed) all influence leukemia presumably through qualitative and quantitative effects on target cell availability.

In studies of restriction by Fv-1, significant differences from host-range restriction in the chicken virus system were noted. In the latter species host-range variations correlate completely with envelope type and restriction occurs at the level of absorption and/or penetration. In the FV-1 system resistance occurs by an intracellular mechanism perhaps implying the presence of restriction enzymes as in bacteriophage. Such a deduction follows from the observations that pseudotypes with vesicular stomatitis virus (VSV) can be prepared which carry the MuLV envelope type and VSV genome. Such pseudotypes bring VSV into cells (based on its ability to replicate) regardless of N or B tropism, showing that the block is intracellular. Analysis of dose-response effects on resistant cells showed multiple-hit kinetics for production of plaques in the XC assay [101]. Using an infectious center assay it was further shown that there was a deficiency in production even from cells which received the requisite number of hits (2 with most viruses, 3 with Gross passage A). As already described there is also a cell to cell transfer problem in producing plaques in a resistant cell population. Among these 3 factors 1 mouse strain, DBA/2, was found to show only the dependence on hitness. Thus either alleic or other genes may come into play by influencing the ability to synthesize virus after the requisite number of hits. Once infected, individual resistant cells produce progeny at the same level as susceptible cells. These most recent data indicate that Fv-1 resistance can be thought of in terms of a positive control which may be overcome by sufficient amounts of some viral gene product. Restriction via repressors or restriction enzymes should be considered. We note again that restriction is dominant even in the case of somatic cell hybrids, thus the system seems optimal for isolation of a "repressor" substance. There are some reports that extracts of N cells added to B cells will allow infection by N-tropic virus. This has not yet been confirmed and would add another complexity to this system—a factor allowing replication as well as 1 preventing replication of a "foreign" virus.

Another genetic system of interest is the hairless mouse (hr) studied by Meier and colleagues. The hr gene must be maintained in heterozygous form because of technical difficulties in breeding homozygotes. hr/hr homozygotes have a high incidence of spontaneous leukemia, the heterozygote intermediate incidence, and the wild type low incidence. All 3 types were positive for virus in earlier studies; thus the incidence of leukemia was thought to be based on a regulatory factor other than virus presence. Recent data of Huebner using quantitative plaque assays of virus in tails at 1-2 months of age, however, show an important dosage effect. Thus the hr/hr have a high level of virus, +/hr an intermediate level, and +/+ barely detectable virus. The correlation of virus dosage with spontaneous leukemia is thus in perfect agreement with the cross-breeding experiments of Meier and Lilly and Rowe, showing a clear correlation between quantitative levels of virus expression and spontaneous malignancy.

This section has presented a brief survey of the multiple types of genetically controlled regulatory activity present in mice which affect naturally occurring disease. Key factors are those controlling spontaneous release and virus spread, ability to respond immunologically, availability of target cells and perhaps others as well. This complex set of controls is superimposed on structural genes such as for Akv-1, v-2, and xenotropic MuLVs.

C. Control of Endogenous Virus Expression in Chickens [18, 19]

In most chickens, viruses of subgroup E are not overtly expressed although their genetic material is readily detected by molecular hybridization and partial expression as the helper factor, *chf* frequently does occur. Several chicken lines have been found (lines 7 and 100) which spontaneously produce RAV-0 and surprisingly these lines are gs- until the time of virus expression. Line 100 contains animals which are either resistant or susceptible to RAV-0 and viremia occurs only in the susceptible birds. Mixed culture experiments established that resistant line 100 cells produce virus but cell to cell spread is limited. This situation is very similar to that seen in mice with the Fv-1 locus. Genes for virus activation in chickens have been described and may be allelic to the gs locus which controls partial expression. Whether these are viral or host genes is not certain; however, based on the presence of the viral genome in all chickens regardless of expression these would seem to be host regulator functions.

The control of susceptibility to subgroup E viruses is exceedingly complex. Results of cross-breeding experiments between 2 resistant lines (C and I) showed that while the F-1 generation was completely resistant, the F-2 and backcrosses gave susceptible animals (19%, F-2; 25% backcross to C). This pattern is interpreted on the basis of 2 independently segregating loci controlling susceptibility. One locus, Tv-e, determines subgroup receptors for subgroup E viruses and has both dominant e^s and resistant e^r alleles; one, designated I^e, which inhibits expression of the e^s gene and the other, i^e, which exerts no control. Thus susceptibility depends on both the presence of a receptor and absence of an inhibitor in these particular crosses. While this has been contrasted with the MuLV system, note that recent evidence indicates an internal block to MuLV synthesis determined by Fv-1 as well as the possibility of a receptor. While the VSV pseudotype experiment described in Section 9.VI.B. argues against the receptor hypothesis, the possibility of confering susceptibility with cell extracts suggests some form of receptor activity may be involved.

In other experiments it was noted that presence of I^e correlated with and was linked to gs+; thus, the possibility that partial expression of the genome results in resistance is raised. This is similar to what would occur in cells deliberately infected with subgroup E viruses and challenged with a RSV (*chf*) pseudotype. Recall that gs+ and presence of *chf* generally correlate and that *chf* determines the envelope glycoprotein, gp85, of subgroup E specificity. Antibody absorption experiments established that gp85 was available at the cell surface, thus the nature of I^e as controlling partial expression of the viral genome seems likely.

Complex interactions between viruses of subgroups B and E have also been noted. Thus all chicken cells resistant to subgroup B (C/B) are also resistant to subgroup E, and preinfection with a subgroup B virus will induce interference to RSV(E). The reciprocal interference pattern is not seen and susceptibility to B and E viruses does not necessarily correlate. The genetic basis of this set of interactions remains to be clarified. Possibilities of multiple or identical loci for e and b subgroups are raised along with differential effects of the regulator I^e gene.

Another poorly understood phenomenon involves facilitation of infection for subgroup E viruses by prior infection of normally resistant cells by viruses of subgroups A and C. Genetic analysis has indicated that facilitation occurs in cells rendered resistant by the I^e gene but which possess the e^s gene for subgroup E receptors. One suggested mechanism involves removal of subgroup E receptors by phenotypic mixing with the infecting subgroup A or C viruses.

As one can readily see, there is a complex of genetic control mechanisms for subgroup E viruses in the chicken. At present there is no clear evidence of controlling genes being structural components as so clearly demonstrated for Akv-1 in the mouse. This is rendered difficult by the ubiquitous presence in chickens of the subgroup E viral genome.

D. Viral Mutants

A classical approach to detecting and mapping of viral genes involves the deliberate production of mutants followed by complementation and functional studies. Such oncornavirus mutants have been isolated after treatment with RNA and DNA mutagens, including 5-azacytidine, bromodeoxyuridine, 5-fluorouracil, nitrosoguanidine, or ultraviolet light. Mutants produced deliberately are generally conditional lethals and readily selected on the basis of growth at 1 temperature (usually low, 37°C), called the permissive temperature, and inability to do so at high temperature (usually 41°C). The altered function(s) of temperature-sensitive (*ts*) mutants can usually be detected by temperature shift experiments at times after infection. In addition to *ts* mutants we have described in the section on recombination (Section 9.VI.B.), nonconditional mutants which spontaneously arise on passage of competent sarcoma viruses. These are transformation defective (*Td*), replication defective (*Rd*), or coordinately defective for both functions (*Cd*). The nonconditional mutants are intrinsically more stable; usually because of deletion as opposed to single base changes. *ts* mutant functions also tend to be "leaky" at the nonpermissive temperature.

With regard to types of mutants seen, 3 broad classes of avian sarcoma viruses are recognized [19] designated: "T," transformation defective; "C," coordinately defective; and "R," replication defective but still capable of transforming. Multiple complementation groups are recognized in these categories; however, since much of this work is in progress we will concentrate on 2 mutants which have yielded important information relating to the unique functions of transformation and reverse transcriptase activity.

Martin [102] derived a mutant of the T class by treatment of SR-RSV-A with nitrosoguanidine. Under the conditions used the mutagenized stock showed a reduction in infectivity of $\sim 10^3$. Surviving virus was cloned by an agar suspension technique in which only transformed cells grow. Of 260 clones analyzed 6 were found to produce foci at 35-36°C and not at 41°C. Of these, 1 (designated T-1) was recloned and used for further analysis. Wild type virus grew equally well at both temperatures as did the T-1 mutant. Inactivation rates at 41°C were also similar for both viruses; thus thermolability could not explain the T defect. Cells infected at 41°C by T-1 were resistant to superinfection by SR-RSV-A but not by a subgroup C virus. Thus, the T-1 mutant at 41°C behaved exactly as a leukosis virus.

A key experiment was then performed which related to the question of a requirement for initiation versus continued production of some product to maintain the transformed state. Cells transformed at 36°C by T-1 were transferred to 42°C, whereupon a rapid loss (4 hr) of transformed properties occurred. These include cell morphology and rate of transport of certain sugars which is also a correlate of the transformed state, and ability to grow in agar suspension. This mutant has thus provided the information that *continued synthesis of a viral gene product is required to maintain the transformed state,* and that this function is completely independent of replication requirements. At least 4 complementation groups have been described for mutants with properties similar to T-1 but whether this indicates 4 essential genes to maintain transformation is uncertain. Downshifts of cells infected at nonpermissive temperature also results in the appearance of transformed cells. This effect requires protein synthesis and at least for some mutants of the T-1 class neither DNA or RNA synthesis. In addition to maintenance functions mutants have been described which appear to be lacking in initiation capability. These are detected by virtue of inability to restore transformation by downshift within 24 hr after infection at nonpermissive temperatures. It should be obvious that identification of these vital functions in terms of specific proteins is a key area of investigation.

Among the initiation-defective *ts* mutants are 2 which are coordinate for replication and transformation. These mutants, designated *ts*335 and *ts*337, were found to have a thermolabile reverse transcriptase activity. For these experiments by Verma, Baltimore, and colleagues [103], both endogenous and exogenous (response to synthetic templates) reactions were studied at various temperatures. The ratio of wild type to *ts*335 activity was ~2-fold greater at 40.5°C compared to equivalent activity at 37°C (endogenous reaction), and ~6-fold greater for the exogenous reactions using poly(rC):oligo(dG). Both crude and purified enzymes from the 2 mutant viruses were inactivated 3-fold more rapidly at 44°C than the wild type. Ribonuclease H activity was also associated with RT (Section 9.III.) and was also relatively heat labile compared to wild type enzyme. The relevant data are summarized in Table 9.21.

Revertants of *ts*335 and *ts*337 have been isolated which are no longer *ts* at 41°C. Enzymes from the revertants were also heat stable, e.g., a 10-fold

Table 9.21 Heat Lability of Reverse Transcriptase in Two Temperature-Sensitive Avian Sarcoma Virus Mutants[a]

Enzyme	$T^{1/2}$ of enzyme activity at 44°C		
	RNA template	DNA template	RNase H
Wild type	11.5	11.5	30.0
*ts*335	2.0	3.0	7.4
*ts*337	1.8	2.8	5.0

[a]Three activities are tabulated, RNA-directed DNA synthesis [poly(C):oligo(dG)], DNA-directed DNA synthesis [poly (dC):oligo(dG)] using as substrate radioactive dGTP. $T^{1/2}$ = time for 50% loss of activity in minutes. From Verma et al. [103].

increase in resistance of RNase H activity, 3- to 4-fold increase in RT activity. This shows a clear correlation between biological and biochemical activity.

In addition to the revertant viruses, recombinants of the *ts* mutants both from subgroup C, Pr-RSV, were prepared with group B leukosis virus, RAV-6. These were selected based on transformation at permissive temperature and host range of the B subgroup. The *ts* lesion is thus unselected. Heat inactivation kinetics of wild type virus enzyme revealed greater stability than did the enzyme from recombinant viruses which possessed *ts* lesions in growth and transformation. These data permit several clear conclusions.

1. Virion reverse transcriptase is viral coded.
2. An essential role for RT in transformation and replication is clearly indicated, obviously acting through synthesis of the DNA provirus.
3. RT and RNase H activity are associated on the same polypeptide chain.

Thus, in 2 distinctive features of oncornaviruses, transformation and reverse transcriptase function, *ts* mutants have provided clear-cut valuable information. Undoubtedly this area of research will prove very rewarding for it will allow a complete mapping of the virus life cycle. Mutants of mouse leukemia and sarcoma viruses have also been described; however, these have not yet given information as crucial as that described in this section.

VIII. Oncornaviruses and Pathogenesis

In the preceding sections we have described many of the more recent findings concerned with the molecular biology of oncornaviruses. Yet the central feature of this group of viruses (considered as a subclassification of the retravirus family) is their ability to induce malignancy in their natural hosts. Patterns of infection and virus spread seem to be highly variable. Thus there can no longer be debate over the hypothesis of an inherited viral genome. This form of transmission, vertical-inherited, is shown for the endogenous viruses of chickens, mice, rats, hamsters, cats, pigs, and baboons. Vertical-congenital transmission via milk, seminal fluid, or uterine secretions is also well established in mice and should not be confused with inheritance. In fact this has been explicitly stated by Gross even though provision for an inheritance phase may now be made in his concept of "obligate communicability." Horizontal transmissions (animal to animal) has also been clearly shown in chickens, more recently and emphatically in cats [104], and in addition transmission between species has been indicated even if not in contemporary time, e.g., SSV-1/GaLV from a rodent species and RD114, to a cat ancestor from an old world monkey progenitor. The situation is obviously complex and to sort out the key factors may be difficult. The evidence for horizontal transmission as a determinant of natural disease is clearest in the cat, where immunoepidemiology has provided the clearest evidence for antibody as a controlling element of natural disease. In this case resistance to disease correlates well with antibody to a cell surface antigen (FOCMA) which appears to be distinct from virion structural antigens. Early statistical data did not show evidence of a horizontal transmission component in FeLV-related disease but this undoubtedly suffered from a compressed time scale used for the data gathering. The change from a state of gs^- (virus negative) to gs+ in large cat households increases risk of leukemia many fold. Yet in this situation we also see evidence for an inherited genome related to FeLV in all cats. Most likely the cat counterpart to chicken subgroup E viruses has not yet been isolated and perhaps this will still be an important determinant of natural disease; yet if one were to attempt to cure malignancy in the cat, the horizontal component is the obvious choice and a vaccine approach is thus reasonable and feasible. In terms of pathogenesis, FeLV is implicated in natural malignancy of the cat population and other diseases, e.g., anemia, as well. Sarcoma viruses have been isolated several times from cats even though as in other species this is a rare event. RD114 has so far shown no relation to disease and we are confronted here with a high level of expression in normal cells as well. For me the nature of feline oncornavirus cell membrane antigen (FOCMA) is 1 of the important questions of viral oncology with potential application to man.

In the chicken, Bauer and colleagues [105] have recently provided evidence for a cell membrane antigen produced by all avian viruses. This antigen seems distinct from structural components and thus may be an analog of FOCMA. The chicken viruses have a bewildering "natural" history and despite the early ease of Rous in isolating sarcoma viruses, contemporary attempts from spontaneous sarcomas do not generally meet with success. Viruses of virtually every pathogenic potential have been described and in some cases induction of a particular neoplasm depends on dosage. Recalling the Friend virus situation in mice, the well-studied avian myeloblastosis virus (AMV) can evidently transform myeloid cell precursors in vitro but seems to be replication defective. A recent publication describes isolation of 2 separate helper viruses from AMV. This points up 1 key problem, namely, that many potential pathogenic viruses have not been described because of lack of appropriate target cells in vitro and the widespread presence in vivo of helper viruses whose own pathogenic potential may appear much more rapidly than the sought for virus. One interesting virus in mice, the Abelson leukemia virus, has been shown to transform both fibroblasts and fetal liver cells in vitro [106]. The fetal cells gave rise to lymphoid tumors which were either of B cell (bone marrow derived, IgG+) origin or contained neither T (thymus derived Θ +) or B cell markers. This type of assay system will provide a model, at least for thought, to apply to assay of leukemia viruses and other pathogenic varients; e.g., how does one isolate a carcinoma virus? In the chicken, ubiquitous subgroup E virus does not appear to be a determinant of natural disease. Viremic chickens of lines 7 and 100 do not develop an unusual incidence of malignancy nor has the virus induced disease in animals. This has led some to question the role of inherited viruses in cancer. Yet in mice one cannot escape the fact that AKR virus is inherited and when expressed is quantitatively related to both onset and incidence of lymphoid leukemia. The xenotropic virus in mice has not yet been reported to induce disease.

In a completely different line of investigation, Freeman et al. [107] have reported that infection with leukemia viruses predisposes cells to transformation with chemical carcinogens. In some cases, e.g., early passage rat cells, chemicals alone are ineffective as transforming agents while at high passage, when endogenous rat virus is synthesized, chemicals are effective without addition of exogenous virus. This type of experiment leads one to speculate that even if certain endogenous viruses are not overtly oncogenic they may interact with other carcinogens to provide essential factors for malignant transformation. This hypothesis would better explain the normal age-incidence patterns found in man than would the existence of a virulent virus of the Gross type. The relationship of many of the viruses described in recent years to malignancy in their natural hosts should be sought with consideration of the complexities of genetic control, mode of assay, and possible cocarcinogenicity before negative conclusions are drawn. Such work is time consuming and tedious but must be done properly to avoid blanket dismissal of the potential malignancy of any virus. All viruses are obviously not alike; e.g., among the few gibbon isolates 1 has induced AML upon inoculation into newborn gibbons and the others have not induced any disease as yet. Thus, as is true of most virus families, pathogenic and benign viruses of the same serotype are frequently found.

Consequences of infection may be tolerance in the case of congenital vertial transmission or immune response as in gibbons or cats with their respective viruses. In mice, evidence for antibody to the envelope components of the xenotropic virus class has been obtained by neutralization and broader specificity by radioimmunoassay. The significance of these responses remains to be shown.

Thus we can safely conclude that there are multiple patterns of virus-host interaction in vivo which show variability from species to species. There are obvious critical questions as to the mechanism of transformation where it does occur and host defense potential for destroying transformed cells. This is an active area of research at present and 1 to which the virus systems may contribute greatly to similar studies in man.

IX. Studies in Man

It should not be surprising that all of the various techniques and reagents developed in model systems have been applied to studies of human materials. Obviously the results and logic derived from the model systems should serve as a guideline for what to expect. At present my bias is that the most convincing evidence for oncornaviral information in man is its presence in many species, including several primates. Reproducible isolation of virus, viral antigens, or viral nucleic acids has not been achieved to the satisfaction of many workers in the field despite publications of positive evidence. This is not to dispute the data but to consider alternate interpretations, including those based on technical artifacts. It seems clear that the critical lack is laboratory to laboratory reproducibility.

A summary of positive findings is as follows.

1. p30 determinants of SSV-1/GaLV, BaLV/RD 114, MuLV, and FeLV in human tissues [108]. One report shows 5/5 leukemic cells positive for SSV-1/GaLV p30 [109].
2. RT related to SSV-1/GaLV in leukemic cells [66].
3. RNA sequences of Rauscher virus in human leukemia and sarcoma [110].
4. RNA sequences of MMTV in breast cancer and normal breast tissue [111]. Particles in human milk.
5. RNA sequences of MP-MV in breast cancer [68].
6. Virus isolation from human leukemic cells–related to SSV-1 [112].

7. Baboon viral sequences in human DNA, related on an evolutionary basis to BaLV [90].
8. Visualization of oncornavirus–like particles in human placenta [114].

Regarding the reports of antigen detection, these have all been made using radioimmunoassay which is a highly sensitive procedure. Most laboratories have found troublesome artifacts using tissue extracts, mainly because of degradation of the labeled antigen which is present in trace amounts. The species-specific p30 determinants are trypsin sensitive and very short incubations will give extensive degradation. Without control this could give the appearance of positive inhibition. One should also be aware of the problem of overt contamination. Since laboratories performing tests generally have initially purified the authentic antigen, accidental contamination of glassware, concentrating devices, centrifuge tubes, etc., is a distinct problem. Since the levels of reported positivity in human material has been low, ~5-20 ng/mg tissue protein, and authentic material is generally available at 10^3-10^5 times higher concentration, this problem should not be dismissed lightly. In several laboratories, applying equally sensitive techniques including purification methodology, no positive reactions have so far been noted. Thus, a definitive conclusion must await wider experience.

The evidence of reverse transcriptase in AML samples which is related to that of the SSV-1/GaLV group has been reported by 1 laboratory on a repeated basis. In contrast, other laboratories report negative findings. This again must await further confirmation, although one wonders about the discrepancies.

Several types of molecular hybridization analyses have indicated presence of viral-related sequences in human material. This is another area of broad contradiction; for example, positive results with Rauscher virus as well as negative results have been reported. Discrepancies in probe specificity seem the most likely explanation. The sighting of virus-like particles in human milk is also a subject of controversy. In a large series, Moore and colleagues report only a small number of particles which by negative-stain techniques are very similar to B particles. Many other particles are seen which may not even be viruses. One problem in this work is the apparent destruction of virus by incubation in milk; thus, negative results may be subject to varying interpretation. These particles are also thought by some workers to be more similar to MP-MV than MMTV; however, in no case has an authentic virus been isolated from human milk. Indirect assays such as for reverse transcriptase have also been controversial and with the recognition that none of the synthetic templates currently in use for screening is absolutely virus specific, any positive results are doubtful. The human placental virus-like particles are also disputed by some workers and may in fact be normal constituents of trophoblastic cells.

Regarding the presence of RNA related to either MMTV or MP-MV in breast tissue, one may attribute this to virus or an intrinsic contribution to viral sequences of sequences related to differentiation of breast tissue. For example, Friend virus isolated from hemoglobin producing "Friend cells" may actually incorporate Hb mRNA into high molecular weight RNA. This perhaps enters the area of exactly what is a virus and what are its origins. The ability to enhance MMTV synthesis by the synthetic steroid dexamethasone via normal steroid receptors seems to indicate an interface between viral and cellular sequences [114]. Thus while the results may be real their interpretation may be based other than on viral involvement.

With regard to direct virus isolation, there are 2 contemporary "isolates" under study which are candidate human oncornaviruses. One has been described in a human breast cancer cell line and has properties similar to MMTV according to published reports [115]. Production of virus is erratic, and visualization by electron microscopy has been difficult. These difficulties have essentially kept this out of the realm of intense investigation which any putative human virus must undergo.

A virus isolated from cells from a single human leukemic patient has been reported by Gallagher and Gallo [112]. This virus is highly related to SSV-1 based on molecular hybridization and presence of type-specific p12 and gp70 determinants (R. Gilden and S. Aaronson, unpublished data). Although the isolation of a virus with these characteristics is not surprising based on earlier enzyme [66] and p30 [109] data, one is surprised based on the current subdivision of the SSV-1/GaLV group into 4 categories (see Table 9.11) to have this virus be essentially identical to SSV-1. The above workers also studied other leukemic cell preparations and so far only this single patient has yielded the woolly-like virus. As noted above SSV-1/GaLV sequences are not found in normal or leukemic cellular DNA, so that one may wonder about the origin of the GG virus. In gibbons, where virus has been repeatedly isolated, unaffected members of a colony reveal antibody to both envelope and p30 antigens [116]. However, so far, humans have not shown antibody to any of the type C viral proteins, even in sensitive radioimmunoassays, except in 1 experimental series which involved direct inoculation of killed RLV.

We also noted that sequences related to the baboon virus have been reported in human DNA. Based on reaction extent and T_m data, these seem related on an evolutionary basis. Does this mean a baboon-related virus is endogenous in humans? If so, its antigenic characteristics should be close to BaLV, closer than the hybridization analyses show the total genome to be. Consistent with this there are reports of baboon/RD114 related p30 determinants in a few human tumors [117]. This result awaits confirmation from other laboratories.

The most balanced summary would have to be that no definitive evidence for a bona fide human oncornavirus has yet been presented. All contemporary reports are subject to alternate interpretations including, unfortunately, overt contamination. While this is a problem for the present, there is no doubt that "truth will out," and that at this stage an open mind is required. We also should add that the progress in the field, aside from the human question, has been rapid and substantive and that ultimately an impact in man should be realized.

Acknowledgments

Work from the author's laboratory was supported by contract from the Virus Cancer Program of the National Cancer Institute, National Institutes of Health, Bethesda, Maryland. I thank my colleagues Stephen Oroszlan, Masakazu Hatanaka, and Howard P. Charman for reading portions of the manuscript and for their helpful suggestions. I am especially grateful to Nancy Hampar for technical preparation, editing, and suggestions concerning organization of this chapter.

References

The reference list is intended to be a mixture of key research papers and review articles. No attempt has been made to be inclusive and thus omissions will hopefully be tolerated by the author's colleagues. For detailed history of the oncogenic virus field Ref. 14 is highly recommended. Reference 19 also contains several excellent chapters on oncornaviruses.

1. D. Baltimore, Nature, 226, 1209 (1970).
2. H. M. Temin, and S. Mizutani, Nature, 226, 1211 (1970).
3. H. Temin, Nat. Cancer Inst. Monog., 17, 557 (1964).
4. J. P. Bader, Virology, 22, 462 (1964).
5. R. J. Huebner and G. Todaro, Proc. Natl. Acad. Sci. U.S., 64, 1087 (1969).
6. P. Bentvelzen, in RNA viruses and host genome in oncogenesis (P. Emmelot and P. Bentvelzen, eds.), North-Holland, Amsterdam, 1968, pp. 309-337.
7. G. J. Todaro, and R. J. Huebner, Proc. Natl. Acad. Sci. U.S., 69, 1009 (1972).
8. H. Temin, Ann. Rev. Genetics, 8, 155-177 (1974).
9. R. J. Huebner, G. J. Kelloff, P. S. Sarma, W. T. Lane, H. C. Turner, R. V. Gilden, S. Oroszlan, H. Meier, D. D. Myers, and R. L. Peters, Proc. Natl. Acad. Sci. U.S., 67, 366 (1970).

10. W. P. Rowe, J. Exp. Med., 136, 1272 (1972).
11. S. K. Chattopadhyay, W. P. Rowe, N. M. Teich, and D. R. Lowy, Proc. Natl. Acad. Sci. U.S., 72, 906 (1975).
12. J. Levy, Science, 182, 1151 (1973).
13. R. M. McAllister, M. Nicolson, M. B. Gardner, R. W. Rongey, S. Rasheed, P. S. Sarma, R. J. Huebner, M. Hatanaka, S. Oroszlan, R. V. Gilden, A. Kabigting, and L. Vernon, Nature, 235, 3 (1972).
14. L. Gross, Oncogenic Viruses, 2nd Ed., Pergamon, London, New York, 1970.
15. W. P. Parks, R. V. Gilden, A. F. Bykovsky, C. G. Miller, V. M. Zhdanov, V. D. Soloviev, and E. M. Scolnick, J. Virol., 12, 1540 (1973a).
16. A. J. Dalton, J. L. Nelnick, H. Bauer, G. Beaudreau, P. Bentvelzen, D. Bolognesi, R. Gallo, A. Graffi, F. Haguenau, W. Heston, R. Huebner, G. Todaro, and U. I. Heine, Intervirology, 4, 201 (1974).
17. P. S. Sarma, H. C. Turner, and R. J. Huebner, Virology, 23, 313 (1964).
18. H. Hanafusa, in Cancer–A Comprehensive Treatise, Vol. 2, 1975.
19. J. Tooze, The Molecular Biology of Tumour Viruses, Cold Spring Harbor Laboratory, New York, 1973.
20. D. R. Lowry, W. P. Rowe, N. Teich, and J. W. Hartley, Science, 174, 155 (1971).
21. S. A. Aaronson, G. J. Todaro, and E. M. Scolnick, Science, 174, 157 (1971).
22. R. A. Weiss, R. R. Friis, E. Katz, and P. K. Vogt, Virology, 46, 920 (1971).
23. V. Klement, W. P. Rowe, J. W. Hartley, and W. E. Pugh, Proc. Natl. Acad. Sci. U.S., 63, 753 (1969).
24. P. S. Sarma and T. Log, Virology, 54, 160 (1973).
25. M. V. Nermut, F. Hermann, and W. Schafer, Virology, 49, 345 (1972).
26. J. T. August, D. P. Bolognesi, E. Fleissner, R. V. Gilden, and R. C. Nowinski, Virology, 60, 1595 (1974).
27. H. Ideda, W. Hardy, Jr., E. Tress, and E. Fleissner, J. Virol., 16, 53 (1975).
28. J. N. Ihle, M. G. Hanna, Jr., W. Schafer, G. Hunsmann, D. P. Bolognesi, and G. Huper, Virology, 63, 60 (1975).
29. H. M. Temin and D. Baltimore, Adv. Virus Res., 17, 129 (1972).
30. M. Green and G. F. Gerard, In Progress in Nucleic Acid Research and Molecular Biology, Vol. 14, Academic Press, New York, 1974, p. 187.
31. W. P. Parks, R. S. Howk, E. M. Scolnick, S. Oroszlan, and R. V. Gilden, J. Virol., 13, 1200 (1974).
32. Y. A. Teramoto, M. J. Puentes, L. J. T. Young, and R. D. Cardiff, Am. Soc. Microbiol., 13, 411 (1974).
33. A. G. Mosser, R. C. Montelaro, and R. R. Rueckert, J. Virol., 15, 1088 (1975).
34. R. V. Gilden, C. W. Long, M. Hanson, R. Toni, H. P. Charman, J. M. Miller, and M. J. Van der Maaten, J. Gen. Virol., 29, 305 (1975).
35. P. R. Murray, and D. P. Nayak, J. Virol., 14, 679 (1974).

36. H. -J. Kung, N. D. Bailey, M. O. Nicolson, and R. M. McAllister, J. Virol., 16, 397 (1975).
37. K. Beemon, P. Duesberg, and P. Vogt, Proc. Natl. Acad. Sci. U.S., 71, 4254 (1974).
38. M. M. C. Lai, P. H. Duesberg, J. Horst, and P. K. Vogt, Proc. Natl. Acad. Sci. U.S., 70, 2266 (1973).
39. N. Tsuchida, C. Long, and M. Hatanaka, Virology, 60, 200 (1974).
40. R. C. Sawyer, F. Harada, and J. E. Dahlberg, J. Virol., 13, 1302 (1974).
41. J. E. Dahlberg, R. C. Sawyer, J. M. Taylor, A. J. Faras, W. E. Levinson, H. M. Goodman, and J. M. Bishop, J. Virol., 13, 1126 (1974).
42. J. M. Taylor and R. Illmensee, J. Virol., 16, 553 (1975).
43. M. M. C. Lai and P. H. Duesberg, Nature, 235, 383 (1972).
44. R. V. Gilden, S. Oroszlan, and M. Hatanaka, Viruses, Evolution and Cancer, Academic, New York, 1974, Chap. 10, p. 235.
45. A. Hellman, P. T. Peebles, J. E. Strickland, A. K. Fowler, S. Kalter, S. Oroszlan, and R. V. Gilden, J. Virol., 14, 133 (1974).
46. R. V. Gilden, in Advances in Cancer Research (G. Klein and S. Weinhouse, eds.), Academic, New York, Vol. 22, 1975, pp. 157-202.
47. M. M. Lieber, C. J. Sherr, G. J. Todaro, R. E. Benveniste, R. Callahan, and H. G. Coon, Proc. Natl. Acad. Sci. U.S., 72, 2315 (1975).
48. R. E. Benveniste, R. Heinemann, G. L. Wilson, R. Callahan, and G. J. Todaro, J. Virol., 14, 56 (1974).
49. H. P. Charman, N. Kim, and R. V. Gilden, J. Virol., 14, 910 (1974).
50. J. Davis, R. V. Gilden, and S. Oroszlan, Immunochemistry, 12, 67 (1975).
51. W. P. Parks, and E. M. Scolnick, Proc. Natl. Acad. Sci. U.S., 69, 1766 (1972).
52. S. Oroszlan, T. Copeland, M. R. Summers, G. Smythers, and R. V. Gilden, J. Biol. Chem., 250, 6232 (1975).
53. S. Oroszlan, M. R. Summers, C. Foreman, and R. V. Gilden, J. Virol., 14, 1559 (1974).
54. M. Strand and J. T. August, J. Biol. Chem., 248, 5627 (1973).
55. G. Hunsmann, V. Moennig, L. Pister, E. Seifert, and W. Schafer, Virology, 62, 307 (1974).
56. W. Schafer, G. Hunsmann, V. Moennig, and F. de Noronha, Virology, 63, 48 (1975).
57. M. Strand, R. Wilsnak, and J. T. August, J. Virol., 14, 1575 (1974).
58. W. P. Parks, M. C. Noon, R. Gilden, and E. M. Scolnick, J. Virol., 15, 1385 (1975).
59. J. R. Stephenson, S. R. Tronick, and S. A. Aaronson, Virology, 58, 1 (1974).
60. W. P. Parks, E. M. Scolnick, M. C. Noon, and C. J. Watson, J. Virol., 14, 430 (1974).
61. R. Callahan, M. M. Lieber, and G. J. Todaro, J. Virol., 15, 1378 (1975).
62. E. Fleissner and E. Tress, J. Virol., 12, 1612 (1973).

63. M. S. Halpern, D. P. Bolognesi, R. R. Friis, and W. S. Mason, J. Virol., 15, 1131 (1975).
64. D. P. Bolognesi, R. Ishizaki, G. Huper, T. C. Vanaman, and R. W. Smith, Virology, 64, 349 (1975).
65. A. C. Herman, R. W. Green, D. P. Bolognesi, and T. C. Vanaman, Virology, 64, 339 (1975).
66. G. J. Todaro and R. C. Gallo, Nature, 244, 206 (1973).
67. M. C. Noon, R. G. Wolford, and W. P. Parks, J. Immunol., 115, 653 (1975).
68. D. Colcher, S. Spiegelman, and J. Schlom, Proc. Natl. Acad. Sci. U.S., 71, 4975 (1974).
69. H. Okabe, R. V. Gilden, and M. Hatanaka, Intl. J. Cancer, 15, 849 (1975).
70. G. J. Todaro, M. M. Lieber, R. E. Benveniste, C. J. Sherr, C. J. Gibbs, Jr., and D. C. Gajdusek, Virology, 67, 335 (1975).
71. E. M. Scolnick, E. Rands, D. Williams, and W. P. Parks, J. Virol., 12, 458 (1973).
72. N. Tsuchida, R. V. Gilden, and M. Hatanaka, J. Virol, 16, 832 (1975).
73. K. Rand, J. Davis, R. V. Gilden, S. Oroszlan, and C. Long, Virology, 64, 63 (1975).
74. M. Hatanaka, T. Kakefuda, R. V. Gilden, and E. A. O. Callan, Proc. Natl. Acad. Sci. U.S., 68, 1844 (1971).
75. T. Takano and M. Hatanaka, Proc. Natl. Acad. Sci. U.S., 72, 343 (1975).
76. G. G. Lovinger, R. Klein, H. P. Ling, R. V. Gilden, and M. Hatanaka, J. Virol., 16, 824 (1975).
77. R. V. Guntaka, B. W. J. Mahy, J. M. Bishop, and H. E. Varmus, Nature, 253, 507 (1975).
78. V. M. Vogt, R. Eisenman, and H. Digglemann, J. Mol. Biol., 96, 471 (1975).
79. W. S. Hayward and H. Hanafusa, J. Virol., 122, 1367 (1975).
80. P. H. Duesberg, S. Kawai, L. -H. Wang, P. K. Vogt, H. M. Murphy, and H. Hanafusa, Proc. Natl. Acad. Sci. U.S., 72, 1569 (1975).
81. R. A. Weiss, W. S. Mason, and P. K. Vogt, Virology, 52, 535 (1973).
82. D. Stehelin, R. V. Guntaka, H. E. Varmus, and J. M. Bishop, J. Mol. Biol., 101, 349 (1976).
83. S. E. Wright and P. E. Neiman, Biochemistry, 13, 1549 (1974).
84. M. Lieberman and H. S. Kaplan, Science, 130, 387 (1959).
85. L. N. Payne and R. C. Chubb, J. Gen. Virol., 3, 379 (1968).
86. H. Meier, B. A. Taylor, M. Cherry, and R. J. Huebner, Proc. Natl. Acad. Sci. U.S., 70, 1450 (1973).
87. B. A. Taylor, H. Meier, and R. J. Huebner, Nature, New Biol., 241, 184 (1973).
88. C. J. Sherr, R. E. Benveniste, and G. J. Todaro, Proc. Natl. Acad. Sci. (U.S., 71, 3721 (1974).
89. R. E. Benveniste and G. J. Todaro, Nature, 252, 456 (1974).
90. R. E. Benveniste and G. J. Todaro, Proc. Natl. Acad. Sci. U.S., 71, 4513 (1974).

91. E. M. Scolnick, W. Parks, T. Kawakami, D. Kohne, H. Okabe, R. Gilden, and M. Hatanaka, J. Virol., 13, 363 (1974).
92. N. Tsuchida, R. V. Gilden, and M. Hatanaka, Proc. Natl. Acad. Sc. U.S., 72, 4503 (1974).
93. H. E. Varmus, N. Quintrell, E. Medeiros, J. M. Bishop, R. C. Nowinski, and N. H. Sarkar, J. Mol. Biol., 79, 663 (1973).
94. D. P. Nayak, Proc. Natl. Acad. Sci. U.S., 71, 1164 (1974).
95. M. Hill and J. Hillova, Virology, 49, 309 (1972).
96. M. O. Nicolson, E. Hariri, H. M. Krempin, R. M. McAllister, and R. V. Gilden, Virology, 70, 301 (1976).
97. A. Karpas and C. Milstein, Eu. J. Cancer, 9, 295 (1973).
98. E. M. Scolnick and S. J. Bumgarner, J. Virol., 15, 1293 (1975).
99. F. Lilly, M. L. Duran-Reynals, and W. P. Rowe, J. Exp. Med., 141, 882 (1975).
100. F. Lilly, and T. Pincus, Adv. Cancer Res., 17, 231 (1973).
101. T. Pincus, J. W. Hartley, and W. P. Rowe, Virology, 65, 333 (1975).
102. G. S. Martin, Nature, 227, 1021 (1970).
103. I. M. Verma, W. S. Mason, S. D. Drost, and D. Baltimore, Nature, 251, 27 (1974).
104. M. Essex, Adv. Cancer Res., 21, 175 (1975).
105. L. R. Rohrschneider, R. Kurth, and H. Bauer, Virology, 66, 481 (1975).
106. N. Rosenberg, D. Baltimore, and C. D. Scher, Proc. Natl. Acad. Sci. U.S., 72, 1932 (1975).
107. A. E. Freeman, P. J. Price, H. J. Igel, J. C. Young, J. M. Maryak, and R. J. Huebner, J. Natl. Cancer Inst., 44, 65 (1970).
108. M. Strand and J. T. August, J. Virol., 14, 1584 (1974).
109. C. J. Sheer and G. J. Todaro, Science, 187, 855 (1975).
110. R. Hehlmann, D. Kufe, and S. Spiegelman, Proc. Natl. Acad. Sci. U.S., 69, 435 (1972).
111. A. B. Vaidya, M. M. Black, A. S. Dion, and D. H. Moore, Nature, 249, 565 (1974).
112. R. E. Gallagher and R. C. Gallo, Science, 187, 350 (1975).
113. S. S. Kalter, R. J. Helmke, R. L. Heberling, M. Panigel, A. K. Fowler, J. E. Strikland, and A. Hellman, J. Natl. Cancer Inst., 50, 1081 (1973).
114. W. P. Parks, J. C. Ransom, H. A. Young, and E. M. Scolnick, J. Biol. Chem., 250, 3330 (1975).
115. C. M. McGrath, P. M. Grant, H. D. Soule, T. Glancy, and M. A. Rich, Nature, 252, 247 (1975).
116. H. P. Charman, N. Kim, M. White, H. Marquardt, R. V. Gilden, and T. Kawakami, J. Natl. Cancer Inst., 55, 1419 (1975).
117. C. J. Sherr and G. J. Todaro, Proc. Natl. Acad. Sci. U.S., 71, 4703 (1974).